BIOLOGICAL DIV

BIOLOGICAL DIVERSITY
Exploiters and Exploited

PAUL HATCHER AND NICK BATTEY

UNIVERSITY OF READING, UK

WILEY-BLACKWELL

A John Wiley & Sons, Ltd., Publication

Library of Congress Cataloging-in-Publication Data

Hatcher, Paul (Paul E.)
  Biological diversity : exploiters and exploited / Paul Hatcher and Nick Battey.
     p. cm.
  Includes index.
  ISBN 978-0-470-77806-7 (cloth) – ISBN 978-0-470-77807-4 (pbk.)
  1. Biodiversity. 2. Nature–Effect of human beings on. 3. Human beings–Effect of environment on. I. Battey, N. H. II. Title.
  QH541.15.B56H38 2011
  577–dc22
                              2010049408

A catalogue record for this book is available from the British Library.

This book is published in the following electronic formats: ePDF: 978-0-470-97964-8; ePub: 978-0-470-97986-0

Typeset in10.5/12.5pt Joanna MT by Laserwords Private Limited, Chennai, India

Printed and bound in Singapore by Ho Printing Singapore Pte Ltd

First Impression   2011

Contents

Acknowledgements

W E THANK OUR families, friends and colleagues for their forbearance and support during the several years that we were writing this book, and without which we would never have completed it. Nick Battey particularly acknowledges Paul Simmonds for early research assistance on some of the topics covered here; and Frank Tallett, David Stack and all in the Department of History, University of Reading for study support during a sabbatical year. At Wiley-Blackwell, our editors Celia Carden and Nicky McGirr, have been unfailingly helpful and supportive throughout this process, and never lost faith in the project. Fiona Woods and Chelsea Cheh from Wiley-Blackwell, and Sangeetha Parthasarathy from Laserwords have been adept at turning our manuscript into the finished book.

Keith Lindsey and Ron O'Dor reviewed the final manuscript. The following reviewed individual chapters: David Morritt, Dennis Greer, Tony Clare, Karen Tait, Claire Helio, Steve Smith, Derek Yalden, Rob Marrs, Sid Thomas, Duncan Cameron, Hilary Ranson, Paul Rodhouse, Peter Maitland, the late Frank Fenner, John Sheail, Ian Carter, Joan Cottrell, Julie Hofer, Chris Drakeley, Rick Titball and John Feltwell. Together they saved us from many inconsistencies, errors and misunderstandings; any that remain, and the opinions expressed here, are the authors'.

All illustrations are by Paul Hatcher, unless stated otherwise; we would particularly like to thank Henry Battey for drawings and Chris Madgwick, Catherine Newell Price and Colin Brownlee for photographs. Steve Appleby has provided not only illustrations but also interest, enthusiasm and commitment.

We have been privileged to teach our Exploiters and Exploited biodiversity course to a bunch of bright and inquisitive students over the last few years. Without them there would be no book, and their enthusiasm and interest have been vital in helping shape our views on the topics we discuss.

Introduction

OPEN ANY BOOK about biology and you will see that interactions are central to the science of life. Mutualism, parasitism, symbiosis, cooperation, collaboration, association: these words describe how organisms interact with one another, in ways that grade from intimate relationships such as parasitism and symbiosis through to looser associations. Perhaps only rarely do species have absolutely no interactions at any stage of their life-cycles, even if that interaction comes only indirectly, through other members of the web of life. Interaction is therefore a ubiquitous feature of the business of life, from the cell-bound mitochondria that once were free-living bacteria, to the tapeworms that gnaw at the guts of their hosts, to ants which farm aphids to gather plant sugars for food.

Open a different book and the topic is human society, with its history, its achievements and injustices, its social structures, its political systems and power struggles, and its economics. Here is interaction of a different kind, one human with another, one social group or community or country vying with another, down the ages; the tense game of balancing mutual cooperation and need against the tendency to dominate, override, eliminate.

Where do these two domains meet? The answer is nowhere; or at least nowhere in terms of academic discipline. Biology is traditionally concerned with how life is organized, how it develops and evolves at both the individual and the population levels. Human society is the sphere of the humanities and social sciences, and tends to deal with humanity apart from other life; in general, history and sociology have humans at their centres. Yet is this separation not weird? The domains of biology and the humanities cannot remain separate; indeed they are already not doing so, rather it is the academic disciplines which lag behind real life. As we look forward we see the human population threatening increasingly to extinguish other life-forms over the coming decades. Yet, paradoxically we will rely more and more upon (some of) these species, as food security becomes less certain and the means of feeding the 9.2 billion people projected by 2050 seems ever more challenging. In reality the spheres of biology and the humanities overlap in fundamental ways. To understand and improve how humans interact with other organisms, and how the raw materials of the Earth are utilized and conserved, we must see our historical, humanistic selves as part of biology in the widest sense. In a complementary fashion, we must understand how other organisms and their societies are organized and will be organized in the future; and this is at least partly a function of where they have come from, their histories on a human-dominated planet.

This book aims to describe biological diversity from a new perspective. Most biodiversity books treat the organisms, the domains of life, as topics largely

Biological Diversity: Exploiters and Exploited, First Edition.   Paul Hatcher and Nick Battey.
© 2011 John Wiley & Sons, Ltd. Published 2011 by John Wiley & Sons, Ltd.

separate from humans—or with humans as rather distant onlookers. We take the opposite approach, attempting to capture the diversity of life with the human connections as integral to it. We revel in the history of interaction between one organism and another, with humans as central players, fashioning and manipulating other species for their own ends. Yet we also see organisms exploiting humans as part of the web of potential. Thus we divide our subjects roughly into exploited and exploiters, aware that, depending on where one is standing, exploiter can be exploited and vice versa. The book is therefore a selection of topics from which we believe the reader can discover general principles about humans and our companion species seeking to sustain themselves here on Earth.

Our choice of organisms and the systems in which they live may seem arbitrary; it is probably idiosyncratic. We talk about things we think are interesting, and this is inevitably a reflection of our background, the students we have taught (their feedback has been crucial), and aspects of the subject of biological interactions we consider important. The first ten chapters deal with animals and plants which seem to us to give more than they take. This is not a precise, scientific determination, but rather an initial judgement that these organisms have been heavily exploited by humans or/and by other species. Thus the *Sargassum* seaweed, floating in giant rafts in the warmer regions of the North Atlantic Ocean, harbours a unique and bizarre fauna that has survived by adapting to life on and around the weed. Cephalopods have extraordinary life-cycles and dispersal around the oceans, and humans have developed sophisticated ways of tracking and catching commercially important members of the group, such as squid. We have also become fascinated with our largest invertebrates, the giant squids. Bee society is one that has grown up in close conjunction with humans; we discuss the factors which lead bee populations in many parts of the world to be either threatened or dangerous, both suggesting an interaction with humans which is not in balance. The silk moth is an organism of intricate beauty, requiring detailed knowledge of its biology as well as particular economic conditions for successful silk cultivation. This has meant that numerous attempts to export the technology around the world have failed; yet silk is a product valued especially, perhaps, because of its biological rather than synthetic origin.

Sugar cane carries on its sharp leaves and sucrose-rich stems a historical burden of slavery which, while well-known, is not very often discussed in relation to the peculiar physiology and agronomy of the crop. Legumes are members of a plant family that have been particularly exploited by humans because of their valuable seed protein and oil; their ability to grow in marginal lands; and for the nitrogen with which they enrich the soil through their symbiotic association with the bacterium *Rhizobium*. The grapevine gives us wine. The development of its cultivation and production technologies (particularly in the New World) is an object lesson in how to exploit to the maximum what is basically a pretty unpromising plant. Salmon is a fish whose complex life-cycle makes it vulnerable at many different steps to the vicissitudes of a human-dominated environment. Yet it has been farmed very successfully, even though this is far distant from the wild salmon pursuing its strange instincts to occupy both the marine and the freshwater waters of the world. There is an unavoidable pathos associated with the fate of many organisms whose natural behaviour and life history we marvel at, while we destroy them. Then we discuss the oak, great bastion of the navy in times gone by, the symbolic heart of many countries and societies around the world, with a special history of exploitation by humans and by a rich variety of animal types. The rabbit is the final exploited organism: both exploiter and exploited, when we had grown tired of it we unleashed upon it the terrible disease myxomatosis.

Then we turn to the exploiters. Mosquitoes, and the malarial plasmodium parasite they carry, are a major target for world health organizations. Total malaria eradication by 2015 is the stated objective, and we examine the biology behind the problem and explore the difficulties associated with meeting this challenge. In the marine environment, the humble barnacle, along with other encrusting organisms, causes serious problems for humans in ships. The solutions to these will increasingly require an understanding of the life-cycle and behaviour of these biofoulers. Back on dry land, bracken is an example of a plant that has changed from being a useful resource, exploited in the manufacture of soap and glass, and for thatch and bedding, to a problematic weed that invades disused land, out-competing other species, and creating a potential health hazard because of its toxic and carcinogenic properties. This example shows how in the course of a few hundred years an exploited resource can itself become the exploiter. Locusts, on the other hand, have been a ravaging exploiter of epic proportions since biblical times. They are still a serious problem, particularly in Africa, and we discuss the fascinating biology of these insects and the way this knowledge can be employed in attempts to predict and control locust plagues. In the chapter on plague we consider modern bubonic plague, which erupted in the nineteenth century, left a legacy of disease foci throughout the world, and has since been exploited as a bioweapon. We also explore the assumption that this was the same plague as the Black Death, which killed up to a third of the population of Europe in the mid fourteenth century.

Red kites show how an organism with a largely unchanging behaviour pattern has, over the past 500 years in Europe, changed its interaction with human society: from valued scavenger and urban waste remover, to pest, to object of conservation focus. Mistletoes are parasitic plants, so there is apparently no question of their status; they exploit other plants for both resource (photosynthates) and habitat (attaching to and sometimes living entirely within their hosts). But recent research has emphasized how mistletoes are actually key members of plant and animal communities, having evolved close associations with birds, mammals and their host plants. There is also evidence that they enrich their habitats disproportionately to their abundance. All in all, while sometimes a significant pest, they are in many cases as much exploited as exploiter, to be valued as mutualists rather than disparaged as parasites. Wolves are creatures with deep psychological ties to humans. We are not sure whether to value or fear them, and they are much misunderstood, their sophisticated sociology perhaps making them uncomfortable reminders of our own animal natures. We arrive at the present day with a human society uncertain as to whether to preserve or eliminate the wolf; but is that the real wolf, or the one we imagine?

We recognize the danger of our approach. We cover a lot of ground, and we have aimed to provide accurate, detailed and contemporary, yet accessible, accounts of our subjects. We have, however, wanted to avoid becoming bogged down in the minutiae. To this end, we give hints for further reading, with references: either reviews or books that expand upon what we talk about, or key papers that have advanced the area. Our main intention has been to avoid cluttering up the text with a rash of citations. Boxes are used to expand on topics of interest, and those that lead on from the main subject of each chapter. Finally, in relation to format, it is not necessary to read the book in any particular order, and we would encourage readers to dip in and out as time and inclination allow. Each chapter is self-contained, although we have ensured that links are made to other chapters wherever possible.

We have been fortunate to be able to work closely with Steven Appleby on his illustrations. The mysterious chemistry of Steven's brain has allowed him, for each chapter, to capture an essence from amongst our cascade of facts. This process of artistic distillation from biology has created an unusual dimension to our project: the concept of exploiters and exploited has been realized via the world of Steven Appleby, and so has, for us at least, become very distinctly etched.

We await the specialists and their criticisms, but we have been heartened by many of the supportive and instructive comments we have had from specialists during the reviewing process. Ultimately, it's a matter of judgement, getting the balance right for our intended audiences: biology undergraduate students, higher level school students, and anyone interested in the world around them and how human beings interact with it. In the end, we believe that the end we seek – understanding, through a broad and detailed picture – justifies our means. We hope you will agree.

PART I
EXPLOITED

Sargassum and the Sargasso Sea — CHAPTER 2

It is well known to seamen, and others who have crossed the Atlantic ocean, that a certain part of that sea is generally covered more or less with a particular species of weed, called gulf-weed, but the reason of its accumulating there, and its origin, have given rise to much difference of opinion. The gulf-weed, or Fucus natans, (which may be translated to floating weed,) of botanists, is said to have been first met with by the early Portuguese navigators, when they extended their voyages to the south, in quest of discovery. It is related, that they found a part of the ocean, to the north and west of the Cape Verd Islands, profusely covered with it; and from the peculiar resemblance which the little nodules on it bear to a small species of grape, they gave it the name of Sargaçao, from their word sarga, the name of this fruit. From thence our term Sargasso is derived, and the part of the ocean in which it is found is usually called the Sargasso sea.

Anon, The Nautical Magazine, 1832

THE SARGASSO SEA, occupying a large region of the subtropical North Atlantic, brings to mind seamen's tales of mystery. Sea monsters, impenetrable clinging weeds, the lost vessels and aeroplanes of the Bermuda Triangle: these may all be considered to reflect the human capacity for imaginative invention. There are, however, many features of the Sea and its real inhabitants that are genuinely unusual or unique. First, the Sea itself: a sea is usually defined by the land surrounding it, as in North Sea, Baltic Sea, Caspian Sea; but the Sargasso Sea is bounded instead by a gigantic system of ocean currents known as the North Atlantic subtropical gyre, which reflects in turn the interaction of the prevailing winds with the rotating Earth. Then there is the *Sargassum* seaweed (referred to in the passage above by its old name *Fucus natans*), an alga from a family whose members normally attach securely to the seafloor by a holdfast. Yet in the Sargasso Sea it floats freely over a vast area, propagating vegetatively, providing a consistent environment over sufficient time to allow members of an abundant fauna to come to mimic the seaweed. One example is the sargassum frogfish, *Histrio histrio* (see Fig. 2.4). At the smaller end of the scale of life, the environmental genome of the Sargasso Sea is being prospected for its potential bounty of undiscovered genes, and unanticipated diversity has been encountered in marine bacteria. There are many transitory inhabitants of the Sargasso Sea, including turtles, barracuda and tuna. Perhaps the most remarkable behaviour is shown by the eel: American and European species of the genus *Anguilla* are believed to migrate thousands of miles to find the Sargasso Sea, where they spawn and die, their transparent, leaf-like

Biological Diversity: Exploiters and Exploited, First Edition. Paul Hatcher and Nick Battey.
© 2011 John Wiley & Sons, Ltd. Published 2011 by John Wiley & Sons, Ltd.

offspring then following the surrounding ocean gyre, eventually to return to the rivers from which their parents emigrated.

In this chapter we present for sober contemplation a survey of the remarkable Sargasso Sea, mindful that the mythology associated with the location reflects its extraordinary physical and biological features. The Sea is a vast vortex, within which lives an exotic cocktail of bacterial, plant and animal life, contained by and carefully adapted to the location. Underlying questions concern the future security of this unique ecosystem and the rights of ownership over the genes of the millions of creatures, both macroscopic and microscopic, that live in it.

LOCATION AND PHYSICAL CHARACTERISTICS OF THE SARGASSO SEA

Christopher Columbus is credited with the discovery of the Sargasso Sea, and he makes frequent reference to the weed in his Journal of the First Voyage (1492) (Box 2.1). The first encounter was on 17 September, nine days after Columbus left the Canary Islands in search of Eastern Asia; and although it would be a further 25 days before his ships made landfall (in the Bahamas), the journal shows the optimism to which the weed gave rise, as an apparent omen of nearby land:

> At dawn on that Monday they saw much more weed which seemed to be a river weed and in which they found a live crab which the Admiral kept. He says that these were sure signs of land, because they are not found 80 leagues from land. They found the water less salty since they had left the Canaries and the breezes more and more gentle. Everyone was very happy and the ships sailed as fast as they could to be the first to sight land . . . .[1]

BOX 2.1	CHRISTOPHER COLUMBUS' FIRST VOYAGE

When Columbus left Spain on his first voyage in 1492, his intention was to reach 'The Indies'—the Far East of the Asian continent—by a western sea route, as an alternative to the known eastern land route. One political motivation for the journey was to make contact with the Mongol Emperor of China (who had actually been deposed in 1368) and to seek a religious alliance for the Catholic Church against Islam. Columbus also sought to win for the Crown of Spain new territories over which he would become viceroy and from which Spain would gain economic benefit. He departed on 3 August 1492 from Palos de la Frontera, near Cadiz in southern Spain, initially sailing to the Canary Islands for ease of subsequent navigation: the Canaries are located at 28°N, the same latitude as the reputed location of 'Cipangu' (Japan), west across the Atlantic at unknown distance. On 12 October Columbus made landfall at Watlings Island,

Bahamas. He visited Cuba and Hispaniola, setting out for Spain again on 16 January 1493, and after being forced by storms to shelter in the Azores, returned to Palos on 13 March 1493, 32 weeks after his departure.

Columbus kept a detailed daily record of his journey, in the form of a journal. The existence of this record has been of major importance in preserving Columbus' reputation and place in history—in the face of vicious dispute over his claims. The original journal has been lost, probably sold by Columbus' grandson Luis. Our knowledge of it, and therefore of the detail of the first voyage, is thus due to the preservation of its contents in note form by Bartolomé de Las Casas, who consulted it extensively in writing his *Historia de las Indias* (begun in 1527). Hence Columbus' journal is in the third person.

---

[1]Ife (1990).

Thereafter, for a six-day period, Columbus experienced another characteristic of the Sargasso Sea – excessive calm, as the journal entry for 22 September shows:

> He steered WNW more or less, sometimes inclining one way, sometimes another. They made about 30 leagues. They hardly saw any weed. They saw some petrels and another bird. At this point the Admiral says: *I was in great need of this head wind because my men were very agitated and thought that no winds blew on these seas that would get them back to Spain.* For part of the day there was no weed; then it was very thick.[1]

The extent of the Sargasso Sea was subsequently defined by Krümmel in the nineteenth century based on the records of German ships plying the Atlantic, and in the early twentieth century by Winge using data obtained by Danish marine biological vessels, as the area of the North Atlantic where floating *Sargassum* weed was encountered, from the Azores in the north-east to the West Indies in the south-west (Fig. 2.1). However, because of the nature of floating weed, this is a shifting region; a better way to define the Sea is by the currents that bound it and create its unique environment. These currents are the Gulf Stream on the west, the North Atlantic Drift to the north, the Canaries Current to the east and the North Equatorial Drift to the south (Fig. 2.1). Together they make up the North Atlantic subtropical gyre, a huge clockwise-circulating movement of ocean water that encloses an area of about 4.4 million square kilometres. This gyre is brought about by the action of wind movements in combination with the rotating earth (Box 2.2).

The currents in the gyre are generally stronger on the west than the east. The Gulf Stream, for example, reaches speeds of 9 km/h off Cape Hatteras in North Carolina, and at its strongest moves 90 million cubic metres of water every second, reaching to the full depth of the ocean in coastal waters. In contrast, the Canaries Current is much more diffuse and shallow. The power of the Gulf Stream is important in the biology of the Sargasso Sea, because it creates local eddies (known as cold-core rings) which spin off into the Sea and lead to mixing as cool, nutrient-rich waters are drawn up from depth (Fig. 2.2). This is significant because the large-scale physical forces operating on the Sargasso Sea (Box 2.3) mean that in general its surface waters are warm, highly saline, and oligotrophic (nutrient-poor). As a consequence, although there is a local abundance of *Sargassum* weed, it is a relatively unproductive ecosystem overall. Cold core rings provide one important means of bringing nutrients into the system.

Fig. 2.1 *Currents surrounding the Sargasso Sea, and distribution of Sargassum weed based on Krümmel's (1891, dark blue) and Winge's (1923, light blue) surveys. Distribution of European eel larvae is given by red dotted lines which show the limits of occurrence of larval sizes (e.g. '25', larvae less than 25 mm only found within this curve) [Based on Ryther (1956) and Schmidt (1923).]*

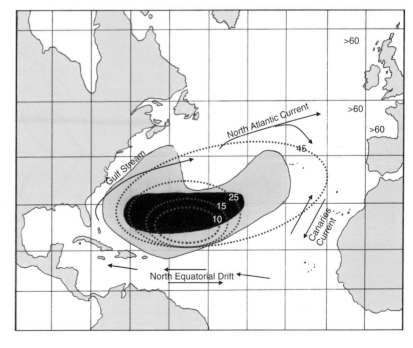

BOX 2.2 THE EARTH'S MAJOR OCEAN GYRES

The surface currents of the major oceans of the world are shown in Fig. 1. The essential cause of these currents is wind movement. This relationship was discovered by the Norwegian explorer Fridtjof Nansen. In 1893, Nansen allowed his wooden ship 'the Fram' to freeze into Arctic pack ice in an attempt to

Ekman transport also leads to upwelling regions along continental margins (see Fig. 1). These and the nutrient-rich waters at high latitudes are the really productive regions of the world's oceans.

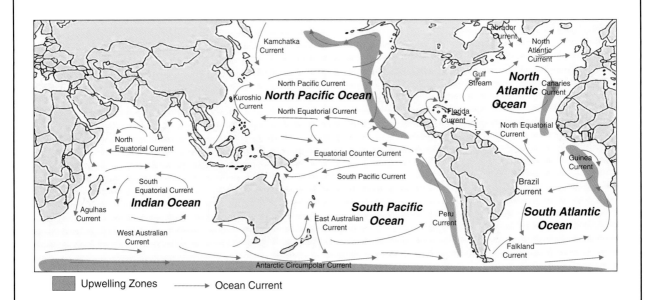

Upwelling Zones   → Ocean Current

Box 2.2 Fig. 1 *The Earth's major ocean gyres* [Adapted by Henry Battey from http://oceanmotion.org.]

drift to the North Pole. He observed that the ship (and the ice) drifted at an angle of 20–40 degrees to the right of the prevailing wind. He realized this was a result of the interaction of the force of the wind with that due to the rotation of the earth (the Coriolis force). The precise relationship between wind movement and water movement at increasing depths was worked out in detail by Walfrid Ekman in 1905 and is therefore known as Ekman transport. The net transport of water to a depth of about 150 m is at 90 degrees to the direction of the wind; below this depth the influence of wind movement is negligible.

The prevailing winds at different latitudes are shown in Fig. 2. Comparison with Fig. 1 shows that the major ocean gyres rotate in the expected relation to these winds, remembering that the Coriolis force acts in opposite directions in the two hemispheres—so the water moves to the right of wind direction in the northern hemisphere, and the gyres therefore move clockwise; whereas water moves to the left of the wind in the southern hemisphere, and the gyres there therefore rotate anticlockwise.

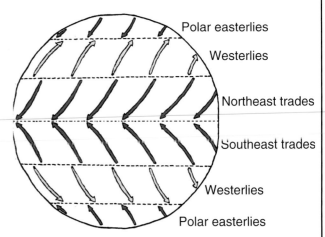

Polar easterlies

Westerlies

Northeast trades

Southeast trades

Westerlies

Polar easterlies

Box 2.2 Fig. 2 *Prevailing winds at different latitudes* [Adapted by Henry Battey from http://oceanmotion.org.]

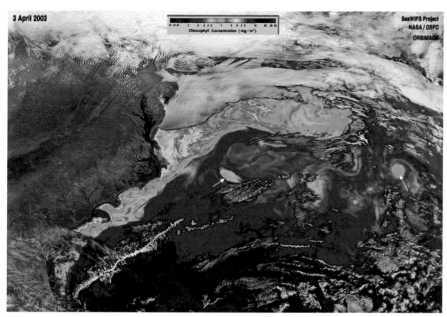

Fig. 2.2   *Cold-core rings in the Sargasso Sea. In this satellite image, chlorophyll content is shown in false colour, blue being low and red high (see scale at top). The Gulf Stream is relatively productive, with green colours grading to orange and red. The Gulf Stream can form warm-core rings to the north and cold-core rings to the south. Cold-core rings (arrowed) contribute nutrient-rich waters to the relatively barren Sargasso Sea, which is indicated by the blue and purple colours in this image [From http://oceanmotion.org.]*

BOX 2.3   LARGE-SCALE PHYSICAL PROCESSES IN THE SARGASSO SEA

The North Atlantic subtropical gyre is a massive eddy, and its net effect is to drive the water within it into a mound as much as a metre higher in the middle than at the edges. This mounding leads to a horizontal movement of surface water parallel to the contours running around the mound; this is called geostrophic flow and tends to reinforce the wind-driven water flow around the gyre. It also means that the Sargasso Sea is effectively a lens of water, about 500 m thick at the centre. It is relatively warm, mostly about 18°C, resting on the huge mass of permanently cold ocean water below (Fig. 1). The point where the temperature changes rapidly at depth is known as the permanent thermocline. Within 100 m of the surface, the water is seasonally much warmer than 18°C, rising to about 28°C during mid-summer; this creates the seasonal thermocline which stratifies the water: as water heats up it expands, becoming less dense and thereby harder to mix with the colder, heavier water below. This surface water tends to be less saline than the 18°C water below, possibly due to a diluting effect of rain (annual rainfall in the vicinity of Bermuda is 150 cm).

Running roughly east–west across the lens of Sargasso Sea water, at latitudes between 20°N and 30°N, is the subtropical convergence zone. Here the surface water of the northern Sargasso Sea, which is cooled in the winter by cold air from the North American continent, meets permanently stratified, permanently warm surface water. This creates fronts, as the colder water slides down under the warmer, lighter water to depths of about 100 m. Along the boundary between these different water masses rapid currents are generated, with surface speeds up to 80 cm/s (Voorhis and Bruce, 1982).

An important detail about the Sargasso Sea eastern boundary is that, because the Canaries current is relatively weak, the effect of the submerged mountain range formed by the mid-Atlantic ridge may be as significant – it is a sufficiently large topographical feature to constrain the water circulation in the ocean (Longhurst, 1995). It is notable that it is on its eastern edge that the hydrographic and biological (distribution of *Sargassum* weed) boundaries of the Sea are least clearly defined and correlate least well.

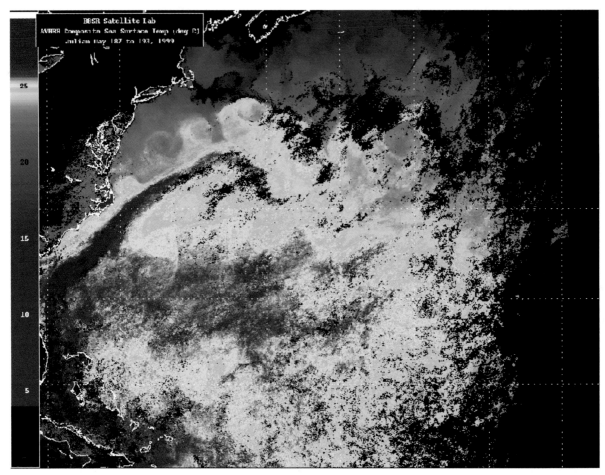

Box 2.3 Fig. 1   *Large-scale physical processes in the Sargasso Sea. Satellite image shows warmer water in red and orange, cooler water in blue and purple, and land and clouds in black. The legend on the left shows sea surface temperatures in degrees Celsius. The Gulf Stream is visible as a 'river' of warm (red) water flowing along the east coast of the USA and then bending seaward at Cape Hatteras in North Carolina. Three cold core rings can be seen between the Gulf Stream and Cape Cod [Reproduced by permission of Bermuda Institute of Ocean Sciences (BIOS).]*

SARGASSUM WEED AND ITS INHABITANTS

The genus *Sargassum* has 577 currently recognized species distributed through the warm-temperate and tropical zones of the world. It is a brown alga (Class Phaeophyceae, Order Fucales), related to the attached genera *Fucus* and *Laminaria*, both important members of biofouling communities familiar from temperate sea shores (see Chapter 13). Only two species, *S. fluitans* and *S. natans*, are considered to be fully pelagic (free-floating), and together they make up the majority of the floating weed in the Sargasso Sea. Their common name, sea holly, reflects the fact that the thallus (the plant body) is highly branched, with leaf-like blades and berry-like floats (Fig. 2.3). There is more *S. natans* than *S. fluitans*, but both are common. These two species are said to grow only by vegetative means, presumably having evolved from attached, benthic (sea bottom-dwelling) species with the accompanying loss of a means of attachment, the holdfast, and of sexual reproduction. Growth occurs at the tips of the branches, being most rapid in the spring and summer; vegetative reproduction occurs by fragmentation, tip regions giving rise to new plants.

The life of an individual plant is probably most often curtailed after a period of some years when the associated fauna, including encrustations, builds up to such an extent that the buoyancy provided by its air bladders is overcome. Gradually the weed sinks and begins a 4000 m descent to the sea floor. Here it makes an important contribution to the nutrition of the deep-sea benthos.

The most striking thing about the distribution of *Sargassum* in the Sargasso Sea is its irregularity. Most typically it is found in clumps of less than a metre across. However, these clumps can form into parallel lines aligned in the direction of the wind, which generates counter-rotating eddies that push the weed into 'windrows'. The horizontal spacing of these rows increases in proportion to wind speed. In addition to the wind, mesoscale circulation effects (current variations on the scale of hundreds of kilometres) also influence distribution of the weed. Where water bodies of different temperatures meet and downwelling currents are generated, *Sargassum* can build into rafts up to 50 m across, giving the appearance of solidity and extension which so impressed early travellers through the region.

Fig. 2.3 Sargassum fluitans
[*From http://algaebase.org.*]

Although accurate data are difficult to obtain, there are an estimated 4−11 million tonnes of *Sargassum* weed in the Sargasso Sea. This sounds a lot; but once the size of the sea is accounted for, it amounts to only 0.9−2.5 g fresh weight per square metre, less than the planktonic biomass in the water beneath. Overall, *Sargassum* is estimated to contribute only about 0.5% of total primary production of the water column, but this rises to 60% in the surface 1 m of water. The relatively low productivity of *Sargassum*, compared to that of attached seaweed in the littoral zone, reflects the nutrient-poor status of the surface water in the Sargasso Sea. In being nutrient-poor, the Sea is typical of tropical pelagic regions (see Box 2.2).

The blue, exceptionally clear water of the Sargasso Sea, in which light can be detected at depths of up to 1000 m, coupled with the high light intensity and temperature at the surface, offer an ideal environment for rapid photosynthesis, were it not for this extreme limitation of available nutrients. It is the consequent low productivity that has led the Sargasso Sea to be described as an ocean desert; it is therefore even more paradoxical that the floating weed itself hosts a miniature ecosystem, teeming with a diversity of life more like a jungle.

This diversity can at first seem daunting − but a good way to grasp the key elements of the community is through its food web. Table 2.1 is derived from detailed studies carried out some time ago, but it shows the principal animal groups involved. The major food sources are plankton; the *Sargassum* itself and the epiphytes growing on it (cyanobacteria, red algae, diatoms); and detritus from the community, which includes organic aggregates (protein and carbohydrates) whose precipitation is aided by phenolic compounds exuded by the *Sargassum*. The filter feeders (hydrozoans, bryozoans, barnacles and the tube worm *Spirorbis* attached to the weed) bring food into the community from the plankton and from particulate material of both plant and animal origin; omnivorous amphipods, isopods and copepods feed on the alga and/or on detritus. The tube worm *Platy-nereis* also includes hydrozoans and bryozoans in its diet. At the next trophic level are the carnivores such as pycnogonids (sea spiders) and nudibranch gastropods (sea slugs). At the top of the food web are carnivorous shrimp, crabs and fish. However, in all but its lowest stratum the community is characterized by blurring of trophic levels, and by a great degree of opportunism in feeding habits, imposed by the limited availability of food. For instance, although the crab *Planes minutes* is a carnivore it has been observed to eat *Sargassum* weed and even lumps of tar in times of need. Similarly the snail *Litiopa* will consume plant as well as animal material. There are apparently no strict herbivores in this community.

Table 2.1 The Sargassum fauna [*Adapted from Butler et al. (1983); Geiselman (1983)*]

Group	Generic names of commonly occurring examples	Principal food sources
Bryozoans	Membranipora	Plankton, detritus
Hydrozoans	Zanclea, Clytia, Obelia	Plankton, detritus
Barnacles	Lepas	Plankton, detritus
Anthozoan: sea anemone	Anemonia	Copepods, worms, plankton
Copepods	Harpacticus, Macrochiron	Detritus
Amphipods	Sunamphitoe, Hemiaegina	Sargassum, epiphytic algae, detritus
Isopod	Bagatus (Carpias)	Sargassum, epiphytic algae, detritus
Polychaete (tube) worms	Platynereis, Spirorbis	Sargassum, epiphytic algae, hydrozoans, bryozoans
Flatworms	Acerotisa	Copepods, bryozoans
Pycnogonid	Anoplodactylus	Bryozoans, hydrozoans
Shelled gastropod (snail)	Litiopa	Bryozoans, hydrozoan remains, Sargassum, detritus, epiphytic algae
Nudibranch gastropods	Doto, Scyllaea, Spurilla	Hydrozoans, anemone, bryozoans
Shrimps	Latreutes, Hippolyte	Worms, amphipods, isopods, hydrozoans, copepods
Fish	Syngnathus, Histrio	Small motile animals and plankton
Crabs	Planes, Portunus	Any of the Sargassum fauna with the exception of shrimps

Fig. 2.4 *Sargassum frogfish, Histrio histrio, showing hand-like fins* [Photograph courtesy of Dr Sung Kim from the Census of Marine Life www.coml.org and Korean Ocean Research and Development (KORDI).]

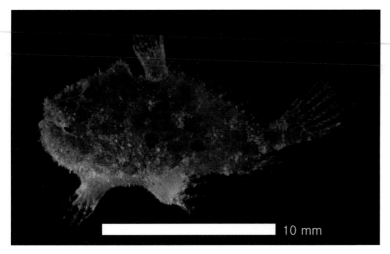

10 mm

A remarkable feature of the *Sargassum* fauna is the high degree of endemism (being peculiar to this particular locality) and specialized adaptation to the surrounding weed rafts. The invertebrate fauna comprises about 100 species, of which nearly 10% are endemic; and two fish species, *Syngnathus pelagicus* (the sargassum pipefish) and *Histrio histrio* (the sargassum frogfish) are considered to be endemic. Individual *Histrio* spend their whole lives in association with a particular raft of *Sargassum*, and not only have surface markings and colouration to make them blend with the weed (cryptism), but also prehensile pectoral fins which allow them to grasp, finger-like on to the fronds as they stalk their prey (Fig. 2.4). In more open parts of the weed they can move extremely rapidly using jet propulsion: frogfish can open their mouths so wide that they create at least a 12-fold increase in the size of the oral cavity, and the gill aperture is restricted to a small opening at the base of the pectoral fin; the normal respiratory cycle can therefore generate powerful pulses of water. The large mouth also creates negative pressure on rapid opening, sucking in prey; and the location of the gill apertures hides respiratory movements, aiding stealth. In addition, the sargassum frogfish has subdermal cavities which act as buoyancy aids.

The pipefish (*S. pelagicus*) shows mimesis, its elongate body providing camouflage against the weed; and the sargassum crab exhibits protective colouration and markings which mimic the seaweed whilst white flecks give the appearance of bryozoan and other encrusting colonies. In general, the fauna has evolved to look like *Sargassum*. These pronounced adaptive features and examples of endemism, the highly integrated nature of the sargassum community, and its widespread occurrence wherever the pelagic weed is found, all point to an ancient origin for the association. The animal community has been described as a 'displaced benthos'—most of the species are derivative of littoral types, suggesting their origin in that environment, but long enough ago to allow close adaptation to the floating life. Interestingly, the *Sargassum* weed itself appears to have changed little, consistent with an apparent lack of sexual reproduction (based on the absence of reproductive structures on thalli).

Jerzmánska and Kotlarczyk (1976) reported a fossil assemblage very like that currently found in the Sargasso Sea (including fucalean algae without holdfasts, and *Syngnathus incompletus*, similar to *S. pelagicus*), from the siltstones of the Carpathian mountains, deposited between 40 and 20 million years ago. This has invited speculation that the *Sargassum* assemblage began to evolve in the Carpathian region of the ancient Tethys Sea, which separated the super-continents of Laurasia and Gondwanaland. With the closure of the Tethys by 17 million years ago, the assemblage is supposed to have moved to its current position where it has been maintained by the action of the North Atlantic subtropical gyre.

What caused the formation of the *Sargassum* community and maintains the association now? Based on current observations, the key factors seem to be nutrition and protection. *Sargassum* releases dissolved organic matter which is metabolized by heterotrophic bacteria. Levels of phosphates in the water surrounding the weed can be two to three times those in the open ocean, and since nitrogen fixation in the oceans is often limited by availability of phosphates and organic matter, the presence of *Sargassum* is beneficial to autotrophs as well. For example, the surface of the weed provides a home for epiphytic nitrogen-fixing cyanobacteria. In general it is thought that the presence of the floating *Sargassum* increases productivity of the micro-organisms at the base of the food web, providing the foundation for the diverse community of larger organisms. Varying degrees of development of epiphytic communities have been documented in samples from different parts of the Sargasso Sea, and this has been attributed either to varying production of antimicrobial compounds by the *Sargassum*, or to varying nutrient availability across the Sargasso Sea. An increase in diversity with age of *Sargassum* rafts is also a feature, and since fragmentation at the growing tips generates new rafts, these have been likened to individual islands, each with its own distinctive resident population.

The weed rafts also attract a more transitory, loosely associated fish fauna. Jacks (*Caranx* and *Seriola* species) gain protection and feed on copepods and larval decapods, using the *Sargassum* as a nursery area for their young. Filefish and triggerfish also feed on copepods, as well as hydrozoans and bryozoans. Large predatory species like barracuda, tuna and swordfish tend to associate with the *Sargassum* because of feeding opportunities. In addition a number of species of sea turtle visit the *Sargassum* weed during their pelagic phase, which lasts for several years, before they return to the beaches on the east coast of North America to lay their eggs. There is particular worry that the turtles, which appear to seek out *Sargassum* rafts, put themselves at increased risk from commercial trawling of the *Sargassum* weed, and from oil and tar pollutants and discarded plastic waste. Wind-driven eddy effects tend to concentrate these pollutants with the weed rafts. The accumulation of plastics, tar and other pollutants within ocean current systems like the North Atlantic subtropical gyre is a topic of great current concern.

THE OPEN OCEAN: THE PLANKTON COMMUNITY AND PHYSICAL DETERMINANTS OF PRODUCTIVITY

The Sargasso Sea itself is the focus of current research interest: rather than the rafts of weed with their extraordinary community, biological oceanographers seek to understand the plankton population of the open water between the rafts. Physical factors (climate, water movements, nutrient input and loss) are seen as determinants of changes in plankton composition and extent through the year (so-called physical 'forcing'). The position of the Sargasso Sea—at the centre of the North Atlantic gyre, at sub-tropical latitudes—condemns it to relative unproductivity, compared to higher latitude ocean waters, and to those at upwelling regions along continental shelves (Boxes 2.2 and 2.3). However, the planktonic life is the most productive element in the Sargasso Sea, and the vast volume of water there is important for global biogeochemical processes. As discussed by Raven and Falkowski (1999), the oceans of the world contain 50 times more inorganic carbon than the atmosphere, so the interaction between the two is an important determinant of atmospheric CO_2 levels.

As a whole, the North Atlantic ocean shows very strong seasonality—through the year the surface layer undergoes large changes in nutrient input due to depth of mixing, and as a consequence of heat and light inputs. This can lead to pronounced spring blooms of phytoplankton, but these blooms vary in both extent and biological composition, according to location and to year. The key factors are the critical depth—that above which, if nutrients are not limiting, phytoplankton show net growth; and the depth of the mixed layer—where significant vertical mixing of water occurs. In the Sargasso Sea, surface waters become thermally stratified during the summer and mixing is then very limited; surface water temperatures rise to 20–28°C. Only during winter does mixing occur, and cooler water of around 18°C (so-called 18-degree water) reaches almost to the surface. This cooler water brings extra nutrients which lead to a late winter/early spring bloom. At more northerly latitudes outside the Sargasso Sea (e.g. the subpolar waters to the south of Iceland) the mixed layer is deeper relative to the critical depth, and therefore phytoplankton productivity is much greater. Spring typically leads to blooming, but anomalously high spring mixing can actually decrease phytoplankton growth by drawing them down to deeper waters, below the critical depth.[2] Some examples of phytoplankton are shown in Fig. 2.5.

The northern Sargasso Sea follows the pattern for subtropical waters, and spring phytoplankton blooms are typical. However, south of the transition zone between latitudes 20 and 30°N (the subtropical convergence zone, roughly from the tip of Florida in the west, to the Canaries in the east—see Box 2.3), it becomes more tropical, surface temperatures never falling below 20°C. The wide band of 18-degree water characteristic of the Sargasso Sea becomes thinner, the surface waters are more permanently stratified, and there is a correspondingly lower input of nutrients from deep water and lower overall phytoplankton productivity.

Current oceanographic research in the Sargasso Sea is focused at the Bermuda Institute of Ocean Sciences; the Bermuda Atlantic Time-series Study (BATS)

[2]For further detail on phytoplankton abundance in relation to critical depth, see for example Dutkiewicz *et al.* (2001).

is a long-term ocean sampling programme designed to characterize the seasonal changes in the ocean community and the physical factors that control these dynamics. The main sampling area to the south-east of Bermuda has been monitored since 1954, and BATS has been in operation since 1988. In a recent modelling study using BATS data, Salihoglu *et al.* (2008) showed that autotrophic eukaryotes and cyanobacteria are responsible for 63% and 33%, respectively, of carbon production. The community is largely dependent on mineral nutrients brought up from depth, or recycled: only 9% of nitrogen comes from nitrogen fixation; and phosphate probably limits production throughout the year, even during the spring bloom. *Prochlorococcus* and *Synechococcus* are the dominant cyanobacteria, but neither appears to fix nitrogen; that activity seems to be principally carried out by another cyanobacterium, *Trichodesmium*.

The dominant autotrophic eukaryotes are prymnesiophytes (golden-brown coccolithophores, with ornate calcium carbonate skeletons) and pelagophytes (chrysophytes), which both significantly contribute to spring blooms. Diatoms (golden-brown algae with silica shells) are also present but bloom less frequently and much later. Prasinophytes (green algae) and dinoflagellates (photosynthetic protists) are less significant members of the autotrophic community in this location.

Microbial heterotrophs use up much of the primary production – leaving relatively little for metazoans. This is reflected in the fact that bacteria and nanozooplankton (protists in the size range 2–5 μm) represent about 70% of the heterotrophic carbon in the surface waters of this part of the Sargasso Sea. The protists include the sarcodines (Acantharia, Radiolaria and Foraminifera), which often have symbionts and therefore also contribute in a minor way to primary production. Overall, zooplankton peaks tend to follow the peaks in phytoplankton production. There is also a pronounced diurnal vertical migration, zooplankton biomass in the upper 200 m of the water column roughly doubling at night. Migrating species include copepods, euphausiids and amphipods.

In the wider Sargasso Sea, bacterial biomass varies but usually equals and can exceed that attributable to phytoplankton. This indicates that heterotrophic bacteria are crucial to carbon processing there. This is in addition to the key role of the photoautotrophic cyanobacteria. The growing realization of the relatively uncharacterized microbial biodiversity in subtropical gyres like the Sargasso Sea is reflected in the Environmental Genome Shotgun Sequencing project, led by Craig Venter. There is also a more general realization that the role of bacteria and viruses in ocean systems has previously been underestimated. But before discussing Craig Venter's project, we need to describe a very ancient association – that of eels, long-distance travellers to the Sargasso Sea.

EXTRAORDINARY VISITORS TO THE SARGASSO SEA – EELS

The eels come from what we call the entrails of the earth. These are found in places where there is much rotting matter, such as in the sea, where seaweeds accumulate, and in the rivers, at the water's edge, for there, as the sun's heat develops, it induces putrefaction.

Aristotle, *Historia Animalium*

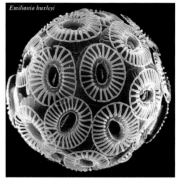

(a) *Emiliania huxleyi*

(b) *Coccolithus pelagicus*

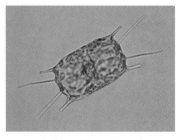

(c) *Odontella mobiliensis*

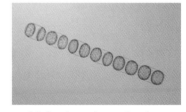

(d) *Stephanopixus palmeriana*

Fig. 2.5 *Coccolithophores and diatoms illustrating the morphologies of some phytoplankton.* **(a)**, **(b)**: *scanning electron micrographs of the coccolithophores* Emiliania huxleyi *(approx 5 μm diameter) and* Coccolithus pelagicus *(approx 30 μm diameter);* **(c)**, **(d)**: *bright field images of the diatoms* Odontella mobiliensis *and* Stephanopixus palmeriana *(approx 30 μm diameter)*[Courtesy of Colin Brownlee, Marine Biological Association, Plymouth, UK, and the Plymouth Culture Collection.]

Still in the pelagic realm, but now at depths of several hundred metres, we can find and follow a different kind of biology, more remote than the weed and its abundant community. Eels are organisms whose origin has fascinated since Aristotle (Box 2.4). At depths of around 200 m, adults are believed – for they have still never been discovered there – to spawn. The eggs – although these have never been found in the wild – are assumed to develop there into larvae which have the appearance of willow leaves, and are known as leptocephali (Fig. 2.6). Reassuringly, these tiny, translucent larvae *have* been found, and their paths back to the rivers and streams of Europe, Iceland and North America have been the subject of almost obsessive scientific study over the past 100 years.

It is more appropriate, however, to start at the rivers of Europe, and to relate how the spectacular migration of mature silver eels back to the sea naturally prompted the quest for the spawning ground of these curious fish.

In 1896, the Italians Grassi and Calandruccio provided the crucial proof that the tiny sea creature *Leptocephalus brevirostris* was in fact the larva of *Anguilla anguilla*. Johannes Schmidt of the Danish Commission, discovering this larva south of Iceland in 1904, and finding it subsequently along the edge of the North European continental shelf, concluded that the freshwater eels of Europe spawn somewhere way out in the Atlantic. By 1922, extensive data collection from both research and commercial vessels had tracked the larvae, based on their ever-decreasing size, to

---

BOX 2.4   EELS

It is frequently stated that Aristotle believed that eels are generated spontaneously from mud, and that this is indicated by the passage quoted at the beginning of this section of the main text. The expression 'entrails of the earth' could, however, be translated from the Greek as 'worms of the earth' – in which case Aristotle may have been aware that eels originated from elvers, the young eels which make their ascent from the sea and often hide in mud during their journey (Bertin, 1956). He certainly knew that adult eels migrate from freshwater back to the sea.

Over the ensuing 2000 years, however, the dominant view seems to have associated the origin of eels (as with snakes) to spontaneous generation, either by abiogenesis (from inanimate matter) or by heterogenesis (from another organism). Hence the myth that eels arise from horse's hair, parodied by the Victorian poet Robert Browning as follows:

Spontaneous generation need I prove
Were facile feat to Nature at a pinch?
Let whoso doubts, steep horsehair certain weeks,
In water, there will be produced a snake;
A second product of the horse...

The Ring and the Book, Book IX Doctor Johannes-Baptista Bottinius, lines 1350–54

A yet more imaginative idea, local to the Fenland of East Anglia, UK, is that in the tenth century St Dunstan in a fit of pious rage transformed sinful monks into eels – giving Ely, cathedral city of the eely fens, its name.

It was not until Francesco Redi's early work on parasitology ('Animals living in Animals', 1684) that scientific clarification about the life-history of the eel was achieved. He states that:

> ... with the first August rains and by night when it is most dark and cloudy, the eels begin to descend from the lakes and rivers in compact groups, towards the sea. Here the female lays her eggs from which, after a variable time depending on the rigours of the sea, hatch elvers or young eels which then ascend the fresh waters by way of the estuaries. Their journey begins at about the end of January or the beginning of February and finishes generally at about the end of April.

Redi's principal interest was in the worms sometimes found living within eels, which were one reason for the widespread belief that eels are viviparous (give birth to live young). He showed that the organisms living within the body of the eel were parasitic worms, not its offspring.

Nevertheless, the belief in eel vivipary persisted; it was perpetuated, for example, by Linnaeus. And it was fuelled by the similarity between eels and the eelpout, *Zoarces viviparus*, a truly viviparous fish known colloquially as aalmutter (eel mother). Peculiarly intense interest therefore attended the sex organs of the eel. Finally, in 1777, Mondini discovered and made the first description of the ovaries of a female eel, which he described as 'frilled organs' lining the body cavity. In 1874, Syrski discovered the presumptive male organs, in the same position as the ovaries in a female, but looped rather than frilled in appearance; but he failed to find spermatozoa. Sigmund Freud searched intensively for definitive proof that these 'organs of Syrski' were male sex organs, and his first published paper (in 1877) acknowledges his failure to achieve this. Eventually, by the end of the nineteenth century such proof was obtained, coinciding with the discovery of the larval leptocephali in the sea.

The life-cycle of the freshwater eel is summarized in Fig. 1. Eels in the genus *Anguilla* are catadromous, which means that they spend their adult lives in freshwater, returning to the sea to spawn. This is the opposite to anadromous salmon (Chapter 9). The leptocephalus stage lasts about 2.5 years and the yellow eel stage from 5 to 50 years. As yellow eels reach near-maturity, they undergo changes (silvering, subcutaneous fat accumulation, enlargement of the eyes, hormonal changes, reduction in feeding) in preparation for their return to the sea. That return, building to a peak in the autumn, and associated with moonless nights and the rise of waters following heavy rain, can be dramatic. The instinct to return to the sea drives the eels frantic: they would rather be caught than turn back. At the Italian eel fisheries of Comacchio it was common practice to light fires at the water's edge to calm migrating eels caught in traps—phototaxis is just one of several exquisite eel sensitivities which include monitoring of current speed, temperature and salinity. It has also been suggested that European eels begin to migrate in anticipation of stormy weather because they can detect

microseisms (ground vibrations due to pressure changes associated with a depression over the North Sea) which reach inland waters before the depression arrives (Sinha and Jones, 1975).

Of most contemporary significance is the massive decline in European and American eel stocks over the last 30 years. Changes to waterways, over-fishing, parasitic infestation, have combined with a small effective population size to make both species of *Anguilla* particularly vulnerable. These effects are exacerbated in what Wirth and Bernatchez (2003) have described as a 'fatal synergy', by the long spatial and temporal life-cycle of the eel. As a consequence extreme strategies are needed to conserve what was once a biological phenomenon characterized by its exuberant abundance.

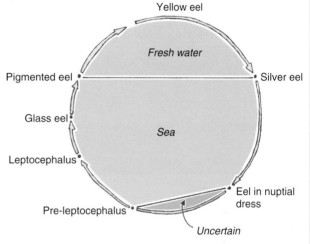

Box 2.4 Fig. 1 *Life-cycle of the freshwater eel* [Adapted by Henry Battey from Sinha and Jones (1975).]

the Sargasso Sea. In Fig. 2.1 we have included information from Schmidt's famous paper 'The breeding places of the eel', showing the distribution of size limits of the larvae of the European eel (*A. anguilla*). Interestingly, the American eel (*A. rostrata*) also spawns in the Sargasso Sea, its larvae making the comparatively short journey to the rivers of the Atlantic coast of North America.

It is hard to know what to say about this story. On the one hand, modern research has identified a comparable pattern of migration for Japanese eels, which apparently spawn near sea mounts in the Pacific Ocean to the west of the Mariana Islands, and migrate on the Kuroshio Current to the rivers of Japan. Population genetic studies confirm that the American and European eels are indeed distinct species (something contested by Denys Tucker in a famous challenge to the Schmidt hypothesis in 1959); even within the European eel species there is some genetic heterogeneity.[3] This suggests that in the Sargasso Sea there are distinct locations for, or timing of, breeding of eel species, and possibly of populations. Nevertheless, overlap has been deduced for the breeding grounds of the American eel and the European eel, and the timing of spawning in the spring. The subtropical convergence zone, where the tropical

[3]For discussion of this vexed issue, see Wirth and Bernatchez (2001), Dannewitz *et al.* (2005) and van Ginneken and Maes (2005).

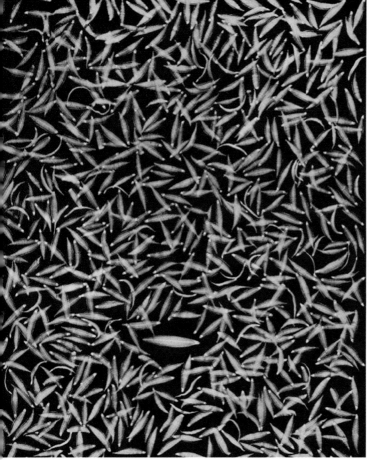

waters of the southern Sargasso Sea meet the subtropical waters from further north (see Box 2.3), may act as an environmental trigger for spawning. The leptocephali of both species move with the surface currents of the Gulf Stream, and the North Atlantic Drift in the case of the European eel, to find their way to continental waters where they metamorphose into glass eels and elvers; further development into yellow and finally silver eels typically occurs in the freshwaters of rivers and streams (see Box 2.4). The journey time of *A. anguilla* from the Sargasso Sea to the rivers of Europe is about 2.5 years.

On the other hand, there is much that is still uncertain and unclear. Schmidt's work can viewed as classic, not as a piece of biological sleuthing but as an example of a scientist over-interpreting his data. So sure was Schmidt of his conclusion that, arguably, his method ensured that he found exactly what he was looking for: he sampled the ocean much more intensively in the area where he believed the eels spawned (Boëtius and Harding, 1985). Thus, while it is clear that the larvae are abundant and small in the Sargasso Sea, the east–west boundaries of the proposed spawning region are uncertain. Furthermore, because the adults are presumed to spawn in the Sargasso Sea but have never been found there, it is still not certain that European silver eels migrate across the Atlantic to their birth place. Perhaps the eggs are spawned in the coastal waters of Europe (where the adults then die) and float on currents until they reach the Sargasso Sea, where conditions most suit their development into larvae. The established view that European eels make the 3000-mile journey back across the Atlantic is therefore still in need of confirmation. Current tagging studies of European eels are likely to provide crucial information to help resolve the issue.[4]

Fig. 2.6 *Leptocephali (from Schmidt, 1923).* **Inset:** *Modern image of a single leptocephalus, scale bar = 1 mm [Courtesy Dr Russ Hopcroft at the University of Alaska Fairbanks.]*

EXTRAORDINARY VISITORS TO THE SARGASSO SEA—CRAIG VENTER

The aim of the Environmental Genome Shotgun Sequencing project, from which data were first published in 2004, was to describe the microbial diversity in the Sargasso Sea environment by analysing the range and character of the genes isolated from filtered seawater samples. The filters were designed to specifically capture the microbial rather than viral or eukaryotic members of the community. Genomic DNA was isolated, cloned and sequenced to produce about 2 million sequences of average length about 800 base-pairs. These were assembled to provide outline information on the genomes of the organisms present. Striking heterogeneity of sequences from *Prochlorococcus* suggests the presence of ecotypes; this finding is backed up by other research which indicates the evolution of ecotypes of *Prochlorococcus* and *Synechococcus* adapted to light and nutrient conditions at specific regions of the water column. In terms of gene discovery, probably the most dramatic finding from the Venter project has been the diversity of proteorhodpsin-like genes. These are bacterial versions of genes which encode

[4]See van Ginneken and Maes (2005), Aarestrup *et al.* (2009), and www.bbc.co.uk/radio4/worldonthemove/species/european-eel/

light-sensing rhodopsin proteins from the retina in the eyes of higher animals, and they offer the potential to couple light energy to ion transport (protons and other ions such as chloride) and thus to allow chlorophyll-independent phototrophy. Their significance to the carbon economy of the open ocean is still unclear, but they indicate that some bacteria previously considered exclusively heterotrophic may have an alternative energy supply. This gives reason to reconsider the conventional view of the oceanic food web; it seems likely that the Sargasso Sea will contribute important information on this question in the future.

The Sargasso Sea is thus a site for gene discovery. It is also subject to accumulation of plastics and other pollutants and, presumably, to a decline in populations of eel larvae (eels are now given conservation status in Europe, indicating the seriousness of their decline there). Ocean currents, like the Gulf Stream, which surround and to some extent define the Sea may be seriously altered by climate change, which may further diminish the boundaries of this unusual Sea. The future of this remarkable region is unclear, but it is perhaps unlikely to continue for long as the home for quite such an exceptional and unique assemblage of organisms.

FURTHER READING

We are not aware of any recent books or articles covering the Sargasso Sea in general. Early descriptions of the *Sargassum* weed and its distribution were made by Winge (1923) and Parr (1939); reviews have been written by Deacon (1942), Ryther (1956), Teal and Teal (1975). The *Sargassum* community was studied intensively in the 1960s, 70s and 80s, and good summaries can be found in Peres (1982), Butler *et al.* (1983), Coston-Clements *et al.* (1991) and Hacker and Madin (1991). The fish fauna was analysed by Dooley (1972) and some of it was beautifully photographed by Sisson (1976); and Pietsch and Grobacker (1987) provide a comprehensive account for those interested in the bizarre and remarkably adapted frogfish. Fish (1987) provides an interesting account of the frogfish jet-propulsion mechanism. The evolutionary origin of the *Sargassum* weed and its community is considered by Lüning (1990). Evidence for antimicrobial compounds from *Sargassum* was provided by Conover and Sieburth (1964), who suggested geographical variation in the extent of microbial colonization of the seaweed could reflect variation in antimicrobial levels; Carpenter and Cox (1974) obtained evidence that nutrient availability was the determinant, with iron suggested as the likely limiting factor. The spacing of *Sargassum* windrows was analysed by Faller and Woodcock (1964) and the contribution of dead *Sargassum* weed to the nutrition of the deep-sea benthos by Schoener and Rowe (1970). Lapointe (1995) has investigated the productivity of the *Sargassum* weed.

Key early works on plankton and productivity in the Sargasso Sea are those written by Menzel and Ryther (1960, 1961) and Ryther and Menzel (1960). A detailed study of the plankton community by Steinberg *et al.* (2001) provides the basis for the description given here. Longhurst (1995, 1998) is an influential overview of worldwide ocean productivity.

Reviews of the biology of the eel are provided by Bertin (1956), Sinha and Jones (1975) and Tesch (2003). Important papers on the breeding places of European and American eels are Schmidt (1923), Tucker (1959) and Kleckner and McCleave (1987). There is a large literature on the population structure of eels, and the location and timing of breeding and its environmental triggers: useful entry points are Lecomte-Finiger (2003), van Ginneken and Maes (2005). Evidence on the breeding place of Japanese eels is summarized in Tsukamoto (2006) and molecular evidence on the evolution of the genus *Anguilla* is assessed in Minegishi *et al.* (2004).

Environmental genomics in the Sargasso Sea has been pioneered by Venter *et al.* (2004); advances in understanding marine bacteria, and the value of sequence data from Venter's project for systematics are summarized by Giovannoni and Stingl (2005). Proteorhodopsins and their implications for ocean phototrophy are discussed by Béjà *et al.* (2000, 2001). For information on BATS see http://www.bios.edu/research/bats.html; and on marine biodiversity in general see the Census of Marine Life at http://www.coml.org/.

REFERENCES

Aarestrup, K., Økland, F., Hansen, M. M. *et al.* (2009) Oceanic spawning migration of the European eel (*Anguilla anguilla*). *Science* **325**, 1660.

Anon. (1832) Sargasso weed. *The Nautical Magazine* **1**, 175–179.

Béjà, O., Aravind, L., Koonin, E.V. *et al.* (2000) Bacterial rhodopsin: evidence for a new type of phototrophy in the sea. *Science* **289**, 902–906.

Béjà, O., Spudich, E.N., Spudich, J.L., Leclerc, M. and Delong, E.F. (2001) Proteorhodopsin phototrophy in the ocean. *Nature* **411**, 786–789.

Bertin, L. (1956) *Eels: a biological study.* Cleaver-Hume Press, London.

Boëtius, J. and Harding, E.F. (1985) A re-examination of Johannes Schmidt's Atlantic eel investigations. *Dana* **4**, 129–162.

Browning, R. (1971). *The Ring and the Book* (ed. R.D. Altick). Yale University Press, New Haven, USA.

Butler, J.N., Morris, B.F., Cadwallader, J. and Stoner, A.W. (1983) Studies of *Sargassum* and the *Sargassum* community. Bermuda Biological Station Special Publication 22, Bermuda Biological Station for Research, USA.

Carpenter, E.J. and Cox, J.L. (1974) Production of pelagic *Sargassum* and a blue–green epiphyte in the western Sargasso Sea. *Limnology and Oceanography* **19**, 429–435.

Conover, J.T. and Sieburth, J.McN. (1964) Effect of *Sargassum* distribution on its epibiota and antibacterial activity. *Botanica Marina* **6**, 147–157.

Coston-Clements, L., Settle, L.R., Hoss, D.E. and Cross, F.A. (1991) Utilization of the *Sargassum* habitat by marine invertebrates and vertebrates: a review. NOAA Technical Memorandum NMFS-SEFSC-296. National Marine Fisheries Service, North Carolina, USA.

Dannewitz, J., Maes, G.E., Johansson, L., Wickström, H., Volckaert, F.A.M. and Järvi, T. (2005) Panmixia in the European eel: a matter of time. *Proceedings of the Royal Society of London B* **272**, 1129–1137.

Deacon, G.E.R. (1942) The Sargasso Sea. *The Geographical Journal* **99**, 16–28.

Dooley, J.K. (1972) Fishes associated with the pelagic *Sargassum* complex, with a discussion of the *Sargassum* community. *Contributions in Marine Sciences* **16**, 1–32.

Dutkiewicz, S., Follows, M., Marshall, J. and Gregg, W.W. (2001) Interannual variability of phytoplankton abundances in the North Atlantic. *Deep-Sea Research II* **48**, 2323–2344.

Faller, A.J. and Woodcock, A.H. (1964) The spacing of windrows of *Sargassum* in the ocean. *Journal of Marine Research* **22**, 22–29.

Fish, F.E. (1987) Kinematics and power output of jet propulsion by the frogfish genus *Antennarius* (Lophiiformes: Antennariidae). *Copeia* **1987**, 1046–1048.

Geiselman, J. (1983) The food web of the *Sargassum* community. Bermuda Biological Station Special Publication 22, pp. 260–270, Bermuda Biological Station for Research, USA.

Giovannoni, S.J. and Stingl, U. (2005) Molecular diversity and ecology of microbial plankton. *Nature* **437**, 343–348.

Hacker, S.D. and Madin, L.P. (1991) Why habitat architecture and color are important to shrimps living in pelagic *Sargassum*: use of camouflage and plant-part mimicry. *Marine Ecology Progress Series* **70**, 143–155.

Ife, B.W. (1990) *Christopher Columbus: journal of the first voyage 1492*. Aris and Phillips, Warminster, UK.

Jerzmánska, A. and Kotlarczyk, J. (1976) The beginnings of the Sargasso assemblage in the Tethys? *Palaeogeography, Palaeoclimatology, Palaeoecology* **20**, 297–306.

Kleckner, R.C. and McCleave, J.D. (1987) The northern limit of spawning by Atlantic eels (*Anguilla* spp.) in the Sargasso Sea in relation to thermal fronts and surface water masses. *Journal of Marine Research* **46**, 647–667.

Lapointe, B.E. (1995) A comparison of nutrient-limited productivity in *Sargassum natans* from neritic vs. oceanic waters of the western North Atlantic Ocean. *Limnology and Oceanography* **40**, 625–633.

Lecomte-Finiger, R. (2003) The genus *Anguilla* Schrank, 1798: current state of knowledge and questions. *Reviews in Fish Biology and Fisheries* **13**, 265–279.

Longhurst, A. (1995) Seasonal cycles of pelagic production and consumption. *Progress in Oceanography* **36**, 77–167.

Longhurst, A. (1998) *Ecological Geography of the Sea*. Academic Press, San Diego, USA.

Lüning, K. (1990) *Seaweeds: their environment, biogeography, and ecophysiology*. John Wiley & Sons, New York.

Menzel, D.W. and Ryther, J.H. (1960) The annual cycle of primary production in the Sargasso Sea off Bermuda. *Deep-Sea Research* **6**, 351–367.

Menzel, D.W. and Ryther, J.H. (1961) Annual variations in primary production in the Sargasso Sea off Bermuda. *Deep-Sea Research* **7**, 282–288.

Minegishi, Y., Aoyama, J., Inoue, J.G., Miya, M., Nishida, M. and Tsukamoto, K. (2004) Molecular phylogeny and evolution of the freshwater eel genus *Anguilla* based on the whole mitochondrial genome sequences. *Molecular Phylogenetics and Evolution* **34**, 134–146.

Parr, A.E. (1939) Quantitative observations on the pelagic Sargassum vegetation of the western North Atlantic. *Bulletin of the Bingham Oceanographic Collection* **6**, 1–94.

Peres, J.M. (1982) Specific pelagic assemblages. In: *Marine Ecology, Volume 5, Part 1* (ed. O. Kinne), pp. 313–372. John Wiley & Sons, New York.

Pietsch, T.W. and Grobacker, D.B. (1987) *Frogfishes of the world: systematics, zoogeography, and behavioural ecology*. Stanford University Press, Stanford, CA, USA.

Raven, J.A. and Falkowski, P.G. (1999) Oceanic sinks for atmospheric CO_2. *Plant, Cell and Environment* **22**, 741–755.

Ryther, J.H. (1956) The Sargasso Sea. *Scientific American* **194**, 98–104.

Ryther, J.H. and Menzel, D.W. (1960) The seasonal and geographical range of primary production in the western Sargasso Sea. *Deep-Sea Research* **6**, 235–238.

Salihoglu, B., Garçon, V., Oschlies, A. and Lomas, M.W. (2008) Influence of nutrient utilization and remineralization stoichiometry on phytoplankton species and carbon export: a modeling study at BATS. *Deep-Sea Research I* **55**, 73–107.

Schmidt, J. (1923) The breeding places of the eel. *Philosophical Transactions of the Royal Society of London Series B* **211**, 179–208.

Schoener, A. and Rowe, G.T. (1970) Pelagic *Sargassum* and its presence among the deep-sea benthos. *Deep-Sea Research* **17**, 923–925.

Sinha V.R.P. and Jones J.W. (1975) *The European Freshwater Eel.* Liverpool University Press, Liverpool.

Sisson, R.F. (1976) Adrift on a raft of *Sargassum. Scientific American* **149**, 188–199.

Steinberg, D.K., Carlson, C.A., Bates, N.R., Johnson, R.J., Michaels, A.F. and Knap, A.H. (2001). Overview of the US JGOFS Bermuda Atlantic Time-Series Study (BATS): a decade-scale look at ocean biology and biogeochemistry. *Deep-Sea Research II* **48**, 1405–1447.

Teal, J. and Teal, M. (1975) *The Sargasso Sea.* Little, Brown, New York, USA.

Tesch, F.-W. (2003) *The Eel.* Blackwell, Oxford.

Tsukamoto, K. (2006) Spawning of eels near a seamount. *Nature* **439**, 929.

Tucker, D.W. (1959) A new solution to the Atlantic eel problem. *Nature* **183**, 495–501.

van Ginneken, V.J.T. and Maes, G.E. (2005) The European eel (*Anguilla anguilla*, Linnaeus), its lifecycle, evolution and reproduction: a literature review. *Reviews in Fish Biology and Fisheries* **15**, 367–398.

Venter, J.C., Remington, K., Heidelberg, J.F. *et al.* (2004) Environmental genome shotgun sequencing of the Sargasso Sea. *Science* **304**, 66–74.

Voorhis, A.D. and Bruce, J.G. (1982) Small-scale surface stirring and frontogenesis in the subtropical convergence of the western North Atlantic. *Journal of Marine Research (Supplement)* **40**, 801–821.

Winge, Ö. (1923) The Sargasso Sea, its boundaries and vegetation. *Report on the Danish Oceanographical Expeditions 1908–1910 to the Mediterranean and Adjacent Seas* **3**, 3–34.

Wirth, T. and Bernatchez, L. (2001) Genetic evidence against panmixia in the European eel. *Nature* **409**, 1037–1040.

Wirth, T. and Bernatchez, L. (2003) Decline of North Atlantic eels: a fatal synergy. *Proceedings of the Royal Society of London B* **270**, 681–688.

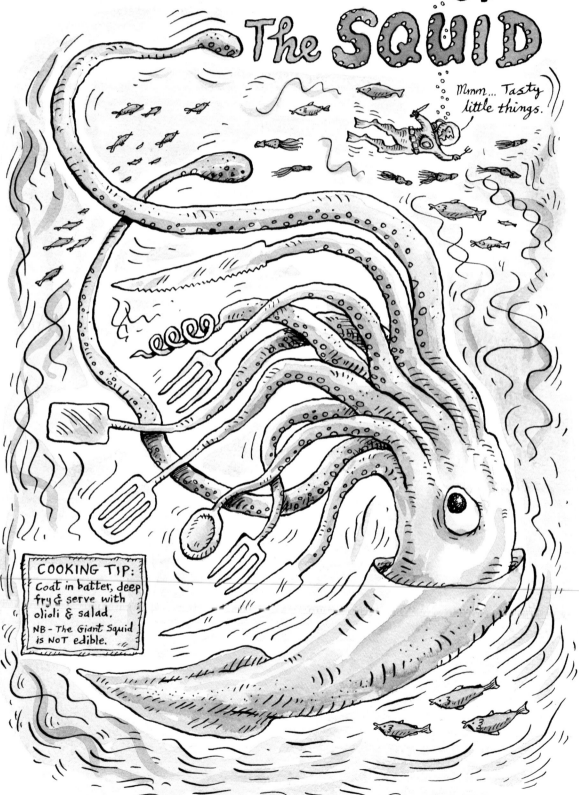

Cephalopods

CEPHALOPODS, FOR EXAMPLE the large oceanic squid and the octopus, may seem like exotic creatures, to be viewed only in nature documentaries. Yet, take a walk along a European beach and you may come across the cuttle 'bone' from a cuttlefish. On some beaches, one of our favourite being the World Heritage Site 'Jurassic Coast' between Charmouth and Lyme Regis in Dorset, UK, you can easily find fossil ammonites and the 'lighting bolts' from belemnites (Fig. 3.1). All are cephalopods: Greek for 'head-footed'.

The cephalopods (class Cephalopoda), which are members of the phylum Mollusca, hold several records: they were probably the first active visual predators to evolve, and contain by far the largest invertebrates alive today. In this chapter we will consider the evolution of the cephalopods, their variety of forms, and convergence with fish. Then we will consider two squid species as representative of those cephalopods exploited by man, and end with discussion of the giant squid, which only recently have become more fact than fiction.

Cephalopods have become the most complex and advanced molluscs (themselves the most diverse group of animals after the insects) and have largely become adapted for a swimming existence. There are only about 650 living cephalopod

(a)

(b)

(c)

Fig. 3.1 **(a)** *Large ammonite fossil from the Lower Jurassic on Monmouth Beach, Lyme Regis, Dorset (the shoe is 33 cm long).* **(b)** *Reconstruction of an ammonite: the colouration is conjectural and based on museum reconstructions.* **(c)** *Saucer full of belemnites, all from Charmouth Beach, Dorset, UK.* **(d)** *Internal shell of a belemnite. The portion represented in (c) is arrowed* [(d) *Redrawn form Lehmann (1981).*]

Phragmocone   Proostracum

Rear   **Front**

(d)

Biological Diversity: Exploiters and Exploited, First Edition.   Paul Hatcher and Nick Battey.
© 2011 John Wiley & Sons, Ltd. Published 2011 by John Wiley & Sons, Ltd.

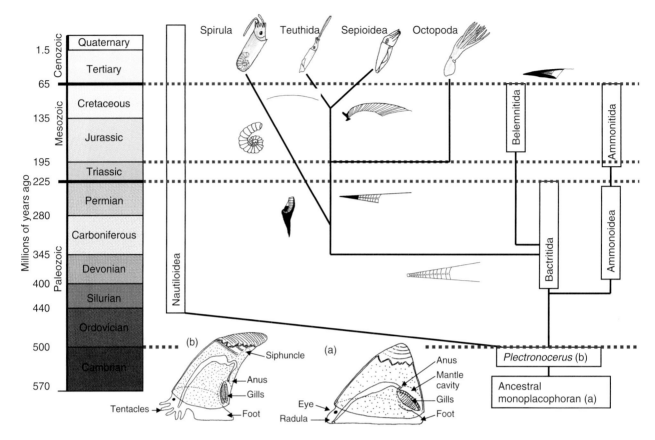

Fig. 3.2 *Evolution of the cephalopods, with some representative internal shells, and putative ancestral monoplacophoran* (a) *and ancestral cephalopod* Plectronocerus (b). *Dashed red lines indicate major extinction events [Partly after Barnes (1980).]*

species, all marine, but the fossil record contains over 10 000 named species. They probably evolved during the Cambrian from early monoplacophoran molluscs, which were somewhat like present-day limpets (Fig. 3.2). The molluscs are characterized by a dorsal body wall covered by a thin layer—the mantle—which secretes the calcium carbonate shell, and a ventral surface shaped as a foot. The earliest known cephalopod, *Plectronocerus cambria* (Fig. 3.2b), already had the distinguishing feature of early cephalopods—a shell enclosing an air-space, providing buoyancy.[1] After neutral buoyancy had been achieved, changes in other molluscan organ systems became possible: the foot developed into tentacles for feeding and a sealed tube for jet propulsion. The mantle cavity became the chamber for this jet pump and for pump-circulation of water over the gills. Ward (2006) maintains that gas exchange rather than locomotion was the main reason for this developing, and suggests that the external shell of early cephalopods evolved as a respiratory structure in response to the low oxygen content of Cambrian oceans. The efficient cephalopod gill could have been a key to their success: cephalopods were the dominant carnivores in the sea for 450 million years. Faster propulsion required improved blood circulation and a nearly closed blood system developed and a complex brain and behaviour evolved.

Molluscan evolution, cephalopods included, has to be pieced together largely from fossilized shells because very few remains of internal organs or fleshy parts have been preserved. The early cephalopods had several phases of explosive radiation, first in the Cambrian of species with generally small, smooth, elongated shells less than 6 cm long. These became all but extinct at the end of the Cambrian, with a re-radiation

[1]This has recently been called into question by Smith and Caron's (2010) reinterpretation of *Nectocaris pteryx* as a primitive, possibly stem group, cephalopod. This is from the Burgess Shale 505 mya, 10–15 million years older than any other cephalopod fossil. It has no shell or radula, but has a large flexible anterior funnel, stalked eyes and a single pair of prehensile tentacles. If categorised correctly, this suggests that jet propulsion evolved in cephalopods before the acquisition of a shell.

BOX 3.1	**AMMONITES**

The ammonites were the most species-rich group of cephalopods, and are the most abundant fossils found in some strata. Their name is derived from the ancient Egyptian god Ammon, who is represented by the head of a ram with twisted spiral horns. Humans have collected ammonite fossils for at least 25 000 years: fossil ammonites were worshiped as the Chakra of the Hindu god Vishnu, and the Greeks created artistic portrayals of ammonite fossils on their coinage (the cult of Ammon was popular with the Greeks: the earliest representation of an ammonite is from 350–250 BC).

Yet, if we did not have a couple of species of living *Nautilus*, a shelled cephalopod with similarities to the ammonites, to examine, it is possible that we would not be able to place the ammonites and fossil nautiloids as cephalopods at all.

Nautiloids appear in the early Ordovician, and have a chambered shell, originally straight. As they grow a wall is secreted between the body chamber and the chamber behind, with a narrow tube (the siphuncle) running along the middle part of the shell connecting the living chamber with the rest of the shell. The empty chambers are gas filled to aid buoyancy.

Ammonoids (subclass Ammonoidea) probably evolved from nautiloid-like ancestors during the early Devonian, about 400 million years ago (mya) (see Fig. 3.2). The suture lines of ammonite shells—the join between the chamber walls and the outer shell—are wavy or crimped, compared to the simple curves found in nautiloids, and the siphuncle generally runs along the outer edge of the whorl. During the late Devonian the ammonoids radiated world-wide.

Some fossils of parts of the ammonite body other than the shell have been found, and indicate that the ammonites had a beak with two jaws; the lower one was up to four times the size of the upper and shaped like a shovel. Ammonoids are thought to have had eight to ten long grasping arms, but it is not known if these had suckers or barbs. Like *Nautilus*, the ammonoids would not have been capable of rapid squid-like movements, and thus were probably not able to actively hunt prey. Furthermore, their mouthparts were probably not capable of cutting or biting. Thus, it is likely that they were bottom feeders, pushing their shovel-like lower jaws along the sea bed like a dustpan.

The ammonoids radiated several times, and seem to have been prone to periodic extinctions: just escaping total extinc-

tion at the Permian–Triassic boundary (when over 90% of all organisms went extinct), and at the Triassic–Jurassic boundary, before a last radiation as ammonites proper (strictly, the term ammonite refers only to these last species, the earlier ones are referred to as ammonoids) in the Jurassic and Cretaceous.

All ammonites went extinct towards the end of the Cretaceous, while nautiloids did not, their species richness declining during the Cenozoic.

Thus, the ammonites seem to have been subjected to boom or bust evolutionary bursts and to die out in times of crisis. A variety of reasons for the final extinction of the ammonites have been put forward (see also Box 3.2). The fossil record suggests that their extinction started before the final late Cretaceous mass extinction, and thus we have to look for causes other than meteorites, or other proposed causes of this extinction for the start of the final decline of ammonites.

Rather, this extinction was probably linked to changes in the late Cretaceous marine ecosystem. During the late Mesozoic there was a radiation of shell-crushing marine animals—clawed crabs, lobsters, skates and rays, carnivorous gastropods, bony fishes, and marine reptiles—which could have preyed upon ammonites. Ward (1983) has suggested that the evolution of ammonites during the Cretaceous reflects this predation pressure, with ammonite shells evolving in three directions. First there were streamlined planispiral shapes with tight coiling, which may have provided increased hydrodynamic stability and speed. Second there was increased shell ornamentation, including spines and ribs in non-streamlined species that could have afforded greater protection. Finally, a variety of partly uncoiled 'heteromorph' shells appeared. These ammonites may have lived near the surface of the sea, or at middle depths, and may have had upward pointing heads. This may have been an adaptation to avoid predators.

Ultimately, we do not know why the ammonites became extinct, or why some nautiloids survived. Was their thicker shell better able to survive attack, or did they have some behavioural or physiological adaptation that aided survival? *Nautilus* has the ability to suppress its metabolic rate drastically and unlike other extant cephalopods is tolerant of hypoxia. Thus it could escape predators by retreating to deep anoxic waters—could ammonites?

from remaining species in the Ordovician, this time throughout the world. The first coiled shells appeared in the lower Ordovician and, during a further range expansion into open water in the middle Ordovician, the greatest differentiation of shelled cephalopods was found; some had shells up to 10 m long. From some of these, the most famous shelled cephalopods, the ammonites, developed (Fig. 3.1 and Box 3.1).

The only cephalopods with external shells alive today are a couple of species of *Nautilus*. Other species, which have only the remains of an internal shell or no shell at all, are placed in the subclass Coleoidea, and are termed endocochleate, in comparison with the externally shelled exocochleate species. As these internal shells are often very reduced (see Fig. 3.2) and fragile, they do not fossilize nearly as well as the externally shelled species, and much remains to be learnt about their evolution. It is thought that all later cephalopods are descended from a small group of nautilus-like cephalopods, the bactrids, which had their main development in the late Silurian. The transition to an internal shell probably occurred slowly, with one of the first coleoid groups, the belemnites (from the Greek for 'lightning arrow'), developing in the late Devonian to early Carboniferous. The portion often found as a fossil on beaches (see Fig. 3.1) is only the most heavily calcified part of the internal shell, which would have also consisted of a chambered phragmocone at the front – this is fragile and easily detached, and is rarely preserved. All belemnites, which were abundant during the Jurassic and Cretaceous, became extinct at the end of the Cretaceous (see Fig. 3.2).

The evolution of an internal shell could have been due to competition or predation pressure from fish (Box 3.2): it would have enabled these cephalopods to float horizontally if it was located centrally; to have outer skin which could become pigmented for camouflage (very important in modern coleoids); and would allow the evolution of lateral fins for propulsion, improved stability and manoeuvrability.

The origin of the modern cephalopods – cuttlefish, squid and octopods, which radiated in the last 100 million years – is uncertain. From the early coleoids such as the belemnites the shell developed in four directions. In Spirula an unthickened coiled shell is retained internally. In squid (order Teuthida) found from the Jurassic and Cretaceous, the shell developed into a horny sword-like rod, the gladius, which comprises only 0.5% of the body weight and serves as an internal support. In the cuttlefish (order Sepioidea) the shell has become the cuttlebone – an internal skeleton and buoyancy aid consisting of large numbers of closely spaced lamellae with narrow-walled air-filled chambers in between. Finally the octopods (orders Vampyromorphida and Octopoda) have lost the shell entirely, except for a tiny stylet.

The squid, order Teuthida, illustrate most of the characteristics of coleoid cephalopods. Some of these characteristics may have evolved either in competition with or to avoid predation from fish (Box 3.2). The head of the squid projects into a circle of large prehensile limbs; this is the foot of other molluscs. Squid have eight short and heavy arms in four pairs, and a pair of much longer tentacles (Fig. 3.3) used to catch prey and also in reproduction. Cuttlefish have the same arrangement, but octopods only have the four pairs of arms. The inner surface of the arms is covered with adhesive discs and sometimes also with hooks and pegs. Muscles on the inner surface of the discs can contract and produce a vacuum. Squid are entirely carnivorous, and prey, which may include fish, crustaceans and other squid, is drawn by the arms and tentacles towards the mouth, where a pair of horny beaks, the lower overlapping the upper, tears off pieces of meat and passes them back to the radula and gut. The radula, a characteristic of molluscs, is a horny, moveable plate containing rows of backward-pointing chitinous teeth; it breaks down the food and transports it to the gut (Fig. 3.4).

Locomotion in squid is mainly by jet propulsion. Water is drawn into the mantle cavity, also containing the gills and reproductive organs. Then the circular mantle muscles contract, the edges of the mantle cavity lock around the head, and thus water can only leave though the narrow ventral funnel opening (Fig. 3.3). This funnel is flexible and can be pointed in different directions

BOX 3.2	CEPHALOPODS AND FISH: CONVERGENCE, COMPETITION AND PREDATION

Eighteenth and nineteenth century anatomists tried to demonstrate the similarity in morphology between cephalopods and fish. This was doomed to failure; morphologically the two phyla are very different. However, both inhabit broadly the same habitat and occupy similar niches. In a very influential paper, Packard (1972) proposed that cephalopods and fish share many functional similarities, which had arisen by convergent evolution. Packard recognized that there were significant differences between fish and cephalopods, for example, mode of locomotion and reproduction, and life-cycle, but he stressed the many similarities. These include:

o mode of life: both groups are marine, utilize all main zones of the seas, are visual predators, and have the same size range;
o have a similar body form, including fins, streamlined, do not bend in a vertical plane, they both shoal and some can 'fly';
o hydrostatic control: both the swim bladder of fish and cuttle-bone of cuttlefish provide about 5% lift;
o feeding: all are carnivorous visual feeders with a similar diet, selectively pouncing on their food;
o physiology: including vascular system, gills, formation of urine;
o central nervous system, with the cephalopod brain in the same size class as the vertebrate brain, similar functional connections, e.g. between retina and brain;
o eyes and vision: lateral position, structure of eye;
o photogenic system: some species of both have light-producing systems;
o colour, displays and behaviour: both can change colour and pattern to mimic surroundings;
o both typically grow continuously and allometrically.

Packard (1972) went on to suggest that the adaptive zone of swimming marine predators was first occupied by the cephalopods and then by vertebrates. Modern fish and cephalopods both evolved and radiated during the Mesozoic, and Packard suggested that cephalopods lost their external shell as an evolutionary response to increase mobility and diving ability (the water pressure would crush the thickest shell at depth) in the face of vertebrate competition. This led to a convergence of features between fish and cephalopods—with the suggestion that fish drove the evolution of the coleoids: cephalopods without external shells. The coleoids were then able to reinvade shallower coastal waters, possibly after the extinction of large fish-like reptiles or a greater differentiation of coastal habitats.

This scenario, for all its teleological rhetoric, is persuasive, but has come under criticism.

Modern squid are probably the cephalopods that most closely functionally resemble a typical fish, such as a salmon, but O'Dor and Webber (1986) have questioned whether squid can compete directly with fish. They suggest that squid have many constraints which preclude successful competition. The main one is locomotion: squid jet propulsion can produce high speed, but it is dependent on accelerating a small mass of water to a high speed through the funnel. This is inefficient compared with moving a large mass of water slowly as the fish does with its fins. Thus, it has been calculated that a squid uses twice as much energy to travel half as far as an average fish. Biochemical factors compound this inefficiency: their inadequate oxygen transport system means that squid have to pump large volumes of blood under high pressure, and also have a limited ability to digest energy-rich lipids, instead metabolizing proteins, of which they have poor reserves.

Overall, this means that squid have a high metabolic rate compared to fish and without reserves must feed all the time. Thus, the squid life-cycle of very fast growth and death after reproduction is contrasted with the slower growth and multiple reproduction of marine fish (a difference acknowledged by Packard). One can contrast the return migration of squid to their breeding ground, which is only achieved by schooling and cannibalism, with the lone return migration of salmon to freshwater, and their ability to survive in freshwater for months without feeding, using up to 50% of their mass as food reserves (Chapter 9). Such differences have great importance for squid fisheries management.

These ideas were extended by Aronson (1991) who suggested, this time comparing the octopus with fish, that predation not competition was the main evolutionary driver. Aronson used recently found fossil evidence which suggested that fish and both shelled and shell-less cephalopods were present at the same time during the late Paleozoic and early Mesozoic, and originally occupied the same shallow inshore and deeper offshore waters, and that the shallow-water shelled cephalopods were eliminated by predators.

Octopuses are examples of modern shallow-water coleoids, and their populations are often limited by fish. They shelter in dens and den availability may also limit their populations. As O'Dor and Webber (1986) conclude 'it is tempting to suggest that squid are no longer so much competing with fish as trying to stay out of their way', and it is probable that octopods are doing the same.

enabling the squid to move forwards or back, and giving it the greatest swimming speed of any invertebrate—some can also leap out of the water and reach over 10 kilometres per hour in the air. Posterior fins are used for slow swimming and also for directional stability under jet propulsion.

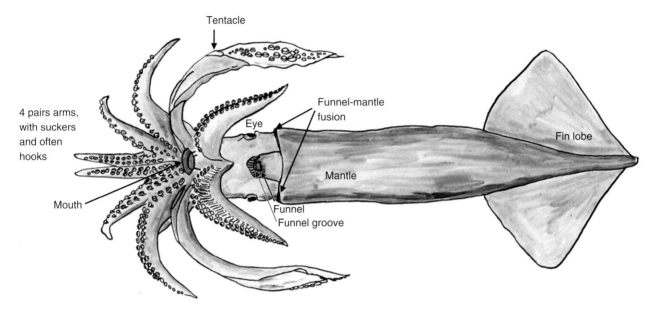

Fig. 3.3 *Ventral view of an adult* Todarodes pacificus, *to illustrate basic squid features (mantle length 30 cm). As most squid can vary their colouration greatly to match surroundings, a standard colouration cannot be given. In this and other squid illustrations in this chapter, the colourations are based on photographs of living examples of the species [Based on Nesis (1987).]*

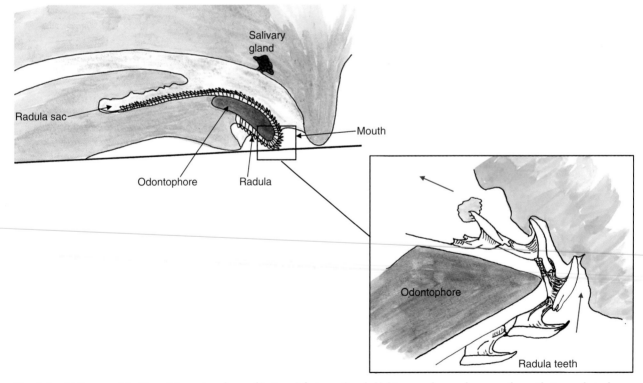

Fig. 3.4 *Molluscan radula. The radula consists of rows of backward-facing teeth embedded in a membranous base, over the cartilaginous odontophore. Muscles enable the odontophore to move in and out of the mouth, moving the radula, which can also move against the odontophore. As the odontophore is projected out of the mouth the radula flattens and the teeth are erected.* **Inset:** *Red arrows show direction of feeding movement [Based on Barnes (1980).]*

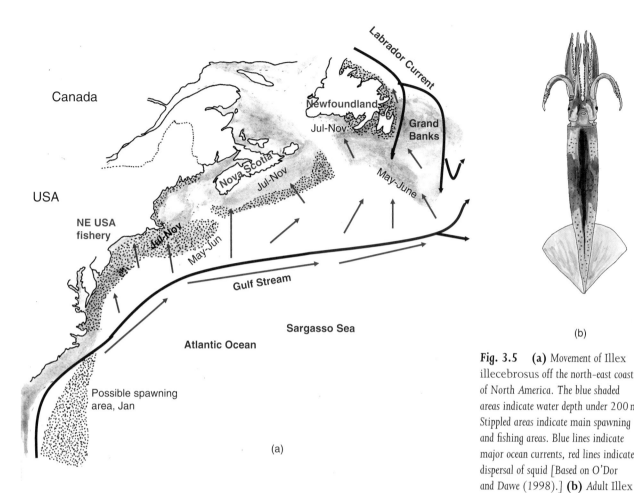

(a)

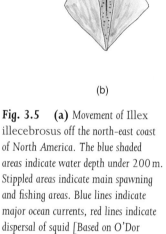

(b)

Fig. 3.5 **(a)** *Movement of Illex illecebrosus off the north-east coast of North America. The blue shaded areas indicate water depth under 200 m. Stippled areas indicate main spawning and fishing areas. Blue lines indicate major ocean currents, red lines indicate dispersal of squid [Based on O'Dor and Dawe (1998).]* **(b)** *Adult Illex illecebrosus [Based on Nesis (1987).]*

Cephalopods have separate male and female sexes. Sperm are packaged into large baseball-bat shaped spermatophores, stored in an organ called Needham's sac, and are passed into the female's mantle cavity to the oviduct, or near the mouth by a modified arm, the hectocotylus, on the male squid. Once in place, a cap on the top of the spermatophore opens and the sperm are released. Eggs are subsequently laid either attached to the ocean floor or floating free.

SQUID FISHERIES

Today, global fisheries catch more than three million tonnes of cephalopods a year, the majority squid. Until recently, two species comprised the majority of world-wide squid catches: *Illex illecebrosus* and *Todarodes pacificus*. These are both members of the Ommastrephidae, the oceanic so-called flying squid, and will serve also as examples of squid ecology and the problems inherent in studying and exploiting this group.[2]

Illex illecebrosus is one of four closely related species in this genus, all present in the western Atlantic, and is common along the north-east Atlantic coast of North America between Newfoundland and Florida (Fig. 3.5). Females produce up to 100 000 eggs in each of several large gel-filled egg masses of up to a metre in diameter. Spawning of these neutrally buoyant egg masses takes place at depth

[2]There are many other interesting commercially fished squid that could be investigated. If we had twice the space available we would also consider the Argentine short-finned squid (*Illex argentinus*) and the Humboldt squid (*Dosidicus gigas*). Both have recently extended their range and increased in abundance, and currently support the most important squid fisheries in the world, with 2008 recorded catches of 608 000 and 856 000 tonnes respectively.

in warmer southern waters off the continental shelf (eggs fail to develop below 12°C, and young larvae have an optimal temperature range of 20–26°C) mainly during the winter.

The newly hatched paralarvae, resembling a small adult, swim quickly to the surface, and while the summer hatchlings tend to remain near their spawning grounds, the winter ones move north and a little inshore on to the continental shelf as they grow (Fig. 3.5). This growth can be rapid, with a daily growth rate of over 5% of body weight being recorded, and they can reach up to 33 cm mantle length, and 1 kg weight as adults. As they grow their prey shifts from plankton to crustacea to fish. In the autumn of their first year they travel south and offshore again to spawn, after which most adults die.

The boom and bust nature of the *Illex* fishery is illustrated in Fig. 3.6. There are three main areas fished for *I. illecebrosus* (see Fig. 3.5); the Newfoundland fishery is the oldest, catching the squid between July and November very near the shore often from small open boats. This Canadian fishery developed from catching squid for export (mainly to China), through providing squid as bait for the cod fishery during the 1950s and 1960s, to lastly, in the early 1970s, for squid as human food, especially for Japan which was experiencing a decline in its squid stocks (see below). The other *I. illecebrosus* fisheries consist of the Canadian Nova Scotia fishery, and the north-east USA coastal fishery. From the late 1960s, other countries started to fish these two areas, with Cuba, Spain and the USSR landing large catches from the early 1970s. By 1979, 16 countries were fishing for these squid, and over 180 000 tonnes total catch was recorded, split 5:4:1 between the Newfoundland, Nova Scotia and US fisheries.

The crash came suddenly after 1979. Within a couple of years the total catch was less than 5% of the 1979 record, and since then the north-east USA coastal fishery has been dominant, but squid catches have been very variable, and usually less than 25 000 tonnes annually. The two fisheries off the Canadian coast have never recovered.

Fig. 3.6 Illex illecebrosus *recorded catches 1950–2008 [From FAO capture production statistics.]*

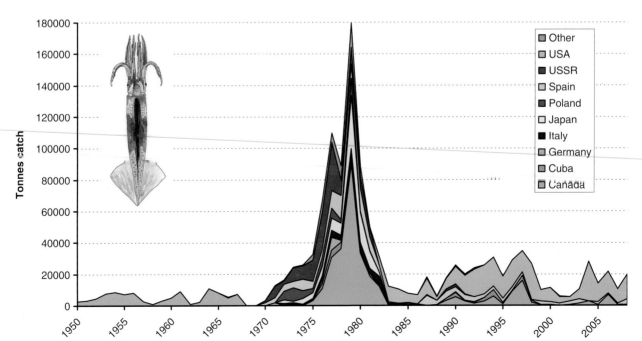

Since the crash of I. *illecebrosus*, fisheries for I. *argentinus* developed off the shelf and slope of the western South Atlantic. Catches increased from 7000 tonnes in 1977 to 700 000 tonnes in 1987, and although decreasing slightly since then still remain above 500 000 tonnes annually.

Todarodes pacificus, the Japanese common squid, or Surume Ika, is the most important commercial squid in Japan, and indeed was subject to the largest cephalopod fishery in the world. It is similar in many ways to I. *illecebrosus*; being largely annual, migratory, producing up to 500 000 eggs, and reaching a maximum mantle length of 30 cm. Its migratory patterns are complex, and still incompletely known. Three populations are now recognized: the winter population (Fig. 3.7a) originates in the East China Sea in January to April, juveniles move north with the Kuroshio Current along the Pacific coast and along the Sea of Japan, and disperse east along the Kuroshio Extension Current during the spring. By August they have reached their north-most limit, are almost adult and are fished in their feeding grounds in summer and autumn. They start to move south in October and November. The autumn population (Fig. 3.7b) originates in the western Sea of Japan in September to October, and spreads all over the Sea of Japan the following summer, before the females mature and start to migrate south by September. These give an early autumn fishery, almost entirely in the Sea of Japan.

Japanese squid fishing has a long history. Between 1661 and 1741, dried squid became a major trade item, and trade increased further after the 1860s when dried squid was exported to China. After the Second World War, squid fishing increased significantly as it was an industry that could be developed with very little capital in a war-ravaged country. At this time fishing was mainly carried out inshore using small unpowered boats with one to three men and simple jigging gear. Catches steadily increased as gear improved and, from the early

Fig. 3.7 *Movement of* **(a)** *winter and* **(b)** *autumn populations of* Todarodes pacificus. *Stippling indicates probable spawning areas. Blue arrows indicate northwards migration of juveniles during spring and summer, red arrows the southwards migration of adults during autumn and winter* [*Based on Okutani (1983).*]

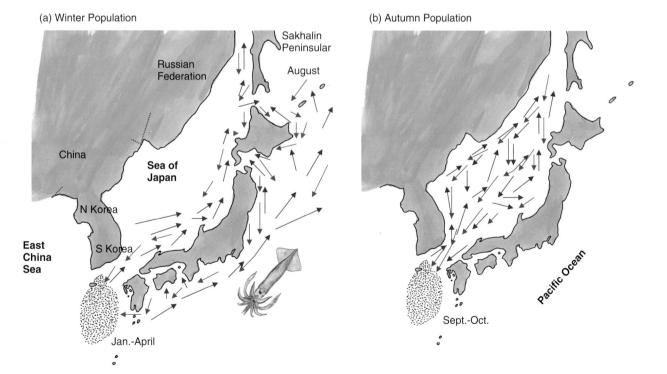

(a) Winter Population

(b) Autumn Population

1960s, automatic jigging machines were introduced (Box 3.3); by the late 1960s almost all boats were using this gear. The Pacific grounds off north Japan had a large catch in 1969 of 330 000 tonnes alone, but then underwent a steep decline, and offshore exploitation of the autumn squid population in the Sea of Japan was promoted. This was aided by the use of larger boats with better freezing capability. Improved navigational equipment now enabled these boats to find and track the migrating squid over long distances, and to select the areas of densest squid to fish: 200 000 tonnes were caught from this population alone in 1970 and 1971. Thus, before 1970, 80% of the catch was from inshore Pacific grounds, but afterwards 60–90% was from the Sea of Japan.

From 1972, the Pacific north coast fishery of the winter population continued to decline, from 154 000 tonnes average in 1970–2 to 6000 tonnes in 1976–9, and this masked the increase in catch from the autumn population in the Sea of Japan. However, the catch from this population also now fell steadily even as new grounds were found, to only 20% of the 1968 peak by 1986. Since the late 1980s

BOX 3.3 SQUID JIGGING

Although oceanic squid are quite vulnerable to high-opening bottom trawls during certain seasons, especially when they move on to the continental shelf and are not actively swimming (this method of trawling has been successfully used for *Illex illecebrosus* in the North Atlantic), a specialized form of squid fishing has developed over many centuries off Japan, largely for *Todarodes pacificus*. This involves fishing at night, with lights and squid jigs.

Squid jigging is one of Japan's oldest coastal fisheries and makes use of the observation that some squid are attracted to lights. From the early seventeenth century simple resin-rich pine-root torches were used, giving way during the twentieth century first to acetylene and paraffin lamps, then from the 1930s to electric lamps. Traditionally, bait of fish or squid would be used on the jig to attract the squid, and hand lines with one jig would be operated by two or three fishermen on a small wooden boat less than 1 km from shore, working for one night at a time. During the 1950s automatic squid jigging machines, each able to reel two lines containing 50 or more jigs were developed, and boats increased in size by the early 1970s to accommodate up to 30 jigging machines and with the capability to fish for over 20 days continuously. Nowadays artificial unbaited lures are used: the jig consists of a flexible plastic stem 4–8 cm long (*T. pacificus* possibly favours red and green jigs) with one to three rings of sharp barbless hooks at the lower end (Fig. 1). Barbed hooks could be used, and would reduce the risk of the squid escaping the jig, but whereas the squid falls from the barbless hooks of the jig automatically when the line is hauled over the side, it would have to be manually removed from a barbed jig, which would be costly in crew time.

The lights and jigs have to be positioned carefully on the boat. Although *T. pacificus* responds to light, it waits in shadow to attack prey visible against the illuminated area: at 1 m from the lure the squid pauses, closes to 60 cm and then vigorously attacks the jig. At this point the jig must be quickly hauled aboard. Therefore, lights are positioned on the centreline of the vessel to give a shaded area below the boat, and lines placed so that they enter the water at the edge of the beam of light.

Echo-sounding is primarily used to locate schools of squid. This is not easy as squid do not differ greatly in density from water and thus do not give a strong echo, unlike fish with swim bladders. Also, compared to fish, squid schools are small, containing only between 100 and 2000 animals. However, use can be made of the knowledge that squid prefer the border between waters of different temperatures, and are generally found in waters between 5°C and 10°C. The echo-sounder will detect this thermocline, partly by the change in density of the water with temperature, and partly because plankton and small fish also collect at this point and return a good echo.

Night-time squid fishing is usually done on dark, moon-free nights; during full moon *T. pacificus* tends to search for mates rather than feeding. Once test-jigging has confirmed the presence of a school of squid, the sea anchor is released (the vessel must be held stationary in the water so that the jigs can fall vertically) and serious fishing begins. The lights are activated and the jigs set to the depth at which the squid is found (*T. pacificus* is very sensitive to light, and can detect at a depth of 40 m the light from 13 kW of lamps from a small boat: larger boats deploy up to 400 kW of light each). After the jig is paid out to the required depth, it is rapidly jerked up 50 cm and allowed to settle again, thus attracting the squid. The fishermen have to find the most efficient jerking pattern to attract the squid and also have to notice immediately when a squid has been caught. Caught squid deteriorate quickly and must be kept on ice or deep frozen.

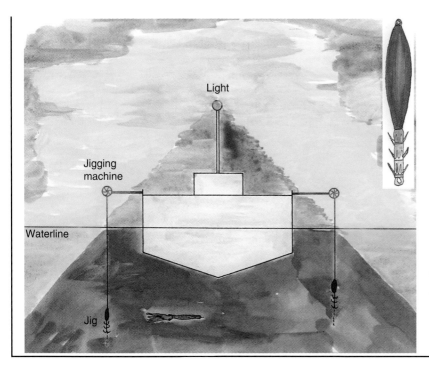

Box 3.3 Fig. I *Squid jigging with light. The highest catch is obtained by setting the jigging machine so that the jig enters the water at the edge of the light.* Inset: *Jig for* Todarodes pacificus, *with three rings of hooks: total length 11.5 cm, of which 7.5 cm is the coloured body* [Based on Hamabe *et al.* (1982).]

stocks have recovered and catches almost reach the 1960s peaks (Fig. 3.8). During the 1970s decline in T. *pacificus* stocks, many larger boats transferred to *Ommastrephes bartramii* 'Akalka' squid fishing. Today, more T. *pacificus* is caught by Korea than Japan.

The histories of these two squid fisheries raise a number of questions. How were these fisheries managed? Why did they suffer such serious population crashes (was overfishing to blame)? How can squid fisheries be managed sustainably? And perhaps most important of all, why are these squid species in these areas able to be so productive during some periods?

To discuss the last question first, knowledge of squid ecology is only just becoming sufficient to provide some clues. It is thought no coincidence that both these major, and at times very productive squid species (I. *illecebrosus* populations are thought to be two to four times as productive as similar ocean-shelf fish species) are associated with major permanent current systems—the Gulf Stream flow in the north-west Atlantic for I. *illecebrosus* and the Kuroshio flow in the north-west Pacific for T. *pacificus*. These are some of the strongest and fastest (up to 6 knots) permanent current systems in the world and are principally wind driven, and arise when waters of different densities (i.e. of different temperatures or salinities) meet—see Chapter 2. At this point a sharp horizontal density gradient is produced by a convergence of flows towards a zone of intersection where the denser water sinks beneath the less dense. This zone is not vertical but tilted, and plankton and small organisms tend to concentrate here, and this is where the young of both squid species feed. The egg masses can find a zone of neutral buoyancy and young can swim vertically to the surface when they hatch, where they will tend to be carried into the zone of food concentration. For I. *illecebrosus*, even if spawning in late winter, egg masses may be picked up by or rise into the Gulf Stream as they develop and be delivered back to sites along the shelf edge in time to meet the next spring plankton bloom.

The squid feed whilst being moved north by the currents, and their growth rate matches the change in size-class of the most abundant food (as the planktonic

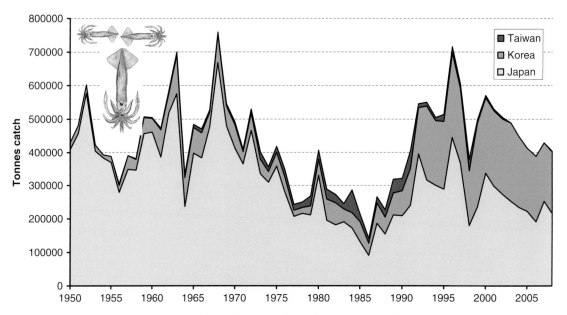

Fig. 3.8 Todarodes pacificus *recorded catches 1950–2008 [From FAO capture production statistics.]*

biomass is passed through the food web by larger organisms eating the smaller ones, so the size of the most abundant food increases) and thus they can feed continuously on the most abundant food. For example, *Todarodes pacificus* young larvae are limited to the temperate surface layers of the current (less than 30 m deep and above 15°C); as their swimming ability and low-temperature tolerance increases with age they show an increased daily vertical migration, spending the day in the deep, down to 600 m, and rising to the surface at night. This migration matches that of their main food during the summer, the Isuda krill, *Euphausia pacifica*, and recent research suggests that the daily vertical migration of the squid enables it to feed continuously throughout the day and night. When this food supply is exhausted, the squid are large enough to feed upon fish which are several years old, for example herring, mackerel and capelin.

From the squid's point of view the only problem with this lifestyle is the need to travel back south against the current to spawn: cannibalism may help them survive this.

For both squid species considered here, fisheries management came rather late. *Illex illecebrosus* management started only in 1974, first by the International Commission for the Northwest Atlantic Fisheries (ICNAF), and from 1977 by the US and Canadian governments separately, with quotas set.

The *T. pacificus* fishery began as a free fishery; regulation and registration of larger vessels started in 1969, primarily to prevent disputes between owners, rather than to regulate catches. The USSR declared a territorial limit of 330 km in 1977, and this prevented Japanese fishing west of the Sakhalin peninsula (see Fig. 3.7), which may have helped squid stocks recover subsequently. In 1998, Japan introduced a separate total allowable catch quota for each of the two main *T. pacificus* fisheries.

Thus, management of these two species is quite different, and we have the situation with *I. illecebrosus* of a species with effectively one population being managed differently by two countries.

If this account had been written ten years ago, all that could have been concluded about the reasons for the population crash of both these species would have been that the highly variable annual populations of T. *pacificus* went through a natural low period in the mid-1970s, possibly due to cooling of waters above the spawning grounds and this coincided with increased fishing pressure. A similar event may have happened to I. *illecebrosus*, with the Newfoundland and Nova Scotia populations being reliant on winter spawning and migration, and thus susceptible to variation in recruitment, coinciding with increasing fishing pressure. The north-east US fishery, which has remained more stable, is in a relatively mild climate, in close proximity to the spawning grounds, and can recruit from several spawnings throughout the year. Bakun and Csirke (1998) concluded that there is a 'near lack of present scientific insight into the mechanisms involved in squid recruitment variability'.

During the last ten years a consensus has emerged that the world's marine ecosystems have undergone a number of climate regime shifts (Box 3.4). One of the best documented of these took place in the North Pacific in the winter of 1976–7. This also took place in the Sea of Japan, which has been described as a little ocean, with many oceanic characteristics including deep water (over 1000 m) and large-scale circulation systems. During this regime shift, the mid trophic level (i.e. organisms that are carnivores, but also prey of other animals), dominated before 1976 by T. *pacificus*, was affected the most. This trophic level is crucial in many highly productive marine systems, is occupied typically by one or two dominant species of small plankton-feeding pelagic species, and exerts a major control on the trophic dynamics of the ecosystem. Before 1976, cephalopods contributed 38% of the energy flow through this trophic level, with walleye pollock (*Theragra chalcogramma*) contributing 19.5%; after the regime shift the situation was reversed with 17.5% and 33% of energy flow respectively being contributed by these species.[3]

The 1976 regime shift in the Sea of Japan was complex, with changes in current strengths, precipitation and thermal fronts, and was characterized by a shift in surface temperatures, particularly in the Kuroshio Current, from about 1.5°C above average to 0.5°C below average, as well as a change in water stratification. Spring-time phytoplankton and zooplankton biomass increased in the northern Sea of Japan, while it decreased to the south. This is likely to have occurred because the plankton in the higher-nutrient northern waters were able to take advantage of the increased light availability caused by the change in water stratification, and the stabilization of the mixed layer of the currents. In the south, however, with lower nutrient levels, the increased stratification of the water is thought to have prevented nutrients reaching the surface and the plankton, leading to decreased biomass. This change was particularly likely to affect young T. *pacificus* in the Kuroshio Current, south, early in the year, and could have led to poor recruitment. The winter spawning areas in the East China Sea consequently contracted.

Todarodes pacificus stocks in the Sea of Japan recovered quite suddenly in the late 1980s (see Fig. 3.8) and this is thought to be due to another North Pacific regime shift that took place during the winter of 1988–9. Whereas the 1976 regime shift is well-characterized by climatic data, the 1988 shift is much better characterized by biological, fisheries data and, although there is still some dispute about its nature, it is largely accepted. The 1988 regime shift was not simply a return to pre-1976 conditions, but was characterized by an increase in surface water temperatures and the strength of the Kuroshio Current, with the temperature at 50 m

[3]Many of these middle trophic level fish and squid support highly productive fisheries, and yet their abundance can be extremely variable. Bakun and Broad (2003) and Bakun (2006) have developed a set of explanations for this, suggesting that these organisms are ideally placed to exploit environmental loopholes that appear in the ecosystem due to changes in weather and climate.

BOX 3.4   REGIME SHIFTS

A regime shift can be defined as an abrupt change between contrasting persistent states of any complex system. Shifts were first identified in terrestrial systems, and first used in marine systems to describe populations of northern anchovy *Engraulis mordax* and pacific sardine *Sardinops sagax* which fluctuated together. Since then, regime shifts have been identified in all the major marine basins, and as well as affecting squid populations are likely also to be affecting salmon in the North Atlantic and North Sea (Chapter 9).

However, the dynamics underlying these shifts remain almost unknown. The drivers are likely to include a combination of abiotic processes including global warming, atmospheric and oceanic oscillations, and biotic processes including overfishing and possibly pollution.

Climate-driven regime shifts in the north Pacific have been suggested to have occurred at least 11 times since 1650. Relevant to the recent exploitation of *Todarodes pacificus*, regime shifts have been observed in 1976, 1988 and possibly 1998. Two climatic indices are important in determining Pacific Ocean regime shifts: the Aleutian Low Pressure Index (ALPI) which is a measure of low surface pressure, and appears to dominate many of the climatic and oceanic indicators of the Pacific; and the Pacific Decadal Oscillation (PDO) which describes an inter-decadal pattern of Pacific climatic variability. These two indices are correlated. The ALPI is associated with precipitation patterns and changes in the up-welling of high-nutrient waters, and changes in the ALPI have occurred together with changes in zooplankton abundance. In the North Pacific, zooplankton levels were highest during 1965–1970, then decreased to the lowest on record in 1989, and remained low until 1997. In the eastern Pacific the sardine–anchovy fluctuations are associated with large-scale changes in ocean temperature: from a warm-water sardine regime of the 1930s and 1940s peaking at 800 000 tonnes in 1936 (this fishery off California and its collapse together with its associated canning industry was described by John Steinbeck in *Cannery Row*): to a cool-water anchovy regime (1950–1975) and back to a warm-water sardine regime (1975 to mid 1990s). The anchovy regime led to the development of the largest single-species fishery in the world – the Peruvian anchoveta (*Engraulis ringens*) fishery – from the late 1950s. This accounted for a quarter of the world's marine landings in 1970, and at its peak could have been landing 12 million tonnes of fish a year.

The Pacific system is so large that it affects most of the world's weather, and is affected by many other systems, increasingly by El Niño and La Niña towards the tropics. This makes the Pacific-wide regime shifts very complex: they are well described by Chavez *et al.* (2003).

Changes in the composition of North Atlantic marine communities in the early 1960s and late 1980s have also been linked to regime shifts. The two primary indicators here are the North Atlantic Oscillation Index (NAOI), a measure of atmospheric pressure distribution at sea level between Iceland and the Azores, and the Gulf Stream Index (GSI), a measure of the north–south shift in latitude of the north wall of the Gulf Stream. Again, these two indices are correlated, with a high NAOI favouring a northward movement of the Gulf Stream two years later. The NAOI influences storms, precipitation, temperature, and drives North Atlantic climatic variability. Changes in the NAOI occurred at the same time as changes in zooplankton community structure and squid and fish abundance. During the negative NAOI regime in the north-east Atlantic during the 1960s there was a high biomass and large mean size of calanoid copepods, and a high abundance of *Calanus finmarchicus* and *Pseudocalanus* species. During the 1980s the mean size of calanoids decreased two-fold; there was delayed emergence, decreases in total biomass and fewer euphausiids. Zooplankton are also strongly correlated with the GSI: a southern GSI giving an earlier spring bloom, and vice versa. These zooplankton fluctuations usually lag 2–3 years behind the shift in the climatic indicators (Fig. 1).

Between 1995 and 1998, both the NAOI and GSI declined to their lowest values since the early 1980s, but North Atlantic zooplankton have not yet increased as expected. Likewise, evidence is emerging that there was a regime shift in the North Pacific in 1998. The effects of these regime shifts are profound on marine ecosystems, and for our exploitation of the sea: witness the effects of the crash of the Peruvian anchoveta industry during the 1970s and the Newfoundland cod industry later (where overfishing exacerbated a widespread regime shift). Yet they are hard to predict, and even to detect when happening, as changes need to be persistent over a number of years: both the 1977 and 1989 North Pacific regime shifts took over ten years to detect. New observation networks, such as the Global Ocean Observing System, will provide more data and may enable regime shifts to be detected in the north Pacific within 2–3 years of occurrence.

The likelihood of climate-driven regime shifts occurring has been shown to increase when humans have reduced ecosystem resilience (the ability of an ecosystem to recover after perturbations), for example from overfishing, waste, pollutants. Flexible and adaptive management is needed to help build up resilience in these ecosystems to prevent these disruptive changes.

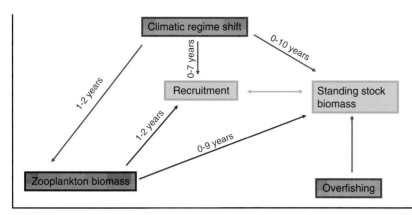

Box 3.4 Fig. 1 *Impacts and lags of climatic regime shifts and overfishing on fish stocks. For squid, standing stock biomass (i.e. adult biomass) is more directly and immediately related to recruitment (i.e. reproduction), given their annual life-cycle and thus lack of a multi-year population structure, and the effects of changes in zooplankton biomass on stock will be felt much quicker* [Redrawn from Lees *et al.* (2006).]

shifting from below long-term average to up to 1°C above average. This led to an increase in zooplankton biomass, a change in its composition, and an increase in larval T. *pacificus* abundance. The squid spawning area increased, expanding over the continental shelf and slope and into the East China Sea. After this regime shift, the total biomass of all species groups increased by 59% and the relative contribution of walleye pollock to the total energy flow decreased from 33% to 4.3% while that of squid increased from 34% to 72%.

The 1970s population crash of I. *illecebrosus* has not yet been linked with a climate regime shift, although it does coincide with a North Atlantic regime shift towards the end of the 1970s (Box 3.4). Recent work has also established that the abundance of this squid is linked to the position of the Gulf Stream and changes in the North Atlantic Oscillation Index (NAOI; see Box 3.4), with high squid abundance linked to a negative NAOI, high water temperatures off Newfoundland and a southwards shift in the Gulf Stream.

Thus, in the last decade the perceived relative importance of exploitation and climate in explaining the population fluctuations of these two squid species has shifted, and in their growth and population sensitivity to oceanographic conditions squid have been suggested as 'ecosystem recorders'. Although we are a little clearer as to the major factors affecting recruitment of these species, we still have no proven direct causative mechanisms linking environmental variables to squid abundance; the conclusions so far are merely correlative.

Therefore, squid fisheries management remains crucial. Squid fisheries have some characteristics which make the application of fish-based management models difficult. Squid have been termed 'marine weeds' and they share many characteristics with the weeds of arable crops: they are usually annual, grow very fast, and die after reproducing once. They are thus able to build up populations quickly in favourable conditions, but equally, populations may crash suddenly with no reserves from different age classes. Compared to slow-growing fish (e.g. salmon; see Chapter 9), with populations composed of different age classes, squid populations are all the same age. Therefore, whereas fish populations can store genetic variability within the age classes, squid cannot, and it is thought that they store variability within populations which have migrated to different areas, and thus have been exposed to different spatial, rather than temporal, conditions. This all affects squid management and it should be taken into account in fishery models. This will require quite a dynamic management system. At the moment only passive management is used: this will have to use closed seasons, closed areas, and a reduction in fishing effort, to have any effect.

With increasing recognition of the importance of climatic variation to squid abundance, predictive models in real time have been developed for these species. However, these are likely to involve a huge investment in scientific resources and run the risk of failing if the environmental indicator variables become decoupled from the as yet unknown processes that regulate abundance. This has been demonstrated recently for north-east Atlantic squid populations. Perhaps it is time to realize that these extremely opportunistic annual species cannot be managed to maintain a stable high population and that it might be better to exploit their short-lived periods of high productivity only when they occur.

THE GIANT SQUID

Cephalopods are ready-made as monsters for our imagination to exploit, with their unusual body plan, many arms and tentacles and remoteness from most of our lives. It is therefore no surprise that the possibility that there may be giant, violent cephalopods lurking in the deep oceans has fascinated humankind, and scientific research has only partially succeeded in replacing fiction with facts about giant squid.

Giant squid are mentioned by Aristotle, and Pliny discusses a large polyp that has long tentacles as well as arms. Thereafter, however, the giant squid is relatively absent from literature until sea monsters appear in northern European accounts from the sixteenth century. The first possible reference to a giant squid occurs in Olaus Magnus's *Historia de Gentibus Septentrionalibus* (1555) (Fig. 3.9):

> A serpent of gigantic bulk, at least two hundred feet long, and twenty feet thick, frequents the cliffs and hollows of the sea-coast near Bergen. It leaves its caves in order to devour calves, sheep, and pigs....It has hairs eighteen inches long hanging from its neck, sharp black scales, and flaming-red eyes. It assaults ships, rearing itself on high like a pillar, seizes men, and devours them.[4]

This is usually taken as a reference to the giant squid, which have been found in these waters. Later authors, including Hans Egede, the first Bishop of Greenland, reported a similar monster in 1734. In 1755, Bishop Pontoppidan published his *Natural History of Norway*, and introduced the term Kraken, meaning sea monster, for this animal. For the

[4]Magnus (1555), p. 1128.

Fig. 3.9 *Woodcut of the sea monster off the coast of Norway described by Olaus Magnus in 1555, possibly the first reference to the giant squid* [From Magnus (1555).]

next hundred years there were many reports of giant worms or sea serpents sighted in various oceans, which Ellis (1998) and others have considered as representing the giant squid. Yet no specimens were caught, and the state of knowledge or lack of it at this time is well represented by Tennyson's poem *The Kraken* (Box 3.5).

BOX 3.5   TENNYSON'S KRAKEN

The Kraken

Below the thunders of the upper deep;
Far, far beneath in the abysmal sea,
His ancient, dreamless, uninvaded sleep
The Kraken sleepeth: faintest sunlights flee
About his shadowy sides: above him swell
Huge sponges of millennial growth and height;
And far away into the sickly light,
From many a wondrous grot and secret cell
Unnumbered and enormous polypi
Winnow with giant arms the slumbering green.
There hath he lain for ages and will lie
Battening upon huge seaworms in his sleep,
Until the latter fire shall heat the deep;
Then once by man and angels to be seen,
In roaring he shall rise and on the surface die.

Alfred Tennyson (1809–1892), later Poet Laureate and Baron, wrote this quasi-sonnet whilst an undergraduate at Cambridge, and it was published in his first collection of poems, *Poems, Chiefly Lyrical*, in 1830. Although some early critics could not understand it, later others were disappointed that it was not included in collections of Tennyson's poetry (it was not reprinted until 1872), and it has become accepted as one of the most important poems in that first collection.

Although Tennyson explained at the time that he was influenced by Pontoppidan's account of the Kraken, this has not stopped a cottage industry developing to ascribe influences, something which Tennyson himself deplored. Thus, academics have found influences of early nineteenth century gothic novels, Shelley's *Prometheus Unbound* and the Book of Revelation within it.

Many also are the interpretations of this poem, but it is clear that for all the Jungian and Freudian analysis and linking to repressed states of consciousness, this poem accurately reflects the lack of knowledge about the giant squid at the time, as well as the sense that it is inherently unknowable. For example, it is not described: only the creatures around it are. In the final lines of the poem Tennyson accurately predicts that the squid is only likely to be seen dead by humans, on the surface of the water.

It is almost obligatory to reproduce this poem when discussing the giant squid in fact or fiction, and yet one of the few novels that explicitly links to the Kraken is not about the giant squid; and the sequence of stories most related to the poem does not mention it.

The first is John Wyndham's 1953 novel *The Kraken Wakes*, where the poem appears in the first chapter. The novel is set in the near future and concerns the alien invasion of the deep ocean. First, ships are lost mysteriously and then aliens appear on land from the oceans in 'sea tanks' which shoot out many tentacles and entrap people. These are likened to the tentacles of sea-anemones. The aliens then induce a warming of the oceans and the melting of the ice caps, and humanity struggles to regroup on the diminished land masses. Although one of the main characters notes 'sometimes I dream of them lying down in those deep dark valleys, and sometimes they look like monstrous squid or huge slugs',[a] and cover illustrations on some editions have portrayed tentacles of a giant squid dragging liners underwater, we never find out what these organisms are, where they came from, or what they want. Indeed there is no communication with them at all. Thus, the link with the Kraken is only as an unknowable organism of the deep oceans.

On the other hand, in H.P. Lovecraft's (1890–1937) series of horror stories about the imaginary Cthulhu mythos written and set during the 1920s and 30s in New England and Massachusetts (both real and imaginary) we closely approach the Tennysonian Kraken. It is also likely that Lovecraft was aware of the other myths and legends surrounding the giant squid, and the specimens found during the 1870s nearby.

In several long short stories, Lovecraft develops the idea that the earth was colonized by aliens from space long before humans, and that they await the opportunity to take over again. The stories usually involve visitations from lesser members of these 'Old Ones', invoked by meddlesome humans, or the discovery of human–alien hybrids. However, in 'The Call of Cthulhu' (1926), the great priest Cthulhu is described: it has an octopus-like head, its face is a mass of feelers, with a scaly rubbery body, claws on its feet, and long narrow wings behind. This squid–dragon 'From his dark house in the mighty city of R'lyeh under the waters, should rise and bring the earth again beneath his sway'.[b]

Most recently, Miéville (2010) produced a post-punk take on the Kraken, with squid-cults worshiping the giant squid and with doubt extending not only to the Kraken but to almost everything else as well. However, unlike the previous authors, and harking back to Melville, Miéville does include a fair amount of giant squid biology in his novel. And so the story continues.

[a]Wyndham (1953), p. 229.
[b]Lovecraft (1982), p. 139.

However, it is likely that some had seen close-up evidence of the giant squid: whalers out of the north-east American ports had associated giant squid with sperm whales (*Physeter macrocephalus*), and Melville has a chapter on the giant squid in *Moby-Dick* including the following:

> A vast pulpy mass, furlongs in length and breadth, of a glancing cream-colour, lay floating on the water, innumerable long arms radiating from its centre, and curling and twisting like a nest of anacondas, as if blindly to clutch at any hapless object within reach. No perceptible face or front did it have; no conceivable token of either sensation or instinct; but undulated there on the billows, an unearthly, formless, chance-like apparition of life.
> … 'What was it, Sir?' said Flask.
> 'The great live squid, which, they say, few whale-ships ever beheld, and returned to their ports to tell of it.'[5]

The first fragment of a giant squid was not caught until 1861 when the French warship *Alecton* spotted one off Tenerife, shot at it, and tried to haul it on board. Unfortunately, it broke in two and only the tail was recovered (information from this was later adapted by Jules Verne for his novel *Twenty Thousand Leagues Under the Sea*, where a giant squid battles with a submarine). In 1871 the Gloucester, Mass., schooner B.D. *Haskins* caught a dead and decomposing giant squid off the Grand Banks and subsequently several dying specimens were caught in nets or washed ashore dead on the Newfoundland and north-east American coast during the 1870s. One of the most famous captures, and still the most important interaction between a human and a living giant squid, occurred in October 1873 off Newfoundland, as told in a letter reproduced in the *American Naturalist* in 1875:

> On or about the 25th of October last, while a man by the name of Theophilus Picot was engaged at his usual occupation of fishing, off the eastern end of Great Bell Island in Conception Bay, his attention was attracted to an object floating on the surface of the water, which at a distance he supposed to be a sail, or the *débris* of some wreck, but which proved upon nearer inspection to be endowed with life. Picot, on observing that the object was alive, to satisfy his curiosity pushed his boat alongside, and I believe struck at it with an oar or boat-hook, whereupon the creature's fury seemed to be aroused, and it struck at the bottom of the boat with its beak, and immediately afterward threw its monstrous tentacles over the boat, which probably it might have dragged to the bottom had not Picot with great presence of mind severed one (or more) of the tentacles with his axe.[6]

This report was later much embellished; other persons, including a small boy, appeared in the boat, and the report undoubtedly formed the basis for H.G. Wells' 1896 story *The Sea Raiders* in which a shoal of luminescent giant squid (we now know that unlike other squid they probably do not shoal and do not have bioluminescent cells) attack and kill several holidaymakers in small boats off the south Devon coast, before disappearing as mysteriously as they arrived.

Luckily, enough portions of giant squid, including the tentacle severed by Picot, were preserved for long enough to be photographed and examined by scientists; and, along with further specimens observed since, we now have a reasonable picture of its

[5]Melville (1851), p. 236.
[6]Anon (1874), pp. 120–121.

morphology. The recent increase in deep-water trawling has provided more specimens, in better condition than before, although the problem of preservation remains acute: squid flesh rapidly decomposes, and even research vessels do not have the facilities to store and preserve the whole of such a large animal. Up to 19 species of giant squid have been described based solely on their anatomy, all placed in the genus *Architeuthis*, but many scientists believe that there are really far fewer species, and probably only three: *A. sanctipauli* in the southern hemisphere, *A. japonica* in the North Pacific, and *A. dux* in the North Atlantic. As it is estimated that a large giant squid would be able to swim around the world in 80 days, there is no *a priori* reason why the species should be geographically separated, and there may be just one species.

The largest specimen recorded is probably one found off New Zealand in 1880 which is reported to have had a total length of 20 metres. However, over half this length comprised its tentacles, which are much longer than the arms (usually up to 3 m long). As these tentacles can contract and extend, and become stretchable when dead, imprecise and exaggerated claims of the size of these squid can be made, and it is far preferable to refer to mantle length. No specimens that have been recently accurately recorded have a mantle length greater than 2.25 m, or a total length more than 13 m (Fig. 3.10). The buoyancy of giant squid is maintained by an isotonic concentration of ammonium ions in the mantle, head and arm muscles: ammonium ions are lighter than sodium ions and float, whereas the rest of the squid is heavier. This is probably the reason they float to the surface when dead. Strandings in Newfoundland and elsewhere have been associated with inflows of warmer water, and it has been suggested that an increase in water temperature may have killed these squid. Their blood has a very low oxygen-carrying capacity, and one that is extremely temperature sensitive, and thus they may have insufficient oxygen in their blood to survive in warmer waters.

However, we know little more about the ecology of the species than Melville and Tennyson. We do know that they are occasionally eaten by sperm whales (but form less than 1% of their diet); small, dead, floating specimens may be eaten by albatrosses; and it has recently been found that they are an important part of the diet of the large (up to 4 m long) sleeper sharks, *Somniosus* species, of the Southern Ocean. The giant squid probably live mainly at 500–1000 m depth, although as they have been caught between 18 and 5000 m deep they can probably range throughout the water column. Their diet consists of fish and other squid, and they appear also to be cannibalistic.

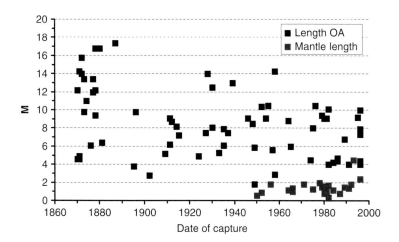

Fig. 3.10 *Giant squid (Architeuthis species) overall (length OA) or mantle length by date of capture. Data from Ellis (1998), excluding pre-1870 records, and some immature specimens. The squid recorded in the 1990s with a mantle length over 4 m is anomalous and unchecked*

It was thought that giant squid were slow and poor swimmers with poorly developed muscles, but in 2004 a Japanese team succeeded in taking the first photographs of a live giant squid (the first live larvae were caught in 2001, but died within days in captivity) which showed that they can be very active. The team lowered a camera off the Japanese coast in a likely giant squid site on a line equipped with squid jigs. A giant squid attacked this at 900 m depth, and the club of one of its tentacles became caught on the jig. Over the next four hours the squid tried to free itself, moving rapidly to and fro, rising to 600 m, and coiling its tentacles in a ball around the line. Eventually the tentacle broke, and the squid escaped.

THE COLOSSAL SQUID

The giant squid is clearly large and may be an active predator, but it falls far short of the size and ferocity reported in legends. Are there larger squid? The answer is almost certainly yes – the colossal squid, *Mesonychoteuthis hamiltoni*. This was first described in 1925 from two tentacles found in the gut of a sperm whale. Since then only 12 reports of this creature have been published, with probably fewer than 20 individuals known. In April 2003, an immature female caught had the greatest mantle length of any squid accurately recorded at the time (2.5 m) and weighed about 300 kg. A mature female in excess of 400 kg was caught and frozen by a fishing vessel in early 2007, also in the Antarctic. After thawing and dissection it was placed on display towards the end of 2008 in a tank in Te Papa museum, Wellington, New Zealand, where it immediately became a popular exhibit.

It is thought that the colossal squid may reach up to 4 m mantle length (Fig. 3.11). Although the giant squid has much longer tentacles, and thus may be longer overall, the colossal squid is undoubtedly heavier. It is also thought to be a much more active predator than the giant squid, with a powerful muscular fin, and along with fixed hooks on its eight arms it has up to 25 rotating hooks on each of its tentacles. The two recent specimens have been caught by trawlers fishing for the up to 2-metre long Patagonian toothfish, *Dissostichus eleginoides*, in Antarctic waters, a species the colossal squid also feeds on. This squid also forms part of the food of sperm whales and sleeper sharks in the Antarctic.

Although the size range of food taken by the colossal squid suggests that it could cope with human prey, there is no evidence at all that it would be a threat to humans. However, increased deepwater trawling and exploitation of Antarctica

Fig. 3.11 *Comparison between the largest colossal squid* (Mesonychoteuthis hamiltoni) *recorded, with 4 m mantle length (in red) with the largest giant squid,* Architeuthis *species, reliably recorded with a 2.25 m mantle length (in blue). Note the much stouter mantle and fin, and relatively shorter arms and tentacles of the colossal squid compared to the giant squid [Based on illustrations at www.tonmo. com.]*

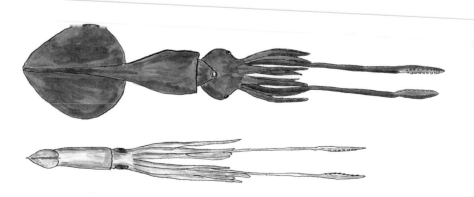

and other remote oceans could pose a threat to the long-term survival of the giant and colossal squid, our largest invertebrates.

FURTHER READING

There are few recent single-volume overviews of cephalopods, although Boyle and Rodhouse (2005) can be recommended for an up-to-date overview. Nesis (1987) has a useful introduction, along with keys to all living cephalopods and Boyle (1983, 1987) gives useful accounts of many species and also general cephalopod topics. For cephalopod evolution, Clarke and Trueman (1988) contains several useful chapters and Kröger *et al.* (2009) is an up-to-date review of their early evolution. For *Illex* fisheries see O'Dor (1983), Rodhouse *et al.* (1998) and Dawe *et al.* (2007). For *Todarodes* fisheries, earlier reviews by Okutani (1983), Osako and Murata (1983), and Murata (1989) can be supplemented by Zhang *et al.* (2000), Kang *et al.* (2002), and Tian *et al.* (2006).

Lehmann (1981) is a useful summary of ammonite structure and evolution, and McMenamin (2007) gives an account of the earliest human representations of ammonites.

Hamabe *et al.* (1982) and Suzuki (1990) give a practical account of Japanese light squid jigging, with Chen *et al.* (2008) discussing the Chinese squid jigging industries. Regime shifts are reviewed by Chavez *et al.* (2003), Lees *et al.* (2006), Zhang *et al.* (2007), and deYoung *et al.* (2008). The history and mythology of the giant squid is covered by Ellis (1998), and Roper and Boss (1982) is a good introduction to giant squid biology. Much accurate information on the giant and colossal squid can be found at www.tonmo.com. Kubodera and Mori (2005) report the first observations of a live adult giant squid.

REFERENCES

Anon (1874) Capture of a gigantic squid at Newfoundland. *American Naturalist* **9**, 120–123.

Aronson, R.B. (1991) Ecology, paleobiology and evolutionary constraint in the octopus. *Bulletin of Marine Science* **49**, 245–255.

Bakun, A. (2006) *Wasp-waist* populations and marine ecosystem dynamics: navigating the 'predator pit' topographies. *Progress in Oceanography* **68**, 271–288.

Bakun, A. and Broad, K. (2003) Environmental 'loopholes' and fish population dynamics: comparative pattern recognition with focus on El Niño effects in the Pacific. *Fisheries Oceanography* **12**, 458–473.

Bakun, A. and Csirke, J. (1998) Environmental processes and recruitment variability. In: *Squid Recruitment Dynamics: the genus Illex as a model, the commercial Illex species and influences on variability* (eds P.G. Rodhouse, E.G. Dawe and R.K. O'Dor), pp. 105–124. FAO Fisheries Technical Paper 376. FAO, Rome, Italy.

Barnes, R.D. (1980) *Invertebrate Zoology*, 4th edn. Saunders College, Philadelphia, PA, USA.

Boyle, P.R. (ed.) (1983) *Cephalopod Life Cycles. Volume I: Species Accounts.* Academic Press, London.

Boyle, P.R. (ed.) (1987) *Cephalopod Life Cycles. Volume II: Comparative Reviews.* Academic Press, London.

Boyle, P. and Rodhouse, P. (2005) *Cephalopods, Ecology and Fisheries*. Blackwell Science, Oxford.

Chavez, F.P., Ryan, J., Lluch-Cota, S.E. and Ñiquen, M.C. (2003) From anchovies to sardines and back: multidecadal change in the Pacific Ocean. *Science* **299**, 217–221.

Chen, X., Liu, B. and Chen, Y. (2008) A review of the development of Chinese distant-water squid jigging fisheries. *Fisheries Research* **89**, 211–221.

Clarke, M.R. and Trueman, E.R. (eds) (1988) *The Mollusca. Volume 12: Paleontology and Neontology of Cephalopods*. Academic Press, San Diego, USA.

Dawe, E.G., Hendrickson, L.C., Colbourne, E.B., Drinkwater, K.F. and Showell, M.A. (2007) Ocean climate effects on the relative abundance of short-finned (*Illex illecebrosus*) and long-finned (*Loligo pealeii*) squid in the northwest Atlantic Ocean. *Fisheries Oceanography* **16**, 303–316.

Ellis, R. (1998) *The Search for the Giant Squid*. Lyons Press, New York.

Hamabe, M., Hamuro, C. and Ogura, M. (1982) *Squid Jigging from Small Boats*. FAO Fishing Manuals, Fishing News Books, Farnham, Surrey, UK.

Kang, Y.S., Kim, J.Y., Kim, H.G. and Park, J.H. (2002) Long-term changes in zooplankton and its relationship with squid, *Todarodes pacificus*, catch in Japan/East Sea. *Fisheries Oceanography* **11**, 337–346.

Kröger, B., Servais, T. and Zhang, Y. (2009) The origin and initial rise of pelagic cephalopods in the Ordovician. *PLoS ONE* **4**(9): e7262.

Kubodera, T. and Mori, K. (2005) First-ever observations of a live giant squid in the wild. *Proceedings of the Royal Society B* **272**, 2583–2586.

Lees, K., Pitois, S., Scott, C., Frid, C. and Mackinson, S. (2006) Characterizing regime shifts in the marine environment. *Fish and Fisheries* **7**, 104–127.

Lehmann, U. (1981) *The Ammonites: their life and their world*. Cambridge University Press, Cambridge.

Lovecraft, H.P. (1982) *The Dunwich Horror and Others*. Arkham House, Sauk City, Wisconsin, USA.

McMenamin, M.A.S. (2007) Ammonite fossil portrayed on an ancient Greek countermarked coin. *Antiquity* **81**, 944–948.

Magnus, O. (1555) *Historia de Gentibus Septentrionalibus* [transl. P. Fisher and H. Higgens, ed. P. Foote]. The Hakluyt Society, London [3 vols 1996–1998].

Miéville, C. (2010) *Kraken: an anatomy*. Macmillan, London.

Melville, H. (1851) *Moby-Dick*. Everyman, Dent, London [1993 edn, ed. A. R. Lee].

Murata, M. (1989) Population assessment, management and fishery forecasting for the Japanese common squid, *Todarodes pacificus*. In: *Marine Invertebrate Fisheries: their assessment and management* (ed. J.F. Caddy), pp. 613–636. John Wiley & Sons, New York.

Nesis, K.N. (1987) *Cephalopods of the World: squids, cuttlefishes, octopuses and allies* [transl. B.S. Levitov]. T.F.H. Publications, Neptune City, NJ, USA.

O'Dor, R.K. (1983) *Illex illecebrosus*. In: *Cephalopod Life Cycles. Volume I: Species Accounts* (ed. P.R. Boyle), pp. 175–199. Academic Press, London.

O'Dor, R.K. and Webber, D.M. (1986) The constraints on cephalopods: why squid aren't fish. *Canadian Journal of Zoology* **64**, 1591–1605.

O'Dor, R.K. and Dawe, E.G. (1998) *Illex illecebrosus*. In: *Squid Recruitment Dynamics: the genus Illex as a model, the commercial Illex species and influences on variability* (eds. P.G. Rodhouse, E.G. Dawe and R.K. O'Dor), pp. 77–104. FAO Fisheries Technical Paper 376. FAO, Rome, Italy.

Okutani, T. (1983) *Todarodes pacificus*. In: *Cephalopod Life Cycles. Volume I: Species Accounts* (ed. P.R. Boyle), pp. 201–214. Academic Press, London.

Osako, M. and Murata, M. (1983) Stock assessment of cephalopod resources in the northwestern Pacific. In: *Advances in Assessment of World Cephalopod Resources* (ed. J.F. Caddy), pp. 55–145. FAO, Rome, Italy.

Packard, A. (1972) Cephalopods and fish: the limits of convergence. *Biological Reviews* **47**, 241–307.

Rodhouse, P.G., Dawe, E.G. and O'Dor, R.K. (eds) (1998) *Squid Recruitment Dynamics: the genus Illex as a model, the commercial Illex species and influences on variability.* FAO Fisheries Technical Paper 376. FAO, Rome, Italy.

Roper, C.F.E. and Boss, K.J. (1982) The giant squid. *Scientific American* **246**(4), 82–89.

Smith, M.R. and Caron, J.-B. (2010) Primitive soft-bodied cephalopods from the Cambrian. *Nature* **465**, 469–472.

Suzuki, T. (1990) Japanese common squid: *Todarodes pacificus* Steenstrup. *Marine Behaviour and Physiology* **18**, 73–109.

Tian, Y., Kidokoro, H. and Watanabe, T. (2006) Long-term changes in the fish community structure from the Tsushima warm current region of the Japan/East Sea with an emphasis on the impacts of fishing and climate regime shift over the last four decades. *Progress in Oceanography* **68**, 217–237.

Ward, P. (1983) The extinction of the ammonites. *Scientific American* **249**(4), 114–124.

Ward, P.D. (2006) *Out of Thin Air: dinosaurs, birds, and earth's ancient atmosphere.* Joseph Henry Press, Washington, DC, USA.

Wyndham, J. (1953) *The Kraken Wakes*. Penguin, Harmondsworth, UK [1955].

deYoung, B., Barange, M., Beaugrand, G. *et al.* (2008) Regime shifts in marine ecosystems: detection, prediction and management. *Trends in Ecology and Evolution* **23**, 402–409.

Zhang, C.I., Lee, J.B., Kim, S. and Oh, J.-H. (2000) Climatic regime shifts and their impacts on marine ecosystem and fisheries resources in Korean waters. *Progress in Oceanography* **47**, 171–190.

Zhang, C.I., Yoon, S.C. and Lee, J.B. (2007) Effects of the 1988/89 climatic regime shift on the structure and function of the southwestern Japan/East Sea ecosystem. *Journal of Marine Systems* **67**, 225–235.

EXAMINING THE SUBTLETIES OF
BEE SOCIETY

THE PRESENT ~ THE QUEEN LAYS EGGS WHICH DEVELOP (ACCORDING TO HOW THE GRUBS ARE FED) INTO:

THE FUTURE ~ BEES NEED TO ADAPT TO NEW SITUATIONS TO ENSURE THEIR SURVIVAL:

WORKER BEES

FUNCTIONS: NURSES, CLEANERS, VENTILATION CONSULTANTS, FORAGERS, GUARDS.

DRONES

FUNCTION: TO IMPREGNATE THE QUEEN.

NEW QUEENS (MURDERED)

FUNCTION: TO REPLACE THE QUEEN.

HYGIENIST WORKER BEES.
FUNCTION: TO REMOVE VARROA PARASITES FROM OTHER BEES & CELLS.

QUEEN

PRIM & PROPER DRONES.

FUNCTION: TO RESIST THE EXOTIC ALLURE OF KILLER BEE QUEENS AND END INTERBREEDING.

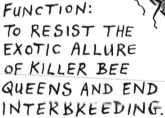

SCIENTIST BEES

FUNCTION: TO SPOT & RESEARCH UNFORESEEN NEW PROBLEMS.

INVENTOR BEES

FUNCTION: TO INVENT WAYS OF TACKLING THE NEW PROBLEMS.

OTHER NEW BEES WHICH MIGHT PROVE USEFUL ONE DAY:

ARTIST BEES

WRITER BEES

MUSICIAN BEES

The Honey Bee

CHAPTER 4

So the industrious bees do hourly strive
To bring their loads of honey to the hive;
Their sordid owners always reap the gains,
And poorly recompense their toils and pains

Mary Collier, *The Woman's Labour*, 1739[1]

THE HONEY BEE, *Apis mellifera* (Fig. 4.1), is the most important insect utilized by humans: important for its honey and for pollination. In the extract above, the proto-feminist Mary Collier (c.1690 to c.1762) likens the work and treatment of the honey bee with that of housewives, and today her words are more relevant than ever.

Although certain characteristics have been selected for by humans, until very recently the honey bee could not be considered truly domesticated, unlike the silk moth (Chapter 5), and many 'feral' or wild honey bee colonies could survive throughout the world (Fig. 4.2). In 2007 there were an estimated 73 million managed honey bee colonies world-wide, an increase compared to 50 years previously. However, the US has experienced a drop of 61% in colonies since 1947 and Europe a 25% reduction since 1970. In this chapter we will consider the history of our exploitation of the honey bee, and possible reasons for these declines.

Although many bees store honey, those of the genus *Apis* are the only species able to store sufficient honey to enable colonies to overwinter, and thus produce perennial rather than annual nests. Thus, *Apis* bees are suitable for management by humans, although honey is collected from other genera in the wild. Although other *Apis* species are managed throughout the world for honey production (Box 4.1 and Fig. 4.3), we shall concentrate on *A. mellifera*, the honey bee, here—by far the most widespread species. Honey bees evolved in the tropical forests of Asia and spread naturally throughout Africa and east throughout Europe. Since 1600 the honey bee has also been moved by humans throughout the world (see Fig. 4.3), and some countries, for example

Fig. 4.1   *The honey bee (Apis mellifera) worker*

[1]See Milne (2001) for a discussion of Collier's use of the beehive metaphor.

Biological Diversity: Exploiters and Exploited, First Edition.   Paul Hatcher and Nick Battey.
© 2011 John Wiley & Sons, Ltd. Published 2011 by John Wiley & Sons, Ltd.

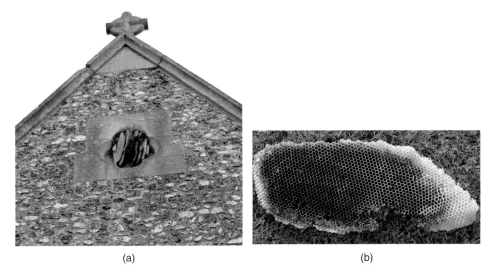

(a)   (b)

Fig. 4.2 **(a)** *Feral honey bee nest established in a recess in the chancel gable end of a Berkshire church, October 2009.* **(b)** *Part of the comb had become detached in recent strong winds, and the colony would not survive the winter in this situation. In* (b) *note the regular spacing of the cells (all are worker cells), the dark brown used cells, and the new, white, unused cells*

BOX 4.1   THE HONEY BEE, *APIS MELLIFERA*

The honey bee is a member of the large insect order Hymenoptera. The aculeate Hymenoptera (those species whose ovipositor has developed into a sting, with eggs being deposited at the base of the sting, rather than travelling through its tubes) are first recorded from the late Jurassic (about 160 mya). The bees, the Apoidea, are characterized, apart from morphological characteristics, as feeding larvae on pollen mixed with floral oils and nectar, and arose during the early Cretaceous (120 mya). The bees are basically vegetarian wasps, to which they are closely related. Advanced sociality developed later.

Seven species in the genus *Apis* are currently recognized (Fig. 1). They are hard to classify and, depending on which characteristics are used, between 4 and 24 species have been recognized previously. *A. mellifera* is especially variable, and is divided into a number of 'races' (see Fig. 4.3 for the distribution of some of the races; over 25 are currently recognized) and more than 90 scientific names have been proposed to describe its species, races or subspecies.

Two branches of the genus *Apis*, the dwarf honey bees *A. florea* and *A. andreniformis* and the giant honey bee *A. dorsata*, live in the tropics and build a single comb nest in the open, often on the branch of a tree. The other four species, including *A. mellifera*, build multiple-comb nests in cavities.

The natural distributions of *A. mellifera*, *A. cerana*, and *A. florea* are given in Fig. 4.3. The range of *A. dorsata* mainly overlaps

with *A. florea*, extends slightly further east, and does not overlap with *A. mellifera*; the range of *A. andreniformis* is similar to that of *A. florea*. *A. koschevnikovi* is restricted to a small part of SE Asia.

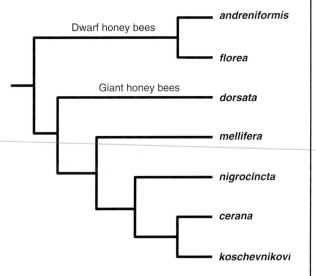

Box 4.1 Fig. 1 *Phylogeny of the genus* Apis [From Grimaldi and Engel (2005).]

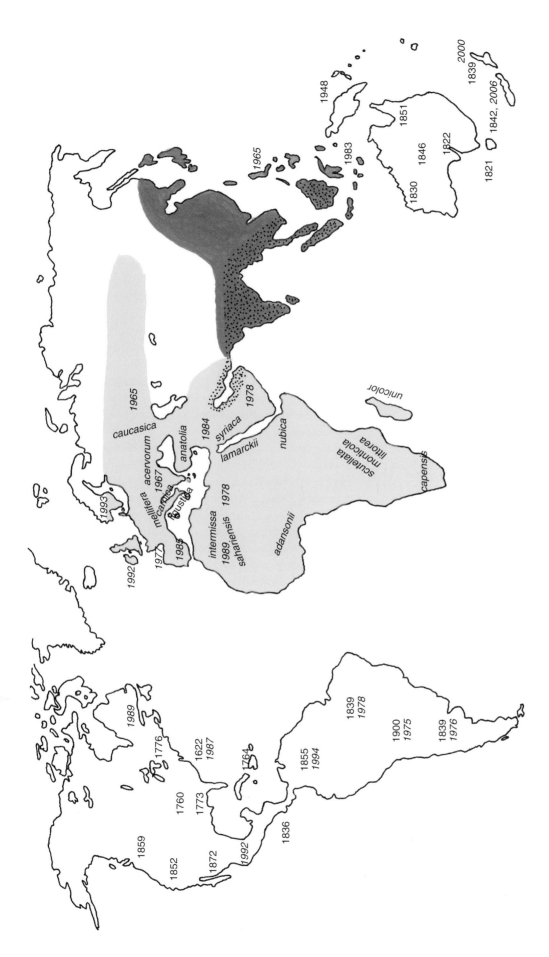

Fig. 4.3 Natural range of Apis mellifera (yellow) and distribution of some races, with dates of introduction elsewhere, and first records of Varroa destructor (in italics). Distribution of A. cerana (orange) and A. florea (dots) [Based on Gould and Gould (1988), with additional from Crane (1999).]

New Zealand and parts of Australia, were without any honey-storing insects until honey bees were introduced in the 1800s.

A honey bee colony at full strength usually consists of one queen, about 1000 drones and around 50 000 workers. In northern Europe the queen begins to lay eggs into brood cells in January; these hatch within three days, and for three days the larvae are fed upon 'royal jelly' produced by glands in the heads of the worker bees. After this, worker and drone larvae are switched to a diet of nectar or dilute honey and nectar, while larvae which will become queens are fed constantly upon royal jelly. By day nine the larvae are fully grown in their cells and these are sealed ('capping'). After about 12 days' pupation the adult emerges from the cell. Thus, the honey bee follows the usual life-cycle for an advanced insect: a larva which has the main function of feeding, and lacks wings; a pupal stage which does not move or feed; followed by the adult stage with wings. However, unlike most insects, different honey bee adults are specialized to carry out different roles.

The workers are female bees, from fertilised eggs, and thus have the full complement of chromosomes (diploid). The drone is a male bee, and its sole function is to mate with queens. In all other respects it does not contribute to the life of the colony, and at times of food shortages will be abandoned. Drone, unfertilized, eggs (thus the drone has half the chromosome complement of the female bees, i.e. is haploid) are laid in March in larger cells at the edges of the outer combs of the nest, and the fully fed drone larva is sealed beneath a distinctive domed cap.

Queens cannot produce brood food or wax, and cannot collect pollen or nectar; they only leave the hive when swarming, and for the rest of the time reproduce. Unlike other larval cells, queen cells are used only once, and are large and free-hanging at the edge of the combs. Queens maintain control over a colony by producing 'queen substance' from her mandibular glands. This is spread over her body and also taken throughout the colony by the workers that feed her. If there is insufficient queen substance in the nest, due to the death or poor health of the queen, workers are triggered into building queen cells and feeding the female larvae appropriately. New queen development is also triggered by overcrowding in the hive.

The first new queen to emerge may kill the old queen and all developing queen larvae and pupae and thus take over the nest. However, more usually, when queens are produced as a result of overcrowding and not the infirmity of the resident queen, the old queen leaves with about half the workers as soon as the first queen cell is capped, and forms a new colony. The first virgin queen to emerge kills any developing queens and leaves the nest and mates with up to 12 drones before returning. This process of colony subdivision, 'swarming', is the means by which the colony replicates. Swarming used to be an important part of beekeeping in the UK, and the formation of a swarm is described by Thomas Hardy in *Far From the Madding Crowd*:

> A process somewhat analogous to that of alleged formations of the universe, time and times ago, was observable. The bustling swarm had swept the sky in a scattered and uniform haze, which now thickened to a nebulous centre: this glided on to a bough and grew still denser, till it formed a solid black spot upon the light.[2]

Modern beekeeping, however, selects for bees that are reluctant to swarm, and they are managed to discourage this; for example by clipping the wings of the queen and by enlarging the hive to prevent overcrowding during the summer.

[2]Hardy (1874), p. 166.

The success of the colony is dependent upon the activities of the workers. As they age so their duties change. During their first day of adulthood they clean cells to make them ready to receive eggs or food stores. Between three and 13 days of adult life they are on nurse duties;

(a)

(b)

their hypopharyngeal and mandibular glands have now developed and they feed larvae with secretions from them, and inspect eggs and larvae in open cells. From seven days of age the wax-producing glands on the underside of the abdomen are developed and until 24 days old the workers seem to concentrate on building new comb. They later receive nectar from foragers, pack down the pollen load deposited into cells, remove debris, and at the end of this period may briefly become guards at the nest entrance whilst taking their first flights outside. The last phase of their life, usually after ten days of age, is to become a forager, collecting nectar and pollen, and the workers do this for the rest of their life of 4–5 weeks.

Foraging workers may collect pollen on one day and nectar the next, or may specialize in one of these (Fig. 4.4). Pollen collectors deposit their load directly in the nest, but nectar collectors give theirs to helpers on entering. By this and other transfers, information is passed throughout the colony: antennae touch, and orientation given, usually from older to younger bees. Thus, the workers are able to respond to changes in external cues such as food availability, as well as internal ones such as the state of the queen.

Fig. 4.4 *The worker honey bee often specializes at any one time in collecting nectar or pollen. At the same time as the worker in* **(a)** *is collecting pollen (note the full pollen baskets on the hind limbs), the worker in* **(b)** *is collecting nectar from the same garden and avoiding contact with the stigmas and style of the flower*

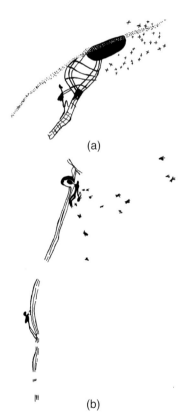

(a)

(b)

Fig. 4.5 *Rock art showing honey hunting from wild nests:* **(a)** *from Eland Cave, KwaZulu Natal;* **(b)** *from La Araña Shelter, Valencia, Spain*

TAMING THE BEE

Our developing knowledge of honey bee biology has gone hand in hand with improvements in its management, but this has been a drawn-out process.

Mesolithic and later rock art in Spain and Africa illustrates hunter-gatherers collecting honey from nests, and honey hunting is probably as old as *Homo sapiens* (Fig. 4.5). In Egypt there are many representations of adult bees from 6000 BC onwards. Honey hunting was practised in Ancient Greece but there is little mention of it in the west after the collapse of the Roman Empire, as by then hives were mainly used around the Mediterranean. However, honey hunting is still practised, and in the 1980s 90% of the honey marketed in Ghana and 80% in Madagascar was from honey hunting. A remarkable honey-hunting partnership between man and the honey-guide bird has developed in Africa (Box 4.2).

Honey bees probably arrived in the UK at or before the arrival of lime trees after the last ice age (e.g. by 5500 BC), with beeswax being found on pottery sherds in England dated 3000–2650 BC. There was also much honey collected from nests in trees during the Anglo-Saxon period (AD 410–1066).

The ownership and tending of nests was the next development, probably due to greater population pressures and a need for more honey. Nests in rocks

BOX 4.2	THE HONEY-GUIDE

For many centuries there have been reports of the peculiar behaviour of a small family, the Indicatoridae (related to the woodpeckers) of African birds, the honey-guides. It was reported that hunting for honey bee nests by the honey badger or ratel (*Mellivora capensis*) was done in partnership with a honey-guide bird. The guiding behaviour was released in the bird by sight or sound of a ratel, and provided it stayed at a distance this guiding continued, leading the mammal to a nest. The ratel opens the nest, and the honey-guide feeds on the remains, taking brood and wax. The bird is one of the few organisms able to digest beeswax.

There were also anecdotal reports from the seventeenth century onwards of the birds guiding human nest hunters, but this had largely been considered a myth. Recently, Isack and Reyer (1989) have demonstrated that this interaction does take place, and that there is a sophisticated interaction between bird and man. Their research involved the nomadic Boran people of Kenya, who claimed that they could deduce the direction and distance to the nest from the greater honey-guide's (*Indicator indicator*) behaviour. The Boran gain the bird's attention with a penetrating whistle, and make noise throughout the interaction to keep the bird interested. The bird flies close to the humans, moves restlessly between perches and emits a persistent

'tirr-tirr-tirr-tirr' call (a sound apparently similar to that made by shaking a partly full matchbox rapidly lengthwise). The honey-guide often then flies directionally above the tree-tops and disappears for a minute or more: when it returns it perches somewhere conspicuous. Repeated leading and following takes the Boran to the nest.

A change in the bird's behaviour signals arrival at the bee nest: it emits a softer call, with longer between notes, and then remains silent. It can fly around the nest from perch to perch, and if the nest is not found by the Boran, the bird looses interest and tries to lead others to another nest.

The birds were able to accurately guide the Boran to nests because they continually monitored the position of bee nests in their locality, peering into nests early in the morning while the bees are still quiescent.

This is an interesting example of a human–bird mutualism: 96% of the nests located in the study in this way were accessible to honey-guides only after being opened by man, and Boran search time per nest of nine hours unguided was reduced to just over three hours when guided. It is also an example of the unformalized ecological knowledge of indigenous peoples, which is liable to be lost.

were often long-lasting and could be owned by a community, whereas a tree cavity nest was usually owned by one person. The Domesday Book (AD 1086–87) mentions some instances, for example at Fladbury and Bredon in Herefordshire, of Bishops receiving as dues all the honey produced from certain areas of woodland; this probably refers to honey collection from wild nests. There are records of honey ownership in Royal Forests and of people being fined for taking honey from these woods. These nests were often tended and sometimes nests were incorporated into stone walls.

The hiving of bees in an artificial cavity in a moveable container came next. This overcame the problem of insufficient natural nest sites in areas with abundant bee forage and enabled an increasing amount of honey and wax to be produced. Hiving probably started in warm temperate regions where bee forage was available for most of the year; in colder regions bee colonies often died out during winter and thus there was little incentive for hiving. Honey bee hives are first

Fig. 4.6   *Part of a stone bas-relief from the Sun Temple of Ne-user-re, Abu Ghorab, Lower Egypt, c. 2400 BC – the earliest known representation of beekeeping*

(a)

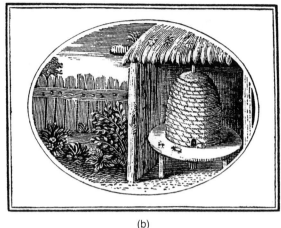
(b)

Fig. 4.7 *Skep beekeeping.* **(a)** *From Magnus (1555), showing skeps under a thatched shelter (top left), a beekeeper in rudimentary protective clothing collecting a swarm from a tree (top right), and transportation of skeps (bottom right).* **(b)** *A skep on a table under a thatched shelter [From Bewick and Bewick (1820).]*

represented in Ancient Egypt (Fig. 4.6) and consisted of a horizontal mud or clay cylinder. This became the dominant type world-wide. Traditionally, these hives were about 140 cm long, 22 cm wide externally and 17 cm internally, made of mud formed around a straw mat. Both ends were closed with mud, and a flight entrance pierced in the front. These hives were formed into a stack of 400–1000. Honey and combs were harvested from the back, and the front opened to inspect the colony or remove brood. This type of hive was widely used in Ancient Greece and the Roman Empire, and the earliest archaeological remains of hives date from almost 1000 BC in the Jordan Valley, Israel.

In northern Europe, hives were first made in a vertical section of tree trunk and in more open areas, log hives gave way to skeps – basket hives made from woven plant stems. The open mouth was placed on a wood or stone stool. These hives were not weatherproof and were usually protected in apiaries, wall recesses (bee boles), or shelters (Fig. 4.7). Many of these structures can still be found in the UK (Fig. 4.8), ancient reminders of this all but forgotten practice.

In the UK, hive beekeeping was probably introduced well before the Romans, originally using wicker skeps. More robust woven coiled straw skeps were introduced by the Saxons after AD 500 and gradually replaced the wicker ones. In the Domesday Book, beekeepers and their hives are sometimes recorded; and in the 'Little Domesday' covering Essex, Norfolk and Suffolk, the number of hives is often noted.

The use of vertical skeps led to a characteristic beekeeping practice in cool temperate countries which was retained until the introduction of modern hives in the later nineteenth century. This practice was termed 'swarm beekeeping', and depended on the production and hiving of swarms in early summer, before the main honey flow (the period during which plants produce the most nectar) later. Swarms were produced easily due to overcrowding in the small skeps and were captured in other skeps and colonies allowed to develop. Four stock hives in spring might easily produce 12 hives after swarming. At the end of the honey flow, some hives were left alone with their honey to overwinter, and the colonies in the heaviest and lightest hives (which would have insufficient honey to survive over winter) were killed and the honey removed.

Fig. 4.8 **(a)** *Seven bee boles in an early seventeenth century wall at Finchamstead, Berkshire, UK. One skep would be placed in each of these recesses.* **(b)** *Two-storey bee bole in the seventeenth century wall of a house, Langdale, Cumbria, UK*

These traditional hives, either horizontal or vertical, with fixed combs made by the bees, were gradually improved but were almost impossible to harvest honey from without destroying the colony. However, greater improvements were made from the 1650s onwards, and after 200 years of experimentation and development an effective 'rational' hive was patented in 1851.

Crane (1999) detailed the development of rational hives, through the following sequence:

1. a modular hive of precision-made tiered wooden boxes, which could fit tightly together;
2. a framework inside the hive to which bees could attach combs, and which enabled the combs to be removed;
3. providing parallel top bars in the hive mimicking the bee's natural comb spacing, so that a comb built down from each could be lifted out;
4. each top bar developed into a rectangular frame;
5. the frame moved away from the hive walls by the space the bees left naturally between their combs and the nest cavity;
6. incorporating these features in a cheap, robust and easily worked hive.

Although stage 1 had been reached by 1649, and stage 3 by 1683 (but failed because the correct spacing was not known), stage 6 was not reached until 1851 in the US with the development of the Langstroth hive. This followed the crucial discovery earlier in the nineteenth century that bees would not wax up spaces around 1 cm wide, through which a bee could pass. Thus, frames separated by this distance (the 'bee space') from the hive case could be removed. The Langstroth hive formed the basis of all modern hives, with many later regional modifications. Although much more honey could be produced in these hives, they also cost more to make, and this led to a change in the nature of beekeeping from a cottage industry to an organized profession: many beekeeping associations were formed around this time.

The modern hive has a 'brood box' with removable combs and two or three 'supers'—similar-sized frames which hold the honey frames. The super can be above or below the brood box, and is separated from it by a queen excluder (Fig. 4.9).

More recent developments have mainly been made in the processing of honey, with the centrifugal honey extractor developed in Europe in 1864–5, and the use of combs made artificially from beeswax pressed into hexagonal cells, so that bees could concentrate on filling the combs with honey.

Along with these technological developments has been selection for optimal traits in the honey bee. Artificial selection, through queen rearing and artificial insemination, is relatively easy as only a few queens and drones reproduce. Some of the characteristics selected for include good colony survival through periods with poor food or weather; resistance to diseases or pests; good storage of honey; tolerance of handling and reluctance to sting; being easily pacified (e.g. by smoke); a low tendency to swarm; and returning to their own hives and not another (so as not to spread diseases).

BEE PRODUCTS

Until recently, honey and beeswax were the only bee products of importance. Bees can make honey out of honeydew, exudations from plant wounds (e.g. the cut stems of sugar cane, Chapter 6), and most importantly from nectar. Three kilograms of nectar becomes 1 kg of honey, and one hive can probably produce 10 kg of honey a year. Nectar is an aqueous solution of various sugars and other compounds including low concentrations of some vitamins. The sugar concentration of nectar varies greatly, from 15% in pear and plum to over 70% in marjoram for example, and bees rarely collect nectar with less than 15% sugar content.

The worker bee collects the nectar into its honey sac, a storage space before the foregut, which when full may comprise 90% of the weight of the bee. The nectar is diluted with saliva, and in the hive is passed from bee to bee, repeatedly regurgitated into a thin film, and drawn up again. This carries on for 15–30 minutes and water is evaporated, producing half-ripened honey. It is then placed into cells for 1–3 days where further evaporation occurs by fanning, before more honey is added, forming the honey store. Ripened honey now contains less than 18% water and is sealed beneath a wax cap. A number of chemical changes occur while the nectar is in the bee and being processed: the enzyme invertase is secreted (this catalyzes the hydrolysis of sucrose to glucose and fructose, the main sugars found in honey, and those most easily digested by the bees); and the enzyme glucose oxidase is secreted from the bees' hypopharyngeal glands into the honey. This enzyme oxidizes glucose, liberating hydrogen peroxide, an antiseptic, and becomes active when the honey is diluted with water.

Honey thus contains 30–35% glucose and 35–40% fructose, in a supersaturated solution. This high sugar concentration (fructose is more soluble in water than glucose and sucrose) and its acidity prevents microbial growth, and along with hydrogen peroxide production accounts for its medical uses, described for the last 4000 years. For example, it has been used in wound treatment, especially burns, and for preserving corpses. Some of these uses are being rediscovered by modern medicine. However, honey's main use was as mankind's primary sweetener in the temperate world until replaced by sugar cane (Chapter 6). Its importance can be gauged by its extensive use as currency: in the Domesday Book there are many examples of dues being paid in honey. Mead, and other alcoholic drinks produced by yeasts fermenting the sugars, were probably produced from honey thousands of years before wine and ale.

The properties of honey, its colour, granularity, taste and smell depends on the flowers from which the nectar was collected. Some nectars add other characteristics as well: neurotoxins. These are found especially in honey made from nectar of rhododendrons (particularly *R. ponticum*) and other plants from the Ericaceae, containing grayanotoxins. While in small amounts such 'mad honey' can be a pleasant pick-me-up, and there was a thriving industry importing such honey from the Crimea in the eighteenth century, eating large amounts has on occasion proved dangerous, as Mayor (2003) relates; and although in general its effects are probably overstated, cases of non-fatal poisonings of humans and animals are still reported, especially from Turkey.

Beeswax, the other original bee product, is a useful substance, being plastic above 32°C, melting at 60–65°C, and of a complex composition. Nowadays, much beeswax is used to make empty combs to put back into hives, but originally it was used to make candles, sealing wax, polish, as a component of cosmetics, and in the lost-wax procedure for metal casting. Like honey, beeswax was used as a means of payment and traded over long distances.

Other commercial products from bees were hardly known before the 1950s, when the price of honey was reduced by excess production, and beekeepers diversified their products. Pollen can be collected from returning bees by placing a

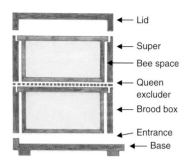

Fig. 4.9   *Cross-section of a Langstroth-type hive. Moveable frames (grey) with comb (yellow) are fitted in tiered wooden boxes. Brood is reared in the lower brood box, honey is stored in the combs in the super. The supers are often shallower than the brood chamber and two or three may be placed above it. The queen excluder prevents drones and queens from passing from the brood box to the supers*

pollen trap across the flight entrance of the hive. Pollen contains 6–28% protein, almost the bee's only protein source, and also contains important amounts of lipids and sterols. A hived colony can collect up to 35 kg of pollen a year, and up to 10 kg of this can be removed without damaging the colony. Royal jelly, the brood food secreted by workers into queen cells, can be collected from colonies manipulated by feeding them sugar syrup and pollen. As the queen bee can live for several years, compared to the life of a worker measured in weeks, people believed that royal jelly had anti-ageing properties. Over 200 tonnes a year of royal jelly are consumed in each of China, Taiwan and Japan.

It is also believed that bee venom has a beneficial effect on some types of rheumatism, and other medicinal uses for it were recorded by the ancient Greeks. Commercial production started in Germany in the 1930s, and since the 1960s a large-scale method of production has been used whereby bees pierce and 'sting' a thin membrane and release their venom; as they can release their sting from this membrane they can sting repeatedly.

Propolis, a sticky substance collected by bees from plant exudates, contains many flavonoids active against micro-organisms. Bees use this to seal cracks in their nests, and humans have used it as glue, polish or a medicine. Finally, the brood itself can be consumed. The fat and mineral content of mature larvae and pupae are about the same as that in beef, and brood contain significant amounts of vitamins A and D. Brood are eaten widely in Asia and Africa (see also Box 15.2).

However, at least as important financially as honey production, and more important overall, is the honey bee's contribution to pollination.

POLLINATION

Pollination involves the transfer of 'male' pollen grain from the anthers of the flower where they are stored to the style at the tip of the ovary. Each pollen grain then grows a germ tube into the flower's ovary and fertilizes an ovule. This pollen transfer can be within a flower, between flowers on a plant, or between plants. Of the more than 250 species of plants grown for food, oil or fodder in the EU, 84% of those studied benefit from or depend upon insect pollination (in some species, such as many of the cereals each flower pollinates itself, and wind pollination is important in some other crops). The understanding of pollination and the role played by bees was realized gradually between the 1670s and the 1880s, and the use of hived bees to pollinate crops has been recommended since the 1890s. Until the late 1980s, honey bees were the only managed pollinators available.

Although all hived bees will contribute to crop pollination, and many amateur beekeepers thus provide a free and ad-hoc pollination service to local crops, hives used specifically for pollination need different management from those used mainly for honey production and are usually moved to the crop. Whereas a honey-producing colony needs lots of foragers collecting nectar, for effective pollination these foragers need to be collecting pollen. This happens when the colony contains lots of unsealed brood (which consume much pollen). To encourage this, beekeepers offering a pollination service can add unsealed brood or feed sugar syrup to the colonies. Hives also need to be moved to the crop plants at just the right time (too early and the bees are likely to forage elsewhere, leave them too long and local nectar sources may be depleted), and in the right numbers to ensure efficient pollination without competition for resources.

If managed properly, this pollination service provides a greater benefit to the crop-grower than to the beekeeper, but commercial beekeepers can obtain a significant part of their income from this service. When the value of pollination in producing extra yield to crops is taken into account, this service becomes more financially valuable than honey production. In 2007 just over 1 million tonnes of honey was produced globally, worth US$ 1.25 billion, yet the value of insect pollination was estimated as US$ 212 billion, and about 80% of that was estimated to be due to honey bees.

Fig. 4.10 *The worker honey bee often has to try quite hard to open a flower in order to extract nectar or pollen from it*

Honey bees are generalist foragers, they are good at exploiting a wide range of flowers and are very good at producing more honey bees, but this does not necessarily make them good pollinators. Indeed it appears that no flowers have co-evolved with honey bees as pollinators. Honey bees are rather small bees, meaning that they loose heat rapidly and they do not forage in cold weather or the rain. Their small size also means that they find it physically difficult to pollinate some plants (Fig. 4.10). For example, honey bees have been promoted as pollinators of alfalfa or lucerne (*Medicago sativa*), a valuable forage crop in the US which needs insect pollination to produce a good seed set. However, in order to pollinate this crop, the bee needs to 'trip' the flower (Fig. 4.11) whereby the stamens and style emerge explosively from the keel and the style becomes abraded and able to receive pollen; pollen is also left on the bee. Honey bees can trip alfalfa flowers, but their small size makes this difficult and dangerous; they can become trapped inside the flower, or killed by the tripping process. Thus, experienced honey bees tend to avoid untripped flowers, and nectar gatherers learn to insert their tongue sideways to collect the nectar without tripping the flower. The European leaf-cutter bee *Megachile rotundata* is a much better alfalfa pollinator and has recently been introduced to the US for this purpose.

Apart from their small size, the relatively short tongue of honey bees means that they will be unable to collect nectar from a number of plants, and thus will

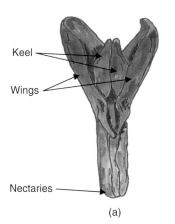

Keel

Wings

Nectaries

(a)

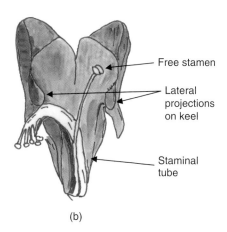

Free stamen

Lateral projections on keel

Staminal tube

(b)

Fig. 4.11 *Alfalfa* (Medicago sativa) *flowers:* **(a)** untripped; **(b)** tripped [Based on Free (1993).]

not visit them for pollination. Also they cannot 'buzz-pollinate', whereby the anthers are grasped by a bee and rapidly vibrated, releasing pollen—essential for pollination of plants from the Solanaceae, such as tomatoes and peppers.

Many of these problems are not shared by bumblebees, many species of which naturally occur with the honey bee. Bumblebees are larger, with longer tongues, thus able to pollinate more species and continue to pollinate in poor weather, and can buzz-pollinate. They are often more efficient pollinators, visiting flowers quicker than honey bees and pollinating a greater proportion of them. In northern Europe bumblebees are probably the most important pollinator; elsewhere, honey bees have this role by virtue of their overwhelming numbers, not because of their pollinating abilities.

Since Darwin, there has been speculation and concern that introduced honey bees may forage to the detriment of other species, out-competing native species for nectar and pollen. Recently there has been particular concern over competition between honey bees and bumblebees. In cold climates this is less likely; the bumblebee is able to start foraging earlier in the day and will be able to accumulate nectar reserves before the honey bee is foraging. When the honey bees start foraging they will deplete the flowers' nectar reserves, making it not cost-effective for the bumblebee to forage on these plants, but bumblebees will still be able to forage on the flowers with deep nectar sources that the honey bees cannot reach, or flowers, such as lucerne, red clover and field bean which honey bees find difficult to enter.

Competition between honey bees and native pollinators has been difficult to demonstrate and virtually nothing is known about the costs to the native fauna of honey bees introduced into new countries. However, increasing information is becoming available about the effects on established honey bee colonies of a newly introduced honey bee race, the Africanized honey bee, or 'killer bee'.

KILLER BEES?

In H. G. Wells' *The Empire of the Ants*, a satire on colonial expansion and the futility of trying to control nature published in the 1890s, a Brazilian gunboat is sent up the Amazon to confront a 'plague' of ants: ants which were especially vicious and intelligent. It ends: 'By 1920 they will be halfway down the Amazon. I fix 1950 or '60 at the latest for the discovery of Europe.'[3] Just over 50 years later a vicious strain of another social insect, the honey bee, did spread, much more quickly, throughout Latin America.

The honey bee is not native to Brazil, and beekeeping started with the introduction of the races mellifera in 1839 and ligustica in 1870 (see Fig. 4.3). These bees, although docile and productive in Europe, did not produce much honey in their new tropical conditions and continued to decrease brood in winter, an unnecessary activity in that climate. Therefore, beekeepers looked for 'better' races of *A. mellifera*, more suited to Brazilian conditions. In the early 1950s the African honey bee, *Apis mellifera scutellata*, was recommended. Although it was known to be more defensive of its nests than the European races, it was thought to be a prodigious honey producer, and that by selective breeding with European races the good traits of both bees could be established in Brazilian populations.

Therefore, in October 1956, 170 queens of *A. m. scutellata* were imported into Brazil from South Africa. About 47 queens survived and were introduced

[3]Wells (1927), p. 16.

Fig. 4.12 *Spread of the Africanized honey bee, from its introduction into Brazil in 1957. The upper dashed line represents the 120 days with maximum temperature below 10°C isotherm, the projected limit of this bee's spread in North America. The lower dashed line represents estimated spread by 1997 from Winston (1992) [Based on Winston (1992), with additional from Crane (1999) and Harrison et al. (2006).]*

into European bee colonies near São Paulo. Unfortunately, a technician mistakenly removed the queen excluders (used to stop colonies swarming) from some of the hives in early 1957, and 27 of the colonies promptly swarmed. Furthermore, African queens were propagated through queen rearing and artificial insemination from 1958 and distributed widely among south Brazilian beekeepers.

It did not take long for the effects to be noted. Many feral African honey bee colonies became established, with the feral bees spreading 300–500 km a year (Fig. 4.12) in a large swarming front of bees probably selected for colonization (long-distance swarm migration, absconding), so that by 1990 there were 50–100 million feral honey bee colonies throughout Latin America.

The European honey bee races live in a seasonal environment defined largely by temperature and thus the colony stores honey and produces a large insulated nest able to contain a large number of workers over the winter. In the African honey bee's natural environment seasonality is mainly determined by rainfall; there is no cold season and so no need to store much honey or to produce large insulated nests. Thus, the nest of the Africanized honey bees is smaller (50% of the European volume and containing 30% of the honey) and often is not formed in a cavity (external nests are more easily cooled).

Periodic lack of nectar or pollen, or increased predator pressure, means that Africanized honey bees often 'abscond', leaving their nest entirely, and travel often a long distance to form another nest. This is not an option for the European honey bee, as the absconding swarm would not be able to collect enough nectar to survive the next winter. The Africanized colonies grow much faster than the European colonies, and thus they swarm up to 12 times a year, compared to the 1–3 times usual for European races.

The major public interest was in the Africanized honey bees' increased defensiveness. An individual sting from one of these bees is no worse than that from a European honey bee, but the Africanized colonies are much more capable of sudden large-scale attacks after minimal disturbance. When they appear in a new area there is often considerable livestock and human mortality, until humans learn co-exist with the bees. For example, in Venezuela there were 400 fatalities from Africanized honey bee stings between 1975 and 1990, but 100 of these occurred in 1978, just after the bee appeared, and only 12 in 1988.

The Africanized honey bees also have a major effect on beekeeping. These colonies produce a large number of workers and drones, and on arriving in a new area as a feral swarm, after increasing their numbers, swamp the European honey bee colonies, and African drones mate with European queens. There is little mating the other way. Therefore, the European colonies became steadily Africanized, while the African colonies did not gain any European bee traits. Thus, within a couple of years of Africanized bees invading an area, beekeepers were faced with colonies of aggressive, absconding, unpredictable bees, that produced little honey.

Until local beekeeping methods changed to cope with these bees (for example, by using greater protective clothing, better pacification, treating the bees very gently, placing colonies away from livestock and people) honey production declined drastically. Some countries managed this transition well: for example, in Brazil beekeeping increased from a low point of 320 000 hives and 5000 tonnes of honey in 1972 after the arrival of Africanized honey bees to 2 million hives and 36 000 tonnes of honey in 1988. Indeed, beekeepers with the technology to cope with these bees often prefer them to the European races; their defensiveness prevents theft and good colonies can always be obtained. However, the honey production per hive at 18 kg is still well below the 60–80 kg per hive claimed to be possible from this bee. Other countries fared less well: Venezuela saw honey production fall from 1300 to 78 tonnes between 1976 and 1981; and in 1987 Panama had 50% of the beekeepers, 40% of the colonies and 19% of the honey production it had five years previously. Thus, management of this invasion has been almost entirely by humans becoming adapted to dealing with the problem. Although breeding out the undesirable traits in the bees was still being suggested into the 1980s, this has not worked. However, as there has always been a great variation in defensiveness exhibited by different Africanized honey bee colonies, it is possible to eliminate queens from the most aggressive colonies. This procedure was widely adopted in Brazil, and by 1990 it was considered that the problem of colony aggressiveness had been practically solved.

The first feral swarm in the US was caught near Hidaglo, Texas, in October 1990 and there were dire predictions of the effect of the spread of the Africanized bee in the US, where its final distribution would only be limited by its poor ability to survive temperate winters. The impact was estimated at US$ 26–58 million per year, with possible reduction in crop yields, due to lack of managed pollination, of the value of US$ 100 million.

The long wait for the Africanized honey bee to enter North America, and its subsequent spread led to a keen national interest, almost obsession, with this insect. Pop groups, children's books and adult fiction featured killer bees, for example Arthur Herzog's 'The Swarm', and several films, including the film version of 'The Swarm' (1978, which used 22 million bees one of the last of the US wave of 1970s disaster movies).

However, the spread of the Africanized honey bee within the US has been slower than in South America (compare in Fig. 4.12 the expected range in 1997 with the actual range in 2005). It reached California in 1994, Nevada in 1998 and Oklahoma in 2004, and was even slower reaching the east coast, being found in Arkansas, Florida and Louisiana only in 2005. Its economic impact has also been much less than predicted. A reduced capacity for winter survival, limited rainfall and a greater European honey bee population, could all have led to a slower spread of African characteristics.

Where will the Africanized honey bee reach in the US? There are two models: one that its range limit would be where the mean maximum temperature falls below 16°C for the coldest month, and the other that it will be where there are more than 120 days with a temperature below 10°C (see Fig. 4.12). The Africanized honey bee has already passed the former limit, and the latter limit extends up to the west coast of Canada. As the Africanization process extends northwards, it is possible that hybrids will gain a greater coldhardiness from their European parents, and thus extend their range further north still, a factor which climate change could exacerbate.

However, while the US was awaiting the invasion of one threat to beekeeping, another had already invaded, one with a far greater impact on beekeeping world-wide: the mite *Varroa*.

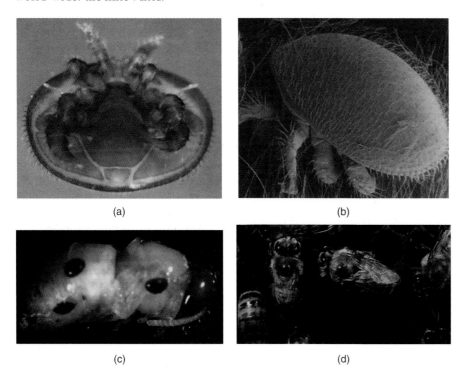

(a) (b)

(c) (d)

Fig. 4.13 Varroa destructor: **(a)** *ventral view;* **(b)** *dorsal view;* **(c)** *on honey bee pupa;* **(d)** *on adult honey bee* [From Kanga, L.H.B., Jones, W.A. and James, R.A. (2005) Enlisting fungi to protect the honey bee. Biologist **52**, 88–94 © Society of Biology, reproduced with permission.]

VARROA!

Of the several parasitic mites of honey bees, *Varroa destructor* is probably of most current importance. The adults are quite large mites; the female is 1–1.75 mm long by 1.5–2 mm wide, and the male 0.75–1 mm long and about 0.75 mm wide. They are flattened, which enables them to fit between the abdominal plates of the adult bees, and avoid being brushed off by them (Fig. 4.13).

The adult female mite seeks out brood cells; drones are greatly preferred, but workers will do. She enters the brood cell of a developing larva about two days before the cell is closed and capped, and hides in the liquid brood food at the bottom of the cell. About 60 hours after capping she starts to lay eggs on the pupae; the first is unfertilized and gives rise to a male mite (*Varroa* has the same haplo-diploid sex determination system as its host); all the rest of the eggs, laid at daily intervals thereafter, are fertilized and give rise to females. The hatched larvae feed on the haemolymph of the pupa, through a hole made and kept open by the adult mite. When mature, the male's mouthparts change into a hollow tube to transfer sperm, and he mates with his sisters in the pupal cell. When the adult bee emerges from the cell, the mated females are carried on its back, and can infest other cells; the male mite and any remaining larval mites die. An average of one mite is produced per worker cell and about two per drone cell.

BOX 4.3   COLONY COLLAPSE DISORDER

Beekeepers have always lost hives over winter: in the US beekeepers would expect to loose 5–10% of their hives each winter, rising to 15–25% annually after varroa became established. Yet in the autumn of 2006 a new phenomenon was noticed in the USA. Many hives, on opening, were found to contain the queen, many untended brood cells, and just a few newly emerged adults. Plenty of food reserves were left, and other insects had not started to rob them, but there were no adult workers, nor were there any dead workers in and around the hive.

This phenomenon became known as colony collapse disorder (CCD), and during the winter of 2006–7 it was estimated that 23% of US beekeepers experienced it, and lost an average of 45% of their hives. The following winter was even worse, with an estimated loss of 36% of the almost 2.5 million US managed honey bee colonies. Similar losses have been reported throughout the world, and have continued in subsequent years.

Although some were sceptical that CCD was a new phenomenon, the overwinter losses were greater than experienced previously, and people began to look for causes. There were many candidates; some such as alien abduction and mobile phones were ruled out early on, and other single causes such as genetically modified crops, pesticides, bad weather, practices associated with the rearing of bees for pollination (in the US 60% of honey bee colonies are used in the pollination of Californian almonds) and climate change could not explain CCD.

One possibility was that a new pathogen had infected bees. An early candidate was the newly found Israeli acute paralysis virus (IAPV), infection with which correlated with CCD.

IAPV was reported to have been introduced into the US in honey bees imported from Australia, a trade that started only in 2004. This caused great consternation among Australian beekeepers, and later studies have shown that this virus had been present in the US since at least 2002, is found in other parts of the world not associated with this trade, and is overall poorly correlated with CCD.

After this hunt for single-factor causes, scientists are gradually realizing that there is probably no one single cause of CCD. Rather, it is likely that many of the factors listed above (with the exception of aliens and mobile phones) combine to weaken the immune system of bees, enabling opportunistic pathogens to multiply. Infected bees often leave the nest never to return, and it is becoming clear that bees in CCD colonies have a larger pathogen load than those from other colonies.

Varroa is implicated in CCD, as it is well known to suppress the immune system of bees and also to vector viruses. The microsporidian (see Box 5.4) parasite *Nosema ceranae* is also implicated. This species has only recently been discovered on *Apis mellifera* (its normal host is *A. cerana*) and appears to be more virulent than *N. apis*, which has been known for a long time. *N. ceranae* can cause colony collapse within two years, and has been implicated in large-scale honey bee losses in Spain.

The only bright side to this latest affliction to hit the honey bee is that predictions of the effect of CCD were so dire that governments on both sides of the Atlantic have been persuaded to significantly increase research into honey bee and pollinator health, a hitherto neglected area.

The mites, as good parasites, do not kill their hosts, but do weaken them, leading to weight loss, deformities and shortened life-span, and increased effects from viruses, some of which are vectored by *Varroa*. It is these viruses that kill the bees, and ultimately cause the bee colony to collapse (Box 4.3). At a mite population of several thousand the bee colony loses cohesion and collapse starts. Bees may migrate from these colonies to nearby ones, spreading the mite.

All this suggests that *V. destructor* is well adapted to the honey bee, and that this relationship has been developing and evolving over many generations. This is not so: the interaction is relatively recent. *Varroa destructor* and its close relative *V. jacobsoni* (the two species were only separated in 2000) originally parasitized the Asiatic honey bee, *A. cerana*. At some point, *V. destructor* started to parasitize the European honey bee *A. mellifera*. The events leading to this are obscure, not least because we can no longer be sure which mite species the original records refer to; but there have been several opportunities for the two bee species to interact, and for mites to spread. One such opportunity occurred in the Philippines, where *A. mellifera* had been introduced and become contaminated with *Varroa* through contact with *A. cerana*, probably through mutual nest robbing, or the introduction of sealed *A. cerana* brood into *A. mellifera* nests to strengthen colonies. *Varroa* was first recognized as a pest of *A. mellifera* in 1965 in the Philippines. In 1971, infected *A. mellifera* were introduced from Japan to Paraguay and from there to São Paulo in Brazil in 1972. *Varroa* was first noted in South America in 1975 (see Fig. 4.3), and it is thought that all of it originated from Paraguay.

Another interaction between bee species occurred in the former USSR. Beekeepers from western USSR moved to the far-eastern province of Primorye in the early twentieth century, and began working with *A. cerana*, the honey bee native to this region. Later they introduced *A. mellifera*. Mites were observed on *A. cerana* here in 1952. In the 1960s, reports of high honey yields from this region led western USSR beekeepers, thinking that there was a 'special' bee here (shades of the African honey bee story), to import *A. mellifera* queens from this region. By 1965 there was widespread *Varroa* infestation of western USSR, and it was noted in Bulgaria in 1967 and Romania in 1975. This did not stop these countries exporting colonies to Libya in 1976 and Tunisia in 1975, respectively, introducing *Varroa* to Africa.

Varroa was detected in the US in 1987 and the UK in 1992. Until the twenty-first century, Australia and New Zealand remained the only countries free of this mite, but *Varroa* was detected in Auckland, North Island, New Zealand in April 2000, and rapidly spread throughout the island. It was first found on the South Island at Nelson in June 2006, and despite an immediate eradication programme it has spread to the surrounding area. In 1996 the global winter kill of *A. mellifera* by *Varroa* was estimated as 13 million colonies, a quarter of global commercial bee colonies, with up to 65% being destroyed in some countries.

The spread of the mite through the UK is salutary. It was first found in Devon in April 1992. The responsible government agency at the time, MAFF, organized a search of colonies in the locality, and heavy outbreaks were found a few days later in a wide area of north Dartmoor, two weeks later in Somerset, and Surrey one week later. A 6 km statutory infected area (SIA) was declared around the first apiary, this was later extended to 40 km, and eventually the whole of Devon was included. MAFF wanted the whole country declared endemically infested, to avoid regulatory disruption to the beekeeping industry. However, the Minister of State for Agriculture consulted with national beekeeping associations and, anxious to limit the perceived spread of the disease, did not recommend this, but gradually increased the SIA to 16 southern counties by May 1992[4]. Over the next five years *Varroa* spread progressively northwards throughout the UK, and by 2000 it was generally accepted that it was endemic.

Control of *Varroa* was difficult, but some pyrethroids (see Box 12.5) were effective. In the UK, *Varroa* was at first relatively easy to control with one autumn application of pesticide. Then, there was a period when the mite population increased rapidly, because it was present in the feral bee population, and it was very difficult to control. When treatments did not work, or were delayed, up to 50% colony loss was experienced. Later, *Varroa* became easier to treat; this probably coincided with the loss of the feral bee populations.

However, *Varroa* became increasingly resistant to pyrethroids, and by 2005 resistance was well established in south-west England, with pockets elsewhere. With only one type of pesticide available, this resistance is set to increase. Belatedly, other control methods are being investigated. Some techniques involve hive manipulation, for example by trapping *Varroa* in brood which is then removed, or forcing an artificial swarm to a clean nest. Some other chemicals show promise, examples being oxalic acid (widely used unofficially in Europe) and formic acid (although concentrated application can kill brood and affect queens). Biological control is also possible, with the fungal pathogen *Metarhizium anisopliae* showing promise (see Box 15.3), and the widely used fungal pathogen *Beauveria bassiana* recently isolated from *Varroa*, and able to cause high mortality (see Box 5.3).

Other researchers have tried to find resistance to *Varroa* within *A. mellifera*, and this work has revealed details about why this species is affected by *Varroa* so much more than *A. cerana*. Apparently, *A. cerana* is a more effective groomer than *A. mellifera*, and many mites are removed this way. *A. cerana* also bites and kills mites, and can detect them in the brood cells, uncap the cells and kill the mites—characteristics rare in *A. mellifera*. *A. cerana* pupae also spend less time capped in cells, which gives less time for mites to develop.

In 1988, Gould and Gould were able to conclude:

> For all their efforts, though, humans have not succeeded in domesticating bees. A swarm escaping from a commercial hive has just as good a chance of surviving in the wild as a feral swarm, and the number of wild colonies living in trees still far exceeds the population living in accommodations designed for them by humans. The history of beekeeping, then, has not been a story of domestication, but rather one of humans learning how to accommodate the needs and preferences of the bees themselves.[5]

Over the last few years, however, *Varroa* infestation has led to the total destruction of feral *A. mellifera* colonies in parts of the US and other countries, and the loss of pollination from these feral colonies has been widely noticed. It seems likely that this parasite is achieving something that man has not managed: the conversion of the European honey bee to a truly domesticated species, unable to live in the wild, and wholly dependant on humans. The long-term effect of colony collapse disorder (Box 4.3) on this newly domesticated species remains to be seen.

FURTHER READING

There are many general books on bees and beekeeping, and more websites, but Ellis (2004) and Kritsky (2010) can be recommended as popular accounts. Gould

[4] Ten years earlier, a widely available review of *Varroa* (De Jong et al., 1982) had noted that the mites can be present in an area for 2–6 years before honey bee colonies show obvious signs, and in nearly every country to that time, the mite was discovered too late for its spread within the country to be controlled.

[5] Gould and Gould (1988), 1995 edn, p. 17.

and Gould (1988) and Winston (1987) give good overviews of the honey bee, concentrating on bee behaviour, little of which we have mentioned here, and vanEngelsdorp and Meixner (2010) give an up-to-date survey of the managed honey bee. Crane (1999) is an encyclopaedic coverage of all aspects of beekeeping; More (1996) is a useful practical introduction to the craft. General aspects of pollination, including substantial sections on the honey bee, are covered in Proctor *et al.* (1996) and for crop pollination see Free (1993). For the Africanized honey bee, Winston (1992) gives a short review of the situation at that time; a wide range of studies is presented in Spivak *et al.* (1991). More recent work is covered in Schneider *et al.* (2004) and Harrison *et al.* (2006). The DEFRA web site gives some useful guidelines for managing *Varroa*, Rosenkranz *et al.* (2010) is a useful summary, and Kanga *et al.* (2005) a summary of *Varroa* and its biological control. For colony collapse disorder, see Cox-Foster and vanEngelsdorp (2009) and vanEngelsdorp *et al.* (2009), and Benjamin and McCallum (2008) is a useful popular account.

REFERENCES

Benjamin, A. and McCallum, B. (2008) *A World Without Bees.* Guardian Books, London.

Bewick, T. and Bewick, J. (1820) *Select fables, with cuts, designed and engraved by Thomas and John Bewick, and others, previous to the year 1784: together with a memoir; and a descriptive catalogue of the works of Messrs. Bewick.* Emerson Charnley and Baldwin, Cradock, and Joy, London.

Cox-Foster, D. and vanEngelsdorp, D. (2009) Saving the honeybee. *Scientific American,* April, 24–31.

Crane, E. (1999) *The World History of Beekeeping and Honey Hunting.* Duckworth, London.

De Jong, D., Morse, R.A. and Eickwort, G.C. (1982) Mite pests of honey bees. *Annual Review of Entomology* **27**, 229–252.

Ellis, H. (2004) *Sweetness and Light: the mysterious history of the honey bee.* Sceptre, London.

vanEngelsdorp, D., Evans, J.D., Saegerman, C. *et al.* (2009) Colony collapse disorder: a descriptive study. PLoS ONE **4**(8), e6481.

vanEngelsdorp, D. and Meixner, M.D. (2010) A historical review of managed honey bee populations in Europe and the United States and the factors that may affect them. *Journal of Invertebrate Pathology* **103**, S80–S95.

Free, J.B. (1993) *Insect Pollination of Crops,* 2nd edn. Academic Press, London.

Gould, J.L. and Gould, C.G. (1988) *The Honey Bee.* Scientific American Library, New York.

Grimaldi, D. and Engel, M.S. (2005) *Evolution of the Insects.* Cambridge University Press, Cambridge.

Hardy, T. (1874) *Far From the Madding Crowd.* [New Wessex edition, Macmillan, London, 1985].

Harrison, J.F., Fewell, J.H., Anderson, K.E. and Loper, G.M. (2006) Environmental physiology of the invasion of the Americas by Africanized honeybees. *Integrative and Comparative Biology* **46**, 1110–1122.

Isack, H. A. and Reyer, H.-U. (1989) Honeyguides and honey gatherers: interspecific communication in a symbiotic relationship. *Science* **243**, 1343–1346.

Kanga, L.H.B., Jones, W.A. and James, R.A. (2005) Enlisting fungi to protect the honey bee. *Biologist* **52**, 88–94.

Kritsky, G. (2010) *The Quest for the Perfect Hive: a history of innovation in bee culture*. Oxford University Press, New York.

Magnus, O. (1555) *Historia de Gentibus Septentrionalibus* [transl. P. Fisher and H. Higgens, ed. P. Foote]. The Hakluyt Society, London [3 vols 1996–1998].

Mayor, A. (2003) *Greek Fire, Poison Arrows, and Scorpion Bombs: biological and chemical warfare in the ancient world*. Overlook Duckworth, Woodstock, USA.

Milne, A. (2001) Gender, class, and the beehive. Mary Collier's 'The Woman's Labour' (1739) as nature poem. *Interdisciplinary Studies in Literature and Environment* **8**, 109–129.

More, D. (1996) *Discovering Beekeeping*. 2nd edn. Shire, Princes Risborough, UK.

Proctor, M., Yeo, P. and Lack, A. (1996) *The Natural History of Pollination*. HarperCollins, London.

Rosenkranz, P., Aumeier, P. and Ziegelmann, B. (2010) Biology and control of *Varroa destructor*. *Journal of Invertebrate Pathology* **103**, S96–S119.

Schneider, S.S., DeGrandi-Hoffman, G. and Smith, D.R. (2004) The African honey bee: factors contributing to a successful biological invasion. *Annual Review of Entomology* **49**, 351–376.

Spivak, M., Fletcher, D.J.C. and Breed, M.D. (eds) (1991) *The 'African' Honey Bee*. Westview Press, Boulder, Colorado, USA.

Wells, H.G. (1927) *Short Stories*. [Folio Society, London, 1990].

Winston, M.L. (1987) *The Biology of the Honey Bee*. Harvard University Press, Cambridge, MA, USA.

Winston, M.L. (1992) The biology and management of Africanized honey bees. *Annual Review of Entomology* **37**, 173–193.

HOW FABRIC IS MADE

WORMS COMBINE SUBSTANCES TO MAKE THREAD. THIS IS THEN WOVEN INTO SHIRTS, DRESSES, CURTAINS & THINGS OF GREAT BEAUTY.

SILK WORM

VISCOSE WORM

NYLON WORM

LYCRA WORM

POLYESTER WORM

RAYON WORM

Sericulture: Silkworms and Mulberries

Does the silk-worm expend her yellow labours
For thee? For thee does she undo herself?

Thomas Middleton, *The Revenger's Tragedy* (1607), act 3, scene 7

WALK THROUGH WOODLAND after a spring or summer storm and you will come across small caterpillars literally hanging in the air. A closer investigation will reveal that they are suspended from a leaf by an almost invisible thread of silk. Some moth caterpillars use these silk threads to move, blown on the wind from tree to tree like miniature Tarzans, but for most the threads are a safety harness with which they can climb up to the leaf from which they have fallen.

Almost all insects and spiders produce silk of some sort but many moth caterpillars (order Lepidoptera) produce copious amounts of silk when mature to form a protective cocoon in which they pupate (Fig. 5.1a). In some cases 'wild' silk is collected from these cocoons (Box 5.1), but most silk is produced from one moth, the silkworm *Bombyx mori*. This has now become entirely domesticated – silkworm caterpillars no longer cling to their food and the adults cannot fly (Fig. 5.1b) – and indeed is probably the only entirely domesticated insect.

Silk has long been a prized and expensive commodity and the secrets of its production were guarded. Countries that somehow obtained sufficient information were eager to set up their own sericulture (the production of silk) industries, and sericulture gradually spread west from its origin in China. However,

Fig. 5.1 **(a)** *Many Lepidoptera, such as this lackey moth (Malacosoma neustria: Lasiocampidae), produce a rough silken cocoon to protect the pupa and anchor it to the substrate.* **(b)** *The adult silk moth, Bombyx mori: note the small wings for the size of body – the moth does not fly*

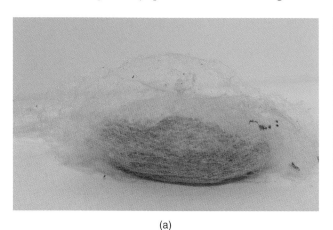

(a)

(b)

Biological Diversity: Exploiters and Exploited, First Edition. Paul Hatcher and Nick Battey.
© 2011 John Wiley & Sons, Ltd. Published 2011 by John Wiley & Sons, Ltd.

BOX 5.1 OTHER SILK MOTHS

Bombyx mori is the only truly domesticated silk moth, and accounts for about 90% of silk production. The other 10% is 'wild' silk made from the cocoons of a number of moth species within the superfamily Bombycoidea. There has been interest throughout the world in using silk from local silk moths feeding on native plants, and in introducing other non-native wild silk producers. Some of the latter have naturalized, and occasionally become pests – the most notorious being the gypsy moth (see Box 5.2) (Fig. 1).

The first silk seen in Europe was from wild silkworms – this is the silk mentioned by Aristotle and Pliny – from an industry which started in at least the fourth century BC centred on the Mediterranean island of Cos. Recently, a cast of a cocoon found in excavations of settlements on the Mediterranean island of Santorini destroyed by a volcanic eruption in about 1470 BC has been identified as that of *Pachypasa otus,* and may demonstrate that wild silk production in the Mediterranean was much earlier than previously thought. This species, common throughout the eastern Mediterranean, is now thought the most likely source of Cos silk, and it continued to be used to produce silk into the nineteenth century.

A variety of moths were used for wild silk production in India long before *B. mori* silk was introduced, and today about a dozen species of moths are still used. *Antheraea paphia* (Saturniidae) accounts for about 7% of worldwide silk production. Strictly, this species produces Tusseh silk, but this term has been used loosely for silk produced by related species, Indian silk in general, or even any wild silk. *Antheraea paphia* is a large moth which feeds on a variety of jungle plants, and in parts of India is regarded as sacred – its wing spots are seen as the chakra of the god Vishnu. The cocoons produce durable dark coloured silk; the moth is not domesticated but is traditionally looked after in the wild and guarded to deter predators: today trees are 'seeded' with eggs from captive moths. *Antheraea pernyi* and *A. yamamai,* respectively Chinese and Japanese oak silk moths, have also been introduced into North America and Europe, have become naturalized and sometimes become pests. There are several other *Antheraea* species used for silk production; one, *A. assamensis,* has been used in India to avoid religious taboos associated with the destruction of life. This silkworm spins a cocoon with a built-in exit hole, and so the adult can be allowed to emerge as it does not damage the silk thread.

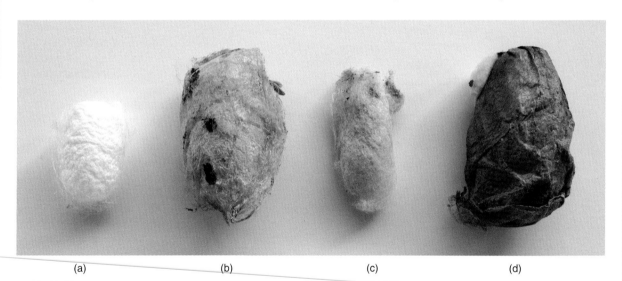

(a)   (b)   (c)   (d)

Box 5.1 Fig. 1 *Silk moth cocoons:* **(a)** Bombyx mori; **(b)** Antheraea pernyi *(Chinese oak silk moth);* **(c)** Samia cynthia *(cynthia silk moth);* **(d)** Attacus atlas *(atlas moth);* **(e)** *adult atlas moth (wingspan 22 cm). All cocoons are to the same scale – (a) is 2.7 cm long. Note that the cocoon and silk of the wild silk moths are much rougher than that of the silkworm B. mori [Specimen (e) courtesy J. Millard.]*

(e)

After Tusseh, Eri silk is the most important wild silk; this is produced from the wild cynthia or ailanthus silk moth *Samia cynthia* as well as the domesticated eri silk moth *S. ricini*. *Samia cynthia* is native to China and is very hardy. Since the mid-nineteenth century it has also been introduced to many parts of Europe and the US for silk production. It feeds on many trees and shrubs, especially the tree of heaven, *Ailanthus altissima*, widely planted in urban environments (introduction of the moth into North America was encouraged to make use of this otherwise 'worthless' tree). *Samia ricini* is used in China and Japan, and is fed on leaves of castor oil plants, *Ricinus communis*. The silk is of variable quality and cannot be reeled: it has to be carded like wool, but it can then be woven and made into tough garments.

Many of the Bombycoidea are large moths, although the silkworm *B. mori* is smaller than most and produces a relatively small cocoon. At the other end of the scale is the atlas moth, *Attacus atlas* (Saturniidae), one of the largest extant moth species, with a wingspan reaching 28 cm. The large rough cocoons are collected from the wild and processed on a small scale in India. Like most of the wild silk moths, it has catholic tastes.

sericulture is not simple; both the silkworms and the mulberry trees they feed on have to be reared, involving a series of complex operations that can be carried out on a large scale in only some economic and social situations. This will become clear as, after discussing the spread of sericulture and what it involves, we discuss the extensive attempts and ultimate failure to establish a sericulture industry in England and North America. Finally, we discuss one of the unexpected outcomes from the establishment of sericulture in Europe–the attempts at diagnosis and control of the many diseases of silkworms were pivotal to the development of insect pathology, and in the training of several famous scientists.

THE SPREAD OF SERICULTURE

Sericulture originated in China, where the ancestor to the silkworm, the wild *Bombyx mandarina* and its foodplant, the white mulberry *Morus alba*, are native. This moth is a pest species and can occur in large numbers. Thus, it would have come to the attention of humans and could have been harvested, similar to wild silk today (Box 5.1). Recent DNA analysis of *B. mandarina* and *B. mori* from China suggests that there was a single domestication event that took place over a relatively short time using many wild worms. However, we do not know when or where this happened.

In Chinese legends, silk was created by Empress Xi Ling, the wife of the Yellow Emperor (2698–2598 BC), who gathered silkworms from trees and reared them in her royal apartments, and she is revered as the Goddess of Silkworms. Archaeology places the origin of sericulture even earlier, with implements and dyed silk gauzes dating to 3600 BC, and more complex woven designs from 2700 BC. Certainly by the Shang Dynasty (c.1750–1100 BC) domestication of the silkworm had become highly developed. Silk was an important part of Chinese trade, especially with the west and Japan, and in the early centuries AD much trade from the west to China was stimulated by the attraction of Chinese silks, and this led to Chinese expansion west.

Silk was an important gift during the Han period (c.200 BC to AD 200), and large state silk-weaving workshops employing several thousand workers were mainly engaged in producing silk gifts. However, export of information, moths, eggs or mulberries was forbidden by severe penalties. For example, the Chinese learnt to boil cocoons in a slightly alkaline solution, killing the pupa and removing the soluble sericin gum, enabling a continuous filament to be unravelled and very fine weaves to be made. This technology did not penetrate the West until the sixth century AD.

In the fourth and third centuries BC, silk from the kingdom of Ch'in (modern Shensi, China) reached India. Chinese silk goods reached the Mediterranean by at least the second century BC, but only became common in the fourth century AD. Along with Chinese silks, the Romans received silk from Asia Minor and Syria, and wild silks from India (see Box 5.1).

Although finished silk goods were traded widely from China, sericulture spread much later: by Chinese peasants to Korea, from where it reached Japan, and also by smuggling to Persia, which was producing silk by AD 500 (Fig. 5.2).

The silkworm was introduced into Byzantium in about AD 553 (some say by Christian monks who had learnt to make silk in China) and the emperor Justinian embarked on enforcing a silk monopoly. However, from the middle of the eleventh century the Byzantine emperors tried to maintain their rule by selling this monopoly piecemeal to foreigners.

From these origins, sericulture spread via the Arabian Gulf to Egypt between AD 800 and 900, along the coast of North Africa by the Moors to Spain and Portugal, and from Greece during the mid-twelfth century to Sicily and Italy, and later in the fifteenth century to France. In 2007, almost 422 000 tonnes of reelable silkworm cocoons were produced in the world, 73% were from China and 18% from India. Yet in 1961 over 50% of world production occurred in Japan. However, between 1975 and 1995 the Japanese sericulture industry almost disappeared, while Chinese cocoon production steadily increased (Fig. 5.3). Sericulture has been promoted as a suitable rural industry for several countries in Africa and South America. Several hundred years ago, sericulture was also being suggested as an ideal industry for Britain and North America. However, before discussing this, we need to consider the component parts of sericulture, namely rearing the silkworms and growing the mulberries.

Fig. 5.2 *The spread of sericulture from Asia to Europe. The century of first record is given; e.g. AD 4 indicates fourth century AD. Green patches indicate recent planting of mulberries for sericulture [Partially based on Tazima (1984) and Lim et al. (1990).]*

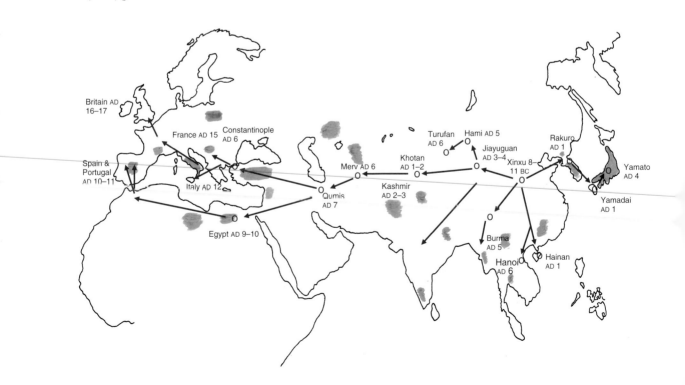

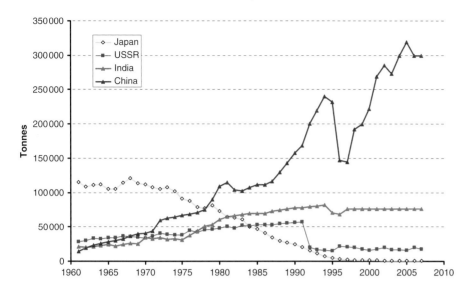

Fig. 5.3 *Production of reelable silkworm cocoons during 1961–2007 by the four main silkworm rearing countries. From 1992, former USSR production has been recorded under the new states (almost all is produced by Uzbekistan). This may account for the drop in production recorded from that time [Data from FAO.]*

SERICULTURE: SILKWORM, MULBERRY AND SILK

The silkworm

The silkworm *Bombyx mori* passes through the same stages as the honey bee (Chapter 4): eggs hatch into larvae that grow through a number of moults, and metamorphose to the adult via a pupal stage, inside a cocoon. However, whereas the honey bee colony looks after its own young, with workers caring for and feeding the larvae, the artificial silkworm colony formed in sericulture has to be tended by humans, who assume the role of the worker bees.

There are two main silkworm breeding operations in sericulture: the formation of cocoons from which silk is reeled, and the separate rearing of eggs, as in order to reel the silk the silkworm pupae have to be killed before the adult emerges. For cocoon production, the freshly hatched larvae (called 'ants') are gently given mulberry leaves, a process called brushing. For the first three larval instars (the growing stage between moults) the young larvae (Fig. 5.4a) grow slowly and require the best quality food, high humidity and the highest temperature. After moulting into the fourth instar the larvae are called 'grown worms' (Fig. 5.4b) and are moved into a new rearing room. Traditionally, these larvae would be reared in open trays housed in racks and would be fed four to six times a day. This is space efficient but labour intensive and thus the grown worms are often reared on the floor, or in shallow ditches sheltered by thatched sheds. The temperature must now

Fig. 5.4 *Silkworm Bombyx mori:* **(a)** *young worms feeding on artificial diet;* **(b)** *grown worms feeding on artificial diet;* **(c)** *fully grown silkworm starting to spin the 'bave' for its cocoon next to a completed one;* **(d)** *completed cocoon, ready for harvesting*

(a)

(b)

(c)

(d)

be lowered, and good ventilation is required. The major concerns are to supply enough food, to make sure all larvae have equal access to it and use it efficiently, and to prevent disease from spreading.

During the fifth (last) instar, larvae feed voraciously for about a week and most of their weight is put on in this instar. After a while they stop feeding and begin to secrete a silk 'bave' for spinning their cocoons (Fig. 5.4c). The larvae are now moved to a separate room with structures (mountages), made of wood, paper, straw or bamboo, on which they can spin their cocoon. A mountage should provide adequate space, so that double cocoons (whereby two larvae spin their cocoons together, making it difficult to reel the silk) are not formed, good air circulation, and should be cheap and re-usable. The larva spins its cocoon from a single fibre of fibroin over 3–4 days, and conditions during this time need to be carefully regulated to obtain good silk. For example, if the temperature is too high the larvae will spin too fast and a loose cocoon with wrinkles will result; the sericin protein (see below) becomes stickier, making the silk harder to reel. If the humidity is too high the larvae may suffer greater mortality and the silk does not dry. Six to eight days after mounting, when the pupa has formed inside the cocoon, the cocoons can be removed from the mountage and harvested.

The cocoon at this stage is 75% by weight pupa and 25% cocoon. First the pupa is killed (to emerge the adult would naturally dissolve a large hole in the cocoon, rendering it unspinnable), by drying. The cocoon can now be stored for up to one year. The sericin that binds the silk fibroin has now to be removed by 'cooking' the cocoons: soaking in water and heating to 100°C for 1–5 minutes followed by rapid cooling. The ends of the fibres from several cocoons are joined and reeled on to a spindle, and standardized skeins of 70 or 140 g of silk are produced.

No by-product from this process is wasted: silkworm excrement ('frass') is used as fertilizer, often on mulberry trees, and the killed protein-rich pupae are used as animal feed. The pupae are cooked with rice powder, dried, water and sugar is added and the mixture is allowed to ferment. The resulting 'pupal rice cakes' are a very popular feed for pigs, fowl and rabbits in Asia.

Considerable numbers of silkworms are required to produce a useable amount of silk: to produce about 15 kg of silk fabric 3.5 to 4 boxes of silkworm eggs (20 000 eggs per box) will be needed. Silkworm egg production is a specialized, usually separate, activity. The silkworm moth naturally produces one generation a year (i.e. is univoltine). The larvae feed during the spring and early summer, the adults emerge during the summer and lay eggs which hibernate through the autumn and winter, hatching the following spring. After several millennia of breeding, however, a variety of races now occur. These can be univoltine, bivoltine (two generations a year) or multivoltine. For example, Chinese races have a short larval duration, with roundish larvae resistant to high temperatures, producing a thin and strong silk which is easily reeled; they are mainly uni- and bivoltine. European races have large fat larvae, usually univoltine, with a long feeding period, and are sensitive to disease. Their cocoons produce a thick silk with good reelability but with high sericin content. Japanese races also have larvae with a long feeding period, but which are resistant to poor conditions, and are uni- or bivoltine, while tropical races have small slender larvae, multivoltine, resistant to high temperatures, but producing poor quality silk.

Egg producers aim to supply good quality eggs, ready to hatch at the time specified by the cocoon rearers. This can be difficult: the univoltine races have hibernating eggs, whereas bivoltines and multivoltines sometimes have hibernating eggs, and sometimes not. Normally, the univoltine eggs hibernate during the year in which they were laid and will not hatch until after overwintering at low temperatures. This restricts silk production to once per year. In bivoltine races, whether the eggs enter hibernation depends on temperature and light during the egg stage (e.g. high temperature and light encourage hibernation). However, larvae that hatch from non-hibernating eggs produce poorer quality cocoons than those from hibernating ones, and thus the latter are used, but the problem now becomes how to break the hibernation early.

In the nineteenth century, methods were developed to break hibernation reliably, for example by using warm water, friction, high temperature, or electricity. A five-minute dip of just-laid eggs in dilute warm hydrochloric acid is commonly used and is probably best, although why it works is not known. This enables a summer/autumn second generation of worms to be produced, and was largely responsible for the increase in annual silk production in Japan from 7300 tonnes in 1905 to 42 500 tonnes in 1930.

Furthermore, in 1905 it was found that an F_1 hybrid between Japanese and Siamese races showed a 30% increase in cocoon yield, and this hybrid had a shorter feeding period, lower mortality and a larger silk fibre. Later it was found that hybrids between Chinese and European races were best. Whereas the weight of silk per cocoon increased in Japan from 161 mg in 1802–30 to a maximum of 226 mg by 1906, after the introduction of F_1 hybrids it rose to 241 mg in 1913–16 and to 551 mg in 1969. However, as the size of the hybrids rose, so the number of eggs laid by the females decreased; this was overcome by creating further hybrids which in turn resulted in greater breeding costs.

A wish list from sericulturalists would include adult silkworms that mate more reliably and lay more eggs, larvae that have greater efficiency at converting mulberry to body weight, have regular growth, are easier to handle, and produce cocoons which are easier to reel and have fewer malformations. Some of these factors have been improved recently, and current silkworm breeding aims to produce races more resistant to high temperatures, humidity and diseases (to enable sericulture to be successful in the tropics) and to produce races adapted to artificial diet. Artificial diet (containing about 25% powdered mulberry leaves) has been successfully used in Japan since 1960, and now is completely used for young larvae, but rarely for later instars (see Fig. 5.4). Larvae feeding on artificial diet have a tendency to grow more slowly and are more disease prone than those reared on mulberry.

The mulberry

Although it may be possible to feed silkworms on other plants, they seem only able to reliably feed on mulberries (35 species of the genus *Morus*, family Moraceae). The favoured species for the silkworm is *Morus alba*, the white mulberry (Fig. 5.5a,c). This is native to China, and has gradually spread west with the silkworm. The black mulberry, *Morus nigra*, is also suitable (Fig. 5.5b,e); it has greater cold hardiness and can ripen fruits as far north as Scotland. This species, probably a native of Persia, was known to the Greeks and Romans and was planted throughout the Mediterranean basin for fruit production long before being used for silkworm rearing. It was probably introduced into Britain by the Romans.

Mulberry cultivation for silkworms needs to produce sufficient quantity of leaf material of the right quality. In many areas the main limiting factor to silk

(a) (b) (c) (d)

(e) (f)

Fig. 5.5 **(a)** *Three-year-old Morus alba seedlings, just ready to have leaves harvested.* **(b)** *Morus nigra leaves and fruit. The two species are hard to distinguish when not in fruit: unlike M. nigra, M. alba usually has leaves without basal lobes* **(c)***, but both species can have similarly shaped irregular leaves* **(d)** *as well. If not pruned, mulberry trees can become large. The black mulberry in* **(e)** *is over 150 years old, its main trunk has largely died, but branches have rooted in the ground and produced an array of new shoots* **(f)**

production is insufficient mulberry leaves to feed the larvae: it has been calculated that 20 kg of mulberry leaves are needed to produce 1 kg of cocoons, which will only produce about 100 g of silk fabric. Tender leaves at the top of the tree are most suitable for the young larvae but would cause too high a mortality in the older larvae which do better fed on middle-aged leaves.

In temperate areas two harvests of mulberry leaves are possible in a year, during May and in mid July, which may support bivoltine larvae. In tropical areas up to five harvests of leaves can be taken, throughout the year. The plants are propagated by grafting on to a resistant rootstock, or by softwood cuttings or 'striking', which involves taking a 15 cm long shoot, trimming off most of the leaves, soaking in root hormone solution and planting up to its tip in wet soil. After 40 days the young plants should have sufficient roots for replanting. Plants are seldom allowed to get large, being pruned by a variety of methods. The low cutting method is the commonest and enables leaf harvesting to start at three years of age (Fig. 5.6). Such intensive pruning and continual harvesting can seriously weaken the mulberry, so rest periods and good fertilization are important. Even so, after 15–20 years the productivity of the trees declines and they need to be replaced.

The mulberry tree has a variety of other uses. The wood withstands rot and is used in boatbuilding. The bark can be processed into artificial leather and made into ropes and nets.

Silk

Understanding how the silkworm spins silk has great practical importance to mankind. For example, nylon is the closest artificial fibre to silk, and to produce the nylon fibre the molten polymer has to be forced through a small jet and cooled. The resulting fibre is soft, as it cannot be run through the jet fast enough to align its molecules. Thus, it now needs to be cold-drawn to six times its original length so that the molecules become oriented parallel to the fibre

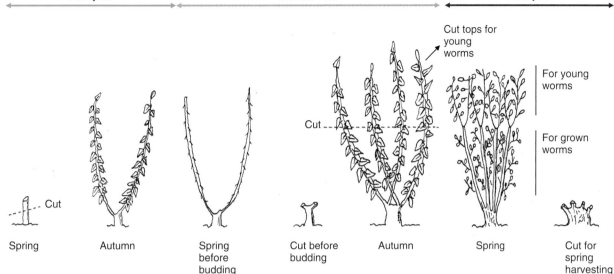

First year | Second Year | Third year

Cut

Spring   Autumn   Spring before budding

Cut before budding   Autumn

Cut tops for young worms

For young worms

For grown worms

Spring   Cut for spring harvesting

axis and solidify. By contrast, the silkworm spins the liquid fibroin with a rapid figure-of-eight movement of the head, with the legs drawing out the protein, into a solid fibre in one action at room temperature (see Fig. 5.4c). It is unclear exactly how this is done, but it appears that in the silk gland the fibroin is already in an aggregated state and with some alignment of molecules. The pressure and tension applied to the silk as it is drawn orients and shears the structure, causing it to polymerize and solidify.

The silk secreted by the silkworm is composed of two main proteins, fibroin (the core) and sericin, which surrounds it. Fibroin is a very large insoluble polypeptide consisting mainly of a repetition of one glycine amino acid with usually either alanine or serine. The protein is complex, with about a third existing in an amorphous state and the rest in a crystalline state. The crystalline portion is formed from β-pleated sheets of anti-parallel amino acid chains maintained by hydrogen bonds (this important type of protein structure was first described in fibroin in 1955). Sericin consists of a mixture of at least five different polypeptides; they are soluble in hot water and form at least three layers around the fibroin core. Up to 20% of the cocoon weight can be sericin.

Silks produced by different species of insects have different structures; for example, silk from *Antheraea pernyi* is more like that of spiders than silk from the silkworm.

Silk is produced in the silkworm from a silk gland, a modified salivary gland (Fig. 5.7). Each arm of the gland is a hollow tube made of two rows of epithelium. The posterior part synthesizes fibroin and this passes to the middle part in a 15% aqueous solution. It is now wrapped in sericin and concentrated to a 30% solution, and stored until spun. Towards the head both arms of the gland fuse and lead through the spinneret. The silk gland is fully formed, but very small, in the 10-day-old

Fig. 5.6 *Establishing mulberries from planting and pruning using the low cutting method. The newly planted mulberry is cut close to the ground in the spring to produce new shoots ready for harvesting in the autumn. By following this method, by the third year the plant is producing leaves for harvest in the spring and in the autumn [Based on Lim et al. (1990).]*

Fig. 5.7 *Silk gland of the silkworm Bombyx mori. (a) Longitudinal section through a 5th instar larva showing extent of the gland. (b) The silk gland dissected from the larva [(a) redrawn from Lim et al. (1990), (b) from Wigglesworth (1972).]*

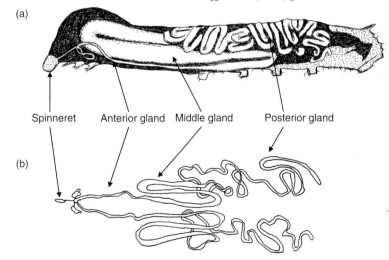

(a)

(b)

Spinneret   Anterior gland   Middle gland   Posterior gland

embryo and produces small amounts of silk during each instar, spun as the insect moults to anchor the old skin to the leaf. During the fifth instar the gland grows and becomes 40–50% of the body weight (Fig. 5.7a); during cocoon spinning the gland starts to degenerate, and has disappeared in the pupa.

SERICULTURE IN BRITAIN AND NORTH AMERICA

Although sericulture was successfully introduced into several countries outside of Asia, for example Italy and France, not all introductions were successful. Notably, attempts to establish sericulture in Britain and North America failed, not for want of trying. Given that sericulture is a complex industry, why was it repeatedly thought that it could succeed, and why did it fail?

The idea of establishing sericulture in England originated in the middle of the sixteenth century, among a range of projects proposed to employ the poor, prevent costly imports and increase exports. One such import was silk fabric and knitted silk stockings, which had become a high fashion item. The first stage in reducing imports was to process the raw silk into garments in England. This was encouraged by the settlement of refugees from foreign religious persecution in England from the 1560s, many of whom, for example Huguenots from France, were silk knitters and weavers, and the development of knitting technology, especially the silk framework knitting machine perfected in England towards the end of the sixteenth century.

Raw silk still had to be imported, however, and increasingly suggestions were made that silk should be produced in England (a not impossible dream given that sericulture had recently started in France); indeed the successful establishment of silk weaving in England increased the demand for imported raw silk. The first published attempt to promote silk production seems to have been Thomas Moffet's poem *The Silkewormes, and their Flies* published in 1599 (Fig. 5.8):

> I sing of little Wormes and tender Flies,
> Creeping along, or basking on the ground,[1]

he promises in the introduction, yet it is not until over half-way through this long poem that he gets around to the silkworm itself. Although he provides little concrete information, and certainly nowhere near enough to instruct someone wishing to begin sericulture, some of his concerns are those that reappear later in the history of English sericulture; for example, the need to feed worms of different ages with different leaves:

> The first three weekes the tend'rest leaves are best,
> The next, they crave them of a greater size,
> The last, the hardest ones they can digest,
> As strength with age increasing doth arise:[2]

and the value of sericulture to the poor

> What multitudes of poore doth it relieve,
> That otherwise could scarce be kept alive?[3]

THE Silkewormes, and their Flies:

Liuely deſcribed in verſe, by T. M. a *Countrie Farmar*, and an apprentice in Phyſicke.

For the great benefit and enriching of England.

Printed at London by V. S. for Nicholas Ling, and are to be ſold at his ſhop at the Weſt ende of Paules. 1599.

Fig. 5.8 *The title page of Moffet* (1599)

[1] Moffet (1599), preface.
[2] Moffet (1599), p. 55.
[3] Moffet (1599), p. 69.

and

> No man so poore but he may Mulb'ries plant,
> No man so small but will a silke-worm feede,[4]

Although some silkworms were reared in the later sixteenth century in England, formal attempts at an English sericulture industry had to await the instalment of James I to the English throne in 1603. Influenced by his French advisors, he became an enthusiastic supporter of sericulture and started a pattern of state intervention that was repeated in the New World. This involved establishing trial enterprises, the provision of silkworm eggs ('seed'), mulberry cuttings, and information. From 1606, James I set up plantations of mulberry at some of his royal palaces (the first one was established at the newly built Charlton House, Greenwich) and had rearing houses for silkworms erected. Early in 1607 he licensed William Stallenge (as Keeper of the King's Silkworms) to publish a book on sericulture and the following year licensed him to import mulberry seeds and distribute them throughout the country. Stallenge established a 2.5 ha mulberry garden in Westminster Palace during 1609, and in the same year James I wrote to his Lords Lieutenants, his county representatives, encouraging them to plant mulberries; 10 000 trees were offered to each and there were 500 000 seedlings in a nursery in France available.

How many mulberry trees were actually planted in England at this time is uncertain, although it was claimed that 100 000 trees were distributed. Certainly no sericulture industry resulted, although a little silk was produced, notably at the king's palaces (Stallenge reported in 1611 that he had 4 kg of silk spun from the cocoons). The king's interest in sericulture continued: he appointed John Boenoil to keep his private collection of silkworms at Oatlands Palace, Weybridge, Surrey, and an ornate two-storey silk-house designed by Inigo Jones was built there. By 1618 silkworms and mulberries were also established at the royal palace at Theobalds, Hertfordshire.[5]

The tree that was planted was the black mulberry, not the white. It has often been reported since that this was a mistake, and that this error led to the failure of the English project. Both are unlikely. At the time, it was thought by many that the black mulberry was the more suitable food for the silkworm. For example, Moffet in 1599 writes:

> For Mulbries though they are of double kind,
> The blacker ones are yet to them most sweete[6]

and it was not until the 1620s that most authors started to insist that the white mulberry must be used in sericulture. The white mulberry may indeed be preferable (it comes into leaf earlier, can provide more leaves in a shorter time, and can be more tender), but silkworms can also feed well on black mulberry and this species may be more likely to survive in England than the white one. More important a reason for failure, and mentioned from the late seventeenth century, was lack of leaf material. The ultimate reason for failure in England, as it was in America later, is most likely to have been the lack of suitable labour. What was always overlooked is that sericulture is a complicated industry, only marginally profitable, and one that needs cheap, yet skilled, labour to succeed. In England and America there were other more profitable crops available.

[4]Moffet (1599), p. 71.

[5]Very little remains of James I's attempts to develop sericulture in England. Although several mulberries in existence are claimed to result from this time, it is more likely that they are descendants of the original trees. Nothing survives of the king's mulberry plantations: Oatlands Palace was mainly destroyed in 1650 and the site is now buried under a suburban housing estate; Charlton House, Greenwich, still survives, and although publicly owned is not open to the public; and Theobalds Palace has been destroyed, although the park is still extant and open to the public.

[6]Moffet (1599), p. 56.

The attempts to promote sericulture in the American colonies subsequently encouraged further trials of sericulture in England in the seventeenth and eighteenth centuries, but with no success. Schemes re-arose at intervals; for example the British, Irish and Colonial Silk Company was formed in 1825 with the usual aims of reducing imports of silk and employing the poor. One of its projects was to introduce sericulture to Ireland, where there was 'an enormous population of women and children; it is they who must perform those delicate operations of reeling, which the more clumsy hands of the other sex are incapable of performing'.[7] Although the company claimed to have mulberry plantations established in Co. Cork and also near Slough, no silk was produced and the company quickly broke up in acrimony.

It is likely that many country estates in England continued with sericulture as a hobby after official exhortations stopped, and some did produce enough silk to make garments. But this was not an industry. However, in one case a small industry did result. Zoë, Lady Hart Dyke, kept silkworms as a child at Lullingstone, Kent, and started rearing them on a larger scale in 1932. Expanding this operation, by 1939 she had almost 20 ha of mulberry trees, millions of silkworms and was able to produce 11 kg of silk a week. There was great interest in this, the only commercial silk farm in the country, from the public, many of whom purchased silkworm kits, as well as from other countries wishing to set up a sericulture industry. This silk farm later relocated to Dorset and was producing silk for royal weddings and state occasions into the 1990s.

At the same time as attempts were being made to establish sericulture in England, English adventurers were exploring and settling on the Atlantic coast of North America. The earliest voyagers to Virginia in 1586 keenly noted the presence of mulberry trees:

> By the dwelling of the Savages are some great Mulbery trees, and in some parts of the countrey, they are founde growing naturally in prettie groves[8]

and

> [I]n many of our journeys we founde silke worms fayre and great; as bigge as our ordinary walnuts.[9]

This suggested that sericulture could be practised here and there will rise as great profite in some time to the Virginians, as thereof doth now to the Persians, Turkes, Italians and Spaniards.[10]

Unfortunately, the silkworms found were not of B. mori, but various species of saturniid moths from which silk cannot easily be ungummed and reeled – although this did not stop settlers trying various methods to prepare and reel these 'gummy bottomes' for the next several hundred years (Box 5.2). Unfortunately also, the mulberry they found was not M. alba nor M. nigra, but the native American red mulberry M. rubra. It is unclear when the settlers realized this. Botanists certainly did early on, and Gerard lists M. rubra in his 1599 catalogue and notes that he had grown three plants in his London garden – settlers were still confusing rubra with alba into the 1670s. Silkworms can be reared on rubra, but it is generally considered inferior to alba and nigra.

[7]Dandolo (1825), p. xiv. The translation of this book from the Italian was sponsored by the British, Irish and Colonial Silk Company, and was printed with a preface extolling the virtues of the company.

[8]Smith (1612), p. 11.

[9]Hariot, T. (1588) *A Briefe and True Report of the New Found Land of Virginia*. London, p. 86 in Clarke (1993).

[10]Ibid., p. 86.

BOX 5.2   THE GYPSY MOTH

Although repeated efforts to introduce sericulture to North America ultimately failed, this process did lead to one successful moth introduction, one that no one wanted, the gypsy moth *Lymantria dispar* (Lymantriidae; Fig. 1).

The French-born artist and amateur entomologist Etienne Trouvelot settled in Medford, a suburb of Boston, during the 1850s. He was interested in the potential use of native moths for silk production, and after trialling several species decided that the polyphemus moth *Antheraea polyphemus* (a relative of the Indian tusseh silk moth, see Box 5.1) was most suitable. From a couple of wild-caught individuals he built up a population by the mid-1860s of over a million caterpillars, feeding on the 2 ha woodland behind his house.

Then in 1868 he returned from a trip to France with some eggs of the gypsy moth. It is thought that he wished to interbreed this wild species with silkworms (presumably *Bombyx mori*) to produce a hybrid that was resistant to diseases such as *Nosema bombycis* that were destroying the European sericulture industry at the time (see Box 5.4). *Lymantria dispar* and *B. mori* are not closely related, and it is unlikely that such a hybridization would have been successful, but before he could attempt this the gypsy moth caterpillars escaped. Some say that a gust of wind upset the indoor cage they were being reared in and they crawled though an open window; others say that they were being reared on a tree in his back garden at the time.

Although Trouvelot realized the seriousness of this escape (the gypsy moth was a well-known forest pest in Europe) and alerted other entomologists, little notice and no action was taken. The gypsy moth population slowly increased and by 1889 it extended over an area of 2.5 by 1 km. A ten-year campaign of eradication then started, and although it nearly succeeded, it was stopped too soon and the moth spread. Since then many attempts have been made to control it using chemical and biological methods as it spread slowly throughout the eastern US. None succeeded and by the 1960s it was realized that the best that could be achieved was to slow the spread of the moth and that eradication was impossible. In 1981, a record 5 million hectares of woodland was defoliated by this moth, which is now common in most of eastern USA.

This species has become one of the most serious forestry pests in the US, and is particularly troublesome as in favourable years it can produce large outbreak populations of caterpillars.

These populations are spread by the young larvae 'ballooning'—they hang from the top branches of trees on a silken thread, and are carried 1–2 km by the wind. The females cannot fly, and do not move far from their cocoon, and thus, unusually for Lepidoptera, most of the dispersal of the population is carried out by the larvae. Although primarily an oak feeder, when hungry the caterpillars can feed on almost all broadleaf and coniferous trees, and thus can devastate urban parkland as well as commercial forests.

The main commercial damage is caused by defoliation: healthy trees can usually withstand one or two consecutive defoliations greater than 50% (they just regrow), but defoliation weakens and stunts the trees and those already weakened, perhaps by drought, can be killed by one defoliation. It is estimated that a severe outbreak of the moth can kill 13% of the defoliated trees. Outbreaks of the gypsy moth are hard to predict, but on oak they have been linked to the reduction in predator abundance that occurs in non-masting years (Chapter 10).

Important also is the destruction the gypsy moth causes in urban environments. This is often more of an inconvenience, with large numbers of caterpillars littering pavements and dropping onto people, but the allergenic hairs on the caterpillars and egg masses make it something of a health hazard.

Trouvelot gave up entomology soon after his gypsy moths escaped, and returned to France at the same time as the moths were starting to become a problem in his neighbourhood. He turned his interest to astronomy and became famous for his astronomical illustrations, which are exhibited to this day. Although now remembered for this, his entomological legacy—as the originator of possibly the only non-native pest introduction that can be traced to an individual—should not be forgotten.

Box 5.2 Fig. 1   *The gypsy moth,* Lymantria dispar, *male (left) and female (right). Wood engravings from Newman (1884). At this time the British form of the moth (much larger than the continental European form introduced into North America) was rapidly declining, and it became extinct in the early twentieth century*

The early Virginian colonists were so keen to start silk production that they imported eggs on the next supply ship in 1609, and

there was an assay made to make silke, and surely the wormes prospered excellent well, till the master workman fell sicke. During which time they were eaten with rats.[11]

[11]Smith (1612), p. 11.

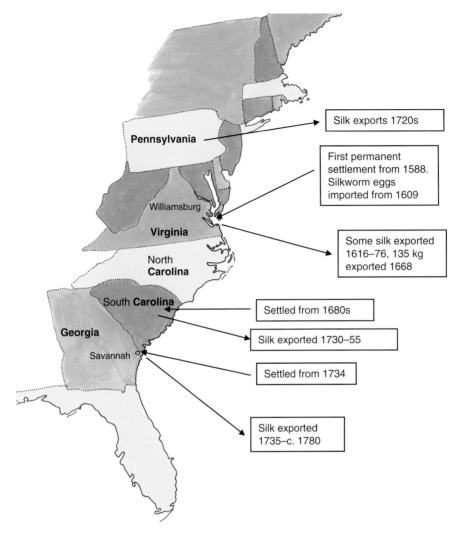

First, local M. *rubra* was planted from fruits, later M. *nigra* was planted, but M. *alba* was not imported until about 1639. Progress was slow, however, due largely to difficulty in transporting eggs across the Atlantic: the voyage was slow and unpredictable (lasting anything from 5 to 12 weeks) and without any means of cooling, the eggs often hatched out before arriving and the larvae died from starvation.

Although renewed efforts at sericulture in Virginia were made in the 1650s, and some silk produced (Charles II was scnt 135 kg of silk in 1668), silk production had to compete with the much more lucrative and easier production of tobacco. Earlier, James I was in a quandary over this; in 1619 he banned the growing of tobacco in England to improve the chances of tobacco growers in Virginia and yet he did not want the colony to be reliant only on tobacco for income. Thus, by 1622 he was writing to the governor of Virginia ordering that farmers work on breeding silkworms rather than growing tobacco. Ironically, by the 1650s colonists were being paid a bounty of 50 kg of tobacco for each kilogram of silk produced. By 1676, however, tobacco had won and sericulture ended (Fig. 5.9).

Fig. 5.9   *Early sericulture in North America: settlement and exports of silk to Britain. Modern state boundaries are given [Partially based on Feltwell (1990).]*

The competition between sericulture and easier, more lucrative crops was repeated in other settlements; for example in South Carolina rice production won out after some successful silk production. One colony, however, was set up specifically for sericulture: Georgia (Fig. 5.9). Again, early explorers here sent erroneous reports that both black and white mulberries were present; for example:

> it must be a weak hand indeed that cannot earn Bread where silk-worms and white Mulberry-trees are so plenty.[12]

Thus, the Ebenezer colony along the Savannah River was settled from 1734 with persecuted Protestants from Salzburg, Austria, and Germany, and later separate settlements of English and Scottish. Land was only granted on the condition that mulberries were planted. The grand aim was to supply England and perhaps the whole of Europe with silk. Good quality silk was exported from Georgia: 3.5 kg in 1735, to 320 kg in 1758 and 4500 kg in 1759. However, production was patchy, and the project had failed by the time of the War of Independence

[12]Martyn (1732), p. 55.

(1775–83). Although many of the problems were due to lack of labour, poor equipment, lack of instruction and poor and variable payment for the silk, there were serious biological problems also. White mulberry, now established as the preferred species for the silkworms, was selected for planting, but this species did not grow well in the swampy coastal lands, which also had unforeseen droughts and unpredictable late frosts. The trees were poorly adapted to the climate: bud burst often started in early spring, only for frosts later in April to kill the new leaves. The few surviving trees had then to be cropped too heavily and were often killed. Black mulberry would have been more suitable here, and did survive the frosts, but was not planted for sericulture as it was erroneously thought to be fatal to the worms.

The climate also affected the silkworms. Those eggs which did not hatch out on the journey from Italy often hatched in response to the early hot weather and ran out of food when the trees were frosted. The colonists were encouraged to breed their own silkworms because of unpredictable supplies of eggs, yet by 1767 these were producing poor quality and weak worms, classic signs of inbreeding depression. Regardless, the enterprise was flawed from the start: while land was cheap, labour was expensive, and to be successful cropping had to take advantage of this.

Attempts at reviving sericulture in North America have been sporadic, usually at times of economic depression and with heavy government subsidy. None has succeeded for long, although in the 1860s there was a lucrative industry in California and Utah supplying silkworm eggs to European sericulturalists devastated by a series of silkworm diseases (of which more below). Once measures had been put in place to stop the spread of these diseases eggs were no longer needed and the industry collapsed.

The greatest collapse and strangest episode in the story of sericulture in North America occurred slightly earlier: the 'multicaulis mania' of the 1830s. In 1821, *Morus multicaulis* (now considered a variety of M. *alba*) was discovered in China and introduced to France, from where in 1826 it was imported to Long Island. Subsequent trials demonstrated that this species grew much faster than white mulberry, the leaves were excellent food for the silkworm (they are now thought acceptable for the small larvae, but not suitable for the older ones), and it was thought to be more hardy than white mulberry. *Morus multicaulis* is a large-leaved multiple-stemmed plant which is best grown in hedgerows and kept below 3 m height. It was suggested that it would be possible to produce two crops of silkworms a year from it: this had been a dream of western sericulturalists for centuries, although there is no evidence that anyone knew how to make their usually univoltine silkworms suddenly bivoltine – they could not even control when they hatched! For example, William Kenrick, a notable proponent of this method, gives very little detail of *how* this could be achieved in his 1835 book on sericulture, but he was clear that *multicaulis*

> Seems destined to replace, everywhere, the common white mulberry, for the nourishment of silk-worms, such is its decided superiority over all others.[13]

From 1829, many US companies were formed for manufacturing silk, based on the virtues of *multicaulis* and encouraged from 1833 by government assistance. Hundreds of thousands of trees were ordered from France, and as the claimed advantages of these trees became more and more outrageous they changed hands

[13]Kenrick (1835), p. 27.

for ever-increasing sums. There was intense speculation and sales of these trees; prices rose 100-fold and many farmers invested heavily in setting up plantations: millions of *multicaulis* were planted in North America in less than ten years.

However, little silk was produced: people were much more interested in buying and selling mulberries than the difficult and messy job of silkworm rearing. Thus, the new silk mills erected were under-utilized – the existing silk mills were already sufficient for the amount of silk being produced. These factors in themselves would have been enough to ensure that the bubble would have burst sooner or later, and this was helped by the fact that *multicaulis* is neither frost-tolerant nor disease resistant, and a poor winter in 1837 led to the loss of many trees. The resulting panic by investors led to a dramatic fall in value of the plantations and many people lost heavily. Eager to recoup losses, more trees were planted in 1838, but a severe winter in 1839 killed further trees. Confidence was now completely lost, and farmers grubbed up and destroyed their remaining trees, and mills turned to other fabrics or were demolished. There was an expression still prevalent in some US states during the 1930s to greet any new get-rich-quick scheme: 'oh, that's just another multicaulis'.

In conclusion, the attempts at sericulture in England and North America were dogged by failures in plant identification, botanical prejudices and unfounded assumptions stemming from a lack of scientific investigation in the area. Such investigation had to await application to silkworm disease epidemics in Europe in the nineteenth century.

BASSI, PASTEUR AND THE DISCIPLINE OF INSECT PATHOLOGY

Rear any species in large numbers in close confinement and it is likely to succumb to pests and diseases, and this is particularly true of moths and butterflies. Silkworm diseases were recorded by the Chinese at least by the seventh century BC, and silkworm diseases were recorded in Europe from 1527. However, it was not until the nineteenth century that scientific research on some silkworm diseases led to measures for their control, and incidentally founded the discipline of insect pathology. This has not only sought to control insect diseases but has promoted the use of some of them for the biological control of pest species.

The first demonstration that a pathogen could cause an infectious disease in an insect was made by Agostino Bassi (1773–1856, later named the father of insect pathology), a lawyer turned farmer at Lodi, Italy. In 1807 he became aware of a serious disease of silkworms, muscardine or *mal del segno*. Originally he was convinced that the disease arose spontaneously in the insect due to poor food quality or fumes from the litter, following accepted theories of the time. He tried to replicate the characteristic features of a caterpillar which had died from muscardine: a dry shrivelled body covered with a whitish efflorescence. However, although some treatments produced corpses that looked similar, none was infectious whereas those from muscardine were.

Thus in 1813 he rejected the theory of the spontaneous generation of disease, and instead hypothesized that the disease was caused by a germ that entered the insect from outside. After many years of experimentation and microscope work he built up an impressive array of information about the disease, its symptoms, the conditions that favour it, and how to control it. He published this in 1835 (Fig. 5.10), concluding:

> The many observations and experiments I have made over a long period of years have proved to me that the mark disease or muscardine never arises spontaneously in the silk worm, or in other insects: that it always derives from an external being, which, by entering the insect and developing, causes the disease, death, and subsequent salification of the corpse: that this being is organised, living, and vegetable: that it is a parasitic plant, a fungous production: that this cryptogamic plant develops, grows, and multiplies only in the living insect, and never in the dead, and only in the genus of caterpillars; and it does not fructify, or at least it does not mature its seeds, until the insect that has nourished it is dead: that the disease produced by this fungus, or rather the insect killed by it, is contagious, but never so long as it is alive … .[14]

Fig. 5.10 *The title page of Bassi (1835)*

He was able to isolate the fungus from silkworms and infect other insect species, along with re-infecting silkworms, thus establishing its infectious nature. This method for proving pathogenicity was re-discovered many years later and became known as Koch's postulates.

Years of microscope study severely weakened Bassi's eyesight and he carried out no further microscope work. He did not describe the fungus, which was later named *Botrytis* (now *Beauveria*) *bassiana* after him. This fungus has since been used successfully as a biocontrol agent against a number of insect species (Box 5.3).

During the middle of the nineteenth century another disease began to attack French silkworms; by 1853 eggs could no longer be produced in France and had to be imported from Lombardy. The disease spread to Italy, Spain and Austria, and finally to China and Japan. This new disease produced small spots on the skin resembling grains of pepper, hence its name *pébrine*, and larvae became arrested at various stages of development. By 1865 the French sericulture industry was in near ruins (*pébrine* had reduced silk output in France from 28 000 tonnes in 1853 to 4000 tonnes in 1865) and the French Ministry of Agriculture appointed a mission for the study of *pébrine*, Louis Pasteur (1822–95) was asked to take charge and he accepted.

Since at least 1849 the Italian biologist Osimo had described small 'corpuscles' in the eggs of silkworms in *pébrine*-infected areas and ten years later recommended that laying moths should be segregated and their eggs be examined microscopically in order to reject stocks that had too many corpuscles. However, this was rejected due to inconsistent results—some eggs initially without corpuscles developed them later and corpuscles could also be found in healthy worms. Pasteur believed for the first three years of his studies that the disease was a physiological, hereditary, disturbance of the caterpillar and the corpuscles were a secondary sign of it produced by the larva (although he was aware of Bassi's work on muscardine, there is no evidence that Pasteur was influenced by it). However, after only two weeks' study he recognized the diagnostic value of the corpuscles, and recommended the adoption of Osimo's egg selection method. This is still used to prevent spread of this disease; for example the method was introduced to Japan in 1885 and the *pébrine* infection rate dropped from 36% in 1890 to 13.3% in 1892 and 0.02% in 1958.

Pasteur continued to seek the cause of the disease. Although his assistants became convinced that the corpuscles were the cause and not just a symptom,

[14]Bassi (1835), p. 39.

BOX 5.3   ENTOMOPATHOGENIC FUNGI AND BIOLOGICAL CONTROL

Over 800 species of fungi, from a variety of groups, have been recorded as infecting insects, and are thus entomopathogenic. Six species are currently commercially available as biocontrol agents against insect pests, including *Metarhizium anisopliae* against locusts (Chapter 15) and two *Beauveria* species. Both these genera are members of the Ascomycota, order Hypocreales, family Clavicipitaceae.

Entomopathogenic fungi all have a broadly similar life-cycle: spores attach to the surface of the insect cuticle, germinate and a germ tube penetrates the cuticle (generally through thinner non-sclerotized areas such as joints) by a combination of physical force and enzymatic degradation. The fungal hyphae multiply in the insect's body cavity and produce yeast-like blastospores which are distributed passively in the haemolymph and invade other tissues. Many species also produce toxins which hasten the death of the insect. Once the insect has been consumed, conidia are released either from the burst cadaver, or from its surface (Fig. 1).

Back in 1836, Bassi suggested that the fungus he found in the silkworm (later to be called *Beauveria bassiana*) could be used as a pesticide by mixing the liquid from putrefying cadavers with water and spraying on to crops. This was not tried in the field until 1884 when the Russian entomologist Metchnikoff mass-produced spores of *Metarhizium anisopliae* and used it against a sugar-beet beetle, causing up to 80% larval mortality.

Beauveria bassiana is considered to be the entomopathogen with the broadest potential. It has a wide host range of over 700 species, is of little risk to beneficial species, and can be sprayed on pests using conventional pesticide sprayers. Considerable advances have been made in mass-producing spores cheaply by solid-state fermentation on clay granules and it has been used on a large scale against a number of pests. A formulation was developed in 1965 and used against the Colorado potato beetle in eastern Europe until recently. More recently, a strain of the fungus has been developed against a wood-boring cerambycid beetle which vectors the pinewood nematode that causes probably the most serious pest problem in Asian pine forests: pine wilt disease. Fibre bands impregnated with spores of the fungus are tied around the branches of dead trees to kill emerging beetles.

Beauveria bassiana has also been used against the European corn borer (*Ostrinia nubilalis*, Lepidoptera) in several countries, especially China. There is a twist here, as long-term suppression of the corn borer on maize achieved by spraying plants with *B. bassiana* seems to be due to the fungus entering the plant directly through its cuticle and establishing as an endophyte (an organism that remains symptomless in its host). The larvae do not become infected by the fungus, but instead have reduced activity possibly due to fungal metabolites causing feeding depression.

Research is continuing into the use of *B. bassiana* against a variety of pests, including moths, whitefly, thrips, aphids, mealybugs and *Anopheles* mosquitoes (Chapter 12), and currently honey bees (Chapter 4) are being studied as potential carriers (vectors) of *Beauveria* to flowers for pest control.

Conidia →

Conidia →

Box 5.3 Fig. 1 **(a)** Beauveria bassiana *and* **(b)** Metarhizium anisopliae *conidia production. Beauveria produces white spores alternately, while* Metarhizium *produces chains of green conidia* [Redrawn from Deacon (2006).]

(a)   (b)

Pasteur was unconvinced, as there were many contradictory results from his experiments. Progress was hindered also by Pasteur's lack of knowledge of type of pathogen he was dealing with: *pébrine* is caused by a microsporidian, *Nosema bombycis*, which at one stage within the insect becomes almost invisible (Box 5.4) and has a rather different life history from the bacteria and fungi he was becoming used to.

During 1866, Pasteur isolated another disease from the same silkworms. This was the non-corpuscular disease *flacherie* (a general term for a silkworm disease known for centuries where the larvae become flabby and weak) and he was able to study it in isolation. This was more virulent than *pébrine* and it attacked all worms of a given age from a batch of eggs. The presence of this disease had confused his earlier work; it attacked *pébrine*-infected larvae quickly and caused death before the corpuscles characteristic of *pébrine* had a chance to develop.

Flacherie is a much more complex disease than *pébrine* or muscardine, and we are still not really sure exactly what causes it. It is still a problem, and is the most important silkworm disease in Japan, accounting in the 1970s for over 70% of diseased cocoons. Pasteur spent 1866 to 1870 studying it and finally concluded that it was due to accumulation of two species of bacteria in the gut, and it could be spread by contaminated excrement or food. It is now thought that these bacteria are secondary invaders, and that viruses are the primary cause of the disease.

In 1870, Pasteur's book on the diseases of silkworms was published, it was very successful and gave the picture now familiar to us of Pasteur the pioneering rational scientist. However, as Cadeddu (2000) has discussed, there is another side to this story. Although Pasteur was originally convinced that *pébrine* was caused by the caterpillar and not by other organisms, from the beginning others thought otherwise. Notably, Béchamp, a professor in the faculty of medicine at Montpellier published from 1865 that the *pébrine* corpuscles were the cause of the disease and carried out experiments to demonstrate this, even before Pasteur had started his work. Later, Béchamp also immediately concluded that *flacherie* was caused by a microbial infection, while Pasteur still thought that it was caused by a disorder of the digestive system and that the microbes were born spontaneously in the silkworm intestine. Pasteur was continually critical of Béchamp's conclusions, yet in his 1870 book had formed the same conclusions. On publication, Béchamp accused Pasteur of having appropriated his ideas without acknowledgement. The truth behind this remains to be investigated.

Silkworm diseases were an initiation for Pasteur into the complexity, unpredictability and variation of infectious diseases of animals, skills he used later in tackling diseases of humans and other mammals (including rabbits, see Chapter 11). In 1874, Pasteur suggested that *pébrine* could be used to control the grape phylloxera (Chapter 8), and subsequently entomopathogenic fungi and microsporidians have been investigated for the biocontrol of many insects (see Boxes 5.3, 5.4 and 15.3). Furthermore, the silkworm has become the model organism for many different fields, including insect nutrition, genomics, endocrinology, sex pheromones and insect genetics. Thus, today the silkworm is as important a scientific organism as an industrial one.

BOX 5.4   MICROSPORIDIA

Nosema bombycis is a member of an unusual group of eukaryotic obligate intracellular parasites – the Microsporidia. Compared to other eukaryotes they are both highly specialized and greatly reduced at every level – morphology, structure, biochemistry, metabolism and genome. They are a very diverse group, consisting of over 1200 species placed in 150 genera, although undoubtedly many more species remain to be identified. Microsporidia parasitize all vertebrate groups and most invertebrate ones, though are most prevalent in arthropods and fish, causing significant losses also to honey bees (see Box 4.3) and some fisheries. Microsporidian infection in humans was rare until the mid-1970s, since when there has been a dramatic increase in line with the rapid rise in numbers of individuals with a suppressed immune system caused by HIV infection.

The spore (Fig. 1) is the only easily recognized and identified stage of microsporidia, and that is viable outside the host. The spore is 1–40 µm long and usually oval. Outside of the normal unit membrane it has two rigid extracellular walls composed of protein and chitin. Inside are three important organs of infection: the polaroplast, consisting of stacked membranes in the anterior of the cell; the polar filament, which is partially coiled and attached to the apex of the spore via an umbrella-shaped anchoring disk; and the posterior vacuole, in which the polar filament terminates.

The germination of this spore is dramatic. Poorly understood environmental cues trigger spore swelling and an increase in osmotic pressure. The anchoring disk ruptures and the polar filament everts and is discharged through the anterior end, like turning the finger of a glove inside out. The polar filament becomes a tube, up to 100 times the length of the spore; this strikes and penetrates (probably by phagocytosis) a host cell, and once fully extended the pressure in the spore forces the sporoplasm through the tube and directly into the cell. Growth and division of the parasite begins, it is now almost invisible in the cell and can be surrounded by host cell organelles. After rounds of binary fission, spores are produced and released either in the host faeces or rupturing cadavers. *Nosema* in silkworms is transmitted both horizontally (between individuals) and vertically (between generations).

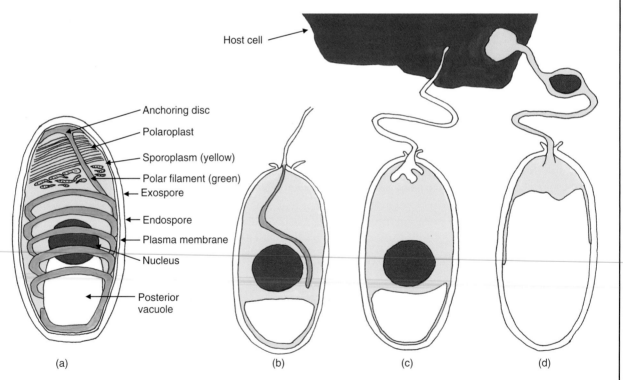

Box 5.4 Fig. 1   *Microsporidia.* **(a)** *Diagram of a typical microsporidian spore.* **(b–d)** *Sequence of spore germination and infection of a host cell: (b) the anchoring disc has ruptured and the polar filament (green) begins to emerge, everting and turning into a tube as it does so; (c) the polar tube is now fully emerged and has penetrated the host cell; (d) the sporoplasm (yellow) and nucleus (red) are passing through the polar tube and into the host cell, leaving an empty spore behind* [(a) based on Franzen (2004).]

Nosema bombycis was the first microsporidian to be discovered, and was first described by Nägeli in 1857 as a fungus. In 1882, Balbiani erected a new group, the Microsporidia, for it. This group has no obvious similarities to other groups of organisms and thus its classification has always been problematic. Over the years it has been allied with a variety of groups: starting from the 1980s the Microsporidia were considered part of the Archaezoa, primitive eukaryotes that evolved before the eukaryotes supposedly gained mitochondria by endosymbiosis.

However, many pathogens appear primitive, when this only reflects the process of reduction from a more complex ancestor. A re-evaluation of the evidence and further studies demonstrated that, although Microsporidia do not have mitochondria now, they retain several mitochondrial genes in their nuclear genome, suggesting that they once did have mitochondria. There is now robust evidence that microsporidia share some relationship with fungi, but it is unclear whether they are fungi, or just close relatives (Fig. 2).

Microsporidia have been investigated as insect biocontrol agents, and one species, *Nosema locustae*, is available for grasshopper and locust control (Chapter 15). Since the 1990s, researchers have been investigating three European Microsporidia species for release against the gypsy moth (see Box 5.2) in North America, and they may be released soon. Although some results with Microsporidia as biocontrol agents have been promising, as obligate parasites they have to be cultured in a living host, which can be expensive, in comparison to entomopathogenic fungi (see Box 5.3), and they often produce only a chronic infection in their host rather than death.

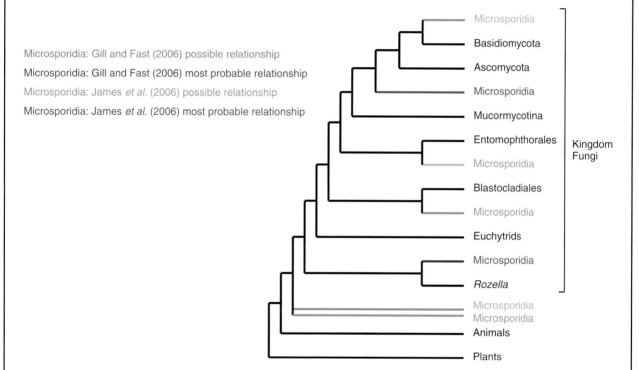

Box 5.4 Fig. 2   *Where are the Microsporidia? Both James et al. (2006) and Gill and Fast (2006) have proposed phylogenies that have placed the Microsporidia as related to the fungi. However, each study has proposed a different most likely relationship, and has not been able to rule out other possible relationships*

FURTHER READING

Feltwell (1990) is a good general account of sericulture and Thirsk (1997) gives a summary of English sericultural attempts in the seventeenth century. Sutherland *et al.* (2010) is a good overview of insect silks. Recent reprints of Moffet (1599), Barham (1719) and Bassi (1835) are useful. Liu (1952) discusses the role of sericulture in Chinese culture, and recent advances in elucidating the origin of sericulture are described by Normile (2009). For practical details of sericulture, see the series of manuals produced by the FAO, including Pang-chuan and Da-chuang (1988), Ting-zing *et al.* (1988) and Lim *et al.* (1990). Sericulture in North America has not been reviewed recently, but Klose (1963) gives a good overview, especially for the nineteenth and twentieth centuries. Older references such as Kenrick (1835) and Wyckoff (1886) now available online are also useful. For the early development of sericulture in Virginia and Georgia, see Hamer (1935), Hatch (1957), Bonner (1969) and Ewan (1969); many of the seventeenth and eighteenth century references mentioning sericulture in these States are available in compilations.

Dubos (1960) provides a good account of Pasteur's scientific career, and devotes a chapter to his study of silkworm diseases. A standard work on insect pathology is Tanada and Kaya (1993), and a recent review of the uses of microorganisms for insect control is given by Hajek *et al.* (2009). Microsporidia are reviewed in general by Keeling and Fast (2002) and Franzen (2004), in silkworms by Kawarabata (2003), and their phylogeny by Bruns (2006), Gill and Fast (2006) and James *et al.* (2006). An introduction to entomopathogenic fungi is given by Deacon (2006), about *Beauveria* by Feng *et al.* (1994) and Zimmermann (2007), *Beauveria* as endophyte by Vega *et al.* (2008), and for bees as possible vectors of *Beauveria* see Al-Mazra'awi *et al.* (2007).

For recent discussions on wild silk production in ancient Greece, see Good (1995) and Panagiotakopulu *et al.* (1997), and for wild silk in general see Peigler (1993). A summary of the gypsy moth problem is given by Marshall (1981), and a history of control by Solter and Hajek (2009); much more can be found by an internet search.

REFERENCES

Al-Mazra'awi, M.S., Kevan, P.S. and Shipp, L. (2007) Development of *Beauveria bassiana* dry formulation for vectoring by honey bees *Apis mellifera* (Hymenoptera: Apidae) to the flowers of crops for pest control. *Biocontrol Science and Technology* **17**, 733–741.

Barham, H. (1719) *An Essay Upon the Silk-Worm*. J. Bettenham and T. Bickerton, London.

Bassi, A. (1835) *Del Mar Del Segno Calcinaccio o Moscardino Malattia che afflugge I bachi da Seta*. Parte Prima. Tipografia Orcesi, Lodi, Italy. [On the Mark Disease, calcinaccio or muscardine, a disease that affects silkworms. Transl. P.J. Yarrow, eds G.C. Ainsworth and P.J. Yarrow. Phytopathological Classics No. 10, American Phytopathological Society, 1958.]

Bonner, J.C. (1969) Silk growing in the Georgia Colony. *Agricultural History* **43**, 143–147.

Bruns, T. (2006) A kingdom revised. *Nature* **443**, 758–761.

Cadeddu, A. (2000) The heuristic function of 'error' in the scientific methodology of Louis Pasteur: the case of the silkworm diseases. *History and Philosophy of the Life Sciences* **22**, 3–28.

Clarke, G. (ed.) (1993) *The American Landscape*. Helm Information, Mountfield, East Sussex, UK.

Dandolo, Count (1825) *The Art of Rearing Silk-Worms*. John Murray, London.

Deacon, J. (2006) *Fungal Biology*, 4th edn. Blackwell Publishing, Malden, MA, USA.

Dubos, R. (1960) *Louis Pasteur: free lance of science*. Charles Scribner's Sons, New York.

Ewan, J. (1969) Silk culture in the colonies, with particular reference to the Ebenezer Colony and the first local flora of Georgia. *Agricultural History* **43**, 129–141.

Feltwell, J. (1990) *The Story of Silk*. Alan Sutton, Stroud, UK.

Feng, M.G., Poprawski, T.J. and Khachatourians, G.G. (1994) Production, formulation and application of the entomopathogenic fungus *Beauveria bassiana* for insect control: current status. *Biocontrol Science and Technology* **4**, 3–34.

Franzen, C. (2004) Microsporidia: how can they invade other cells? *Trends in Parasitology* **20**, 275–279.

Gill, E.E. and Fast, N.M. (2006) Assessing the microsporidia–fungi relationship: combined phylogenetic analysis of eight genes. *Gene* **375**, 103–109.

Good, I. (1995) On the question of silk in pre-Han Eurasia. *Antiquity* **69**, 959–968.

Hajek, A.E., Glare, T.R. and O'Callaghan, M. eds (2009) *Use of Microbes for Control and Eradication of Invasive Arthropods*. Springer, The Netherlands.

Hamer, M.B. (1935) The foundation and failure of the silk industry in provincial Georgia. *North Carolina Historical Review* **12**, 125–148.

Hatch, C.E. (1957) Mulberry trees and silkworms: sericulture in early Virginia. *The Virginia Magazine of History and Biography* **65**, 3–61.

James, T.Y., Kauff, F., Schoch, C.L. *et al.* (2006) Reconstructing the early evolution of Fungi using a six-gene phylogeny. *Nature* **443**, 818–822.

Kawarabata, T. (2003) Biology of microsporidians infecting the silkworm, *Bombyx mori*, in Japan. *Journal of Insect Biotechnology and Sericology* **72**, 1–32.

Keeling, P.J. and Fast, N.M. (2002) Microsporidia: biology and evolution of highly reduced intracellular parasites. *Annual Review of Microbiology* **56**, 93–116.

Kenrick, W. (1835) *The American Silk Grower's Guide; or the art of raising the mulberry and silk on the system of successive crops in each season*. George C. Barrett and Russell, Odiorne and Co., Boston, USA.

Klose, N. (1963) Sericulture in the United States. *Agricultural History* **37**, 225–234.

Lim, S.-H., Kim, Y.-T., Lee, S.-P., Rhee, I.-J., Lim, J.-S. and Lim, B.-H. (1990) *Sericulture Training Manual*. FAO Agricultural Services Bulletin 80. Food and Agriculture Organisation of the United Nations, Rome, Italy.

Liu, G.K.C. (1952) The silkworm and Chinese culture. *Osiris* **10**, 129–194.

Marshall, E. (1981) The summer of the gypsy moth. *Science* **213**, 991–993.

Martyn, B. (1732) *A New and Accurate Account of the Provinces of South-Carolina and Georgia*. J. Worrall, London.

Moffet, T. (1599) *The Silkewormes, and their Flies*. Nicholas Ling, London. [V. Houliston, (ed.) (1989) Medieval and Renaissance Texts and Studies 56, Renaissance English Text Society, Binghamton, New York, USA.]

Newman, E. (1884) *An Illustrated Natural History of British Moths*. W.H. Allen, London.

Normile, D. (2009) Sequencing 40 silkworm genomes unravels history of cultivation. *Science* **325**, 1058–1059.

Panagiotakopulu, E., Buckland, P.C., Day, P.M., Doumas, C., Sarpaki, A. and Skidmore, P. (1997) A lepidopterous cocoon from Thera and evidence for silk in the Aegean Bronze Age. *Antiquity* **71**, 420–429.

Pang-chuan, W. and Da-chuang, C. (1988) *Silkworm Rearing*. FAO Agricultural Services Bulletin 73/2. Food and Agriculture Organisation of the United Nations, Rome, Italy.

Peigler, R.S. (1993) Wild silks of the world. *American Entomologist* **39**, 151–161.

Smith, J. (1612) *A Map of Virginia, with a description of the countrey, the commodities, people, government and religion.* Joseph Barnes, Oxford.

Solter, L.F. and Hajek, A.E. (2009) Control of gypsy moth, Lymantria dispar, in North America since 1878. In: *Use of Microbes for Control and Eradication of Invasive Arthropods* (eds A.E. Hajek, T.R. Glare and M. O'Callaghan), pp. 181–212. Springer, The Netherlands.

Sutherland, T.D., Young, J.H., Weisman, S., Hayashi, C.Y. and Merritt, D.J. (2010) Insect silk: one name, many materials. *Annual Review of Entomology* **55**, 171–188.

Tanada, Y. and Kaya, H.K. (1993) *Insect Pathology.* Academic Press, San Diego, USA.

Tazima, Y. (1984) Silkworm moths. In: *Evolution of Domesticated Animals* (ed. I.L. Mason), pp. 416–424. Longman, London.

Thirsk, J. (1997) *Alternative Agriculture: a history from the Black Death to the present day.* Oxford University Press, Oxford.

Ting-zing, Z., Yun-fang, T., Guang-xian, H., Huaizhong, F. and ben, M. (1988) *Mulberry Cultivation.* FAO Agricultural Services Bulletin 73/1. Food and Agriculture Organisation of the United Nations, Rome, Italy.

Vega, F.E., Posada, F., Aime, M.C., Pava-Ripoll, M., Infante, F. and Rehner, S.A. (2008) Entomopathogenic fungal endophytes. *Biological Control* **46**, 72–82.

Wigglesworth, V.B. (1972) *The Principles of Insect Physiology.* Chapman & Hall, London.

Wyckoff, W. C. (1886) *American Silk Manufacture.* John L. Murphy, Trenton, NJ, USA.

Zimmermann, G. (2007) Review on safety of the entomopathogenic fungi *Beauveria bassiana* and *Beauveria brongniartii*. *Biocontrol Science and Technology* **17**, 553–596.

Sugar Cane

CHAPTER 6

For clothing, we are given a pair of canvas drawers twice a year. Those of us who work in the factories and happen to catch a finger in the grindstone have a hand chopped off; if we try to escape, they cut off one leg. Both accidents happened to me. That's the price of your eating sugar in Europe.

Negro slave near Surinam, Candide or Optimism, Voltaire, 1758

W E LIVE IN a complex and contradictory society. It is hard to discern clear chains of cause and effect and harder still to judge right from wrong in social history. Historians may eventually make such judgments, but biologists need not normally confront the issues involved. And yet here we must, because the peculiar physiology and agronomy of the sugar cane plant is associated with one of the great injustices of the modern era: the enslavement of an estimated 12 million people. This is the number of Negroes taken from Africa, many of them because of a European fad for sugar.

After 1492, when Europeans discovered America, the early colonists of its tropical regions soon found that sugar cane grew very well there. They built the market for sugar at home, but they lacked a workforce to labour in the tropical heat and boil the sugar from the cane. So they enslaved blacks in Africa, shipped them across the Atlantic, and for three centuries worked them in the sugar plantations of Brazil, the Caribbean and Central America, and Louisiana. The horror of this is compounded by the juxtaposition of two types of exploitation – of black people in the New World, and of white Europeans who were made to believe they needed sugar. The only relief lies, perhaps, in the most recent chapter of the story: the conversion of sugar cane plantations from sources of sugar to sources of biofuel.

At the crux of it is the sugar cane (Fig. 6.1), a plant with an unusual combination of stem storage of sucrose and rapid, perennial growth. This combination provided a platform for the development of sugar consumption and the slave trade, because although people could easily be addicted to the instant energy hit of sugar, the sugar cane plant was confined to tropical regions and to be economic required plantation-scale production and rapid processing near the point of harvest. Thus an easy consumption belied a fearsome production system. Slavery offered a solution which perfectly suited the needs of the entrepreneurs who sought to expand the European sugar market in the sixteenth and seventeenth centuries: an apparently endless supply of black labour was imported from Africa

Fig. 6.1 *The sugar cane,* Saccharum officinarum. **(a)** *Sugar cane plantation – traditional harvesting.* **(b)** *Modern methods [Courtesy of Dr Raymond Maughan, Barbados.]*

(a)

(b)

Biological Diversity: Exploiters and Exploited, First Edition.   Paul Hatcher and Nick Battey.
© 2011 John Wiley & Sons, Ltd. Published 2011 by John Wiley & Sons, Ltd.

to the Caribbean and surrounding regions, where the slaves died in the service of sugar.

In recent history, sugar slavery has disappeared, but sugar cane production and sugar extraction still demand hard and unpleasant manual labour, which is provided in many parts of the world by poorly paid migrant workers. This and the complex politics behind the sugar market mean that social injustice persists in association with the sugar cane plant. Yet the need for alternative sources of fuel has led to a focus on sugar cane sucrose as a source of ethanol, and its stem fibre as a direct source of energy. It is an irony that in a country with a long historical association with sugar slavery such as Barbados, sugar cane breeding is now re-orientating itself towards high-fibre varieties that may no longer be grown for their sugar. Instead, cane will be harvested and burnt to provide a renewable source of electricity for the island.

In this chapter we outline the features of the sugar cane plant that make it so controversial: stem storage of sugar; C4 photosynthesis, giving a prodigious ability to assimilate carbon under tropical conditions; and a vigorous perennial habit. With this information in harness, we explore the potential of 'total cane' – a source of biofuel that could reduce dependency on the non-renewable energy sources coal, gas and oil. If the economics of total cane production stack up, this may provide the means of rehabilitation of the sugar cane plant.

THE ORIGIN OF SUGAR CANE

In 1753, Linnaeus named the cultivated sugar cane *Saccharum officinarum*, 'officinarum' – meaning 'of the apothecaries' shops' and reflecting the medieval use of sugar as a medicine. The cane he described was the 'Creole' cane of the New World plantations, which had been taken to Hispaniola (now Haiti/Dominican Republic) in the Caribbean by Columbus in 1493, and which dominated commercial sugar production for about 300 years after that. This cane was a hybrid that had originated in northern India, where Alexander the Great had seen sugar cane growing when he reached the Indus valley in 325 BC. It was brought to Persia, where documents suggest a sugar cane industry existed by about AD 600, and was transmitted relatively quickly around the Mediterranean littoral during the Arab conquests following the death of Mohammad in AD 632. Thus, by 1400, sugar industries had been established in Syria, Egypt, along the Barbary coast of North Africa, in southern Spain, Sicily, Crete, Rhodes and Cyprus (Fig. 6.2), all using what was simply known as 'sugar cane', and is assumed to have been the 'Creole' type. More ancient sugar industries existed in India and China: based on references in Sanskrit literature the extraction of some form of sugar was going on in India by 500 BC, and in China reference to 'stone honey' (loaves of hard sugar) can be found by AD 300. These older industries involved hybrids of the local species *S. barberi* and *S. sinense*.

Creole cane, although responsible for about 2000 years of sugar production, was only named 'Creole' following the arrival in the Caribbean of varieties that displaced it there and elsewhere in the world. These were the 'noble canes' of *S. officinarum* which had previously been in use for millennia in the islands of the South Pacific, Java and New Guinea, selected by the natives for their sweetness and vivid colours (Fig. 6.3). The main 'noble canes' were the Otaheite (or Bourbon) cane, which was used to establish the Mauritian sugar industry in the eighteenth

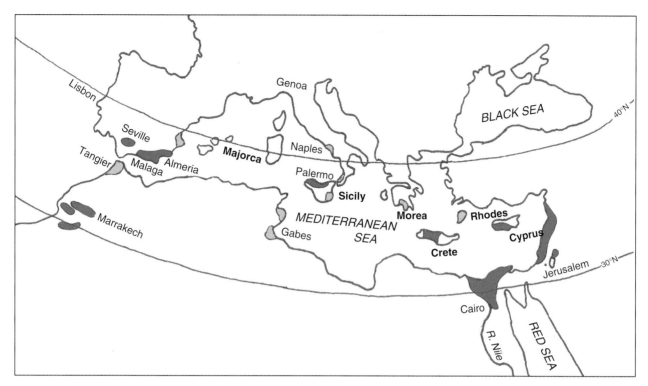

century and which displaced Creole cane in the West Indies following its introduction there by Captain Bligh in 1793; and the Cheribon (or transparent) cane, which refers to a whole series of striped, light and dark types that became important when the Otaheite cane failed through susceptibility to disease. In the twentieth century, following the realization that sugar cane was not sterile, and the discovery of natural hybrids growing wild in Java, 'noble canes' were used to breed new varieties by crossing with the wild species S. *spontaneum* and S. *robustum*, and with S. *barberi*.

The breeding of sugar cane has thus been dominated by the discovery of the 'noble canes' selected by the natives of Polynesia. The beginning of this process of selection is placed in New Guinea about 8000 BC, where it is surmised that the indigenous peoples selected the sweeter forms of wild S. *robustum*, a species that is now noted for being tough and relatively low in sugar, and therefore finds local use as a building material rather than an edible crop. As the sweeter forms of this species spread through the South Pacific islands, hybridization with the wild grass *Erianthus maximus* probably also occurred. By this combination of natural hybridization and human selection, the 'noble canes' of S. *officinarum* are believed to have arisen.

Fig. 6.2 *Sugar production areas in the Mediterranean region, AD 700–1600. Blue, major regions; yellow, minor regions [Redrawn by Henry Battey from Galloway (1977).]*

SUGAR STORAGE

It is worth noting that there are two schools of thought concerning the function of stem sucrose storage. One school argues that it gives a competitive advantage to the plant. In the natural habitat of sugar cane, those mature canes which store most sucrose during the dry season would be able quickly to respond to the onset of the monsoonal rains by sustained rapid growth of new suckers. This would

Fig. 6.3 *Sugar cane stalks [Stevenson GC, 1965. Genetics and breeding of sugar cane. Longman, UK.]*

enable the young canes rapidly to reach the top of the canopy and begin to photo-synthesize themselves. It is therefore envisaged that the sucrose storage trait would have been positively selected during evolution.[1] The other school of thought attributes the high level of stem sucrose to human selection. During the domestication of sugar cane by the native people of New Guinea and Polynesia, those types with highest sugar content would have been selected and maintained. According to this theory, the trait of sucrose storage is not considered to be beneficial in evolutionary terms. Indeed, on this view it is unlikely that the high sucrose-storing genotypes would survive outside domestication.[2]

Either way, stem storage of sucrose is quite unusual. For long-term future use, many plants store carbon as the polysaccharide starch in modified plastids known as amyloplasts. In trees adapted to temperate climates, such as oak and apple, the onset of the dormant period is associated with starch deposition in the roots. In springtime, starch is mobilized to sucrose to drive renewed growth. In plants from hotter climates which need to survive periods of drought, it is the rains that signal the remobilization of stored starch. Starch is also stored in the seeds of many plants, providing a source of stored energy that has the potential to last for many years, until germination.

Sugar cane, however, accumulates high levels of the disaccharide sucrose in its stems after transport from the leaves via the phloem of the vascular tissue. The sucrose concentration in the stem internodes can reach about 20% of fresh weight; this means that the vacuoles of the storage parenchyma cells are able to accumulate sucrose concentrations as high as 900 mM. This formidably sweet solution means the stems are good to chew raw. To achieve it the plant must organize sucrose uptake from the cell wall compartment (where sucrose is transported from the phloem), into the cytoplasm and then, against a massive concentration gradient, across the vacuolar membrane and into the vacuole. Sucrose appears to be broken down into glucose and fructose before uptake into the cytoplasm, and these monosaccharides are then phosphorylated before being reassembled into sucrose-phosphate for transport into the vacuole. Some aspects of this process are still uncertain, but the accumulation of sucrose clearly requires significant energy input and is therefore likely to be of functional importance to the plant.

C4 PHOTOSYNTHESIS: WHERE THE SUGAR AND BIOMASS COME FROM

One of the surprising things about plants is how inefficient they are at harvesting the energy of the sun, and sugar cane is no exception. In fact, this inefficiency is particularly noticeable in sugar cane because, one way or another, exploitation of solar energy conversion (usually into sucrose, but more recently into fibre) is the end at which all sugar cane breeding and cultivation technologies have been aimed. About 1% of sunlight is used in photosynthesis by sugar cane and of this only about a fifth is converted into dry matter during most of the growth cycle of the crop.[3]

And yet, as plants go sugar cane is impressive as a dry matter accumulator. A yield of 212 tonnes of cane stalks per hectare after 12 months growth has been recorded in Australia. This is about the theoretical maximum, given the known photosynthetic efficiency and required usage of fixed carbon in respiration and the

[1]Bull and Glasziou (1963).
[2]Discussed in Stevenson (1965).
[3]Alexander (1973), p. 123.

growth of leaves and roots. If 20% of cane stalk is sucrose (see above), this should give 50 tonnes of sucrose per hectare per year; in fact 33 tonnes were recorded.[4] More typical annual cane stalk yields are 80 tonnes dry matter per hectare, compared to dry matter production of 25 tonnes per hectare by wheat grown in the UK.

This ability of sugar cane to grow rapidly and accumulate dry matter in the high-light, high-temperature environment of the tropics is related to a special form of photosynthesis which it operates. Photosynthesis is the capture of light energy and its use in the reduction and fixation of CO_2 into carbohydrate. Light harvesting is done by chlorophyll in the chloroplasts, and the energy is stored as ATP and reducing power (NADPH) and used to drive the CO_2-fixing reactions. Carbon dioxide is acquired from the atmosphere via the stomata, pores in the leaf that have the disadvantage that they allow water vapour to escape; this means that photosynthesis always involves a compromise between water loss and CO_2 gain.

Most plants (about 85% of all species) carry out C3 photosynthesis, in which the CO_2 is initially fixed into a three-carbon compound before being converted to more complex carbohydrates (glucose, sucrose, starch). Current atmospheric concentrations of CO_2 are sufficiently low that the enzyme which catalyses this initial fixation step, Rubisco (ribulose bisphosphate carboxylase/oxygenase), can be quite inefficient if the CO_2 concentration within the leaf starts to fall. Rubisco then goes into reverse and oxidizes the C3 substrate via a process known as photorespiration, leading to a loss of CO_2 rather than its fixation. The decline in CO_2 concentration which leads to photorespiration can occur at high light intensity as the opportunity for rapid photosynthesis becomes available, particularly if water availability is limited and the plant has to close its stomata.

Sugar cane, along with a number of other mainly tropical species, uses C4 photosynthesis, which is a cunning method of concentrating CO_2 around Rubisco and thus maximizing the efficiency of the enzyme. C4 photosynthesis involves two main changes from the C3 type. First, CO_2 is fixed into a C4 compound (4-carbon organic acid) by a different enzyme, Pepc (phosphoenolpyruvate carboxylase), which has a much higher affinity for CO_2 than Rubisco. Second, the C4 compound is shuttled from its place of synthesis (the mesophyll cells of the leaf) to cells surrounding the leaf vascular bundle, where it is decarboxylated and the CO_2 re-fixed by Rubisco (Fig. 6.4).

This elaborate mechanism enables plants that use it to operate photosynthesis with less water loss, and respond much more effectively to high light conditions, than plants that use conventional C3 photosynthesis. It is therefore more efficient in tropical regions when atmospheric CO_2 concentration is low − as now, when it is about 0.04% of the atmosphere. The shuttling mechanism employed in C4 photosynthesis is energy demanding, so the conventional C3 mechanism would be preferable if CO_2 concentrations were high enough to ensure Rubisco always operated in the forward (CO_2 fixing) direction. In evolutionary terms, the current low atmospheric CO_2 concentration is relatively recent. Photosynthesis began to evolve about 2−3 billion years ago, and at that time atmospheric CO_2 would have been great enough to saturate Rubisco activity. As plant life took over the surface of the earth, oxygen levels steadily increased, whereas more and more CO_2 was locked up − for instance during the Carboniferous period about 360−280 mya, when the coal, gas and oil in use today originated. In the last 50 million years CO_2 concentrations have declined to the point where Rubisco inefficiency means C4 plants can be at an advantage (under high light), and geological evidence indicates that plants of this type have appeared over this relatively recent period. A particularly dramatic

[4]Moore et al. (1997), p. 143.

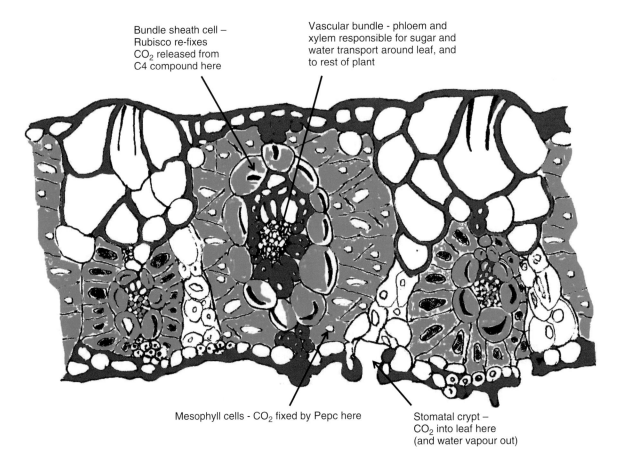

Bundle sheath cell –
Rubisco re-fixes
CO_2 released from
C4 compound here

Vascular bundle - phloem and
xylem responsible for sugar and
water transport around leaf, and
to rest of plant

Mesophyll cells - CO_2 fixed by Pepc here

Stomatal crypt –
CO_2 into leaf here
(and water vapour out)

Fig. 6.4 *Transverse section of a sugar cane leaf [Drawn by Henry Battey from an original photograph in Taiz and Zeiger (2006) Plant Physiology, 4th edn. Sinauer, MA, USA.]*

increase in C4 biomass occurred 7 to 5 million years ago, associated with further decline in CO_2 concentration; it was at this time that the world's major grasslands were formed, with a significant component being C4 plants.[5]

Thus, at least part of the explanation for sugar cane's phenomenal ability to assimilate carbon as sugar and biomass lies in its use of C4 photosynthesis; another part lies in its capacity to sustain active leaf production, and therefore growth, over a prolonged growing season.[6] These features conspired, with the stem storage of sugar and the sweet tooth of humans, to create the potential for exploitation of furious intensity.

SUGAR CANE AND SLAVERY

The facts of the transatlantic slave trade have been extensively documented. It began with the colonization of the Caribbean and surrounding regions by the major European powers. Early in the sixteenthth century attempts were made to use the native Amerindians as slave labour in the fledgling sugar cane plantations. The natives proved poor slaves, however, and through compassion for them Bartolomé de Las Casas suggested a scheme for importing Spanish immigrants, each of whom was to be allowed a quota of 12 African slaves. But it was Judge Alonso de Zuazo of Hispaniola who realized the need for a dependable labour supply in order for

[5]See Ehleringer *et al.* (1991), Cerling *et al.* (1993); also Beerling (2006), particularly Chapter 8, for an evaluation of the 'CO_2 starvation' hypothesis.

[6]Lawlor (2001) and Bull and Glasziou (1975) provide interesting further reading on these two aspects.

the Caribbean sugar industry to flourish, and who saw the potential of Africa to provide this labour. Thus Zuazo wrote to the King of Spain:

> It would be necessary that ships should leave this island (Hispaniola), and go to Sevilla to obtain the things necessary for this trade, such as stuffs of various colours, and other articles that are used as barter at the Cape Verde Islands. The King of Portugal should be approached to grant a licence to this end, and all negroes between the ages of fifteen and eighteen years that can be obtained should be taken.[7]

What followed was capture, transport and forced labour on such scale, and of such brutality, that Las Casas came to bitterly regret his suggestion and to campaign vigorously against the slave trade. But the seventeenth century saw the massive development by the British and French of the transatlantic sugar trade which the Spanish and Portuguese had dominated in the previous century; and with it grew slavery. Sugar produced in the West Indies was transported to Europe, refined and sold; the same ships that had carried the sugar loaded up with finished goods – textiles, guns, trinkets, alcohol – and carried them to the west coast of Africa, where they were sold or exchanged for slaves. In ports from Dakar in the north to Benguela in the south, Negroes captured from the interior of the continent were bartered, loaded on to the ships and transported across the Atlantic in conditions whose inhumanity has become legendary. Those who survived the crossing (and it is estimated that about 13% did not[8]) were sold at market and driven to the sugar cane plantations. The anti-slavery poem 'The Black Man's Lament' by Amelia Opie (1769–1853) emphasizes, by comparing the lot of the slave with that of an English peasant, the relentless toil of the plantation labourer. Here is the question put by the naïve British interlocutor, and the slave's response:

> "But, Negro slave, some men must toil.
> The English peasant works all day;
> Turns up, and sows, and ploughs the soil.
> Thou wouldst not, sure, have Negroes play?"
>
> "Ah! No. But Englishmen can work
> Whene'er they like, and stop for breath;
> No driver dares, like any Turk,
> Flog peasants on almost to death.
> Who dares an English peasant flog,
> Or buy, or sell, or steal away?
> Who sheds his blood? Treats him like dog,
> Or fetters him like beasts of prey?
> He has a cottage, he a wife;
> If child he has, that child is free.
> I am depriv'd of married life,
> And my poor child were *slave* like *me*."

The 'triangular trade' of sugar, goods and slaves continued through the eighteenth century until the abolition of slavery in the early nineteenth century,

[7] Deerr (1950), p. 265.
[8] Thomas (1997) provides a comprehensive description of the Atlantic slave trade.

although slaves continued to be used in sugar plantations in parts of the Caribbean (Cuba, for instance) until almost the end of that century. Vivid accounts of the history of sugar and its association with slavery can be found elsewhere (see Further Reading). There are two aspects of particular importance here: why *sugar* and slavery? And why *sugar* at all? The former question seems to be answered by specific features of the sugar cane plant; the latter by human psychology.

The association of slavery with sugar cane pre-dates the transatlantic triangular trade: the various regions of the Mediterranean involved in sugar production (see Fig. 6.2 and Box 6.1) appear to have independently adopted slavery. Although the historical evidence is fragmentary, it seems that where sugar cane cultivation became extensive, it tended to acquire slave labour. Thus in Egypt in the Mamluk era (1250–1517), although corvée (unpaid labour provided by a vassal) was unusual, it was used in sugar cane cultivation; and in Morocco evidence from place names indicates an association of sugar and slaves.[9] In Crete and Cyprus, the response to the shortage of labour consequent on the plague epidemics of the fourteenth century was to turn to slave labour, sourced from Greece to the shores of the Black Sea. Later, on the Atlantic Islands (Madeira, the Canaries, and Sao Tomé), which provided for Spain and Portugal during the fifteenth and sixteenth centuries a stepping stone between the old Mediterranean and the new Caribbean sugar industries, slavery, now of Negroes stolen from Africa, was the rule. In both Madeira and the Canaries, as sugar production declined due to competition from new sources across the Atlantic, vines took the place of sugar cane; and viticulture used no slaves.

Why the general association of slavery with sugar? The answer lies at least partly in the biology of the sugar cane plant. Frost sensitive, it is confined to the

---

BOX 6.1   MEDITERRANEAN SUGAR PRODUCTION

Fundamental to the development of the sugar industry around the Mediterranean is the contribution of the Arabs, who within a century of the birth of Islam had established, from Arabia to Southern Spain, an administrative and cultural continuity which allowed the development of the technology required for successful production of the novel sugar cane crop. One such technology was irrigation – as a plant adapted by origin to the humid tropics, sugar cane required year-round irrigation in the dryer environment of the Middle East and Mediterranean. Water management was an area in which Islamic engineers excelled and this made sugar cane cultivation possible well outside the plant's natural range. Their overall enthusiasm for technologies of irrigation, production and processing, as well as an exceptional administrative organization, allowed the Arabs to promote the development of sugar cane cultivation in southern Europe. They expanded production to Marrakech in the south, and Valencia and Palermo in the north, devising methods of sugar production even where the growing season was curtailed by frosts, necessitating early harvest. Water power was employed for milling, and a variety of sugars was produced.

Thus the Arabs established an industry that was adopted by subsequent incomers: the first crusaders in the eleventh century reported that on the plains around Tripoli the local inhabitants cultivated in abundance the 'honey reed' or 'Zuchra'. Sugar cane cultivation was supervised by crusaders in the kingdom of Jerusalem (1099–1187) and by the Knights of Malta at Acre until 1291 when it fell to the Saracens. The later Mediterranean trade was coordinated by Venetian merchants, with Antwerp a major centre for sugar refining. The industry declined through a combination of factors, including war, plague and competition from the Atlantic industry. Its wide distribution around the shores of the Mediterranean has almost been forgotten, but ruins of sugar mills can still be seen in the deserts of southern Morocco; and the Gate of the Sugar Workers in the walls of Syracuse, Sicily, reflects its local importance. For further details on the Mediterranean sugar industry, see Galloway (1977) and Mintz (1985).

---

[9]Galloway (1989), pp. 41–42.

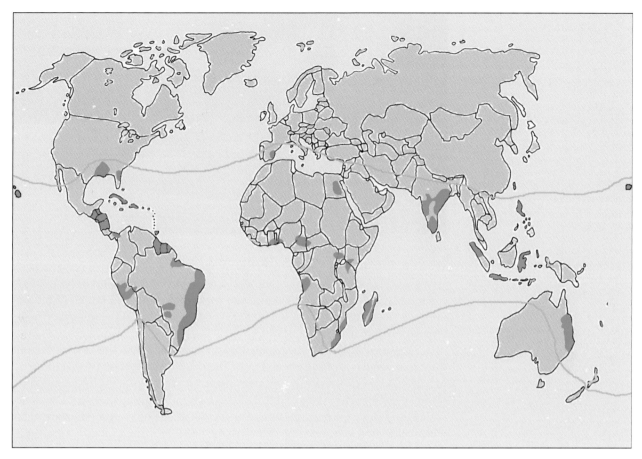

Fig. 6.5 *World distribution of sugar cane cultivation (green shading), with palm tree line shown in yellow [Redrawn by Henry Battey from James (2004).]*

tropics and warm subtropics, the present-day boundaries of production following the palm tree line in both hemispheres (Fig. 6.5). From the seventeenth century onwards, the economic success of the European powers was underpinned by global expansion, colonization and trade. Hobsbawm emphasizes how fundamental this change was to the establishment of modern industrial economies:

> The shift was not merely geographical, but structural. The new kind of relationship between the 'advanced' areas and the rest of the world, unlike the old, tended constantly to intensify and widen the flows of commerce. The powerful, growing and accelerating current of overseas trade which swept the infant industries of Europe with it – which, in fact, sometimes actually *created* them – was hardly conceivable without this change. It rested on three things: in Europe, the rise of a market for overseas products for everyday use, whose market could be expanded as they became available in larger quantities and more cheaply; and overseas the creation of economic systems for producing such goods (such as, for instance, slave-operated plantations) and the conquest of colonies designed to serve the economic advantage of their European owners.[10]

[10]See Hobsbawm (1968), p. 35–36; see also Mintz (1997).

Fig. 6.6 *Sugar cane burning. This dramatic image from Queensland, Australia, dates from 1959 [From www. flickr.com.]*

Crops of tropical origin, particularly those with an addictive or narcotic aspect, such as tobacco, tea, coffee and sugar, contributed importantly to economic development in the temperate regions of northern Europe. In the case of sugar, the confinement of the plant to the tropics offered the opportunity of international trade based on slavery.

The other aspects of sugar cane biology combined with this tropical requirement to make slavery fundamental to it. Then, as now, the plant was entirely propagated by vegetative means – stem segments readily produce roots and new plants when buried in the ground. Each segment or 'sett' (up to 45 cm long and with 3–4 buds) requires a pit, trench or furrow to be dug to accommodate it, and when cultivation areas become extensive the need for manual labour is greater than with a seed-raised crop. There was, and is, a need for constant maintenance and management in the crop – weeds, rats, pests and disease all need to be combated. For optimum yields, high levels of fertilizer, and irrigation (only rice requires more amongst annual crops), are required. In the early days of Caribbean cultivation virgin land, which had previously been heavily forested, was cleared by burning and therefore had high levels of inorganic nutrients. But the perennial nature of sugar cane means that several successive 'ratoon' crops are taken over a number of years. The initial nutrients were therefore quickly exhausted and had to be supplemented by manual application of manure and other fertilizers.

A further key factor was that the desired product (sucrose) is stored in a vegetative part of the plant – the stem – rather than in the reproductive structure (flowers or fruit or seed). This means that prodigious vegetative growth is needed for high sugar yield; further, before harvest, cane burning was practised. This is a sight that can still be seen in some sugar cane plantations (Fig. 6.6). It has the advantage that snakes, poisonous insects and spiders are driven out, but requires constant supervision. After burning, harvesting must be done quickly – the whole cane is cut down, trimmed and delivered to the sugar mill quickly to minimize the breakdown of sucrose, a process that begins rapidly on harvest. Typically, cane needs to reach the sugar mill and the sap be extracted within 48 hours of harvest (Box 6.2).

Extraction and basic refinement of the sugar was achieved by grinding the
cane stems between rollers, and mill slaves often had missing limbs – a machete
was kept nearby for the occasions when hands or arms became caught.[11] The
cane sap was then boiled, crystallized and fractionated in a boiling house, where
the intense heat created conditions which it was considered only Negro labour
could tolerate. In addition, cane extraction led to a heavy demand for fuel, much
of which was obtained locally through manual labour, at least until supplies ran
out and importation became necessary.

In short, sugar cane physiology and agronomy, and the process of sugar
extraction, required huge labour inputs under harsh conditions. The possibility of
obtaining slaves from Africa led to a business opportunity based upon the sugar
cane plantations of the New World; the final piece of the commercial jigsaw was
to convince the population of Europe to believe they needed the product, sugar.

THE CREATION OF DEMAND FOR SUGAR

Before 1600, sugar was confined to a minority of the privileged. Of the subsequent rise in sugar, tea and coffee consumption, Sidney Mintz (1997) writes:

BOX 6.2   MAKING SUGAR

A sugar factory (Fig. 1) does not make sugar. It uses a series
of operations to extract the sucrose that is made by the plant.
Sugar factories are designed to receive either the beets (swollen storage tuber) of the sugar beet *Beta vulgaris*, or the canes
(stems plus some leaf material) of the sugar cane *Saccharum
officinarum* (Fig. 2). Sugar can also be extracted on a more local
basis from sources such as the Palmyra palm (*Borassus flabel-
lifer*), the Kitul palm (*Caryota urens*) and the silver date palm
(*Phoenix sylvestris*).

Cane sugar extraction (summarized in Fig. 3) begins with
washing of the cane stalks. The canes are then crushed in roller
mills and the expressed juice clarified by adding burnt limestone
powder as 'milk of lime'. The lime sludge (including aggregated
impurities from the cane juice) is filtered out; the resulting 'thin
juice' contains about 15% sucrose. Water is evaporated to

Box 6.2 Fig. 1   *Sugar factory in Barbados* [Courtesy of
Dr Raymond Maughan.]

Box 6.2 Fig.2   *Sugar cane factory in operation. The cane
is loaded into a shredder and moved on a conveyor belt to the
processing plant. St Andrew's Sugar Factory in St Joseph, Barbados*
[Courtesy of Moira Mulvey.]

[11]Description of William Mathieson nineteenth century abolitionist–see O'Connell
(2004), p. 62.

produce a syrup containing at least 68% sucrose. Sugar crystals then form as the solution is rapidly cooled, and these are separated from the remaining solution (molasses) by centrifugation. The sugar is dried and stored. The by-products can be recycled (water), or used in agriculture (lime sludge, molasses), fermentation (molasses) or energy production to power the sugar mill (the spent cane stalks or 'bagasse'). The sugar can be further refined by processes that include decolorization (using activated carbon), melting and partial inversion, for example in syrup production (Fig. 4).

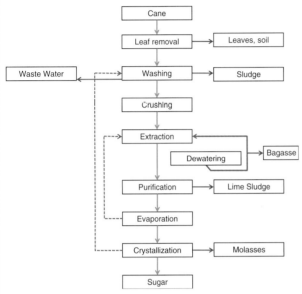

Box 6.2 Fig. 3 *Sugar extraction* [Redrawn by Henry Battey from an original at www.ifc.org.]

Box 6.2 Fig. 4 *Lyle's Golden Syrup, which tastes intensely sweet because of partial inversion of sucrose to fructose. This classic British brand depicts on the tin a dead lion and a swarm of bees, with the slogan OUT OF THE STRONG CAME FORTH SWEETNESS. This is an adaptation of the Biblical riddle (Judges 14:18) 'Out of the eater came forth meat, and out of the strong came forth sweetness', referring to the lion slain by Samson and the formation in the carcass of a bees' nest containing honey. Golden Syrup is a honey-mimic; the conflation of sugar with honey is the end point of a long-standing historical trend. Galen and Dioscorides used theriacs (treacles made from honey) to treat poisonous bites, but in England treacle had come to mean molasses by the seventeenth century; the word molasses itself derives from the Latin mel, meaning honey. See Mintz (1985) for further discussion*

[T]he general spread of these substances through the Western world since the seventeenth century has been one of the truly important economic and cultural phenomena of the modern age. These were, it seems, the first edible luxuries to become proletarian commonplaces; they were surely the first luxuries to become regarded as necessities by vast masses of people who had not produced them; and they were probably the first substances to become the basis of advertising campaigns to increase consumption. In all these ways, they, particularly sugar, have remained unmistakably modern.

Mintz, focusing on the British case, traces the growth in the use of sugar over four centuries, from a treat and exotica of the rich to a staple in the diet of the poor. The change is very striking. The average sugar intake per capita increased over 20-fold from 1700 to 1900, and it continued to increase during the last century. On one level Mintz attributes this rise to an increase in availability and decrease in price, the striking thing about the British sugar market before the abolition of slavery in the early nineteenth century being its protectionism: sugar produced in the colonies was sold within Britain and the proceeds served to make the plantation owners and shipping companies richer. There was very little true trade between Britain and the Caribbean, making the relationship most similar to that between town and country. Thus profit was fuelled by only one input: the consumption of black slaves. After abolition, sugar availability continued to increase and the price to fall, due to the removal of protectionism and the benefits of free trade at a time when beet sugar from temperate climates was becoming increasingly competitive. In fact, abolition can be seen as serving the most fundamental needs of the emergent capitalist British economy, by driving down prices and allowing consumption to increase.

On a second level, Mintz attributes the rise in sugar usage to the manipulation of its social meaning. The instant gratification provided by sucrose (the 'sugar hit') was by a process of social encouragement used to exploit the desperately poor. Thus, for example, Arthur Young, in a social commentary from the late eighteenth century, expresses his surprise at the request of poor people to be allowed to spend their alms money on tea and sugar to eat with their bread and butter dinner, rather than something more nutritious.[12] Mintz accounts for this as part of a process of systematic incorporation of sugar into the everyday lives of the working classes. It was associated with industrialization and the change from an agrarian lifestyle to a factory-based one. Meals were transformed by the inclusion of hot stimulant drinks sweetened with sugar and accompanied by tobacco, in a trend whose endpoint can be seen today in the consumption of fast, junk food. A key point is the 'meaning' associated with the assimilation of apparent luxury items into affordable everyday rewards: the symbolism of a successful system inhabited by an instantly gratified population.

Mintz's argument is hard to prove, and it would be wrong to suggest that the incorporation of sugar into our social framework was deliberately exploitative – but it was encouraged by those with vested interest in the slave trade, plantation-based colonial economies, and later in the profit to be derived from increased mass consumption. It is interesting to note that in France, at least until the First World War, working-class refusal to adopt sugar and sweets in spite of a prolonged hard sell by the French sugar lobby, formed part of an identity-based resistance to what were perceived as bourgeois habits and objectives (Bruegel, 2001). This undermines the generally accepted idea that humans have an innate sweet tooth, and suggests that we acquire, through conditioning and association, even our most basic desires.

SUGAR CANE AND POWER

We have seen that the sugar cane plant has been at the root of two vast global industries: slavery and sugar production. For the last 400 years its product has been

[12]Mintz (1985), p. 172.

used to manipulate the population of the developed world, initially at the expense of black slaves. In the later half of the period, the sugar market has continued to be a source of controversy, as temperate sugar beet has competed with tropical sugar cane. European Union over-production and the sugar quota system have been particularly problematic (Coote, 1987). Yet in the last two decades the possibility has emerged that the sugar cane plant could play a more constructive social role. An urgent need has emerged to find and employ renewable energy sources, both because of the finite size of fossil fuel reserves and because of political instability in the oil-producing regions of the world. Sugar cane offers potential because the sugar can be converted to ethanol, and the cane stalks, trash and other waste material (the 'bagasse') used as fuel to drive the fermentation process. The excess can be used to generate power.

The Brazilians have been engaged since the 1970s in large-scale ethanol production from sugar cane, to supplement oil in so-called 'flex-fuel' cars. The more recent trend has been to look at the biomass potential of sugar cane – making the sugar component less decisive.[13] Hence 'total cane' or 'fuel cane' or 'energy cane' breeding programmes have been established in Puerto Rico, Hawaii and Barbados. In crossing programmes with S. officinarum, clones of S. spontaneum are now being exploited for high fibre, which is crucial in biomass generation. Additionally, consistency of cane yield over several ratoon generations, and the ability rapidly to generate a vigorous crown, are characters of increased significance in sugar cane breeding. There is an ongoing debate over exactly how sustainable energy cane is, relative to fossil fuels and to other biofuels.[14] Questions remain about input levels (land, water, fertilizers, distribution costs) and environmental costs (e.g. soil erosion). But the prospect of a redeemed future for sugar cane gives grounds for genuine optimism.

FURTHER READING

Galloway (1977, 1989, 2000) is a great source for the history of the sugar industry; so also is the comprehensive and detailed and work of Deerr (1949, 1950), though his referencing can present a difficulty. Hobhouse (1999) provides a useful historical commentary on the sugar industry, and O'Connell offers further discussion of the impact of sugar. Stevenson (1965) is a good book for the history of sugar cane breeding. Mintz (1985, 1997) has made an influential analysis of the development of the sugar industry and the sociality of sugar. Details on physiological research on the plant can be found in Alexander (1973), on photosynthesis in Lawlor (2001); and the biosynthesis of sugar is treated clearly and concisely by John (1992). Cogeneration – the use of sugar cane for sugar and fuel – is covered in Payne (1991), and debates about sustainability of energy cane can be accessed through Marris (2006).

[13]Alexander (1985, 1991).
[14]See Dias de Oliveira et al. (2005) and Marris (2006).

REFERENCES

Alexander, A.G. (1973) *Sugarcane Physiology: a comprehensive study of the Saccharum source-to-sink system.* Elsevier, Amsterdam, The Netherlands.

Alexander, A.G. (1985) *The Energy Cane Alternative.* Elsevier, Amsterdam, The Netherlands.

Alexander, A.G. (1991) High energy cane. In: *Cogeneration in the Cane Sugar Industry* (ed. J.H. Payne), pp. 233–242. Elsevier, Amsterdam, The Netherlands.

Beerling, D.J. (2006) *The Emerald Planet: how plants changed Earth's history.* Oxford University Press, Oxford.

Bruegel, M. (2001) A bourgeois good? Sugar, norms of consumption and the labouring classes in nineteenth-century France. In: *Food and Identity: cooking, eating and drinking in Europe since the Middle Ages* (ed. P. Scholliers), pp. 99–118. Berg, Oxford.

Bull, T.A. and Glasziou, K.T. (1963) The evolutionary significance of sugar accumulation in *Saccharum. Australian Journal of Biological Sciences* **16**, 737–742.

Bull, T.A. and Glasziou, K.T. (1975) Sugar cane. In: *Crop Physiology: some case histories* (ed. L.T. Evans), pp. 51–72. Cambridge University Press, Cambridge.

Cerling, T.E., Wang, Y. and Quade, J. (1993) Expansion of C4 ecosystems as an indicator of global ecological change in the late Miocene. *Nature* **361**, 344–345.

Coote, B. (1987) *The Hunger Crop: poverty and the sugar industry.* Oxfam, Oxford.

Dias de Oliveira, M.E., Vaughan, B.E. and Rykiel, E.J. (2005) Ethanol as fuel: energy, carbon dioxide balances, and ecological footprint. *BioScience* **55**, 593–602.

Deerr, N. (1949) *The History of Sugar, Volume I.* Chapman & Hall, London.

Deerr, N. (1950) *The History of Sugar, Volume II.* Chapman & Hall, London.

Ehleringer, J.R., Sage, R.F., Flanagan, L.B. and Pearcy, R.W. (1991) Climate change and the evolution of C4 photosynthesis. *Trends in Ecology and Evolution* **6**, 95–99.

Galloway, J.H. (1977) The Mediterranean sugar industry. *Geographichal Review* **67**, 177–194.

Galloway, J.H. (1989) *The Sugar Cane Industry: an historical geography from its origins to 1914.* Cambridge University Press, Cambridge.

Galloway, J.H. (2000) Sugar. In: *The Cambridge World History of Food, Volume I* (eds K. Kiple and C. Orvelas Kriemhild), pp. 437–449. Cambridge University Press, Cambridge.

Hobsbawm, E.J. (1968) *Industry and Empire: an economic history of Britain since 1750.* Weidenfeld & Nicolson, London.

Hobhouse, H. (1999) *Seeds of Change: six plants that transformed mankind.* Papermac, London.

James, G. (2004) *Sugarcane,* 2nd edn. Blackwell, Oxford.

John, P. (1992) *Biosynthesis of the major crop products.* John Wiley & Sons, Chichester, UK.

Lawlor, D.W. (2001) *Photosynthesis*, 3rd edn. Bios, Abingdon, UK.

Marris, E. (2006) Drink the best and drive the rest. Nature **444**, 670–672.

Mintz, S.W. (1985) *Sweetness and Power: the place of sugar in modern history*. Penguin, Harmondsworth.

Mintz, S.W. (1997) Time, sugar and sweetness. In: *Food and Culture: a reader*, (eds C. Counihan and P. van Esterik), pp. 357–369. Routledge, New York.

Moore, P.H., Botha, F.C., Furbank, R.T. and Grof, C.P.L. (1997) Potential for overcoming physio-biochemical limits to sucrose accumulation. In: *Intensive Sugarcane Production: meeting the challenges beyond 2000*, (eds B.A. Keating and J.R. Wilson), pp. 141–155. CAB International, Wallingford, UK.

O'Connell, S. (2004) *Sugar: the grass that changed the world*. Virgin, London.

Payne, J.H. (1991) *Cogeneration in the Cane Sugar Industry*. Elsevier, Amsterdam, The Netherlands.

Stevenson, G.C. (1965) *Genetics and Breeding of Sugar Cane*. Longman, Harlow, UK.

Thomas, H. (1997) *The Slave Trade: the history of the Atlantic slave trade 1440–1870*. Picador, London.

Legumes

Legumes

THE FABACEAE (LEGUMINOSAE) is the third largest family of angiosperms and the most important economically of the dicotyledons because so many of its members are food crops. The traditional family name refers to its characteristic fruit, the legume (Figs 7.1 and 7.2), which develops from a single carpel (the female part of the flower, enclosing the ovules) and splits along two sutures when dry, to give two valves each containing seeds. The word legume is derived from the Latin *legere* meaning to gather, emphasizing that these are fruits that can be gathered by hand. In French, *légume* means 'vegetable', but the English borrowed the word and then restricted its meaning. Other words associated with legumes include pulse, derived from the Latin *puls* and meaning porridge; the Food and Agriculture Organisation (FAO) use pulse to describe any legume harvested for its dry grain, thus excluding oil, forage and green vegetable legumes. Pease is an ancient word which was treated as a plural to generate 'pea'.

In the Fabaceae there are 650 genera and about 18 000 species, and the family is divided into three sub-families: the Caesalpinoideae, Mimosideae and Papilionoideae. The Caesalpinoideae includes *Cassia* (senna or ringworm bush), *Ceratonia* (carob) and *Tamarindus* (tamarind); while Mimosideae includes *Acacia* and *Mimosa*. Most of the economically important legume species are in the Papilionoideae and representative genera are listed in Table 7.1, grouped according to Tribe. These genera can be divided functionally into forage crops (*Medicago*, *Trifolium*), oil crops (*Arachis*, *Glycine*) and grain crops (the rest).

Fig. 7.1 *Assorted garden legumes showing flowers and fruits (the legume, or pod):* **(a)** *garden sweet pea* (Lathyrus sativa); **(b)** *mangetout* (Pisum sativum); **(c)** *runner bean* (Phaseolus coccineus); **(d)** *climbing French bean* (Phaseolus vulgaris)

(a)   (b)   (c)   (d)

Biological Diversity: Exploiters and Exploited, First Edition.   Paul Hatcher and Nick Battey.
© 2011 John Wiley & Sons, Ltd. Published 2011 by John Wiley & Sons, Ltd.

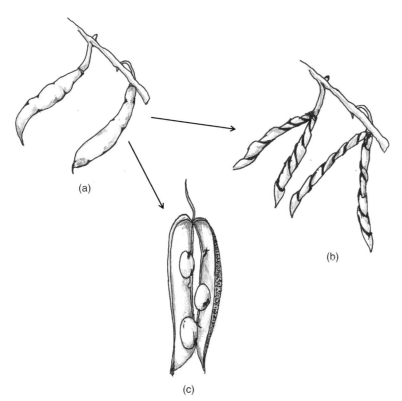

Fig. 7.2 **(a)** *Bean pod before dehiscence.* **(b)** *Wild (undomesticated) bean, in which the valves of the pod roll up at maturity, scattering seed.* **(c)** *Domesticated bean, in which the valves of the pod do not roll up and seeds remain in place [Drawing by Henry Battey.]*

As well as being important as forage, oil, food and fodder crops, many of the Fabaceae have an exceptional ability to enrich the soil because they carry out symbiotic nitrogen fixation in root nodules. These nodules are specialized structures that accommodate rhizobial bacteria and create an environment in which they can fix atmospheric nitrogen.

In this chapter we will follow a journey – from domestication of leguminous wild species in the Near East, Central America and China, to the use of legumes as a widely valued (but sometimes also mistrusted) food source of the ancient Greeks. We will see them feature as a key crop in Roman rotational agriculture and in the 'maize–beans–squash' *milpa* cropping system pivotal to the civilizations of South and Central America. Biblical sources give us a glimpse of the way legumes were generally considered in the West – as important but poorly regarded food. This reputation has followed them down the ages and is epitomized in the expression 'poor man's meat': as people get richer, they tend to discard legumes and move to meat as a source of protein. Nevertheless, diversification of legume cultivation coincided with the discovery of the New World. *Phaseolus* beans were taken to Africa; broad beans to eastern USA. In India, where meat-eating is relatively limited, legumes have remained a central part of the diet. The diverse potential of legumes is confirmed most spectacularly by the development of soybean from a cherished food source of the Far East to the most important world oil and fodder crop.

Finally, we will describe how longstanding awareness of the beneficial effect of legumes on the soil eventually culminated in the discovery of symbiotic nitrogen fixation. Present-day agronomic studies confirm a convincing capacity for soil enrichment which will be crucial in the drive to reduce the use of inorganic nitrogen fertilizer worldwide. Molecular research focuses on the mechanisms behind the bacterium–plant symbiosis and the potential to confer the nitrogen-fixing trait on other crops, such as cereals. When looked at in the round, the ability of legumes to facilitate nitrogen fixation is probably the most remarkable feature of a fascinating group of plants.

DOMESTICATION OF LEGUMES

The histories of settled civilization and plant cultivation are intimately entwined. Legumes appear very early as domesticated crops associated with a sedentary way of life based around agriculture; the process by which they were taken into cultivation is therefore a significant aspect of our own cultural origins.

Table 7.1 *Some economically important legumes*

Tribe	Latin name	Common name	Region of domestication
Aeschynomeneae	*Arachis hypogaea*	Groundnut (peanut)	South America
Cicereae	*Cicer arietinum*	Chickpea	Near East
Vicieae	*Lens culinaris*	Lentil	Near East
	Pisum sativum	Pea	Near East
	Vicia faba	Fava bean (field bean, broad bean, horse bean)	Near East
	Lathyrus sativus	Grasspea	Near East
Phaseoleae	*Glycine max*	Soybean	Far East
	Phaseolus acutifolius	Tepary bean	Central America
	Phaseolus coccineus	Runner bean	Central America
	Phaseolus lunatus	Lima bean (butter bean)	Central and South America
	Phaseolus vulgaris	Common bean – includes green (snap) and dry beans (dwarf, red kidney, haricot, pinto, navy)	Central and South America
	Vigna angularis	Adzuki bean (azuki bean)	South-east Asia
	Vigna unguiculata	Cowpea (black-eyed pea)	Africa
	Vigna mungo	Black gram (urd bean)	South-east Asia
	Vigna radiata	Green gram (mung bean)	South-east Asia
	Vigna subterranea	Groundbean (Bambara groundnut)	Africa
	Cajanus cajan	Pigeonpea	South-east Asia
Trifolieae	*Medicago sativa*	Lucerne, alfalfa	Near East
	Trifolium repens	Clover	Near East

For most of their evolutionary history people have not been sedentary, but lived as hunter-gatherers. Humans are thought to have begun to diverge from chimpanzees about 6 million years ago; 'Lucy' (*Australopithecus afarensis*), an upright biped, dates to 3.2 million years ago. *Homo habilis* and *Homo erectus* follow, and anatomically modern *Homo sapiens* appeared in Africa about 200 000 years ago. Hunter-gatherers, exhibiting complex social behaviour and depictional art, had spread from Africa around the world by 20 000 years ago (Chapter 9). We know that hunter-gatherers often had sophisticated knowledge of plants and animals, and many techniques of plant management were familiar to them. Yet, relatively suddenly, human behaviour changed about 10 000 years ago, and people started farming. It is the combination of sedentism with agriculture that is striking: what made people settle down?

Harlan (1992) listed the following among the most significant theories of the origins of settled agriculture: the need to provide plants and animals for religious ceremony; the adoption of a sedentary lifestyle (associated with, for instance, fishing); population pressure/food shortage; and the creation of a perceptually 'safe' environment. He concludes that there is no theory that works universally; all the above factors may have contributed at different times and in different places. Information gathered on the Natufian peoples of the Levant indicates, however, that at least some of these hunter-gatherers may have been sedentary by around 12 000 BP; and that agriculture may have developed by a process of 'wild gardening': nurturing of wild species and their gradual domestication (Mithen, 2002). What appears universal to all the proposed scenarios is that, once embarked upon, it is very difficult to escape from settled agriculture: it requires additional labour to generate extra productivity, which drives up population growth, which in turn drives up demand for food. This circularity is a problem with which humans have been trying to cope ever since.[1]

Regardless of the exact way in which it arose, one of the most important features associated with the development of settled agriculture is the domestication of plants. The two main groups involved early on were wild grasses (which were domesticated as cereals), and legumes. In different parts of the world different members of these groups were domesticated: maize and beans in Central America; wheat, barley, rye, lentils, peas and chickpeas in the Near East; rice, millet and soybean in the Far East. These three regions were key foci for the development of agriculture, but domestication of plants also went on in more diffuse regions nearer the equator—for example, potatoes were domesticated in South America; yams and sorghum in sub-Saharan Africa; taro and sugar cane in south-east Asia (Chapter 6).

For the story of legume exploitation it is the features which change under domestication that are of particular interest. What makes a plant more 'domestic'? How does this aid cultivation and promote yield? Some key features typically associated with the domestication of seed crops are an increase in reproductive effort—more flowers in larger inflorescences; self-pollination; larger seeds and fruits; a determinate growth habit; an annual life-cycle rather than a perennial one; non-dehiscence of fruits and seeds; reduced seed dormancy; and more even germination. The legumes exhibit changes of this kind. There is a general increase in size in the domesticated forms and, associated with determinacy and annuality, a greater emphasis on flower and fruit production. Wild legumes typically release their seeds explosively when the pod is mature, due to the contraction of a specialized inner layer. The two halves of the pod roll back and the seeds are fired away from the plant. In many domesticated legumes, in contrast, the mechanism for explosive dehiscence is suppressed (see Fig. 7.2). This feature is considered by some to be diagnostic of domestication and has been used to identify domesticated beans (*Phaseolus vulgaris*) in archaeological remains in Central and South America.

The timing of legume domestication in relation to the development of settled agriculture varies in Central America, the Near East and the Far East; the principal legume species domesticated in these regions are also different (Fig. 7.3 and Table 7.1). In Central America, squash (*Cucurbita pepo*) is the earliest domesticated plant, first recorded about 10 000 years ago, next is maize (*Zea mays*), with non-shattering cobs dating from Oaxaca, Mexico, 6250 years ago. The earliest evidence for bean cultivation in this region (derived from a bean pod from Tehuacán in Mexico) is much later, dating to 2285 years ago. The information currently available therefore suggests a significant gap between cereal and legume domestication in this region. Although the wild progenitors of maize and beans are both found in west-central Mexico, recent evidence suggests that they may have been domesticated separately and then brought together into a single cropping system. In Peru, the earliest domestication date for beans (4300 years ago) coincides more closely with the appearance of maize.[2]

In the Near East, people were collecting and processing wild cereals by 15 000 years ago; it seems that in this case a relatively cool, dry interval in the overall process of climate warming following the last glaciation could have

[1]There are contemporary examples of tribes that have successfully reverted to hunter-gathering, but in general settled agriculture seems to generate a self-perpetuating cycle. Recent discussion of the origins of agriculture has placed emphasis on the relative ease with which some plants—such as wild grasses and legumes—could be domesticated, as a contributory factor. This point of view suggests that such 'domestication potential' may be more important than sedentism or population pressure for the development of agriculture (see Murphy, 2007, p. 25).

[2]These beans were recovered at the Guitar-rero cave site in Peru. They were originally thought to be much older (because of dating based on associated material); the revised radiocarbon dates of the beans themselves are presented in Kaplan and Lynch (1999).

provided the stimulus to settled cultivation. This interval was the Younger Dryas and it began about 13 000 years ago. The suggestion is that this led to such depletion of wild food resources that by about 11 000 years ago the occupants of one key site, at Abu Hureyra, located near the Euphrates in modern Syria, had begun to cultivate domesticated rye (*Secale cereale*) and lentils (*Lens culinaris*). Elsewhere in the region the sequence of events may have varied, with barley (*Hordeum vulgare*), einkorn wheat (*Triticum monococcum*) and emmer wheat (*T. turgidum*) being important players, but a very early domestication of legumes is general.

In China, agriculture developed around millet cultivation in the north and rice in the south. Foxtail millet (*Setaria italica*) and proso millet (*Panicum miliaceum*) were being cultivated 10 500 years ago, and rice cultivation had probably begun by 10 000 years ago, being established in the Yangtze valley 8000 years ago. The major legume domesticated in the Far East was soybean (*Glycine max*), but evidence for its domestication is not found until much later than that for the cereals. Written evidence indicates that it was cultivated 3000 years ago during the Chou dynasty: according to Hymowitz (1970), the earliest Chinese character for soybean was 'shu' (see Fig. 7.3) and can be traced back to this period. Soybean was probably domesticated during the preceding Shang Dynasty (*c*.1500–1027 BC). There is, however, still some uncertainty about exactly where this domestication occurred. The centre of soybean genetic diversity provides one indication, and lies in a zone stretching from south-west to north-east China, connecting two important centres of early agriculture – the Yellow River and Yangtze River valleys.

This brief outline indicates that legumes were an important early domesticated crop associated with the development of settled agriculture around the world. Usually secondary to cereals, they are nevertheless an essential component. In the next section, we explore the role of legumes in early civilizations. Essentially, this concerns the methods by which humans reconciled the nutritional (and general agricultural) value of legumes with their unwanted, often toxic properties.

Fig. 7.3 *The earliest domesticated legumes from* **(a)** *the Americas (Phaseolus vulgaris),* **(b)** *the Near East (Lens culinaris) and* **(c)** *the Far East (Glycine max).* **(d)** *Earliest Chinese character for soybean ('shu') according to Hymowitz (1970), from which this is reproduced. The vertical lines are said to represent stem and root, separated by the horizontal line (earth); the three lines below the root are said to represent root nodules. See also Qiu and Chang (2010)*

ANCIENT LEGUME CULTIVATION AND TOXIN AVOIDANCE

In general, legumes have protein-rich seeds, an adaptation to help them germinate and establish quickly. The cotyledons absorb the nutritive endosperm early on in seed development, so that, by maturity, the cotyledons fill the seed and are packed with protein and, in some cases, lipids. Naturally enough, animals seek out these highly nutritious seeds, so legumes often synthesize defence compounds to deter them. For example, in wild-type and tropical varieties of Lima beans (*Phaseolus lunatus*), bruising liberates hydrogen cyanide from a glucoside precursor, linamarin, due to the action of a glucosidase. Beans also contain varying levels (according to species and variety) of haemagglutinins, lectins which are highly toxic but which can be inactivated by soaking followed by boiling. Cases of poisoning by red kidney beans (*P. vulgaris*), attributable to haemagglutinins, have been reported quite regularly in the UK. These are usually due to inadequate care in preparation of raw beans.[3]

Many legumes contain significant levels of inhibitors which, because they reduce the activity of the digestive enzyme trypsin, lower the nutritional value of their protein. Soybean trypsin inhibitor is the most prominent of these inhibitors. Another potentially toxic legume is fava bean (*Vicia faba*), which can cause anaemia (known as favism) in susceptible individuals, due to the action of the glucosides vicine and convicine. Such individuals are deficient in glucose 6-phosphate dehydrogenase, a deficiency which makes their red blood cells susceptible to these compounds. Favism is a genetic disorder that can also confer tolerance to the malarial parasite; hence its frequency in Mediterranean areas where malaria has been prevalent (Chapter 12). Lathyrism is a serious, permanent paralysis of the lower limbs which can occur following prolonged consumption of the grasspea (or chickling vetch), *Lathryus sativus*. Although not encountered much in the West, this species can be important famine food in parts of the developing world. Finally, probably the most notorious problem associated with legume consumption is flatulence. This is caused by cell wall carbohydrates and oligosaccharides that are indigestible to human enzymes, but which are metabolized by bacteria in the lower intestine, generating gas (hydrogen, methane, carbon dioxide) in the process. Not all legumes cause this problem, however; groundnuts and mung beans, for instance, do not.

In spite of these negative qualities, legumes were of great importance in the ancient world. The agricultural benefits of legumes were appreciated by Pliny the Elder (AD 23–79), who in his *Historia Naturalis* repeatedly emphasizes the benefits of growing legumes for the soil, praising vetch, beans, lentils and particularly lupins:

> We have stated that fields and vineyards are enriched by a crop of lupines; and thus it has so little need for manure that it serves instead of manure of the best quality ....[4]

But probably the greatest historical discussion of legumes in the Old World centres around beans. The bean of antiquity was *Vicia faba*; *Phaseolus* beans did not arrive until discovery of the New World. Most present-day production of *Vicia faba* now occurs in China, but the species has a Near Eastern origin. Its remains are to be found in archaeological sites all around the Mediterranean by 2000 BC, indicating that it was extensively cultivated in the region. By the time of the Romans, 'fat bacon and a steaming mess of beans and barley' were the recommended menu for 1 June, simple food in honour of the old-fashioned goddess Carna.[5] These fava beans of the Greek and Roman era were quite small, like a large pea; and probably black, in contrast to the pale, large flatter-seeded *Vicia faba* of the present day.

One way the unwanted properties of fava beans were avoided was simply to ban them. The meaning of such bans is, however, controversial:

> From classical antiquity to the present, Pythagoras' rule that his followers must 'abstain from beans'—not only from eating them, but even from walking through fields where they were growing—has been a source of both popular and scholarly wonder. Humanistic scholarship,

[3]See the survey by Rodhouse *et al.* (1990).

[4]Pliny, *Historia Naturalis* Book XVIII, XXXVI, 133–135.

[5]Nagle, *Ovid's Fasti* 6, 169–182.

with free association as its main guide, has offered explanations that range from the mildly ridiculous to the extremely ridiculous.[6]

This comment from Brumbaugh and Schwartz (1980) refers to the ban on consumption of beans attributed to Pythagoras, who established a Greek sect or brotherhood around 530 BC at Croton, in southern Italy. This was a philosophical and religious society, at least one section of which was focused on the execution of rituals based around the sayings of Pythagoras, sayings that were designed to protect the individual from the dangers of the magical world. The bean ban appears to have related to a belief in the sacredness of all living things and to prohibitions on the consumption of flesh.

Brumbaugh and Schwartz, however, believe that humanist scholars have over-emphasized such symbolic reasons for the bean ban, and argue that in fact the ban was simply a common-sense injunction to protect the inhabitants of the region from favism (see above). Yet the analysis by Simoons[7] assesses the grounds for connecting the Pythagorean bean ban to favism, and finds little evidence that favism existed in the region at the time of Pythagoras. In addition, some elements of the accounts of Pythagoras and his cult are probably not well-founded: for instance, that they would not walk through fields of beans. Perhaps most significant for the current context, bean bans were widespread in the ancient world, extending well away from areas where favism is presently found. An example is the taboo on the urd bean (*Vigna mungo*) in India. It is therefore most likely, despite the current enthusiasm for an explanation based on modern understanding of favism, that the prohibition on beans was based on a superstitious association of them with both death and life. Death: because of the black spot on the flower, and the hollow stems of the plant, which were believed to provide a path for the soul to and from the underworld; life, because of associations with fertility. The discomfort that could be caused by consumption of beans (flatulence, bad dreams) was probably also a factor in rendering them subject to superstitious belief. Perhaps we should leave the final word on the matter to Pliny the Elder:

> Moreover in ancient ritual bean pottage has a sanctity of its own in sacrifice to the gods. It occupies a high place as a delicacy for the table, but it was thought to have a dulling effect on the senses, and also to cause sleeplessness, and it was under a ban with the Pythagorean system on that account—or, as others have reported, because the souls of the dead are contained in a bean, and at all events it is for that reason that beans are employed in memorial sacrifices to dead relatives.[8]

In the New World, archaeological evidence from Mexico and Peru indicates that four main species of *Phaseolus* bean were cultivated—tepary, runner, lima (or butter) and common bean (Fig. 7.4 and Table 7.1). These were typically grown as part of the *milpa* system of co-cultivation of beans, maize and squash. The lysine-rich bean complemented maize, which if treated with lime (as during the traditional production of tortillas), together represented a balanced basic diet which could be supplemented with carbohydrates, vitamins and protein from fruits, vegetables and meat as available. Selection of *Phaseolus* species for co-cultivation meant maintenance of their natural climbing habit, somewhat reduced to allow them to twine around the vertical maize stems. In this system, the indeterminate habit of the beans gives a prolonged flowering and fruiting period, providing a sustained period of cropping. In more modern,

[6]Brumbaugh and Schwartz (1980), p. 421.

[7]Simoons (1998).

[8]Pliny, *Historia Naturalis* Book XVIII, xxx, 118–120.

  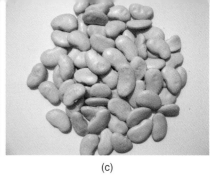

(a)   (b)   (c)

Fig. 7.4 *New World beans:*
(a) Phaseolus vulgaris (*common bean; haricot, pinto and red kidney are shown*); **(b)** P. coccineus (*runner bean*); **(c)** P. lunatus (*Lima bean*)
[*Courtesy of Cathy Newell Price.*]

intensive systems the dwarf, determinate habit is favoured, providing a single (potentially mechanized) harvest of a bushy plant.

One of the most impressive developments of the *milpa* approach is the system of swamp gardens, or *chinampas*, employed by the Aztecs over several centuries before the arrival of the Spanish in 1521. These were raised beds of relatively small dimension (about 3 m wide), but great overall extent, reclaimed from the swamps and lakes in the valley of Mexico. They provided maize, beans, amaranth and other crops for the Aztec capital Tenochtitlan, which was built on an island in Lake Tezcoco. The city's population is estimated to have been 300 000, and the *chinampas* played a crucial role in ensuring its food supply. The plants were raised as seedlings on floating nurseries which could be towed from one *chinampa* to the next, and once transplanted grew in earth taken from the surrounding land supplemented with rich silt dredged from the lake bottom. Crucially, this system meant the plants were provided with a consistent water supply. The extent of *chinampa* horticulture was greatly increased following the famine in the region which resulted from four years of successive crop failures between 1450 and 1454. By the early 1500s there were around 9000 hectares of *chinampas* capable of supplying seven crops a year. Their productivity is one of the reasons for believing that the cannibalism for which the Aztecs are so renowned had a ritual and religious significance, rather than a nutritional one.

We have seen that the story of soybean cultivation began in China. During the Chou Dynasty (1027–221 BC) soybean spread across North China to become a major food crop there, alongside millet. By the end of the Han Dynasty (AD 220) it had reached Korea, and by the fifteenth century had become established in Japan and south-east Asia. During its spread, Chinese farmers selected genotypes suitable to their environments and uses; because soybean is very sensitive to photoperiod, this particularly involved selection of types with maturation dates appropriate to latitude. This process of farmer selection created a large reservoir of genetic diversity, so that by the beginning of the twentieth century there were an estimated 20 000 soybean landraces in China.

The exceptional qualities of the soybean were clear to the ancient Chinese: its high protein content (39–45%) made it a much more economical use of land than meat production. This was particularly important in China, where good pasture and arable land were at a premium; and where Buddhism, with its vegetarian focus, encouraged the production of meat substitutes. Soy protein is also very complete, being rich in the important amino acids. The high oil content (20–23%) of the soybean gives further uses, for example in cooking and soap-making; and the nitrogen-fixing capacity of the root system made the plant central to Chinese crop rotation systems. The pictographic symbol 'shu' (Fig. 7.3) emphasizes the root nodules, rather than the fruit and seed, suggesting that early in its cultivation soybean may have been valued for its effects on soil fertility, rather than as food. This is readily understandable because the mature, unprocessed beans are bitter and cause serious flatulence, as well as containing trypsin inhibitors (see earlier).

Therefore, the Chinese, Japanese and other south-east Asian cultures developed processing techniques in order to exploit the soybean as a food. Some of these are soy milk (an aqueous extract of soybeans); tofu (curdled soy milk); yuba (skin formed on boiled soy milk); edamame (immature soybeans cooked in the pod); and, after fermentation, miso, soy sauce and tempeh. The central importance of just one of these soy foods, tofu, is captured by the words of the Spanish missionary Friar Domingo Navarrete, who in 1665 described it as follows (for Kidney-Beans read soybean):

> It is call'd Teu Fu, that is Paste of Kidney Beans. I did not see how they made it. They drew the Milk out of the Kidney-Beans, and turning it, make great Cakes of it like Cheeses … All the Mass is as white as the very Snow … It is eaten raw, but generally boil'd and dress'd with Herbs, Fish, and other things … It is incredible what vast quantities of it are consum'd in China, and very hard to conceive that there should be such an abundance of Kidney-Beans. That Chinese who has Teu Fu, Herbs and Rice, needs no other Sustenance to work … Teu Fu is one of the most remarkable things in China … .[9]

Yet, although soybean was known to European explorers and its qualities as a food plant were described by Engelbert Kaempfer in 1712, it remained unexploited in the West until the twentieth century. In the next section we discuss the final, most dramatic phase of its development, to achieve global significance not as a human food, but as an industrial crop, yielding oil and animal feed.

'POOR MAN'S MEAT' TO GLOBAL CASH CROP

The rather mixed reputation of legumes in Europe and the Near East is expressed in the biblical account of Esau and Jacob, in which Esau exchanges his birthright for a 'mess of pottage' – a red lentil soup.[10] The implication is that Esau has exchanged something of great value for food which, although sustaining, is of relatively little worth. This reputation is captured in the phrase 'poor man's meat' coined for legumes: that is, a meat substitute (suggesting a recognition of their high protein value), but one associated with poverty. This association has remained to the present day, so that poorer people tend to consume more legumes and less meat, while the rich do the opposite.

Nevertheless, the legumes have been moved around the globe and become established as crops of major local, regional or national importance. Portuguese slave traders are believed to have taken *Phaseolus vulgaris* beans to eastern Africa, where their diversity has been valued and perpetuated. In Malawi, for instance, they are maintained as genotypic mixtures so that dwarf and climbing types, with a range of seed shapes, patterns and colours, germination rates, and flower colours are cultivated together. This variability probably makes the crop more resilient to unpredictable climatic events (such as time of onset of the first rains), and to pests and diseases. Locally, particular combinations of bean shapes and colours have come to be characteristic, suggesting that there may also be cultural, aesthetic reasons for maintaining this diversity: the farmers may value it for its own sake (Fig. 7.5).

[9]Cummins (1962) pp. 195–196.

[10]*Genesis* 25: 29–34.

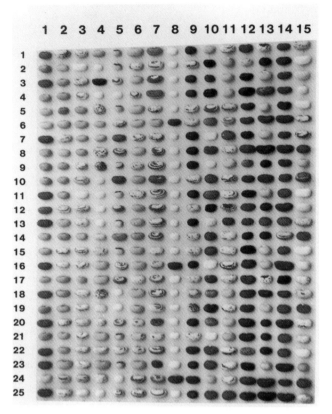

```
    1  2  3  4  5  6  7  8  9 10 11 12 13 14 15
 1
 2
 3
 4
 5
 6
 7
 8
 9
10
11
12
13
14
15
16
17
18
19
20
21
22
23
24
25
```

Fig. 7.5 Phaseolus vulgaris *seed diversity in Malawi [Martin, G.B. & Adams, M.W. (1987). Landraces of Phaseolus vulgaris (Fabaceae) in Northern Malawi. I. Regional variation. Economic Botany* **41**, *190–203 © Society for Economic Botany.]*

A further remarkable feature of the genus *Phaseolus* is its wide latitudinal distribution: it can be grown from cool temperate climates to tropical regions. Genotypes from its region of origin in Central/South America are strictly photoperiodic, and only complete their life-cycle in relatively short days. But as the genus has been cultivated further from the equator, photoperiod-insensitive lines have evolved or been bred. In this respect, *Phaseolus* is similar to soybean (see below) but markedly different from species of *Vigna*, which have a much narrower range, being found mainly in tropical and subtropical climates from Africa to the Far East. *Vigna unguiculata*, the cowpea, is also cultivated in warmer regions of the USA, where it is known as the black-eyed pea, and illustrates another aspect of the exploitability of genetic diversity in legumes. It was domesticated in Africa, where it is consumed mainly as a bean pulse and has a trailing growth habit (Fig. 7.6a). It was probably moved to Asia about 2000 years ago, where the need was for a pod which could be consumed as a green vegetable, and selection for this character has produced, in the most extreme case, the 'yard-long bean' of the Far East: this is the *sesquipedalis* cultivar which produces beans up to a metre in length. Because of the size of the pod, a climbing habit is needed and this has been co-selected (Fig. 7.6b). In India, on the other hand, the need was for a fodder crop, so selection has produced a bush type with short erect branches (Fig. 7.6c).

From its origins as a crop of great regional significance in the Far East, but one which was almost unheard of in the West, the soybean has during the twentieth century taken on major international importance. Its value as an oil and processed feed crop meant that in the US alone the production area increased from 700 000 hectares in 1924 to over 30 million in 2008. Annual production there now stands at over 70 million tonnes, and the value of soybeans is second only to maize. Brazil and Argentina are close to rivalling the US (each producing around 50 million tonnes annually); other major producers are India, China and Canada, but they lag well behind. The latitudinal range of these countries shows that, as with *Phaseolus* beans, there exists in soybean the genetic diversity to allow production from the tropics to cool temperate regions. This is manifested by the wide range of photoperiod sensitivity, allowing the breeding and selection of cultivars which will flower and mature at the required latitude. The fact that there are over 100 (commercial and public) soybean breeders in North America shows the scale of the business opportunity now represented by the crop. The diversity of its potential uses is illustrated by the soybean car developed by Henry Ford in 1941. The plastic paneling of the car was derived from soybean, and this experimental project was designed to reduce metal usage and exploit the products of agriculture for industry.

NITROGEN FIXATION: THE BENEFITS OF COOPERATION

We come now to the discovery of biological nitrogen fixation, and our current understanding of the process. This is a special part of the story of legumes and humans, first because nitrogen fixation is a profoundly important biological process; and second because its proper understanding is of such potential significance for our sustainable future on Earth. The essence of it is that legumes, by creating the appropriate environment within their roots, allow bacteria to catalyse the conversion of inert atmospheric nitrogen to the ammonium ion. This is then made available to the plant in exchange for carbon, and supplies the nitrogen essential for amino acids in proteins, purines and pyrimidines in nucleic acids, and alkaloids, polyamines and other secondary compounds. When the plant dies the root system, in particular, enriches the soil with nitrogen.

(a)

There are alternative routes by which nitrogen can be reduced or oxidized into forms that living organisms can use: there are nitrogen-fixing bacteria that are free-living; there are actinomycete bacteria of the genus *Frankia* that associate symbiotically with non-legumes such as alder (*Alnus*) and *Caeanothus*; there are even animals like termites that can accommodate nitrogen-fixing bacteria. Nitrogen can be made available non-biologically by lightning, which creates nitrogen oxides that reach the soil as rain; and by the combination of nitrogen and hydrogen at high temperature and pressure in the Haber process, the route by which inorganic fertilizer is made.

(b)

(c)

Fig. 7.6 Vigna unguiculata (*the cowpea*): **(a)** *prostrate;* **(b)** *climbing— 'yard-long bean';* **(c)** *bush type* [*From Smartt, J. (1990). Grain Legumes: Evolution and Genetic Resources © Cambridge University Press, Cambridge.*]

Yet the legume root nodule is a vitally important natural route for nitrogen assimilation, and because legumes are, as we have seen, valuable crops in their own right, the root nodule features centrally in agriculture and in the future provision of food. Nitrogen fixation by this method uses no fossil fuel for energy and hydrogen, as the Haber process does. Neither does it cause problems with run-off and cost as do artificial nitrogen fertilisers. The effective exploitation of legume nitrogen fixation can be achieved through judicious use of crop rotations; while, in the long term, better understanding of the process will enable legume productivity to be improved, and may suggest ways by which other crops can be enabled to fix their own nitrogen.

The beneficial effect of legumes on soil fertility was known to the ancient Chinese and to the Romans, but the demonstration that this arose because legumes could somehow gather atmospheric nitrogen had to wait until the end of the nineteenth century. Early stirrings of understanding may have been shown by Sir Humphry Davy, a pioneer in agricultural chemistry, who commented in 1813 that peas and beans contain albumen (protein), and that the nitrogen in this seemed to be derived from the atmosphere. However, it is most likely that Davy believed that the ultimate source of plant nitrogen is ammonia in the atmosphere.

This was certainly the view of Justus von Liebig (1803–73), who felt that nitrogen gas was just too unreactive to be used biologically. Liebig, because of his eminence in nineteenth-century chemistry, was able to insist that all plant nitrogen is derived from ammonia, and that 'the crops of field diminish or increase in exact proportion to the diminution or increase of the mineral substances conveyed to it in manure'.[11] He envisaged a form of nitrogen cycling, but ammonia was at its heart, moving from the atmosphere into soil through rainwater, into plants and thence animals, and back to the soil by decomposition. He maintained this position even in the face of evidence from Jean-Baptiste Boussingault (1802–87) that legume crops (e.g. clover) could gain nitrogen in excess of that provided by manures and that this became available to other non-legume crops (wheat, turnips, oats and potatoes) via the process of crop rotation.

For a long time Liebig prevailed, but experiments conducted by Gilbert and Lawes in the field at Rothamsted, England, during the 1840s and 50s also pointed to the ability of clover, in particular, to assimilate nitrogen. When Gilbert and Lawes grew their plants under completely artificial conditions, however, in which no contamination (including that by bacteria) was possible, this property of legumes disappeared. It was not until 1886 that the conundrum was resolved, when Hellriegel and Wilfarth showed that it was the nodules that were the key, that these needed a soil extract to develop, and that when they did the plants showed vigorous growth in the absence of nitrogen fertilizer. They went on to show that there was some specificity in the relationship between soil extract and legume and that the nodule-inducing agent was inactivated by heat. This was consistent with the demonstration that the nodules contained bacteria responsible for nitrogen fixation, and that these rhizobia show some selectivity in the legumes with which they associate.

It is now over 120 years since the secret of the soil-enriching property of legumes was revealed, and in that time a detailed understanding of the rhizobium–legume symbiosis has been achieved. The rhizobia turn out to be a taxonomically very diverse group, with different genera and species showing varying degrees of specificity in their legume associations. Thus, *Rhizobium* nodulates peas, chickpea and *Phaseolus* beans, while the quite distantly related *Sinorhizobium* nodulates soybean; but soybean can also be nodulated by another genus, *Bradyrhizobium*. Such diversity on the bacterial side probably reflects horizontal gene transfer (via a plasmid which can transfer itself from one bacterial genus to another, unrelated one) of the key genes needed to allow nodulation of the host legume. These genes, known as *nod* genes, allow the bacterium to sense legume species-specific flavonoids secreted by the roots, recognition which in turn causes other *nod* genes to be activated whose products (known as nod-factors) induce the root cells to divide and form the nodule. They also allow the process of infection to take place via root hair cells (Fig. 7.7). Once the bacteria have successfully invaded the plant cells, they differentiate into bacteroids which are active in nitrogen fixation and repressed in their ability to assimilate ammonium. These bacteroids are contained in a structure (termed the symbiosome) by a membrane with specialized transport proteins which move ammonium out to the surrounding plant cell where it is converted to amides or ureides for further transport. These transporters also move carbon (in the form of acids) into the symbiosome to support the metabolism of the bacteroids.

The key rhizobial genes for nitrogen fixation are the *nif* genes, encoding the nitrogenase enzyme. This is a complex of two proteins which operates as shown

[11]Nutman (1987).

in Fig. 7.8. The nitrogenase enzyme is irreversibly inactivated by oxygen, so the plant cells within the nodule synthesize leghaemoglobin, which is related to haemoglobin and imparts a pink colour to the nodules (Fig. 7.9). Leghaemoglobin binds oxygen, maintaining it at a sufficiently low level to allow nitrogen fixation. In a sense this is the crucial contribution of the legume; but it is also notable that the nitrogen-fixing reaction is very costly in terms of energy (16 ATP used for every molecule of nitrogen gas reduced), so the process is not a 'free lunch'. The legume must generate extra photosynthate to drive its nitrogen fixing factory, and this is one reason why yields can be erratic and cultivation can require considerable expertise. The need to establish a good symbiosis (by supplying the appropriate rhizobial strain), the stage of plant development when nodules form, the nitrogen status of the soil and other abiotic factors, can all influence the effectiveness of legume cultivation and nitrogen fixation. Thus, although the opportunity offered to sustainable agriculture by legumes is beyond doubt, there is an accompanying agronomic challenge posed by the need to cultivate both the bacterium and the plant efficiently.

We have described in this chapter how legumes have been a nutritionally important group of plants associated with humans since the development of

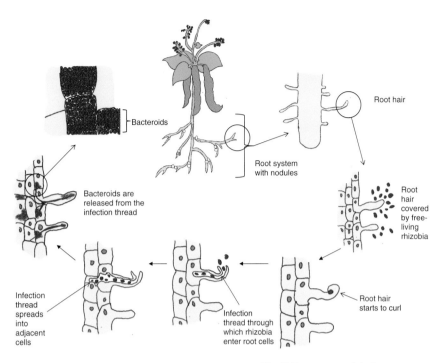

Fig. 7.7   *Invasion of the legume root by rhizobia [Redrawn by Henry Battey from an illustration at www. microbiologyonline.org.uk.]*

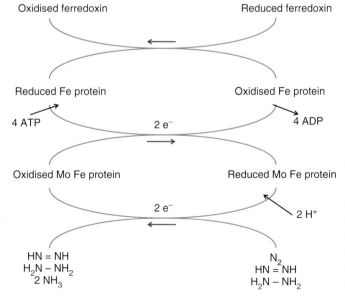

Fig. 7.8   *Biological nitrogen fixation. The reaction is summarized as follows:*

$$N_2 + 8H^+ + 8e^- + 16ATP = 2NH_3 + H_2 + 16ADP + 16Pi.$$

It is performed by the nitrogenase enzyme, a complex of two proteins. The Fe protein is reduced by electrons donated by ferredoxin, binds ATP, and in turn reduces the molybdenum–iron protein. This donates electrons to nitrogen producing HN = NH. Two further cycles result in the synthesis of ammonia. Hydrogen is a by-product of the reaction and could have potential for use as a fuel

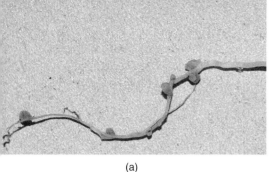
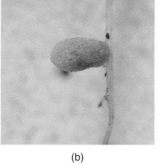

(a)   (b)

Fig. 7.9 **(a)** *Part of a wild vetch root system with nodules of Rhizobium.* **(b)** *Detail of a single nodule. Each nodule is about 2–3 mm long. The pink tinge is due to the presence of leghaemoglobin*

settled agriculture. Their benefits extend to soil nutrition and indirectly to the nutrition of other crops. The significance of this role cannot be overstated. To give one example, a review by Köpke and Nemecek (2010) of the ecological services offered by faba bean concludes that among its key environmental benefits are: biological nitrogen fixation, benefiting the whole crop rotation; high-quality protein provision; positive pre-crop effects (nitrogen and other effects); increased biodiversity following from increased agroecosystem diversity; provision of feed for pollinators and beneficial insects; mobilization of soil nutrients other than nitrogen; reduced tillage options; and reduction in greenhouse gas emissions and energy demand when introduced into cereal-based intensive crop rotations. In general, it is the potential for introduction into crop rotations (which often just use cereals and brassica crops) that is perhaps most striking.

Could cereals, whose global production dwarfs that of legumes, be made to fix their own nitrogen? The contribution to the goals of sustainable agriculture would be enormous: over 40 million tonnes of fertilizer is applied annually to wheat, rice and maize; 25% of this is lost, contributing to environmental pollution through eutrophication of water systems and damage to the ozone layer. The engineering of nitrogen-fixing cereals would markedly reduce this fertilizer use, although potassium, phosphorus and micronutrients would still need to be supplied. The complexity of the nitrogen-fixing process and bacterial association in legumes probably makes its transfer to cereals unrealistic. It may, however, be possible to genetically alter soil nitrogen-fixing bacteria that already associate with cereals—such as *Azospirillum* does with rice and wheat—so that they contribute directly to the plant nitrogen economy. The benefits of this cooperation would work with the more immediate practical opportunities arising from enhanced production of leguminous crops to address the pressing issues of global food security.

FURTHER READING

The origins of agriculture in the Near East are usefully discussed by Harlan (1992), Evans (1998) and Murphy (2007). Zohary and Hopf (1993) deal in detail with the domestication of Old World crops. Chapters in Harris and Hillman (1989) give a good overview of the subject, but these need to be complemented by more up-to-date information. The sequence of events, from gathering of wild species; through their cultivation, including collection and sowing of seed; and on to full-blown domestication of crops that cannot survive without the aid of humans, has been the subject of much debate and change of emphasis. The account given by Hillman (2003) describes how an understanding of the Abu Hureyra site developed and illustrates how views change as new evidence comes to light. The summary by Balter (2010) shows how the debate continues unabated. Smith (2001) offers an accessible summary for the New World, while the same author tackles both New and Old World agriculture in Smith (1995); a comprehensive, global overview of the subject is given by Barker (2006). For the development of agriculture and soybean domestication in the Far East, see Hymowitz and Newell (1981), Carter *et al.* (2004), Murphy (2007), Bettinger *et al.* (2007) and

Qiu and Chang (2010). For *Phaseolus* domestication, see Gepts (1998) and Kwak *et al.* (2009); and for beans in the context of Mesoamerican *milpa* farming, Smartt (1990); the Aztec *chinampas* are discussed by Armillas (1971), Ortiz de Montellano (1978) and Smith (1998). The origin and development of crops in general are covered by Hancock (1992). Mithen (2002) gives an inspired account of the development of human civilization since the last glaciation.

An early but still useful survey of legumes in human nutrition is given by Doughty and Walker (1982); a more recent review focusing on the nutritional benefits of grain legumes is provided by Tharanathan and Mahadevamma (2003). For an interesting discussion of Malawian bean diversity, see Martin and Adams (1987). *Phaseolus*, *Vigna* and *Glycine* genetic resources, including environmental adaptation, are excellently described by Smartt (1990); detailed information on genetic resources of soybean can be found in Carter *et al.* (2004), and of grain legumes in Singh and Jauhar (2005). The availability of the soybean genome sequence (Schmutz *et al.*, 2010) allows key crop production traits such as disease resistance and seed nutritional quality, many of which are quantitatively inherited, to be defined by association genetics. The history of the discovery of nitrogen fixation is described in Nutman (1987), and a useful historical account of nitrogen in agriculture is provided by Leigh (2004). Biological nitrogen fixation is discussed by Chrispeels and Sadava (2003) and Buchanan *et al.* (2000), and an accessible entry to the evolutionary genetics of the legume–rhizobium symbiosis is provided by Brewin (2002). Oldroyd and Downie (2008) give a comprehensive account of signalling during establishment of the symbiosis. Agronomic aspects are discussed by Smartt (1990); a useful account relating mainly to lentil agronomy is given by Quinn (2009). The possibilities associated with nitrogen fixation in non-legumes are discussed by Saikia and Jain (2007). A useful account of the potential of African legumes is given by Sprent *et al.* (2010).

REFERENCES

Armillas, P. (1971) Gardens on swamps. *Science* **174**, 653–661.

Balter, M. (2010) The tangled roots of agriculture. *Science* **327**, 404–406.

Barker, G. (2006) *The Agricultural Revolution in Prehistory: why did foragers become farmers?* Oxford University Press, Oxford.

Bettinger, R.L., Barton, L., Richerson, P.J., Boyd, R., Hui, W. and Won, C. (2007) The transition to agriculture in northwestern China. *Developments in Quaternary Sciences* **9**, 83–101.

Brewin, N.J. (2002) Pods and nods: a new look at symbiotic nitrogen fixation. *Biologist* **49**, 1–5.

Brumbaugh, R. S. and Schwartz, J. (1980) Pythagoras and beans: a medical explanation. *The Classical World* **73**, 421–422.

Buchanan, R.B., Gruissem, W. and Jones, R.L. (2000) *Biochemistry & Molecular Biology of Plants*. ASPP, Rockville, USA.

Carter, T.E., Nelson, R.L., Sneller, C.H. and Cui, Z. (2004). Genetic diversity in soybean. In: *Soybeans: improvement, production, and uses*, 3rd edn (eds H. R. Boerma and J. E. Specht), pp. 303–416. American Society of Agronomy, Madison WI, USA.

Chrispeels, M.J. and Sadava, D.E. (2003) *Plants, Genes and Crop Biotechnology*, 2nd edn. Jones & Bartlett, Sudbury MA, USA.

Cummins, J.S. (1962) *The Travels and Controversies of Friar Domingo Navarette 1618–1686, Volume II*. The Hakluyt Society, 2nd series, no. 119, Cambridge University Press, Cambridge.

Doughty, J. and Walker, A. (1982) *Legumes in Human Nutrition*. FAO, Rome, Italy.

Evans, L.T. (1998) *Feeding the Ten Billion: plants and population growth*. Cambridge University Press, Cambridge.

Gepts, P. (1998) Origin and evolution of common bean: past events and recent trends. *HortScience* **33**, 1124–1130.

Hancock, J.F. (1992) *Plant Evolution and the Origin of Crop Species*. Prentice-Hall, Englewood Cliffs NJ, USA.

Harlan, J.R. (1992). *Crops and Man*, 2nd edn. American Society of Agronomy, Madison WI, USA.

Harris, D.R. and Hillman, G.C. (1989) *Foraging and Farming: the evolution of plant exploitation*. Unwin Hyman, London.

Hillman, G.C. (2003) Investigating the start of cultivation in western Eurasia: studies of plant remains from Abu Hureyra on the Euphrates. In: *The Widening Harvest* (eds A. J. Ammerman and P. Biagi), pp. 75–97. Archaeological Institute of America, Boston, USA.

Hymowitz, T. (1970) On the domestication of the soybean. *Economic Botany* **24**, 408–421.

Hymowitz, T. and Newell, C.A. (1981) Taxonomy of the genus *Glycine*, domestication and uses of soybeans. *Economic Botany* **35**, 272–288.

Kaplan, L. and Lynch, T.F. (1999) *Phaseolus* (Fabaceae) in archaeology: AMS radiocarbon dates and their significance for pre-Colombian agriculture. *Economic Botany* **53**, 261–272.

Köpke, U. and Nemecek, T. (2010) Ecological services of faba bean. *Field Crops Research* **115**, 217–233.

Kwak, M., Kami, J.A. and Gepts, P. (2009) The putative Mesoamerican domestication center of *Phaseolus vulgaris* is located in the Lerma-Santiago basin of Mexico. *Crop Science* **49**, 554–563.

Leigh, G.J. (2004) *The World's Greatest Fix: a history of nitrogen and agriculture*. Oxford University Press, Oxford.

Martin, G.B. and Adams, M.W. (1987) Landraces of *Phaseolus vulgaris* (Fabaceae) in Northern Malawi. I. Regional variation. *Economic Botany* **41**, 190–203.

Mithen, S.J. (2002) *After the Ice*. Weidenfeld & Nicholson, London.

Murphy, D.J. (2007) *People, Plants and Genes*. Oxford University Press, Oxford.

Nagle, B.R. (1995) *Ovid's Fasti: Roman holidays*. Indiana University Press, Indianapolis, USA.

Nutman, P.S. (1987) Centenary lecture. *Philosophical Transactions of the Royal Society of London B* **317**, 69–106.

Oldroyd, G.E.D. and Downie, J.A. (2008) Coordinating nodule morphogenesis with rhizobial infection in legumes. *Annual Review of Plant Biology* **59**, 519–546.

Ortiz de Montellano, B.R. (1978) Aztec cannibalism: an ecological necessity? *Science* **200**, 611–617.

Qiu, L-J. and Chang, R-Z. (2010) The origin and history of soybean. In: *The Soybean: botany, production and uses* (ed. G. Singh), pp. 1–23. CABI, Wallingford, UK.

Quinn, M. A. (2009) Biological nitrogen fixation and soil health improvement. In: *The Lentil: botany, production and uses* (eds W. Erskine, F. J. Muehlbauer, A. Sarker and B. Sharma), pp. 229–247. CABI, Wallingford, UK.

Rodhouse, J. C., Haugh, C. A., Roberts, D. and Gilbert, R. J. (1990) Red kidney bean poisoning in the UK: an analysis of 50 suspected incidents between 1976 and 1989. *Epidemiology and Infection* **105**, 485–491.

Saikia, S. P. and Jain, V. (2007) Biological nitrogen fixation with non-legumes: an achievable target or a dogma? *Current Science* **92**, 317–322.

Schmutz, J., Cannon, S.B., Schlueter, J. *et al.* (2010) Genome sequence of the palaeopolyploid soybean. *Nature* **463**, 178–183.

Simoons, F. J. (1998) *Plants of Life, Plants of Death.* University of Wisconsin Press, Madison WI, USA.

Singh, R. J. and Jauhar, P. P. (eds) (2005) *Genetic Resources, Chromosome Engineering, and Crop Improvement. Volume 1: Grain Legumes.* Taylor & Francis, Boca Raton FL, USA.

Smartt, J. (1990) *Grain Legumes: evolution and genetic resources.* Cambridge University Press, Cambridge.

Smith, B. D. (2001) Documenting plant domestication: the consilience of biological and archaeological approaches. *Proceedings of the National Academy of Sciences of the USA* **98**, 1324–1326.

Smith, B. D. (1995) *The Emergence of Agriculture.* Scientific American Library, New York.

Smith, M. E. (1998) *The Aztecs.* Blackwell, Oxford.

Sprent, J. I., Odee, D. W. and Dakora, F. D. (2010) African legumes: a vital but under-utilized resource. *Journal of Experimental Botany* **61**, 1257–1265.

Tharanathan, R. N. and Mahadevamma, S. (2003) Grain legumes: a boon to human nutrition. *Trends in Food Science & Technology* **14**, 507–518.

Zohary, D. and Hopf, M. (1993) *Domestication of Plants in the Old World,* 2nd edn. Clarendon Press, Oxford.

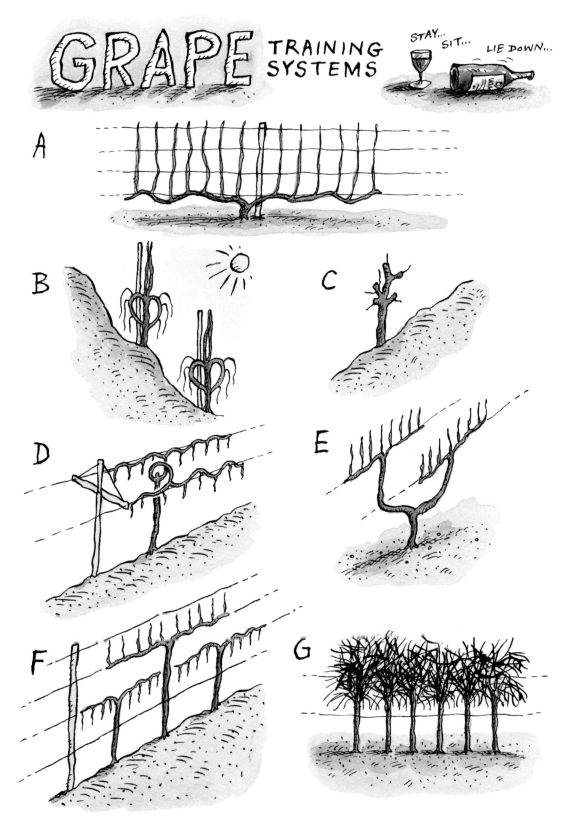

TRAINING SYSTEMS
A - Vertical trellis.   B - Vines on poles.   C - Goblet.
D - Geneva double curtain.   E - Lyre.   F - Scott Henry.   G - Minimal.

The Grapevine

CHAPTER 8

What colour are they now, thy quiet waters?
The evening star has brought the evening light,
And filled the river with the green hillside;
The hill-tops waver in the rippling water,
Trembles the absent vine and swells the grape
In thy clear crystal.

Evening on the Moselle, Ausonius(c.310–395)

THE LATE TWENTIETH CENTURY witnessed a number of revolutions resulting from advances in the ability of mankind to modify and control plants. The most significant was the green revolution of the 1960s and 70s in which the productivity of cereal crops (principally wheat and rice) was dramatically increased following selection for shorter-stemmed, higher-yielding varieties in international breeding programmes. This transformed food production, reducing the proportion of malnourished people in the world, and confounding predictions of widespread famine. A second revolution, which met with less general appreciation than the first, was that based on the development of genetic modification technology in the 1980s and 90s. This allowed existing traits to be altered and new ones, such as herbicide and insect resistance, to be incorporated into the whole range of crop plants. The course of this revolution is still being charted; but, notwithstanding public concerns, it seems certain that genetic modification will be crucial to the major challenge ahead: to feed a world population predicted to reach 9.2 billion by 2050.

There was a third plant-based revolution, one which was more modest and less fundamental than either of the two described above, but which illustrates the potential offered by innovative development of a plant product: wine. The revolution in wine production was led by Australia, California and other New World countries, and relied upon vision, marketing, exploitation of niche opportunity, and the rigorous application of scientific principles to grapevine cultivation (viticulture) and wine production (oenology). The trends in grape and wine production world-wide are summarized in Box 8.1, the key point being the growth in prominence of New World wines over the last 20–30

Biological Diversity: Exploiters and Exploited, First Edition.   Paul Hatcher and Nick Battey.
© 2011 John Wiley & Sons, Ltd. Published 2011 by John Wiley & Sons, Ltd.

| BOX 8.1 | FACTS AND FIGURES ABOUT GRAPES AND WINE |

In 2007, 71% of grape production was for wine, 27% for table grapes, and 2% for dried use (raisins, sultanas, currants). These proportions have been fairly constant in the recent past, with a slight increase in table grapes, principally because of the recent expansion of China, and to a lesser extent the USA, in this market. The major producers of grapes for wine are as they have always been: the Old World European countries of Italy, Spain and France, each producing 35–50 thousand hectolitres per annum. New World producers (USA, Argentina, Australia) are significant but well below these three, each producing 10–20 thousand hectolitres per annum. Wine consumption has traditionally been dominated by the major producers, but in the last ten years there has been a very marked increase in consumption in the US and the UK. Most striking is the change in market share of traditional EU exporters, compared with that of the New World (southern hemisphere and the US) (Fig. 1). Since 1986 the New World share has increased about ten-fold, at the expense of the Old World; this trend has been roughly matched by changes in the planted area of vines. Most impressive of all is the trend for grape production for wine in Australia over the last 20 years (Fig. 2).

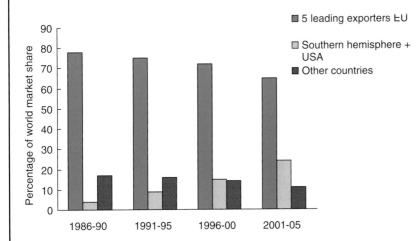

Box 8.1 Fig. 1 *Share of the world market for wine exports* [From 'State of the vitiviniculture world report in 2007' by the Director General of OIV–presentation at http://news.reseau-concept. net.]

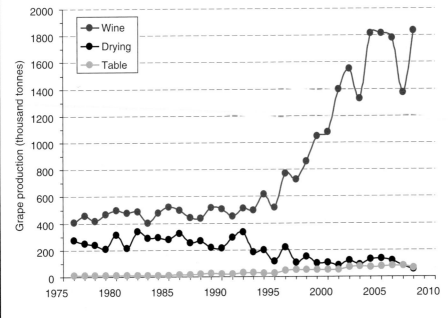

Box 8.1 Fig. 2 *Grape production in Australia, 1976–2008* [Redrawn from Iland, 2004, with additional data from the Australian Bureau of Statistics, Vineyards Estimates–see www.abs.gov.au.]

years. Here is what the *Oxford Companion to Wine* has to say about one of the key players, Australia:

> [In 1983–4, Australian exports] were 8.9 million litres, worth $9.6 million. Australia simply did not rate a mention in the list of world wine exporters. By 2003–4, its exports were 575 million litres worth $2.55 billion dollars, placing it behind only France, Italy, and Spain in value of wine exports. It is not too fanciful to suggest that the wines have an openness, a confident, user-friendly style which reflects the national character (and climate).[1]

In this chapter we discuss the biology of the grapevine and illustrate the range of factors that must be managed to allow its successful exploitation.

It is its combination of contrasting characters which makes the grapevine such a striking plant (Fig. 8.1a). If allowed, it will straggle and smother, weed-like, over whatever offers it support, yet it produces bunches of luscious fruits to attract bird and beast. At ambient temperatures it will grow more-or-less continuously, and rapidly: 2–3 cm per day at its maximum means shoot growth can be actually observed in real time by the patient onlooker. This is in contrast to many other perennials, such as the apple, which grow more sedately, producing distinct flushes of leaves. The grapevine is warmth-loving, deep-rooting and drought tolerant, often being content on relatively thin, poor soils; yet it is a pretty typical temperate-zone perennial in its deciduous habit and its requirement for a period of dormancy, broken by cold winter temperatures, for synchronous flowering and fruiting over the summer season.

As a crop plant, the grapevine offers significant challenges to the grower. Its perennial cycle must be managed, regulating vegetative growth and flowering so as to optimize quantity and quality of fruit. The fact that the optimum fruit character is expressed only after the extensive processing that constitutes wine production adds a further complication. Vines can remain productive for decades: most commercial production is from vines up to 50 years old, and some of the most valued wines are derived from vines much older than this. An example is the Shiraz vines of Langmeil winery in the Barossa Valley, South Australia, which were planted in 1843. The contributions of vine age, soil type, training, irrigation, and especially the weather, lead to local and yearly

Fig. 8.1   *Grapevine, Vitis vinifera. Vigorous growth* **(a)** *in summer and* **(b, c)** *after pruning in winter. Spur pruning* **(b)** *involves cutting back growth made during the previous season to short spurs on a permanent framework of branches (known as 'cordons'). It is usually associated with warmer climates in which the fruitfulness of the buds at the base of shoots (those buds which remain after spur pruning) is not limiting. Cane pruning* **(c)** *is a type of replacement pruning, in which selected shoots (or shoot—in this example, the Single Guyot system, there is only one) which grew during the current season act as the basis for next year's growth. The replacement shoots are tied down from the vertical to the horizontal during winter pruning. In both types of pruning, the quantity of vegetative material removed is immense—up to 90% of the plant every year [(a) courtesy of Cathy Newell Price; (b) and (c) courtesy of Duncan McNeill (Farm Advisory Services Team, Kent, UK).]*

(a)   (b)   (c)

[1]Robinson (2006), p. 42.

variations in wine quality that can be a prize for the connoisseur or a headache for the supermarket looking for consistency of product.

The grapevine is in the Vitaceae, a family that is widely distributed, its members typically being woody shrubs or climbers, including Virginia creeper (*Parthenocissus quinquefolia*). There are two genera within the Vitaceae valued commercially for their grapes: *Muscadinia* and *Vitis*. The muscadine grape (*Muscadinia rotundifolia*) is important locally in south-eastern USA as fresh fruit, in preserves, and as wine. The musky flavour and thick skins make the fruit something of an acquired taste, but the widespread cultivation of muscadine grapes in the region reflects good resistance to local diseases and pests. 'Scuppernong' is the oldest cultivar and is bronze-skinned; 'Fry' is historically interesting because it was used to make Virginia Dare, a popular wine in the US before Prohibition in 1919.

Within the genus *Vitis*, the Eurasian species *V. vinifera* is the grapevine of modern commerce. Species of North American origin are of great significance as rootstocks, which provide pest control (see later); but in general their fruit have been found to be unsatisfactory, due to small size, pronounced seediness and strong flavours. However, in some cases exotic flavours, strawberry- or pineapple-like, can create a more interesting fresh juice than that of *V. vinifera*. The most important North American species grown for its grapes is *V. labrusca* 'Concord', and this supports a significant industry in the eastern states of the US. The fruit, however, contain high levels of methyl anthranilate which, along with other volatiles, imparts a 'foxy' aroma usually considered unacceptable in wine. The East Asian species *V. amurensis* has been hybridized with *V. vinifera* to create cultivars such as 'Rondo', which is dark-skinned but matures early and is resistant to frost, and can therefore be successfully cultivated in northern Europe where dark-skinned *V. vinifera* cultivars often fail to ripen effectively.

Domestication of *V. vinifera* has involved the selection of hermaphrodite types, the wild grapevine typically being dioecious (bearing male and female flowers on separate plants). Genotypes with larger, sweeter fruits suitable for fresh use, drying, or wine production have also been selected. The wild grapevine (*V. vinifera* ssp. *sylvestris*) co-exists with the domesticated grapevine (*V. vinifera* ssp. *sativa*) over its range from Europe to central Asia, and this has led to a debate over the way in which domestication occurred. One hypothesis is that it happened once, in mountainous Transcaucasia, between the Black Sea and the Caspian Sea. This is consistent with archaeological evidence for the origin of large-scale wine making in this area, and in the neighbouring Zagros Mountains (western Iran) and Taurus Mountains (Turkey) during the Neolithic period (Fig. 8.2). The second hypothesis, of independent domestication events over the range of the wild grapevine, is supported by the existence of morphological groups within domesticated *V. vinifera* ssp. *sativa*: the *occidentalis* group from western Europe; the *pontica* group from eastern Europe; and the *orientalis* group from western and central Asia (illustrated in Fig. 8.2). Analysis of chloroplast DNA polymorphisms also suggests independent domestication events corresponding to these broad groupings. Further research may reveal more such events, as wild genotypes suited to local conditions were brought into cultivation over the last 8000 years.

DNA analysis has helped to resolve longstanding questions about the parentage of important modern cultivars. One example is Chardonnay, a distinguished white wine cultivar, which turns out to have originated from a cross between the equally reputable Pinot Noir, and the decidedly ignoble Gouais Blanc, which is of such poor quality that its planting has been banned in France on more than one

Fig. 8.2   *Domestication of the grapevine. Large-scale wine making began in the region delimited by Transcaucasia, the Taurus and Zagros Mountains. Also shown are the approximate areas where the occidentalis, pontica and orientalis groups are believed to have originated (purple grape symbols). Brown colour: approximate position of land over 2000 m [Based on Unwin (1991).]*

occasion. Equally striking is the range of significant modern cultivars that have been derived from these two parents by different crosses at different times, including Auxerrois Blanc (used in Pinot Blanc), Gamay Noir (the Beaujolais grape) and Melon de Bourgogne (Muscadet).

THE GRAPEVINE PLANT

> Of all the plants which cover the surface of our globe, there is, per-haps, none more sensible of the action of the numerous causes which influence vegetation, than the vine. Not only do we see it varied under different climates, but even in the same climate, we see its products changed in the most astonishing manner, in consequence of a dif-ference in the nature of the soil, the exposure of the vineyard, or the system of cultivation pursued in it.
>
> *A Treatise on the Culture of the Vine, and the Art of Making Wine*, James Busby
> (1825)

Wild grapevine is found in temperate forests. It typically grows from seed, making rapid growth up to heights of 30 metres, climbing over other plants, using its tendrils to grip and for support. In this natural setting, flowering is associated with the attainment of the top of the canopy and higher light intensi-ties, where birds are attracted to the dark, sweet fruit of the female plant, which they help to disperse. The expansive vegetative growth, the investment in ten-drils rather than a solid supporting trunk, and the delay in flowering for the first few years of the plant's life, are all adaptations to this habitat and lifestyle. The grapevine that most of us will encounter, however, whether in gardens or in the vineyard, has the same basic features as its wild relative but these appear in a very different context. First, cultivated grapevines are propagated vegetatively, from cuttings. Second, they are managed to produce a regular succession of canes over many years; hence the appearance of gnarled, stumpy stems with straggly canes arising from inconspicuous buds (see Fig. 8.1). For plants like these, which have intrigued observers of nature since ancient times (Box 8.2), the following summarizes the developmental cycle over the year. An appreciation of this cycle is crucial for successful practical management of the grapevine.

Renewed growth in the spring leads to bud break, and emergence of the leaves, tendrils and inflorescences constructed within the bud during the previous season (Fig. 8.3). To produce this suite of structures the shoot apical meristem (a group of dividing stem cells) at the centre of the bud grows in a way that is unique to the Vitaceae, alternating between leaf primordia (groups of cells partitioned from the meristem and destined to become leaves) and lateral primordia

---

BOX 8.2   ANCIENT INSIGHTS FROM THE VINE

For Theophrastus (c.371–287 BC) there was an important distinction between cultivated and wild plants. Cultivation best expressed the dominant nature of a cultivatable plant: to provide food for humans. Wild plants tended to resist cultivation–for instance medicinal plants, which he thought lost their useful properties when cultivated. The nature of a grapevine is to be cultivated; in making its fruit it has the goals of producing the seed for the next generation, and providing the flesh for the benefit of humans. Growers of seedless grapes breed to favour the latter. Lack of cultivation leads to a change in the nature of the grapevine–it degenerates, the wild part of its nature coming 'unnaturally' to dominate. Thus the wild or 'mad' vine has lost its way: it fails to ripen its fruit, favouring the seed because of the neglect of mankind.

Albert of Bollstädt (known as Albertus Magnus because of the esteem in which he was held) was a bishop and a scholastic philosopher of the thirteenth century. He wrote extensively, and with a critical insight into the natural world which at times seems uncannily modern. This is shown by his comments on the grapevine: he observed that the tendril can occur in the place of the bunch of grapes, and that it should therefore be viewed as an incompletely developed grape bunch. The modern interpretation of the tendril is that it is closely related to an inflorescence.

John Gerard's *Herball* was first published in 1597 and was very widely known. Gerard appears to have been a plagiarist, stealing his work from a translation of the Belgian botanist Dodoens' final work. Yet his illustration of 'the manured vine' is attractive (Fig. 1), and his words on wine are memorable:

Almighty God for the comfort of mankinde ordained Wine; but decreed withall, That it should be moderately taken, for so it is wholsome and comfortable: but when

measure is turned into excesse, it becommeth unwholesome, and a poyson most venomous. Besides, how little credence is to be given to drunkards it is evident; for though they be mighty men, yet it maketh them monsters, and worse than brute beasts. Finally in a word to conclude; this excessive drinking of Wine dishonoreth Noblemen, beggereth the poore, and more have beene destroied by surfeiting therewith, than by the sword.

Vitis Vinifera
The manured Vine

Box 8.2 Fig. 1 *Grapevine illustration from Gerard's Herball, edition of 1636, reprinted by Gerald Howe (1927) [Illustration from Gerard's Herball, 1636.]*

(also known as *anlage* or uncommitted primordia, because of their ability to develop into either a tendril or an inflorescence, or less commonly, a complete new shoot). The fate of the lateral primordium depends on the position of the bud on the stem, on temperature and on light intensity, the response to these factors being cultivar dependent. The developmental decision between tendril and inflorescence is crucial for productivity of the vine, because bunch number per vine is the major determinant of yield in grapevine.

The axillary bud (see Fig. 8.3) is the site of intense activity from the moment the parent shoot emerges in the spring.

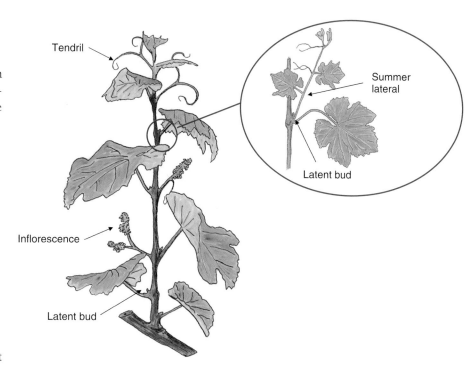

From the bud a shoot begins to emerge almost immediately. This is known as the 'summer lateral', because it grows out in the same season as the parent shoot. This lateral shoot contributes (sometimes excessively) to the leaf canopy, but the really important structures are the buds in the first two or three reduced leaves at its base. These remain in what appears to be a single bud, which by the end of the season becomes very swollen and obvious. This 'latent bud' contains the shoot, and associated tendrils and inflorescences that will grow out the next season. It is because the inflorescences for next year are formed within the latent bud during the current season that the grower has to consider the effect of light and temperature not just on this year's crop, but also on the potential crop for the following year (Fig. 8.4).

The inflorescences bear flowers in early summer, and these transform at fruit set into bunches of young fruit (see Fig. 8.4). Self-fertilization is of major importance in grapevine, although wind movement of pollen, and dry conditions, are required at flower opening (known in viticulture circles as 'capfall') for good fruit set. This takes place in June/July (northern hemisphere) or December/ January (southern hemisphere). Depending on the weather, locality and cultivar, the whole of berry development, from fruit set to full ripeness, takes 3–4 months. Thus for many cultivars consistently warm, dry weather is ideal until early October in the

Fig. 8.3 *Key parts of a grapevine. The axillary bud is circled, and in enlargement shows the summer lateral which develops from it, and the latent bud containing the inflorescences that will grow out next year*

Fig. 8.4 *Two-year cycle of grapevine reproductive development from flower initiation to fruit ripening [Adapted from Carmona et al. (2008).]*

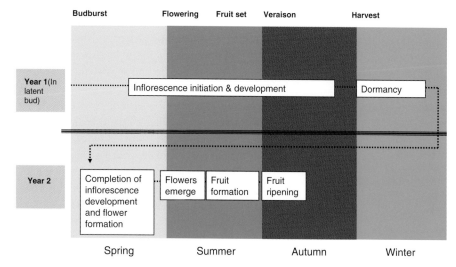

northern hemisphere. Growth of the berries has a rapid initial phase, lasting 3–4 weeks, followed by a pause as seeds mature. It is during this phase that many of the cultivar differences in maturity (harvest) date originate. In some (e.g. seedless types) it takes only a week; in other, late maturing cultivars, six weeks. There then follows a further rapid period of expansion, accompanied by ripening changes: colour development in red and black-fruited cultivars, softening, sugar accumulation, acid breakdown and the development of secondary compounds. The onset of this ripening phase is known as véraison, judged in practice by the beginning of colour change; about 4–6 weeks later flavour ripening or 'engustment' begins. After another three weeks, flavour maturity is reached.

One great debate in the wine industry is over quality of wine in relation to the quantity produced – the yield of wine per hectare. Wine quality is strongly influenced by the composition of the grapes at harvest. Sugar, principally present as glucose and fructose, is important because it represents the potential for fermentation to alcohol, and the proportion of sugar left after fermentation is a large contributor to the 'sweet' or 'dry' character of the wine. Acid is also significant; malic acid is largely degraded during ripening, leaving mainly tartaric acid. Secondary compounds include tannins which impart astringency to red wine; anthocyanins (responsible for colour); flavonols (co-pigments that intensify the colour due to anthocyanin); and terpenes, which are important contributors to aroma. It is this array of compounds, and their variation according to cultivar, locality, and weather conditions during ripening, that provides the foundation for the distinctive properties of different wines. The question of whether too much yield reduces quality is hotly (and continuingly) debated in wine circles: high yields (of around 100 hL per ha) can be achieved in both the New and the Old World, but it is generally agreed that with yields greater than 50 hL per ha fine wine production becomes increasingly challenging.

To return to the grapevine plant: during fruit development and ripening, shoots continue to produce leaves, branches and tendrils, leading to potential competition between vegetative and fruit growth and shading of the fruit. As well as being the time of fruit ripening, autumn also signals the onset of dormancy and leaf-fall; cool temperatures and shortening photoperiods act synergistically to induce dormancy and cold-hardiness. The post-harvest, pre-leaf-fall period is critical because carbohydrates and nitrogenous reserves are replenished. These reserves provide the basis for early spring growth and influence yield the following season. Cold winter temperatures are important because they help to break dormancy and lead to synchronized growth in the spring.

Underpinning all the above-ground activity is root growth, originating in an analogous way to shoot growth, but from the activity of root meristems. An increase in root activity, leading to the development of root pressure and resultant flow of sap to the shoots, is an early event in spring, typically preceding bud burst by some weeks. It results from rising soil temperature, and removal of the dormant state of the plant. Flow of sap is important to clear the xylem vessels of air that has accumulated during the winter, and to provide sugars needed to drive bud break. Thus, wild grapevines (*V. labrusca*, *V. riparia*) growing in the forests of New England achieve high root pressures during the month before bud burst in late May. Sap flow to heights of 18 m is enabled by this process, before any leaf emergence has occurred to allow transpirational flow of water.

One further impressive characteristic of grapevine, which enables viticulture to thrive where many other types of agriculture fail, is its ability to grow and fruit productively in arid environments on soils of poor fertility. The root system is a vital element in this—it can penetrate deeply using 'sinker' roots, and also exploit the surface soil effectively using fibrous absorbing ('spreader') roots which are frequently renewed.

Probably the most important of all the developmental changes discussed in this section is the two-season cycle underlying reproductive growth (see Fig. 8.4). The initiation of the inflorescence within the latent bud during the spring, and the slow development of the flower initials, their overwintering and emergence the following season, dictate many vine management practices and lead to a prolonged and often unpredictable impact of environment, in particular climate, on the crop. It is to the crucial role of climate in viticulture that we now turn.

CLIMATE, WEATHER AND THE GRAPEVINE

Climate is measured or described in long-term averages; weather is the day-to-day variation of those averages. Weather may make or break a vintage, and may explain the particular characteristics of a given wine, but it is the climate that determines which grapes (if any) will grow and ripen well in any given locality.[2]

The major areas for viticulture lie in the region of the world with average annual temperatures between 10°C and 20°C (Fig. 8.5). Thus the Mediterranean region, and the southern hemisphere countries well known for their wine (Chile, Argentina, South Africa, Australia and New Zealand), all have significant areas with this warm-temperate climate. So too do major parts of North America and Asia, although areas with a fully continental climate are unsuitable for grape cultivation because *V. vinifera* is intolerant of extreme winter cold. There are also local variations in climate which permit viticulture in unexpected places: for example, in the Rhine

Fig. 8.5 *World distribution of viticulture (black). Note distribution is largely in the temperate zone (yellow) [Based on Jackson and Schuster (1987).]*

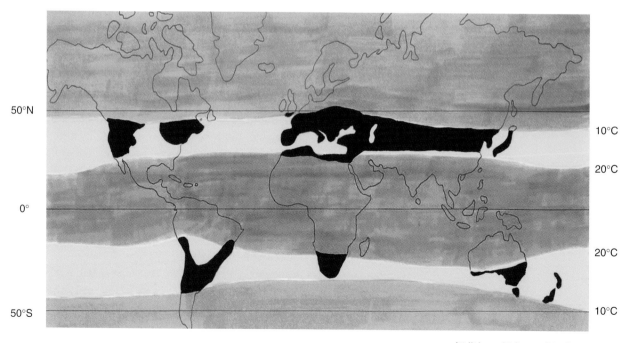

[2]Halliday and Johnston (2006), p.27.

valley, at 51°N, where the effects of the Gulf Stream lessen the extremes of continental weather; and in subequatorial Bolivia, where viticulture occurs at 3000 m, altitudes at which the tropical climate is sufficiently ameliorated.

Its temperate-zone origin means that the grapevine thrives where summers are long, warm and dry, and winters are cool. Cultivation in regions more equatorial than the 20°C isotherm is generally difficult because in the absence of winter cold (and shorter photoperiods) leaf-fall and dormancy do not occur, and flowering becomes poorly synchronized; in addition, excessive humidity makes pest and disease problems more acute. Move pole-ward from the 10°C isotherm and winters may be sufficiently severe to kill the plant. Even where this is not the case, summers are usually too short, terminated by excessive rainfall and early frosts as well as decreasing temperatures, and delayed in their onset by cool temperatures. For this reason southern England is only marginal for viticulture.

The example of southern England brings to the fore another important dimension to climate and viticulture: the idea that the best quality wines come from the cooler of the viticultural regions shown in Fig. 8.5. This is a complex topic and we will first explain the traditional view, moving on to qualify it later in this section. Champagne in northern France is famous for its high-quality white wines derived from the cultivars Chardonnay and Pinot Noir. Similarly, the Central Otago region of New Zealand, right on the 10°C isotherm, is gaining a reputation for exceptional quality Pinot Noir. In contrast, the North African countries of the Mediterranean region are not well known for quality wines, the higher temperatures favouring the production of grapes for drying as sultanas, raisins and currants. An almost equivalent generalization, that the best quality wines are produced in cool years in hot regions, and warm years in cool regions, accounts for the fact that, occasionally, vineyards of southern England can produce a vintage to rival that of the Champagne region of neighbouring France. In these years, a long, dry autumn may achieve an optimal balance between sugar and acid in the grape. This balance is generally a key factor linking temperature to wine quality. If summers are too hot, too little acid remains at harvest and the grapes are too sweet; too cold and acidity dominates. Modern methods of blending, combining musts (juice) of different origins, can overcome the problem, but the tendency then is to produce wines that lack local character and distinctiveness.

James Busby discussed the difficulties associated with high growth temperatures a long time ago:

> Thus, in Burgundy, where the sun's rays do not act so powerfully in the production of saccharine matter, the wines are distinguished by a richness and delicacy of taste and flavour, while those produced under the burning sun of Languedoc and Provence, possessing no virtue but spirituosity, are generally employed in distillation. Not that the existence of a large portion of saccharine matter is incompatible with that of flavour and perfume, but it generally happens, that the volatile principles on which these depend, are dissipated during the lengthened fermentation necessary to convert into alcohol, the excess of saccharine matter.
>
> James Busby, *A Treatise on the Culture of the Vine, and the Art of Making Wine*
> (1825)

There are several extremely important caveats needed here. First, today's equipment and techniques can provide very exact control over wine production, allowing much more flexibility than in Busby's day. There may also be cultural reasons why quality wine is not associated with certain regions—for instance the prohibition on alcohol in Islamic states. The need to ensure stability during prolonged sea transport is another factor which led to the association of hot climates with the addition of brandy to produce 'fortified' wines, such as port. Thus, the interplay of political, social and economic factors can be as important as climate in determining the reputation of particular regions for high- or low-quality wines. One example is the history of wine production in the Upper Douro valley of northern Portugal, where in the early eighteenth century feuding between the British and the French led to a politically inspired British preference for Portuguese wines; another important factor was the need for cheap wine for mass consumption in Britain. The hot climate made achievement of quality wine a challenge, due to excessively sweet grapes at harvest; expediency thus led to the 'fortification' solution.[3] The technical difficulties associated with the production of quality wines in warm climates can often now be avoided, and red wines of renowned quality are produced in Portugal, the Lebanon and the hot regions of Australia.

Gladstones (2004) recognizes the flawed nature of the generalization about the superiority of wines from cool climates, but acknowledges that very high temperatures during ripening can have negative effects on wine quality. He argues that, according to the wine style preferred, there are different temperature optima—particularly during the final phase of ripening. Here's how he puts it:

> Delicate, fresh and aromatic (especially white) table and sparkling wines need cool, equable late ripening to maximize and conserve the more volatile components in the grapes, and retain acid, with only moderate sunshine hours and high humidities to keep alcohol low. Full bodied, red table wines need ample total flavourants, pigments, tannins and alcohol, but usually less acids, and hence more ripening warmth and sunshine. Their essential quality constituents are more temperature-stable and less fugitive than in delicate white wines, so temperature equability is less critical if still important. Fortified wines have the requirements of full-bodied table wines but with still more sunshine to maximize berry sugar content.[4]

The critical influence of fruit ripening temperature on wine quality is part of a bigger picture, in which temperature is the major factor regulating grapevine development. Little growth occurs below 10°C; above this temperature growth increases linearly to approx 17°C, then less rapidly to a maximum at between 22°C and 25°C. The relationship between temperature and vine productivity (yield) is more complex, because photosynthesis depends on light intensity and duration, and CO_2 concentration, as well as on temperature. Net carbon assimilation (photosynthesis minus respiration) is at a maximum between 25°C and 30°C. At higher temperatures, respiration continues to increase while photosynthesis does not; because the rate of development also quickens at high temperature, the time for vegetative growth (and assimilate gain) before fruiting is reduced. In very hot regions light is therefore crucial for good cropping—not so much light intensity (photosynthesis in grapevine saturates at about one-third the light intensity typical of bright sunlight), but duration, or sunshine hours.

Where flower number in the inflorescence is limiting to yield, temperature has an additional effect, because of its role in promoting the initiation of flowers. The optimum temperature for flower initiation is cultivar dependent: for

[3]Note, however, that Gladstones (1992) has advanced a climate-based explanation for the adoption of Portuguese wine by the British, arguing that the 'Little Ice Age' in the late seventeenth century, which reduced average temperatures across much of Europe, would have made the Portuguese climate ideal for table wines, while those from northern Europe would have become poor and acid. The adoption of fortification as a procedure for Portuguese wines in the middle of the eighteenth century, and the resumption of British trade in table wines with France and Germany, coincided with the elevation of growing temperatures.

[4]Gladstones (2004), p. 108.

example in 'Muscat Gordo Blanco' very little initiation occurs below 20°C, and the maximum occurs close to 35°C. This effect of temperature operates, it should be remembered, the season before the flowers emerge and form fruit, when inflorescence development is occurring in the bud (see preceding section).

While climate determines whether and where grapevines can be grown, the weather is the main factor behind the quality of wine produced in any particular year – the vintage. For example, one study of the quality of Burgundy vintages over 17 years, correlating good, average or bad vintages with the weather, suggested that plenty of sunshine hours in the spring and warm temperatures during mid-summer were both key to a good vintage.

Water supply is another important factor in vine growth, a feature of regional climate which varies locally and annually with the weather. Crucially, and somewhat controversially in wine circles, water supply can be controlled in dry regions by irrigation. Grapevines need a minimum annual rainfall of 500 mm, and 750 mm for good productivity. Where rainfall during the growing season is limiting, and irrigation is not used – as in many traditional vineyards of Europe – reduced vine numbers per hectare and hard pruning limit the number of shoots demanding water. Where irrigation is practised – as for example in Australia – the water supply is very carefully controlled, forming the basis for the techniques of partial root-zone drying and regulated deficit irrigation (see later). Excess rainfall can be disastrous for grapevine cultivation, in the spring if it causes the soil to become waterlogged and in the summer and autumn as the likelihood of fungal diseases is increased. Direct rainfall on the fruit can also cause it to swell rapidly and burst. This is another reason why the current climate of England makes viticulture a difficult challenge.

Thus a range of climatic factors influence grape cultivation, temperature being the most critical. There has always been a great need to predict which localities would be best suited to viticulture, because it is a long-term commitment requiring good infrastructure for the associated process of wine production. Prediction based on temperature profiling has therefore been extensively researched. Initially, in California, heat summation methods were developed to allow prediction of the suitability of different regions for various wine styles. This built on the observation that grapevine shows little growth below 10°C, summing day-degrees above this base temperature over a seven-month growing season. Based on this kind of approach, one can say that localities which offer more than 1950 day-degrees are best suited to fresh and dried grapes; 1500–1950 day-degrees for table and fortified wines; and 950–1500 day-degrees for table wines. Below 700–900 day-degrees ripening will be insufficient for any purpose. Other methods have also been useful – for example the mean temperature of the warmest month; and the average temperature over the growing season. A further development is the 'homoclime' method, in which computer simulations are used to find areas of similar climate to places of recognized viticultural merit.

Clearly, future climate will be crucial for sustained success in viticulture in any particular locality. Climate change research suggests, globally, that wine-producing areas will experience an average warming of 2°C during the next 50 years. For those localities producing high-quality fruit at current climate margins, this means that things will get progressively more difficult: at the least, the use of new cultivars will be required, along with the production of alternative types of wine. The predicted additional days of extreme heat (>35°C) in many parts of the US could, however, lead to the elimination of quality wine production in those regions. Despite some compensation by an overall northward movement of viticulture, one projection is of a 50% decline in the highest quality US wines. Similarly alarming forecasts have been made for parts of Australia. This emphasizes the need for long-term breeding programmes selecting cultivars suitable for quality wine production under more extreme conditions. Given the great age of the vines responsible for some of the best quality wines, the opportunities are obvious for speculative plantations in latitudes currently too cool for sustainable viticulture. But the predictions need to be accurate and reliable!

MANAGING THE GRAPEVINE

To exploit a plant we humans impose our will upon its natural inclinations. This requires study of the behaviour of the plant over its yearly cycle and application of discipline to that cycle so as to maximize the productivity we require. In the case of the grapevine, we curb the tendency to excessive vegetative growth – vigorous production of stem and leaf – so as to promote fruit production. But the flowers, and in turn the fruit, are borne by these vegetative structures, and depend on them for sustenance, so great care must be taken to achieve the right balance. The 'right balance' is not a straightforward issue: in contrast to other crops, simply maximizing fruit number and weight is not necessarily the objective in grapevine cultivation. Some of the most prized wines are derived from vines whose productivity is deliberately confined to just a few bunches. This is why detailed methods of training, pruning and managing grapevines have been worked out over many years, to suit the production of grapes and wine in a whole range of different situations. There is also a fair amount of local tradition, which runs alongside the scientific rationalization of methods adopted. Nevertheless, the topic is of great commercial significance because grapevine management absorbs a much greater proportion of costs in comparison to other crops.

There are some important basic principles behind the management of grapevines. First, vine support structures will be used for many years, so the investment involved is large and needs to be carefully planned. Second, training and pruning are closely linked; if, for example, mechanical pruning is intended, factors such as access to specialized machinery may influence decisions about row spacing and training method. Third, pruning is carried out in the winter when the vines are dormant, and is intended to ensure a regular supply of buds that will form flowers and fruits. These must be supported by an architecture sufficient for satisfactory yield and fruit quality, but not so great as to shade fruit and compete excessively for assimilate. Summer pruning can also be employed, to remove excess shoots and reduce associated problems with disease, but it is of secondary importance to winter pruning. There are two principal kinds of winter pruning, spur and cane pruning (see Fig. 8.1).

Further insight into training and pruning can be gained by reviewing the factors that influence vine growth and performance, and in doing so illustrating the systems used under different circumstances. Probably the most fundamental factor is the growing environment. In a cool climate, with a deep, fertile soil and plentiful rainfall, vines are likely to be vigorous, making excessive vegetative growth; flowering and fruitfulness will probably need to be maximized. A typical support system would be a vertical trellis, with a replacement pruning regime (illustration A opposite and on page 134); localities would be Burgundy in France, Germany,

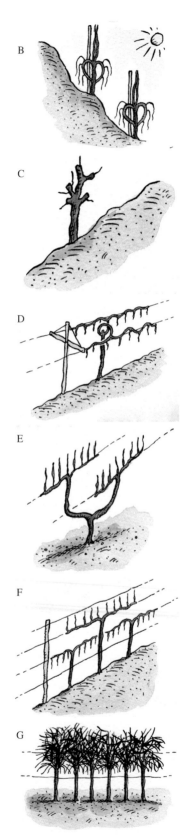

or New Zealand. However, vines are sometimes deliberately grown on steep slopes facing the sun, in order to maximize light interception and warmth, as in the Moselle region of northern France and Germany. Here the topography makes elaborate training systems impractical and the vines are planted densely, supported by single poles (illustration B). At the other extreme from northern Europe, for example in many parts of Spain, with low rainfall, high light and very warm temperatures, vines can be grown without support (other than a small stake), and spur pruned (illustration C). Here the environment does much of the work, limiting vegetative growth. The problem may instead be productivity per hectare. Such 'bush' or 'goblet' vines are also a common site in southern France and Australia.

A radically different approach was developed in Geneva in eastern USA, to optimize productivity of the local *V. labrusca* 'Concord'. In this case the main stem grows up to a wire support system at 1.5–1.8 m; there the canopy is split between two wires and hangs down–hence the name 'Geneva Double Curtain' (illustration D). The system provides good light interception, and copes with the vigorous growth of *V. labrusca*. The importance of tradition is illustrated by the story of the Lyre training system, developed in Bordeaux. This divides the canopy in a similar way to the Geneva Double Curtain; the difference is that the canes are trained upwards, rather than hanging down (illustration E). The method offers the opportunity for improved productivity (due to reduced shading), but the French Appellation d' Origine Contrôlée rules deplore such departure from the norm.

In many cases the overriding factor affecting choice of training system is cost. Cost of labour–or labour availability–mean that mechanical pruning, and increasingly mechanical harvesting, are being adopted in many major viticultural regions. In Australia, for example, about 80% of the annual harvest of grapes for wine is now carried out mechanically. Thus support systems that facilitate machine pruning and harvesting, or can be adapted to allow this, are popular. The Scott Henry system, named after the grower in Oregon who developed it, is one such system. The vines are trained on wires at about 1–1.2 m; when used with a spur pruning regime, cordons from alternate vines are pruned to send their shoots upwards and downwards (illustration F). Such a system is relatively expensive to establish, because of the investment in post and wire support, but offers long-term savings on labour. Because of the urgent need to save labour, conversion (or 'retrofitting'), of existing vines to new training systems allowing mechanization is now relatively common.

Lastly, minimal pruning options have been pioneered in Australia. In the system developed there by Clingeleffer, the vines are grown on a vertical trellis, cut back below the cordon by machine to allow access to sprayers, but otherwise left to proliferate shoots upwards and outwards (illustration G). In this situation some self-regulation of growth occurs as a proportion of terminal shoots abscises naturally; but there is still a thicket of growth and mechanical harvesting is needed. The system has had some success in hot regions of Australia, where water comes through irrigation. In other environments excess shoot growth in the summer can cause shading.

The preceding discussion has focused on training and pruning as methods by which the vegetative/reproductive balance can be managed in grapevine. Scheduled application of water can also complement these methods, and the development of practical techniques of irrigation management stands out as an example of innovation, associated particularly with progressive thinking in New World viticulture. Regulated deficit irrigation was developed in the late

1970s for peach trees, providing control of vegetative growth and benefits to the fruit crop. The technique involves reduced water supply at a specific developmental period, and has subsequently been tested in a range of other tree fruit crops. It is thought to alter assimilate partitioning, reducing shoot growth without penalty to the fruit.

For grapevine, there has been a longstanding awareness that superior wine quality is associated with water stress: hence the traditional ban on irrigation in Spain and France. There is also a pressing need to conserve water. Regulated deficit irrigation has therefore found widespread application in vineyards; a typical schedule would supply water until fruit set, limit it during early fruit development, and ensure adequate water during ripening. A development involving spatial variation in supply is the partial root drying technique, in which drip irrigation is applied alternately (roughly every two weeks) either side of the row of vines. Soil drying causes root-derived signals, including abscisic acid, to be transmitted to the shoot, where they are believed to be responsible for reduced shoot and leaf expansion.

EXPLOITATION OF GRAPEVINE BY PESTS AND DISEASES

As we have seen, the history of viticulture is the history of the transmission of one species around the globe. Wherever discovery of new lands by Europeans was followed by colonization, and the environment was suitable, *V. vinifera* was soon planted. But on the heels of the colonisers came other exploiters of the vine: insect and nematode pests, and disease-causing fungi, bacteria and viruses. One fungus, botrytis (*Botrytis cinerea*), has long been troubling grape growers in Europe, probably since Roman times; it can be very damaging, particularly to fruit where it causes rotting and off-flavours. But its effects can also be marshalled under the right ripening conditions (warm, humid mornings and dry afternoons in late autumn) to produce 'noble rot'. In this way the partially rotted berries of the cultivar Semillon (with a little Sauvignon Blanc) are used to make French Sauternes. The presence of the fungus on the fruit helps to shrivel the grapes, concentrating the sugar, and an intensely sweet white wine is the result. Some other Noble wines are German Trockenbeerenauslesen (made from the cultivar Riesling), and Hungarian Tokay Aszu (made principally from the cultivar Furmint), although the details of the process used to create the end-product are very different. Tokay Aszu, for example, is made by addition of the shrivelled (*aszú*) grapes to previously fermented wine; a second fermentation then follows.

New fungal pathogens of grapevine appeared in Europe during the nineteenth century. The year 1846 saw the introduction to France of grape powdery mildew (*Uncinula necator*). Then in 1878, downy mildew (*Plasmopara viticola*) was brought into south-western France on vines from the US. This disease proved very destructive, until in 1885 the efficacy of Bordeaux mixture (copper sulphate and lime) was accidentally discovered by a French grower who used it as a paint to discourage local scrumpers. But the destruction due to downy mildew was limited compared to the devastation wreaked by the insect phylloxera (*Daktulosphaira vitifoliae*). This is an aphid-like insect which causes tuberosities on *V. vinifera*, so weakening the plant that eventually it dies. It was introduced to France in the 1860s on vines from the US; in 1868 a commission established to discover the

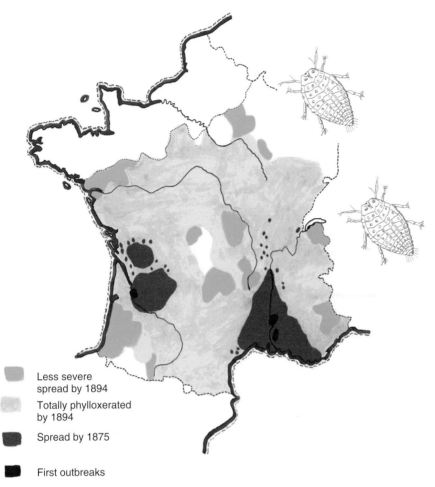

Less severe
spread by 1894

Totally phylloxerated
by 1894

Spread by 1875

First outbreaks

Fig. 8.6 *Spread of grape phylloxera through France [Based on Campbell (2004).]*

cause of vine death in the Rhône valley described how on the roots there were infestations of a 'plant louse of yellowish colour, tight on the wood, sucking the sap'.[5] For the next 30 years arguments raged over the role of the insect, and solutions to the problem were attempted—from flooding the vineyards, to treatment with the lethal chemical carbon disulphide. In the meantime the epidemic threatened to destroy French viticulture—by 1900, 75% of French vines were affected, covering an area of 2 million hectares (Fig. 8.6). It spread progressively through Europe and then to much of the rest of the world. The solution eventually adopted came from North American vines.

The native North American species *V. riparia*, *V. rupestris* and *V. berlandieri* were found to tolerate grape phylloxera, forming apparently harmless galls on their leaves, and being unaffected (or forming only minor nodosities) on their roots. American vines were therefore tried in three ways—directly as fruit producers; as fruit-producing hybrids with *V. vinifera*; and through grafting as rootstocks. Rootstocks proved effective, hybrids of the North American species offering not only tolerance to grape phylloxera, but also other benefits including cold-tolerance, and the ability to grow on lime-rich or sandy soils. However, the battle against the insect will probably never be fully complete. In California's Napa Valley in the 1980s an infestation was discovered and it became clear that the hybrid rootstock AxR#1 was unable to provide protection; one suggested explanation was that a new form of *Daktulosphaira vitifoliae* had emerged, capable of living on AxR#1. About 20 000 ha of vines had to be replaced; probably all vines grafted on AxR#1 will eventually be subject to the same fate. In some parts of the world there has been success in controlling the spread of the insect—for example in Australia, Victoria and New South Wales were early affected, but strict quarantine measures and vigilant monitoring have so far kept South Australia, an area rich in vineyards, free of grape phylloxera. Chile is, however, the only major wine-producing country without the insect. The complex life-cycle of grape phylloxera, responsible for its prodigious rate of increase and its difficulty in elimination, is summarized in Fig. 8.7.

The tragic irony of the ravaging of nineteenth century Europe by new pests and diseases of grapevine is that each successive epidemic was a consequence of efforts to prevent the preceding one. North American vines carrying grape phyl-

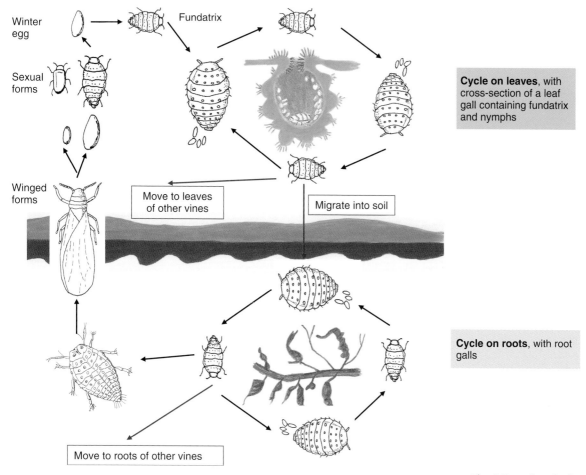

Winter egg

Fundatrix

Cycle on leaves, with cross-section of a leaf gall containing fundatrix and nymphs

Sexual forms

Winged forms

Move to leaves of other vines

Migrate into soil

Cycle on roots, with root galls

Move to roots of other vines

Fig. 8.7 *Life-cycle of grape phylloxera [Partially based on Campbell (2004).]*

loxera were introduced in efforts to control grape powdery mildew; those carrying downy mildew were part of the effort to solve the phylloxera problem.

Many other pests and diseases afflict the grapevine, to varying degrees. Nematodes can be a serious problem, particularly in irrigated vineyards on sandy soils. The bacterium *Xylella fastidiosa* was shown in the 1970s to be the cause of Pierce's disease, which had badly affected Californian viticulture 100 years earlier, destroying 14 000 ha by 1906. *X. fastidiosa* originates in south-eastern USA, where it limits commercial grapevine cultivation to those species native to the region. It is transmitted by leafhoppers, the most recent outbreaks in southern California being associated with the glassy-winged sharpshooter *Homalodisca vitripennis*, itself a new species to California.

THE NEXT REVOLUTION IN VITICULTURE?

In this chapter we have described the biology and commercial exploitation of the grapevine. What are the main challenges for the future? In the medium term, rising labour cost (or reduced labour availability) is likely to be coupled with a developing wine market in which more people are more discerning, demanding authentic, local wines rather than a blended, mass-produced offering. Since the answer to the former problem is mechanization, it will be necessary to ensure this does not conflict with the demand for authenticity. In the longer term, climate change threatens to marginalize currently favoured regions; the areas that will replace these are not easy to predict.

Most generally, it is the fact that the grapevine is a perennial vine that raises costs and makes production so complex, requiring at least a biennial view on climatic effects. Is there room for further innovation to tackle this issue? Perhaps conversion to an annual habit is possible, with shoots growing from the stem bud and flowering and fruiting in one season. This would allow reduced investment in support, and simplify pruning and other management operations. The characteristic which stands in the way in grapevine is the prolonged period of time (14 months) allocated by the plant to flower initiation and development. Rapid progress in understanding flowering in annuals has occurred over the last 20 years, and although for perennials the problem is more complex, many important flowering genes have now been identified in grapevine. One example returns us to the theme at the beginning of this chapter: a homologue of the wheat 'green revolution' gene turns out to control the developmental choice between tendril and inflorescence in grapevine. In the dwarf mutant which reveals the function of this gene, inflorescences are initiated and grow out in the same season.[6] Given this kind of progress, it seems possible that, with appropriate breeding, flower initiation and fruiting could be focused into a single year in commercial cultivars. The autumn-fruiting raspberry provides an example of a commercially useful perennial which does this, in which the costs of support, training and pruning are greatly reduced compared with biennial raspberries. For grapevine, the mysteries of the 'old vine' could still be enjoyed, the perennating stem and root system being retained but with strictly annual shoots simplifying, and making more economical, the task of the viticulturalist.

FURTHER READING

Viticulture and oenology are renowned for their specialized terminology. A good website on wine, with a glossary covering most aspects, is at www.thewinedoctor.com/. For viticulture, start with Wikipedia (http://en.wikipedia.org/wiki/Glossary_of_viticultural_terms). General books on the grapevine and viticulture are Winkler (1962), Mullins et al. (1992), Jackson and Schuster (1987), Creasy and Creasy (2009). An Australian perspective, but with much wider relevance, is provided by Dry and Coombe (2004) and Coombe and Dry (1992); the chapter by Thomas and van Heeswijck (2004) is of particular general interest. Useful books on wine and grapes are Halliday and Johnson (2006), Robinson (2006), Johnson and Robinson (2007).

There are useful chapters in Sandler and Pinder (2002) on wine and health, while the social and economic history of viticulture and wine is covered by Unwin (1991). For the process of domestication of grapevine, see McGovern (2003) and Arroyo-Garcia et al. (2006); for an example of molecular analysis of the parentage of important wine cultivars, see Bowers et al. (1999). Both these topics are succinctly covered in the review by This et al. (2006). For descriptions of grapevine development, see Mullins et al. (1992); for flowering and its relation to yield and fruit quality, see Carmona et al. (2008). For the importance of the post-harvest, pre-leaf-fall period in relation to next season's yield, see, for example, Holzapfel et al. (2006). The relationship between climate and wine quality

[6]Boss and Thomas (2002).

is discussed by Jackson and Schuster (1987) and Gladstones (2004). For detailed discussion on the effects of temperature and light on grape cultivation see Gladstones (1992, 2004). Jones and Davis (2000) analyse the influence of climate on grapevine production and wine quality in Bordeaux. The homoclime method of matching climatic regions is described by Smart and Dry (2004). For the potential impacts of climate change on viticulture, see Jones *et al.* (2005) and White *et al.* (2006). Jackson (2001) is excellent on pruning and training; and the overviews by Freeman *et al.* (1992) and Tassie and Freeman (1992) are also very useful. Irrigation technologies are discussed in Fereres and Soriano (2007) and Creasy and Creasy (2009). For pests and diseases see Mullins *et al.* (1992) and Creasy and Creasy (2009); a review of grape phylloxera is provided by Granett *et al.* (2001). An accessible account of the topic is provided by Campbell (2004). For rootstocks see Howell (1987). The role of the 'green revolution' (gibberellin-response) gene in grapevine is described by Boss and Thomas (2002). The possibilities associated with altering the perennial habit (to an annual one) are discussed further in Battey (2005). For discussion of the views of Theophrastus, see French (1994); see Theophrastus (372–287 BC) for pertinent writings on grapevine. Albertus Magnus is discussed by Arber (1986).

REFERENCES

Arber, A. (1986) *Herbals*, 2nd edn. Cambridge University Press, Cambridge.

Arroyo-Garcia, R., Ruiz-Garcia, L., Bolling, L. *et al.* (2006) Multiple origins of cultivated grapevine (*Vitis vinifera* L. ssp. *sativa*) based on chloroplast DNA polymorphisms. *Molecular Ecology* **15**, 3707–3714.

Battey, N.H. (2005) Applications of plant architecture: haute cuisine for plant developmental biologists. In: *Plant Architecture and its Manipulation* (ed. C.G.N. Turnbull), pp. 288–314. Blackwell, Oxford.

Boss, P.K. and Thomas, M.R. (2002) Association of dwarfism and floral induction with a grape 'green revolution' mutation. *Nature* **416**, 847–850.

Bowers, J., Boursiquot, J-M., This, P., Chu, K., Johansson, H. and Meredith, C. (1999) Historical genetics: the parentage of Chardonnay, Gamay, and other wine grapes of northeastern France. *Science* **285**, 1562–1565.

Busby, J. (1825) *A Treatise on the Culture of the Vine, and the Art of Wine Making*. R. Howe, Sydney, Australia.

Campbell, C. (2004) *Phylloxera: how wine was saved for the world*. HarperCollins, London.

Carmona, M.J., Chaib, J., Martinez-Zapater, J.M. and Thomas, M.R. (2008) A molecular genetic perspective of reproductive development in grapevine. *Journal of Experimental Botany* **59**, 2579–2596.

Coombe, B.G. and Dry, P.R. (eds) (1992) *Viticulture. Volume 2: Practices*. Winetitles, Adelaide, South Australia.

Creasy, G.L. and Creasy, L.L. (2009) *Grapes*. CABI, Wallingford, UK.

Dry, P.R. and Coombe, B.G. (eds) (2004) *Viticulture. Volume 1: Resources*, 2nd edn. Winetitles, Adelaide, Australia.

Fereres, E. and Soriano, M.A. (2007) Deficit irrigation for reducing agricultural water use. *Journal of Experimental Botany* **58**, 147–159.

Freeman, B.M., Tassie, E. and Rebbechi, M.D. (1992) Training and trellising. In: *Viticulture. Volume 2: Practices* (eds B.G. Coombe and P.R. Dry), pp. 42–65. Winetitles, Adelaide, Australia.

French, R. (1994) *Ancient Natural History: histories of nature*. Routledge, London.

Gale, G. (2003). Saving the vine from *Phylloxera*: a never-ending battle. In: *Wine: a scientific exploration* (eds M. Sandler and R. Pinder), pp. 70–91. Taylor & Francis, New York.

Gerard's Herball: the essence thereof distilled by Marcus Woodward; from the edition of Th. Johnson, 1636. (1927). Gerald Howe, London.

Gladstones, J.S. (1992) *Viticulture and Environment*. Winetitles, Adelaide, Australia.

Gladstones, J.S. (2004) Climate and Australian viticulture. In: *Viticulture. Volume 1: Resources*, 2nd edn (eds P.R. Dry and P.G. Coombe), pp. 90–118. Winetitles, Adelaide, Australia.

Granett, J., Walker, M.A., Kocsis, L. and Omer, A.D. (2001) Biology and management of grape phylloxera. *Annual Review of Entomology* **46**, 387–412.

Halliday, J. and Johnson, H. (2006) *The Art and Science of Wine*. Mitchell Beazley, London.

Holzapfel, B.P., Smith, J.P., Mandel, R.M. and Keller, M. (2006). Maintaining the postharvest period and its impact on vine productivity of Semillon grapevines. *American Journal of Enology and Viticulture* **57**, 148–157.

Howell, G.S. (1987) *Vitis* rootstocks. In: *Rootstocks for Fruit Crops* (eds R.C. Rom and R.F. Carlson), pp. 451–472. Wiley-Interscience, New York.

Iland, P.G. (2004) Development and status of Australian viticulture. In: *Viticulture. Volume 1: Resources*, 2nd edn (eds P.R. Dry and P.G. Coombe), pp. 1–16. Winetitles, Adelaide, Australia.

Jackson, D. (2001) *Monographs in Cool Climate Viticulture. 1: Pruning and training*. Daphne Brassell Associates and Lincoln University Press, Canterbury, New Zealand.

Jackson, D. and Schuster, D. (1987) *The Production of Grapes and Wine in Cool Climates*. Butterworths, Wellington, New Zealand.

Johnson, H. and Robinson, J. (2007) *The World Atlas of Wine*, 6th edn. Mitchell Beazley, London.

Jones, G.V. and Davis, R.E. (2000) Climate influences on grapevine phenology, grape composition, and wine production and quality for Bordeaux, France. *American Journal of Enology and Viticulture* **51**, 249–261.

Jones, G.V., White, M.A., Cooper, O.R. and Storchmann, K. (2005) Climate change and global wine quality. *Climatic Change* **73**, 319–343.

McGovern, P.E. (2003) *Ancient Wine: the search for the origins of viticulture*. Princeton University Press, Princeton, USA.

Mullins, M.G., Bouquet, A. and Williams, L.E. (1992) *Biology of the Grapevine*. Cambridge University Press, Cambridge.

Robinson, J. (2006) *The Oxford Companion to Wine*, 3rd edn. Oxford University Press, Oxford.

Sandler, M. and Pinder, R. (eds) (2002) *Wine: a scientific exploration*. Taylor & Francis, New York.

Smart, R.E. and Dry, P.R. (2004) Vineyard site selection. In: *Viticulture. Volume 1: Resources*, 2nd edn. (eds P.R. Dry and P.G. Coombe), pp. 196–209. Winetitles, Adelaide, Australia.

Tassie, E. and Freeman, B.M. (1992) Pruning. In: *Viticulture. Volume 2: Practices* (eds B.G. Coombe and P.R. Dry), pp. 66–84. Winetitles, Adelaide, Australia.

This, P., Lacombe, T. and Thomas, M.R. (2006) Historical origins and genetic diversity of wine grapes. *Trends in Genetics* **22**, 511–519.

Thomas, M.R. and van Heeswijck, R. (2004) Classification of grapevines and their interrelationships. In: *Viticulture. Volume 1: Resources*, 2nd edn. (eds P.R. Dry and P.G. Coombe). Winetitles, Adelaide, Australia.

Theophrastus (371–287 BC) *De Causis Plantarum* Harvard University Press 1976.

Unwin, T. (1991) *Wine and the Vine: an historical geography of viticulture and the wine trade.* Routledge, London.

White, M.A., Diffenbaugh, N.S., Jones, G.V., Pal, J.S. and Giorgi, F. (2006) Extreme heat reduces and shifts United States premium wine production in the 21st century. *Proceedings of the National Academy of Sciences USA* **103**, 11217–11222.

Winkler, A.J. (1962) *General Viticulture*. University of California Press, Berkeley CA, USA.

The Salmon

CHAPTER 9

The Salmon is accounted the King of fresh-water-fish

Izaak Walton, *The Compleat Angler*[1]

PERHAPS THE TITLE of Netboy's (1980) book, *Salmon: the world's most harassed fish*, is more apposite than the quote from Isaak Walton's *The Compleat Angler* given above: for the salmon has been hardly treated as a king, unless it be Charles I. In this chapter we will explore the exploitation of the salmon, discussing how its life-cycle makes it particularly vulnerable, and how the once-unlikely success of salmon farming may also be adding to the threats to wild salmon populations.

The salmon, *Salmo salar* (meaning 'salmon the leaper'), is one of about 55 species of freshwater fish found in Britain and is a member of the family Salmonidae. (In this chapter we will use 'salmon' to refer to *S. salar* alone; there are also five species of Pacific salmon.) In Britain this family contains over 35 species in seven genera, including the rainbow trout *Oncorhynchus mykiss*, the arctic charr *Salvelinus alpinus*, brook charr *S. fontinalis*, and the brown trout *Salmo trutta*. The latter is the only other member of the genus *Salmo* present in Britain and can hybridize with the salmon. The salmon is the largest of the British Salmonidae, with the British rod-caught record of 29 kg set in 1922, and the world record 36 kg salmon caught in Norway during 1925.

The salmon once occurred naturally along both the east and west coasts of the North Atlantic (Fig. 9.1), and has a complex life-cycle split between hatching and early growth in fresh water, a move to the sea for weight gain and return to fresh water for reproduction. Such a life history is termed anadromous (Box 9.1) − the move from fresh water to salt is the opposite of that seen in the eel (Chapter 2).

The sexually mature salmon return from the sea and assemble in freshwater pools close to their spawning area by October. By this time their morphology has changed − the male now has a hooked lower jaw (the kype) and both sexes have a thick and spongy skin (Box 9.2). Towards the end of November the female salmon selects a shallow, fast-flowing stream with a gravel bottom (so that flowing water can percolate around the eggs and oxygenate them) for egg-laying (Fig. 9.2). She takes considerable effort to dig and shape an egg-pocket in the gravel, excavating a saucer-shaped hole up to 30 cm deep: this can take up to 48 hours. When this is completed the male (which has been waiting, quivering increasingly, nearby) and female lie side by side in the hole and the eggs and sperm (the milt) are released almost simultaneously into the bottom of the hole, where the current is much weaker. The female covers the eggs with gravel before moving

[1]Fifth edn, 1676, p. 274 (Bevan, 1983).

Biological Diversity: Exploiters and Exploited, First Edition.   Paul Hatcher and Nick Battey.
© 2011 John Wiley & Sons, Ltd. Published 2011 by John Wiley & Sons, Ltd.

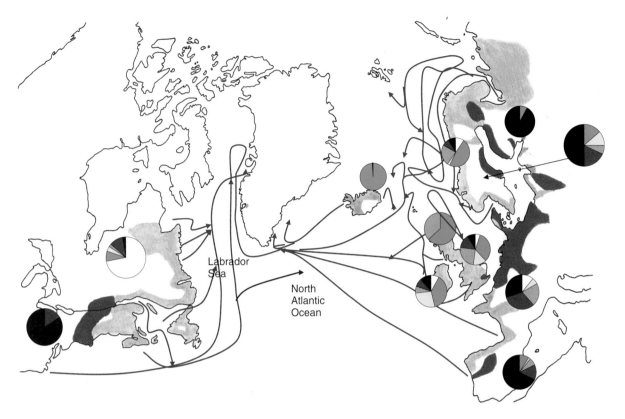

Fig. 9.1 *Distribution and ocean migration of the salmon,* Salmo salar. *Distribution [from Verspoor et al. (2007)]: red, formerly present, now extinct; orange, extant. Blue lines give some approximate migration routes at sea. Pie charts represent status of salmon rivers [WWF (2001)]: white, no data available; green, healthy; yellow, vulnerable; orange, endangered; red, critical; black, extinct*

BOX 9.1	THE ORIGIN OF ANADROMY

Why do salmon move between fresh and sea water, and how did this life-style evolve? Early views favoured a marine salmon ancestor which later invaded fresh water. More recently there has been a shift of view towards a freshwater ancestor which invaded the sea during the last glaciation and the subsequent restriction of its freshwater habitats. The return to fresh water to spawn is therefore thought to indicate the salmon's ancestral biome.

To resolve the question we need to know the ancestors of the salmon, and thus workers have tried to identify the closest living group to the salmonids. Unfortunately, there have been at least nine different phylogenies of the group of bony fish similar to the salmon (the Euteleostei) in the last 35 years, with the closest relative to the salmon appearing in four different groups, depending on the analysis. McDowall (2001) suggested that the closest ancestor also moved between sea and freshwater, and that we would probably never know whether the ancestor originated in fresh or salt water.

However, in 2003 Ishiguro et al. published the complete mitochondrial DNA sequences from salmon and 33 related species. This revealed that the Esociformes, the pikes (all freshwater

fish), were the sister group to the salmonids, and that the two groups probably evolved in the northern hemisphere, with the two lineages diverging about 95 million years ago.

This is consistent with the hypothesis articulated by Gross et al. (1988) that anadromy enabled the salmonids to exploit the more productive saltwater environment. These authors noted that anadromy was common only in the northern hemisphere, while the opposite, catadromy (exhibited by eels, see Chapter 2) was more common in the southern. They linked this to productivity in the sea – it is higher than fresh water in northern latitudes, but generally lower in the southern. Thus, in the northern hemisphere fish that could migrate to sea to grow could be at a selective advantage (growing faster and larger and thus producing more offspring more quickly) than fish that remained in fresh water.

Recently, McDowall (2008) has questioned aspects of this theory, suggesting that the cold-tolerating northern latitude fish were preadapted to capitalize on habitats that became available with the retreat of the ice sheets, and that anadromy enabled them to find these habitats, now separated by sea. Escape to the sea may also have enabled the fish to avoid harsh winter conditions.

BOX 9.2   **A LIFE-CYCLE IN POETRY**

Ted Hughes (1930–98, British poet laureate from 1984) was interested in fishing from an early age and started to fish for salmon seriously in his forties. In 1986 he wrote 'fishing is a substitute symbolic activity that simply short-circuits the need to write',[a] but during the 1980s he wrote many poems involving salmon, from which a complete salmon life-cycle in poetry can be constructed. Hughes' interest in the occult and astrology developed throughout his life along with his interest in natural history, and by the time of writing the poems, many concerning salmon, collected in *River* (1983), he reflected a growing awareness of a force animating all living creatures, in this case driving salmon back upriver to reproduce. The salmon is also a mythological and totemic fish in these poems. Hughes was also concerned with the more down-to-earth areas of salmon conservation and water quality.

Thus, grab a copy of the *Collected Poems* of Ted Hughes (Keegan, 2003), and within its almost 1200 pages of poems you will find much from which to construct a salmon life-cycle. Why not start with *The Best Worker in Europe* about salmon smolt and saltwater salmon producing biomass for human consumption, yet poisoned by their pollution, and continue to *September Salmon* and *October Salmon* for the return of the mature salmon to fresh water and an evocative description of the physical state of these fish. This can be followed by *Salmon Eggs* for spawning and *Orts Poem 49* for a description of the eggs.

[a]Letter to Anne Stevenson, Autumn 1986, pp. 521–522 (Reid, 2007).

---

upstream to prepare another egg-pocket: the completed nest, termed a redd, may consist of up to six egg-pockets. Salmon fecundity is very variable and associated with body size, but fewer than 20 000 eggs are laid, and often less than 2000. This is a low number for such a large fish, and also in comparison with the squid discussed in Chapter 3.

The salmon does not feed during its return to fresh water, even though it may spend ten months there, but strangely will take a fishing bait or lure. The female can lose up to 45% of her body weight during spawning, bringing her close to physiological death, and many die, along with most males. The survivors return to the sea to feed, usually drifting downstream tail-first, being too weak to move against the current. At this stage they are called kelts; only about 5–10% of salmon, predominantly females, survive to breed a second time in Britain and very few breed a third time. Those that do survive spend 6–18 months at sea before returning to fresh water to breed again.

The fertilized eggs hatch in the spring into alevins, some 70–160 days after being laid. These still have the egg yolk sac attached (having consumed only about 20% of the yolk) and feed from this within the gravel for 4–5 weeks. Once this has been exhausted the young, now about 2.5 cm long (and called fry or fingerlings) emerge: several thousand may emerge from a few square metres of gravel. From now on they feed upon invertebrates and grow slowly. At about 6–7 cm length the fry develop dark blotches along their sides, like thumb prints, and are now termed parr. During winter they conserve energy, entering spaces between or under stones, sometimes in redds from previous years.

Most young salmon migrate to sea between April and June, many after 2–3 years but some up to six years after hatching. The salmon (now called a smolt) is still small, no more than 15 cm long and weighing a few grams. During its travel to the sea the salmon has to become able to survive in salt water (Box 9.3).

Fig. 9.2 *Cassley Falls, on the river Cassley, Sutherland, an excellent Scottish salmon river [Courtesy of Chris Madgwick.]*

BOX 9.3 SMOLTIFICATION

The development of the salmon from the freshwater parr to the smolt able to survive in salt water is termed smoltification. The study of this phenomenon has intensified recently as it is a key developmental stage needing to be controlled in the salmon farming industry.

Smoltification involves behavioural, morphological and physiological changes to the fish. The developing smolt changes from an aggressive territory-holding individual who orients upstream against the current, to a shoal member facing downstream and moving with the water flow. Morphological changes include the loss and restructuring of fat tissue within the elongating body; the skin loses the parr marks and becomes more silvery, the fins become transparent and acquire black margins, and the scales become looser and shed more easily.

Physiological changes include a change in eye pigments to make the eye more blue-sensitive, matching light in the open oceans, and an increase in metabolic rate and scope for growth. One of the most important changes the smolt will have to cope with is salt water. Almost all bony (teleost) fish maintain their plasma osmotic concentration at about a third that of salt water, regardless of whether they are freshwater or saltwater species. Thus, in fresh water the fish will be more concentrated than its surroundings, and will tend to take in water through its skin (Fig. 1). The fish can compensate for this by not drinking water and excreting copious urine, but however efficient its kidneys some salts will be lost. Therefore, it takes up ions (e.g. sodium and potassium) actively from the water through the gills.

In salt water the fish has the opposite problem. It is now less concentrated than its surroundings, and will tend to lose water. Therefore, the fish drinks water, but as a consequence will tend to collect too many salts. These have to be actively secreted by the gills and kidneys. The salmon changes its ion and water regulation during the smoltification process. This is thought to be under the control of a number of hormones.

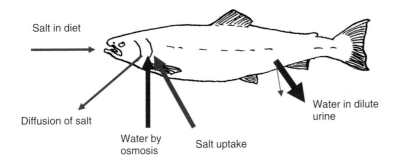

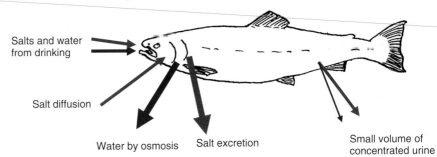

Box 9.3 Fig. 1 *Osmoregulation by the salmon in fresh and sea water. Arrows represent direction and magnitude of flow: blue, water; red, salts*

Smoltification is initiated the autumn before the fish move to the sea. Salmon parr that have reached a critical length of 7.5–8.5 cm in September to October start the smoltification process and put on a spurt of growth, reaching 12 cm length by the following spring, while smaller parr will grow more slowly and remain in fresh water for at least another year. Once smoltification is under way it is regulated by photoperiod, with temperature speeding up or slowing down the process. Until the role of photoperiod was realized, hatcheries tended to rear parr under a constant day length, which could result in unsynchronized smoltification – external signs of smoltification that were not matched by physiological changes. Fish farms now mimic changing winter and spring day lengths, and produce smolt that also can be transferred to sea in the autumn and at regular intervals throughout the year. This spreads out freshwater productivity and removes peaks and troughs of production: 30–40% of Norwegian smolts are now produced out of season.

As the smoltification process concludes, a series of factors such as increased temperature, light, and water flow interact with the fish's physiological state and initiate downstream movement. There is a 'smolt window' during which the salmon is physiologically ready to go to sea; this corresponds to an environmental window of best sea conditions and lack of predators. If for some reason – such as an obstruction in the river or low water level – the fish misses this window and does not make it to sea in time, a process of de-smoltification starts, and the fish has to wait another year to reach the sea.

The change in shape and colour of the salmon as it matures in fresh water has led to confusion as to its identity. Until the 1830s parr were regarded as either a variety of the brown trout, to which they look similar, or as a separate species, *Salmo salmundis* or *S. samulus*. This confusion arose as some male parr become sexually mature (a process called paedogenesis) and can fertilize the eggs of the returning females – the parr lie at the bottom of the egg-pocket, and in some cases may be more effective fertilizers than the mature males. This is probably an adaptation to ensure some egg fertilization in the absence of mature males. Old male salmon were also thought a separate species, *S. hamatus*, and the kelts another, *S. argenteus*, in the nineteenth century.

THE SALMON AT SEA

Jones (1959) wrote at the start of a short chapter on the salmon in the sea:

Once the smolts leave our rivers, we see little or nothing of them: we do not know where they go to; there is no evidence to show that they stay near our coasts. ... The first we usually see of our Salmon is when they enter our rivers again after at least a year and a half in the sea.[2]

This is no more informative than Izaak Walton could be 300 years previously. This lack of information was due to the salmon not shoaling, and avoiding the main fishing grounds; for example, there are only 78 Scottish records of salmon taken at sea between 1888 and 1954.

However, in 1956 a tagged Scottish salmon was caught in a west Greenland inshore fishery, and subsequently tagged salmon from Canada, Britain, Ireland and Sweden were all caught off Greenland. Changes in fishing areas in the North Atlantic led to further discoveries, and today we have a fair idea of the preferred feeding grounds for the salmon in the Atlantic (see Fig. 9.1): at the edge of the Arctic Ocean and along the east and west coasts of Greenland. The salmon concentrate in food-rich eddies where surface temperatures are between 4 and 8°C, associated with prominent ocean currents and continental shelf features. There are at least three races of salmon: a North American one in the Labrador Sea off

[2]Jones (1959), p. 57.

Greenland (also containing fish from parts of Europe); a European race off the Faeroes, which move far into the northern and north-eastern Atlantic; and a smaller more segregated Baltic Sea race. The eastern and western Atlantic populations are genetically different and have probably been diverging for over 500 000 years, with only limited gene flow between them. This has led some to suggest that these salmon populations should be distinct subspecies, *salar* and *sebago*, respectively.

While at sea the salmon behave as individuals and do not shoal. They show little directional migration and feed upon crustaceans and other large zooplankton, small squids, sand eels (*Ammodytes* species) and fish including sprats (*Sprattus sprattus*) and herring (*Clupea harengus*). They now grow far more quickly than in fresh water: after a year of sea feeding they may be 50 cm long and weigh over 2 kg, reaching 20 kg after four years.

Some salmon return to fresh water to breed after one year at sea (these are called grilse), but most return after two or more winters at sea. Salmon return at different times of the year and are classified as 'spring fish', 'summer fish' or 'autumn fish'. Some British rivers are 'early' (e.g. the Welsh Dee with salmon moving upstream in January and large numbers in early March); some are 'summer' (e.g. the Cumbrian Derwent, where most fish move upstream from early summer to autumn); and some are 'late', with no salmon run until late summer. This variation in return time may be a form of 'bet hedging' to try to ensure that some fish return at the most suitable time.

The salmon do not just return to fresh water, but tend to return to the same river they were born in: less than 3% stray to other rivers. This was stated by Izaak Walton in 1653:

> [I]t has been observed by tying a Ribon in the tail of some number of the young *Salmons*, which have been taken in *Weires*, as they swimm'd towards the salt water, and then by taking a part of them again with the same mark, at the same place, at their returne from the Sea...which hath inclined many to think, that every *Salmon* usually returns to the same River in which it was bred....[3]

It was not until the early twentieth century that this was confirmed by careful experimentation. But how do the salmon find their way back? Here we can be less sure.

The cue to start the return to fresh water is probably increasing sexual maturity. Navigation to their river must involve a couple of stages. First the salmon have to find the coast; this migration appears to be active and directional, with salmon following an almost constant compass heading and not always following the currents. The directional information required may be obtained from the sun, polarized light, or the Earth's magnetic field. It has also been proposed that salmon can detect the orbital movement of swell using sensors in their inner ear. To use such information, however, the salmon needs a map against which to relate these changes – could this be a magnetic map?

Once in coastal waters, more physical cues are available to the salmon, including changes in salinity, currents and tides. It is thought that olfaction plays an important part in finding the right river, and that the young smolts may learn characteristic odours of their home river while moving downstream to the sea, and that this becomes imprinted – a long-term memory. Salmon could learn the odours of more than one place along the river, and thus use this to guide them home. However, it is not clear which odours they are learning.

[3]Izaak Walton, *The Compleat Angler*, 1st edn, 1653, p. 119 (Bevan, 1983).

EXPLOITATION

As the glaciers retreated and the climate ameliorated slowly and fitfully in the northern hemisphere from about 15 000 BP, both salmon and humans would have returned (long before this, salmon bones found in Spanish caves inhabited 40 000 BP are some of the first records of fish consumption by humans). Stone Age populations in much of northern Europe lived along the coastline, a high-biodiversity high-stability environment, and characteristic fishing communities developed such as the Vedbæk (8000–6500 BP) and Ertebølle cultures (6500–5100 BP) in present-day Denmark. Fishing appears to have been a summer and early autumn occupation, and along with bone fish hooks remains have been found of fish nets made from plant fibres, wood net floats and stone sinkers, and wooden leisters – fish spears with a hazel shaft and hawthorn barbs, lashed together with cord, possibly from nettle: used especially by salmon poachers to the present day (Fig. 9.3).

Much of the fishing was carried out from fixed installations – fish traps or weirs, also called 'fixed engines'. Numerous fish traps in estuaries and along the coast have been found in north-western Europe from the late Mesolithic onwards. One of the earliest was recently excavated in the Liffey estuary, Dublin, where the remains of two 25 m long wattle fences, a C-shaped fish trap and a basket trap, dating from 6100–5700 BC were found. The fences were made mainly from hazel, *Corylus avellana*, and give an important indication also of the nature of woodland management at the time – the hazel was carefully cut and fashioned probably from coppice on an 8- to 9-year rotation (Chapter 10). Salmon have probably always been caught using these structures, and the reliable appearance of migratory salmon at their estuary at certain times of the year could have been an important event for early humans.

Other evidence also attests to the importance of salmon. Carved pictures of salmon are found on reindeer bones left in caves at Altamira, north-western Spain, dating from the Palaeolithic, and appear on rock paintings 6–10 000 years old in Scandinavia (Fig. 9.4). Although salmon appear only occasionally in the Norse myths, they are important to Celtic mythology in both Welsh and Irish records. In these myths, salmon inhabited pools or wells along important rivers, and in the Irish myths salmon gained wisdom by eating the nuts of the hazel tree. As in Norse myths when Loki (Chapter 19) changes into a salmon to escape detection, in Celtic myths various people are transformed into salmon. Salmon-like fish and fishing images are also found at the fourth-century temple at Lydney, above the Severn estuary, to the Romano-Celtic god Nodens. Furthermore, a fish, thought to be the salmon, is the second most common animal representation in the nearly 200 symbol stones erected by the Picts in eastern Scotland mainly during the eighth century AD.

Fig. 9.3 *Catching salmon by hand.* (**a**) *Catching salmon in Scandinavia in the sixteenth century with an unusual single-pronged leister, also drying and smoking salmon [from Magnus (1555).]* (**b**) *Salmon spear from Radnorshire, Scotland; the prongs are about 25 cm long [Redrawn from Jenkins (1974).]*

(a)

(b)

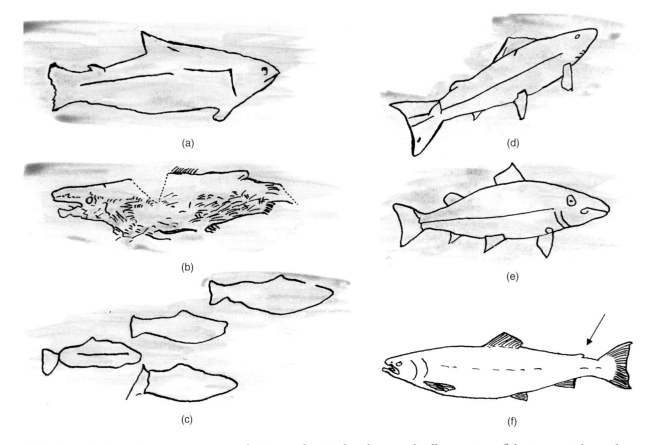

Fig. 9.4 *Early salmon rock art. In all cases the identification is conjectural, but these drawings all show signs of the adipose fin, characteristic of salmonids.* **(a)** *Gravid female, Palaeolithic, from Labastide, France.* **(b)** *Male (with kype), Palaeolithic, from Trois Frères, France.* **(c)** *Rock paintings from Tingvoll, Norway, 6000 BP.* **(d)** *From Glamis, Scotland, symbol stone, Pictish, eighth to ninth century.* **(e)** *From Easterton of Roseisle, Moray, Scotland, Pictish, eighth to ninth century.* **(f)** *Modern drawing of a salmon, with the small adipose fin arrowed [(a) and (b) redrawn from Guthrie (2005); (c) redrawn from Verspoor et al. (2007). Not to scale.]*

The Domesday Book only sporadically mentions fisheries, mainly on the major rivers, for example the Thames (which had weirs also) and Severn, but explicit reference to salmon is only occasional. For example, the church of St Peter, Gloucester, received a payment of sixteen salmon, a payment of six salmon was made at Much Marcle, Herefordshire, and a fishery at Eaton, Cheshire, had six fishermen and paid 1000 salmon. These fisheries, either 'fixed engines' or weirs, were a source of dispute and conflict. Many of the fishing rights were owned by monastic houses, and they clung tenaciously to them in the face of confiscation by various sovereigns. Local fishermen were also concerned that the monasteries were taking too many fish, so reducing their livelihood. As many of these disputes took place on excellent salmon rivers such as the Tyne, Derwent and Severn, we can be sure that salmon were involved. Acts were put in place in the late thirteenth century specifically to prevent fishing for salmon at certain times of the year, and in 1384 forbidding the use of a 'stalker' – a fine meshed net that would take salmon fry and smolt.

For larger riverside towns, fishing traps or weirs more importantly impeded navigation, and from the early thirteenth century efforts were made to control their use. Acts against fish weirs in the Thames, Severn, Ouse and Trent were passed in 1346, and repeated Acts were passed over the next 150 years, probably a sign of their ineffectiveness.

The salmon fishery at Wareham, on the west of Poole Harbour in Dorset, UK was probably typical of many small fisheries throughout the country. Wareham means 'wear settlement' or farmstead, and is mentioned from the seventh century. The early fishing wear, or kidel (from which the expression 'a pretty kettle of fish' comes), was probably removed during the thirteenth century, as the Magna Carta (signed in 1215) made provision for the removal of all such weirs from rivers in England. However, a fresh channel was cut into the river Frome at Wareham (Fig. 9.5) probably to get around this (the original course of the river has subsequently silted up) and a wooden weir was in place by 1234. This reached across the river and records from 1760 show it to have had five sections. By the nineteenth century the weir comprised wooden frames with 3-cm diameter wooden bars to prevent salmon moving upriver. In the middle of the weir was an opening with a purse-shaped hoop net leading to a bag, in which the salmon were caught.

In the early seventeenth century another cut, 1.5 km long and 3.5 m wide, containing two weirs was made by a tenant farmer seaward of the town (Fig. 9.5) to catch salmon before they reached the original weir. This was declared illegal, and the farmer had to remove the weir and fill in the cut.

During the first half of the nineteenth century the weir brought the salmon almost to extinction in the river Frome. The Salmon Fishery Act 1861 made this weir illegal. It was removed, and numbers of salmon gradually increased from only seven taken in the Frome in 1900, to 200 in 1920 and 600 in 1962, allowing an increase in rod fishing.

A variety of salmon weirs and traps were used on rivers in Britain and Ireland (see Figs 9.6 and 9.7), and were gradually regulated, reduced, and in some cases banned during the nineteenth and early twentieth centuries. Under the 1861 Act and the 1923 Salmon and Freshwater Fisheries Act, any weir used for catching salmon and trout and extending more than half way across a river at lowest water had to have a free gap of at least 10% of the width of the river at the

Fig. 9.5 *Salmon weirs at Wareham, Dorset, UK. Sketch map of present day river, with, inset, a conjectural mid-eighteenth century view.* **(a)** *The river Frome today, looking upstream from the bridge at Wareham Quay. The original course of the river is almost entirely silted up, the entrance indicated by the arrow.* **(b)** *The salmon weather-vane on St Mary's Church, Wareham, indicates the former importance of salmon to the town* [*Inset based on Clark (1950).*]

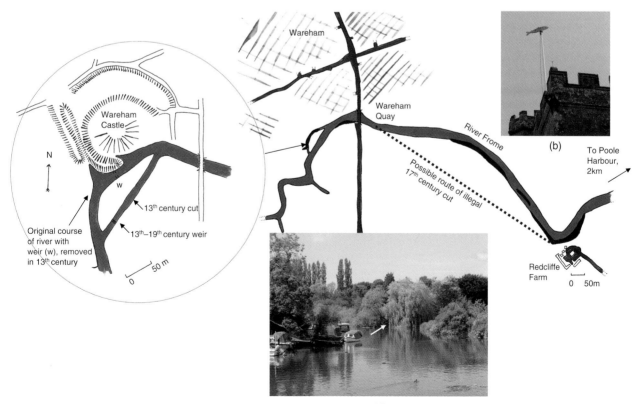

(a)

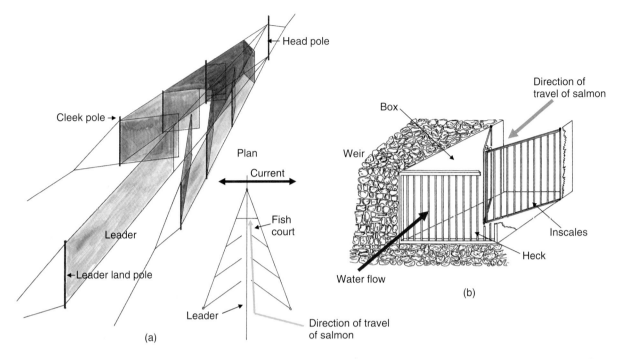

Fig. 9.6 (**a**) *Scottish bag net for coastal or estuarine salmon fishing.* (**b**) *Cut-away diagram of a salmon box trap, one of several designs once common in river weirs throughout Britain and Ireland. Salmon swimming upriver find it obstructed and move along the weir to the trap, passing through the inscales (arranged horizontally in some traps) into the box. Although salmon could swim through the heck they generally do not and can be netted from the box*

deepest part. This led to many salmon weirs being abandoned. The 1923 Act also ordered that any dam or obstruction to fish migration placed across a river had to have an approved fish ladder.

In estuarine and coastal waters, too wide to span with weirs, 'fixed engines' little different from Neolithic fish traps were used. On rocky coasts the bag net was used – a leader net was placed across the route the salmon took along the coast; the fish would see this and be forced into a bag (Fig. 9.6). In the Severn estuary a distinctive form of 'fixed engine' salmon fishery developed over 1000 years using putchers – conical open woven wicker baskets 60 cm in diameter at the mouth, 150–200 cm long and tapering to a point, placed in three to five parallel rows to form a weir or rank between the high and low tide levels, set at an angle to the current. Some ranks faced upstream and some downstream to catch fish on both the ebb and flow of the tide. The visibility is so poor in the silt-laden Severn that salmon cannot see these obstructions: this method would not work in clearer waters. Over 11 000 putchers and the larger and more complicated putts were in operation in the Severn estuary in 1865, and the intertidal area of the Severn is littered with their remains (Fig. 9.7). Since the 1940s putchers have been made from metal, and today fewer than half a dozen ranks remain on the Severn, and only a couple still operate.

Fixed engines were expensive to erect and maintain and relatively easy to regulate. Far more portable and harder to regulate were salmon nets, and again many estuaries had their own netting methods. For example, in the Lune estuary and Solway Firth, UK the haaf net is still used. This is a net on a 5-m wide e-shaped frame; the fisherman wades into position with the net, facing the current and twisting and lifting the net to catch salmon individually. In other rivers and estuaries nets were paid out from small craft. In many cases a coracle was used – a very small keel-less bowl-shaped boat made from animal skins stretched over a wicker frame and cartable by one person. In other cases specially designed fishing boats were developed, including the Severn stop-net boats (Fig. 9.7c).

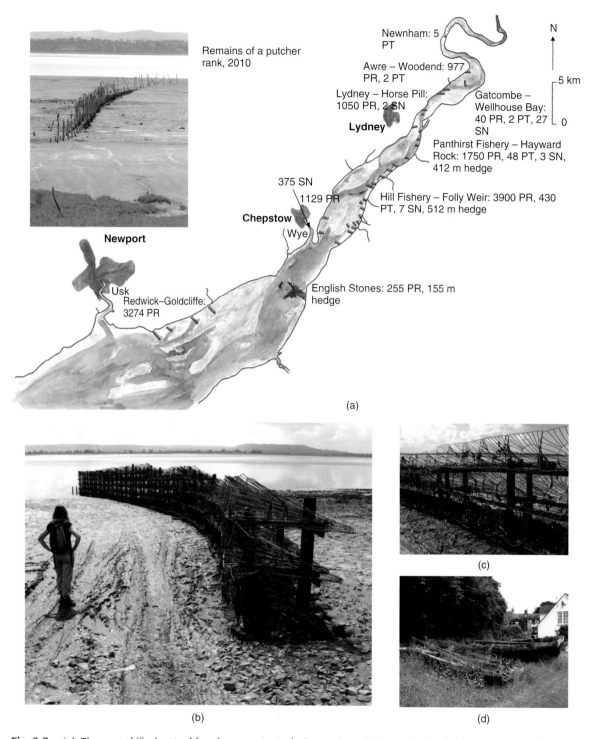

Fig. 9.7   (**a**) *The extent of 'fixed engines' for salmon trapping in the Severn estuary, UK in 1865. Grey indicates mud at low tide. PR, putcher; PT, putt; SN, stop net. Hedges were wickerwork fences placed to guide salmon into the putchers and putts.* (**b, c**) *A working putcher rank in the Severn Estuary in 2009, at low tide. It uses metal putchers, the open mesh of which will catch only larger salmon.* (**d**) *Remains of the last Severn salmon stop-net boats at Gatcombe in 2009. Stop-net fishing involved anchoring the boats to a cable fixed to the river bed. The boats carried a large U-shaped net suspended from two poles; this net was placed under the boat, counterbalanced by weights and held open by a support stick controlled by the lone fisherman in the boat. When a salmon hits the net the stick is dislodged, and the net rises and closes. Stop-net boats worked off Chepstow and Lydney until the mid-1980s* [Data in (a) from Jenkins (1974).]

Coastal sea fishing for salmon was also practised. Much of this was by drift-netting, using a net in which salmon were caught by the gills. Although used in British estuaries and in the Baltic for many years previously, the use of this net offshore by Norwegian and Scottish fishermen from 1960 was viewed with alarm by salmon anglers – for the first time large numbers of salmon could be caught away from their rivers. In 1960 a few boats started fishing for salmon using drift nets off the Northumberland coast, and when reports of good catches were received several Scottish boats took part in this fishery later in the year. More joined in 1961 and salmon were drift-netted up the east coast of Scotland from up to 40 boats. In the following year former herring drifters joined in as well, with 160 boats involved and 115 000 drift-netted salmon landed in Scottish ports. This salmon fishing stampede was halted in Scotland later in 1962 with the banning of drift-netting and the landing of fish so caught at any British port.

Salmon drift-netting relies on intercepting fish returning up the coast to their spawning grounds. It involves shooting an 8-m-deep net suspended on a rope with floaters from a boat. Originally hemp nets were used; these are visible to the fish, and thus work best in the disturbed seas produced from winds of force 3–5. By the late 1960s, light, artificial monofilament nets had been developed; these are almost invisible to the fish and thus can be used in calm seas. These nets not only made legal fishing more efficient and safer but made illegal fishing much quicker and easier, and considerable effort has been made by the coastguard and navy to eradicate illegal drift-net salmon fishing off the British coast: in 1986 it became an offence even to carry a monofilament drift net on a British fishing boat in Scottish inshore waters. Today regular helicopter patrols help control salmon poaching, but during the 1970s and 80s there were many reports of violent incidents during the seizure of illegal fishing boats and nets.

The ease of catching salmon with monofilament nets caused further alarm. Their use was banned in some countries, and in 1989 salmon drift-netting was banned altogether in Norway. However, in England and Ireland salmon drift-netting was still legal, although licences were controlled. On the north-east coast of England between Filey and Berwick-upon-Tweed, salmon fishing increased rapidly from the 1960s: 140 nets were licensed in 1992, catching on average that decade 34 000 fish a year, and dominating the English and Welsh reported salmon catch. After decades of negotiations, the fewer than 100 remaining salmon drift-net fishermen accepted voluntary compensation to quit in 2003, at a cost of over £3 million. The even larger Irish salmon drift-net fishery was not banned until 2007.

The discovery in the late 1950s of where salmon fed at sea led to the development of deep-sea salmon fishing. In 1964, Faeroese and Norwegian fishermen started to fish for salmon off Greenland, and were later joined by boats from Denmark and Greenland. These used free-floating drift nets of up to 5 km in length. Monofilament nets were introduced here as well, which enabled daylight fishing. This soon became the largest salmon fishery of all, and at its peak, in 1971, 2700 tonnes of salmon was caught – representing about a fifth of the world catch of salmon (Fig. 9.8).

For the first time, salmon were being caught outside of their country of origin (60% of the salmon caught in the Greenland fishery had originated in

North American rivers, the rest from Europe), and any regulation would have to take place by international agreement. Since the late 1970s, regulation of this fishery has been difficult and only the threat of an embargo of Danish imports by the USA (Greenland is a self-governing province of Denmark) was effective in reducing the fishery. In 1998, commercial fishing stopped and a quota of 20 tonnes for internal use was set. Bar a couple of glitches, this has been maintained since, and in 2007 a new seven-year moratorium on commercial salmon fishing in Greenland waters was signed.

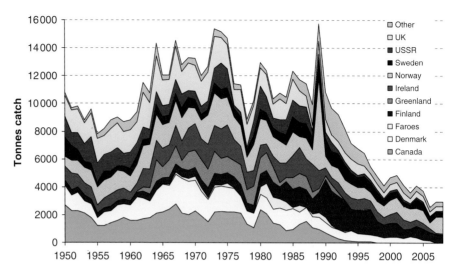

Fig. 9.8 *Annual catch of wild salmon between 1950 and 2008. Note that the variation in catch is dependent on changes in quotas as well as salmon abundance. A better measure of the latter would be catch per unit effort [Data from FAO fishery statistics.]*

After the declaration of a 320 km Faeroese Exclusive Economic Zone in 1977, deep-sea salmon fishing by Faeroese vessels steadily increased. They had experimented during the 1970s with catching salmon on floating sprat-baited long-lines, similar to those used for squids (Chapter 3), and although initially thought to be unsuccessful, the number of boats using this method increased greatly during the 1980s, and up to 30 km of line could be fished at once. This fishery became a valuable industry after the collapse of the Faeroese herring fishery and the reduction in cod stocks. It peaked in 1982 (Fig. 9.8) and was then regulated; in subsequent years the quota was often not caught. Since 2000 there has been no commercial Faroese salmon fishing.

With the regulation of sea fishing, an increasing proportion of the salmon catch now comes from rivers by anglers: about 65% of the catch in North America and 40% in Europe, with increasing use of catch and release.

SALMON FARMING

In his 1932 novel *Brave New World*, Aldous Huxley described a dystopian future in which most humans were bred artificially and natural reproduction was outlawed. Depending on conditioning and treatment, the human embryo could develop into an 'epsilon-minus semi-moron' or an 'alpha-plus', the ruling caste. This world has parallels with the beehive (Chapter 4) as well as with salmon farming, not least in the domestication of a species. In the Brave New World only a few 'wild' humans are left in reservations: the same has happened to the salmon, where over 99% are now farmed.

Although freshwater salmon hatcheries have been used since the mid nineteenth century to augment the salmon stock in rivers, farming salmon to produce mature fish for food started, in Norway, only during the 1950s when young salmon were taken into salt water. Salmon farming began in Scotland during the 1960s in a similar manner, and since then has increased significantly (Fig. 9.9). It is now a highly developed industry with many distinct processes and it is perhaps

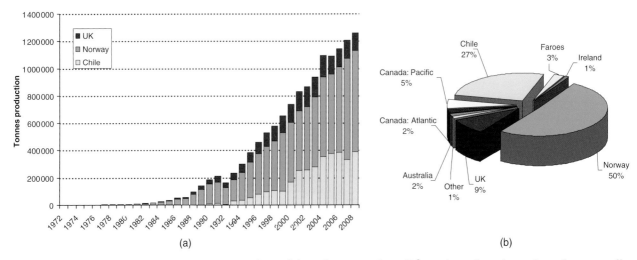

(a)   (b)

Fig. 9.9 **(a)** Tonnes of salmon produced by the three main salmon farming countries. **(b)** Proportional contribution of the main salmon farming countries to farmed salmon production in 2008 [Data from FAO fishery statistics.]

surprising that a fish with as complex a life-cycle as the salmon has taken so well to being artificially reared.

The salmon farming cycle begins with the production of eggs and sperm. These are 'stripped' from a reared, selectively bred broodstock of mainly Norwegian origin. The males (cocks) and females (hens) are separated as they become sexually mature, and when 'ripe' they are stripped by hand. After mixing, the fertilized eggs are moved into a hatchery.

Hatcheries are now specialized operations, some producing up to 30 million fish in a year. The eggs are incubated in the dark in high-quality flowing fresh water. On hatching, the alevins drop to the bottom and, to prevent bunching and to reduce activity, gravel, astro-turf, or rubber mats are used as a substrate. When the yolk sac has been consumed the fry are moved to tanks and kept at a high density (e.g. 10 000 per square metre) to encourage feeding by learning and competition. This is a critical period, and water quality, temperature and flow, and light (now a stimulus for growth) needs to be controlled. The fry are fed a carefully formulated starter diet; as they grow the water level in their tank is increased to reduce fish density, and water flow is adjusted so that the fish can maintain station without too much movement. As the fish develop at different rates they are subsequently graded and separated into different tanks based on size.

When the fish have developed into parr, at about 2 g weight, they are transferred into freshwater pens in sheltered lochs (Fig. 9.10a), or moved into larger tanks on land. In the latter, by controlling temperature and day length, smolt can be produced at different times of the year (see Box 9.3). These fish are fed a formulated dry diet and can grow quickly: a 5 g parr can double in weight in 14 days, and double again in another 21 days, far faster than they would grow in the wild. After smoltification the fish are acclimatized to salt water and are moved into the sea for their final growth (Fig. 9.10b). In 2008, over 36 million smolt were produced in Scotland.

At sea the fish are placed in cages. The first cages were simple floating wooden cradles with a net hanging below. These structures were small (under 10 m³) and could be used only in the sheltered waters of fjords, bays and lochs. These areas could only support small farms, and the slow mixing of the water could cause problems in waste dispersal. Therefore, from the early 1960s offshore salmon farms were developed. These became common in the 1980s, allowing for a dramatic increase in salmon farming (Fig. 9.9), with large-volume

(a)

(b)

Fig. 9.10 *Scottish salmon farms.* **(a)** *Freshwater rearing in Loch Migdale, Sutherland, Scotland.* **(b)** *Saltwater rearing in Badcall Bay, Sutherland* *[Courtesy of Chris Madgwick.]*

5000–10 000 m³ cages being used. Offshore salmon cages have to be able to withstand much greater wave and wind action than the sheltered inshore ones. Furthermore, disturbance adds to the stress on the salmon and they will not feed during a storm. Several solutions have been developed, including flexible breakwaters around the cages to reduce wave action, and cages that automatically submerge when waves reach a predetermined height.

The cost of the smolts represents about 20% of the total cost of salmon production at sea, but the cost of feed represents 40%. At sea the fish need to be fed as much as they will eat. Although fish were originally fed frozen blocks of fish, by the late 1980s almost all farms used a manufactured food. Protein is the largest part of this feed and fish-meal is the main protein component, coming largely from Chilean and Peruvian herring, menhaden or anchovy (see Box 3.5). It takes roughly three tonnes of these 'industrial fish' to produce one tonne of salmon. Owing to the high cost of fish-meal, attempts have been made to replace some of it with soya and maize meals. Bone and poultry meals were also used until banned after the BSE scare in Europe during the late 1980s.

Salmon in the wild have pink flesh caused by carotenoids present in the krill and shrimp they eat. These carotenoids are not found in the manufactured diets of farmed salmon and have to be added as supplements to the farmed fish's diet, or the resulting flesh would be a drab grey colour. After up to two years' growth at sea the salmon weigh up to 5 kg and are harvested.

With the rapid increase in production of farmed salmon since the early 1990s and its consequent fall in price, salmon has gone from being a luxury food of the few to a more everyday food of the many. The addition of this healthy food to more diets can only be a good thing. The increase in local employment has also been positive; for example, in 2008 the salmon farming industry supported about 5000 jobs in Scotland and produced salmon worth £324 million. However, the farmed production of salmon is not without its environmental consequences and there has been considerable, often acrimonious, argument over the regulation of fish farms, particularly in Scottish coastal waters. Until recently, the Crown Estate (which owns most of the foreshore and seabed around Scotland) was effectively both landlord and regulator of salmon farming in Scotland; this dual role was widely criticized.

In both freshwater and sea salmon rearing, the three main environmental problems are effluents, parasites/diseases, and escaped fish. Effluent, in the form of faeces, uneaten food, chemicals used as disinfectants, anti-foulants (Chapter 13), pesticides and antibiotics will be washed out of cages. At the turn of the millennium it was estimated that the organic waste entering the sea from all existing Scottish and Norwegian salmon farms was equivalent to the post-treatment sewage discharges from 52 million people. Some of this effluent sinks to the sea bed around the cage where it can smother the community and lead to an increase in bacteria. Some of these consume oxygen, leaving less oxygen in the water for other organisms; other anaerobic bacteria within the waste can produce hydrogen sulphide and methane. Much of the waste nitrogen and phosphorous remains in the water column, and although rapidly dispersed from the area, may be linked with the production of toxic algal blooms. These have killed several million farmed salmon as well as countless wild organisms, and have rendered some cultured and fished mussels unfit for human consumption.

Farmed salmon suffer from many diseases and parasites and infection is probably aided by stress and overcrowding. Although treatments for many of these afflictions have been developed, from antibiotics and pesticides through to

BOX 9.4   SEA LICE

A number of ectoparasitic organisms are associated with the salmon. Most, such as the freshwater pearl mussel (see Box 9.5), occur in low densities and do not harm the fish. The saltwater sea louse, *Lepeophtheirus salmonis*, a crustacean (Copepoda: Caligidae), also caused few problems until the advent of salmon farming. The first sea louse was noted in farmed salmon soon after farming began in the 1960s, and by the 1990s was one of the greatest threats to the sea-farming of salmon: the cost to salmon farming of sea lice in Scotland alone was estimated at £15–30 million in 1998.

The sea louse *L. salmonis* is found exclusively on salmonids and is an obligate parasite – it cannot survive without its host, and cannot survive for long in fresh water. The adult females produce pairs of strings from their abdomens, containing 100 to 1000 eggs each, and these hatch into non-feeding free-swimming planktonic nauplius larvae. After another naupliar stage the larvae moult into an infective copepodid stage (Fig. 1) which attach to the host by their antennae. They then develop through four chalimus stages immobile on their host: during the subsequent pre-adult stage they can move between them. After a second pre-adult stage the adult stage is reached. An individual sea louse can live for up to seven months. This life-cycle has certain similarities to that of another crustacean, the barnacle (Chapter 13).

The damage caused by the sea louse depends on the number present on the salmon. The lice are epithelial feeders and as this tissue is continually replaced a couple of lice will cause little problem. In fish farms, however, large populations can build up and over 500 lice have been found on one fish. High densities cause significant erosion around the head, causing osmotic shock and increased infection by pathogens, and leads to sluggish fish which do not feed and are more likely to die.

Treatment of sea lice is problematic. The fish are not in a closed system, and any prophylactic used will need to be applied in large amounts, will spread into, and could affect, the surrounding sea. Treatment originally involved one of a number of organophosphate insecticides, to which sea lice soon developed resistance. The fish needed to be repeatedly treated and also suffered increased mortality from the chemicals, which were accumulated by the salmon. This was the main reason for their reduced use during the late 1980s.

Hydrogen peroxide treatment superseded organophosphates in the UK. This is effective against the lice, but also very toxic to the salmon. Other insecticides, including pyrethroids (very toxic to aquatic invertebrates; see Box 12.5) and insect growth regulators, have also been tried and are used under licence in some situations. During the 1990s there was increased use of the veterinary 'wonder drug' ivermectin (a broad-spectrum invertebrate pesticide) for sea louse control. This again has a low margin of safety to the fish, and accumulates within it (in the mid-1990s, 10% of farmed salmon sampled in British shops contained residues of ivermectin). It was used illegally on salmon in several countries, and was licensed for a time in Scotland and Ireland. This was despite the threat it posed to other marine fisheries, for example shellfish, which are very sensitive to the chemical.

Cultural control methods – such as fallowing, selecting stock for resistance and careful monitoring of sea lice in fish farms – should be part of the management of the sea louse problem. In Norway in particular, several species of wrasse cleaner fish, caught wild and introduced into the fish farms, have been effective in removing sea lice from the salmon.

As well as the increased pollution caused by chemical sea louse treatments, there is increasing concern over the escape of sea lice from fish farms into the wild, and the potential effect this could have on wild salmonids. The nauplius larvae tend to float to the surface and are carried into coastal waters by onshore winds. Thus, wild salmonids waiting in estuaries to enter rivers are particularly susceptible to infection. Sea trout are more susceptible to the effects of sea lice than salmon, and the escape of sea lice has been suggested as one of the main reasons for the severe decline in Irish sea trout numbers since the late 1980s. Sea trout have been observed returning to fresh water early, possibly to kill the lice, before returning to sea again.

It is now accepted that sea lice can move between farmed and wild salmonids, and that sea lice densities are significantly greater around coastal fish farms. However, the link between this and the decline in wild salmonids is difficult to prove as there is much variation in sea lice densities in the wild away from salmon farms. Since the late 1990s one of the aims of fish-farm sea louse management programmes has been to greatly reduce the number of sea lice escaping into the sea. But, a decade or more on, it is unclear whether these plans are having the desired effect.

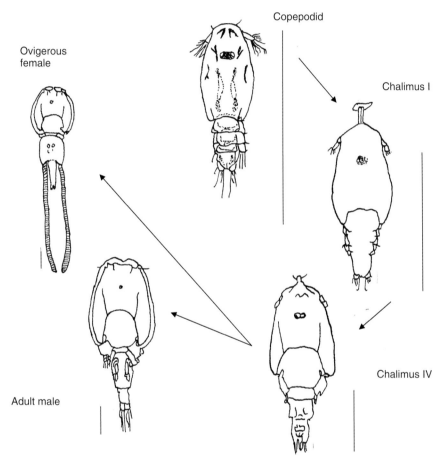

Box 9.4 Fig. 1   *Life-cycle of the sea louse* Lepeophtheirus salmonis. *Bar = 1 mm* [Redrawn from Pike and Wadsworth (1999).]

vaccination, the stimulation of natural immunity, and use of probiotics, salmon farming has led to an increase in infection of wild salmonids. Salmonid-specific pests such as the sea louse (Box 9.4), the monogean fluke *Gyrodactylus salaris* (a freshwater pest), and diseases such as furunculosis (caused by the bacterium *Aeromonas salmonicida*) and the infectious salmon anaemia virus[4] have been increasingly detected in wild fish.

        Many farmed salmon escape into the sea from cages damaged by storms or collisions with vessels. In 1999, 250 000 salmon escaped from one farm alone in the Orkneys when cages broke in a storm, and 20–40% of the fish caught by the

[4]A severe outbreak of this virus in Chilean salmon farms is likely to lead to an 85% reduction in salmon production in 2010 compared to 2008.

Faeroese offshore salmon fishery are thought to be escaped farm salmon (in some Norwegian fjords over 80% of the salmon caught in 2003 were escaped farmed fish). Overall, it is likely that two million farmed salmon escape into the Atlantic each year; this is about 50% of the apparently wild salmon found in this location. This loss from sea cages is one reason why transgenic salmon, developed since the 1990s (with higher growth rate and tolerance to lower temperatures), should not be approved for release.[5]

The farmed salmon are genetically dissimilar to the wild fish, having been selected during 30 years of breeding for rapid growth in fish-farming conditions. Unwittingly, selection may have also taken place for increased competitiveness and aggression for food and for decreased anti-predator responses. In contrast, wild salmon exist in genetically distinct populations adapted to particular rivers – body shape and fin size vary with water flow characteristics. There is concern that these characteristics may be lost in wild salmon from interbreeding with farmed fish, a loss that may reduce the ability of the wild salmon to cope with environmental change. Farmed female salmon seem to hybridize with brown trout more than wild salmon do, but these resultant hybrids are rarely fully fertile.

Strong evidence for the effects of fish farming on the environment has been hard to obtain. Part of this is due to the nature of the marine environment and salmon life history, but it has been suggested that the unwillingness of some organizations to sponsor research which may lead to greater regulation of this industry has also had an effect. The results can be seen in current Scottish Government reports. For example, in June 2009 a report on restoration guidance for west-coast salmonid fisheries stated that a restocking solution could not be recommended as the cause of the decline had not been clearly identified. This report was widely seen as a whitewash, especially as there was little mention of aquaculture effects and no mention of sea lice. A month before, 'A Fresh Start', a renewed strategic framework for Scottish aquaculture, was launched by Marine Scotland (itself established on April Fool's Day that year). This did mention sea lice, but stated that its wish to control them was hampered by lack of knowledge of the extent of the problem.[6] It was not until June 2010 that proposals were made to establish an engineering standard for Scottish Aquaculture.

THE END OF THE WILD SALMON?

The life-cycle and habits of the salmon make it particularly vulnerable to the actions of humans. Wild fish can only be profitably harvested when mature, and before breeding. Coupled with a slow growth rate, this means that it is easy to over-fish – the reason why recent developments in deep-sea salmon fishing were controversial. Furthermore, the rapid growth of fish farming is likely to be adding to the problems of the wild stock. This will have knock-on effects on other members of the ecosystem (Box 9.5).

Until recently, the major threats to the salmon were found in fresh water: fishing, obstructions and pollution. Fish weirs were not only a hindrance to navigation, but a barrier to salmon returning upstream to spawn. Other weirs (e.g. for mills), coupled with locks for navigation and, increasingly, dams for hydroelectric schemes have also inhibited the movement of salmon. Increasing water extraction has lowered river levels and caused parts of salmon rivers

[5]However, in July 2010 it looked likely that a transgenic salmon would be authorized by the end of the year for release in the US. This would be the first transgenically engineered animal for human consumption, and has a gene inserted from the Chinook salmon (*Oncorhynchus tshawytscha*) and one from ocean pout, *Zoarces americanus* which enable it to grow to size in half the time currently taken. The fish are all females and are bred sterile to prevent interbreeding with wild salmon.

[6]Insert your own proverb here; suggested are 'seek and ye shall find' or 'none so blind as will not see'.

BOX 9.5   THE FRESHWATER PEARL MUSSEL

The decline of the wild salmon will have an effect on many other organisms, including food species and competitors in fresh water and sea. One species, however, has a more intimate relationship with the salmon and is being affected directly by its decline – the freshwater pearl mussel *Margaritifera margaritifera* (Bivalvia: Unionoida). This mollusc is considered a keystone species (one that has profound effects on a range of species in an ecosystem) of salmon rivers and was once common throughout much of the salmon's range: it is now threatened and endangered.

This filter-feeding mussel lives partly or completely buried in sand and gravel in fast-flowing unpolluted rivers (Fig. 1). Large mussel beds can develop and individuals can live for well over 100 years (reaching 12–15 cm length), making them one of the longest-lived invertebrates. The species has separate sexes and the male sheds sperm into the water in early summer; some is inhaled by the females and fertilized eggs develop in brood pouches on their gills. In late summer the eggs hatch and the larvae – called glochidia, and looking like tiny (0.6 mm long) mussels with their shells held apart – are released into the water. These need to be inhaled by salmon or trout, usually fry in their first year after hatching; the shells snap shut on a gill filament and the mussels encyst and grow parasitically. During the following spring the cysts rupture and the juvenile mussels fall to the bottom and burrow into the sand or gravel and mature over 10–15 years.

This is probably an example of a symbiotic relationship, whereby both organisms benefit. Certainly the pearl mussel is dependent on salmonids for part of its life-cycle, and the fish will also enable the mussel to colonize upstream. The relationship is not obligatory for the salmon, but they could benefit from cleaner water resulting from the mussel's filter-feeding, and the mussel beds could provide a suitable habitat for the invertebrates on which the salmon feed. The cysts do not damage the salmon, and there is some evidence that they could be beneficial, inhibiting senescence and stimulating resistance to stress and some diseases. Pearl mussels are subject to the same threats as salmon to their habitat, and also have suffered from a destructive fishery.

The abundance of freshwater pearl mussels in Britain was known to the Romans, and Suetonius suggested that this was one reason why Julius Caesar ordered the first Roman invasion of Britain in 55 BC. The mussels have been fished intermittently since, with a large-scale industry developing in Britain and Ireland in the sixteenth century, and again in the 1860s when a German merchant expressed interest in purchasing freshwater pearls. It is generally impossible to search a mussel for its pearl without killing it and this, along with their slow growth rate and very poor glochidial survival, means that this industry cannot be sustainable and fishing declines after a few years.

Into the 1970s a handful of full-time freshwater pearl fishers operated, travelling Britain and the continent to find mussel beds, which they usually searched on a 15-year rotation. As mussels under 8 cm in length are unlikely to contain a pearl, these were generally not sampled. It is probable that this low-level exploitation would have been more sustainable, but there also has been considerable damage to mussel beds from less-discerning amateurs looking for pearls.

(a)

(b)

Box 9.5 Fig. 1   *The freshwater pearl mussel* Margaritifera margaritifera. **(a)** *Freshwater pearl mussel being reared in captivity, Windermere, UK.* **(b)** *Freshwater pearl mussel fishing in Scandinavia in the sixteenth century* [(b) from Magnus (1555).]

Freshwater pearl fishing was licensed from 1991 and in 1998 was made illegal in Britain. Under the Wildlife and Countryside Act 1981 it is an offence to harm freshwater pearl mussels or their habitat, and similar protection has been afforded throughout most of their range. Furthermore, the selling of pre-1998 caught Scottish pearls is strictly licensed. In 2008, the first British prosecution took place: a contractor was fined for excavating a stretch of the River Irt in Cumbria and damaging over 1 km of the river bed, destroying mussels and thousands of salmon and trout eggs. In 2005, the Scottish Police and Scottish Natural Heritage started 'Operation Necklace' to prevent fishing of Scottish sites – thought to represent over half of the 100-odd known viable freshwater pearl mussel populations in the world, and the only viable populations in Britain.

The mussel unfortunately continues to decline, and 2008 was possibly the worst year yet with several Scottish rivers targeted by thieves. In Wales, where very few populations remain, all the mussels from some sites have been removed into captivity, and there is now a programme of hatchery rearing of juvenile mussels and their subsequent release. Likewise, at the Freshwater Biological Association's laboratory on the banks of Windermere, freshwater pearl mussels from threatened northern English populations are being reared using fish-farm technology and farmed salmon and brown trout. However, captive rearing has to go hand-in-hand with restoration of the salmonid populations of rivers where the mussels live.

to dry out at times, and during the twentieth century increasing canalization and simplification of river systems for effective flood prevention has destroyed salmon habitats.

During the nineteenth century, increased untreated sewage from the rapidly growing British population was discharged into rivers, along with waste and chemicals from developing industries, agriculture and forestry. The salmon is a fish requiring clean water and one of the first to be affected by this pollution.

The above effects rapidly reduced salmon numbers during the nineteenth century, and by the middle of the century many of the richest salmon rivers had declined markedly: the Thames was one of the first British rivers to lose the salmon, during the 1830s. Publicity, for example by Charles Dickens, drew attention to this decline and a Royal Commission of Enquiry was set up in 1860. The Salmon Fisheries Act 1861 resulted, and laid down a uniform closed season for nets and rods throughout the country, and attempted to regulate the discharge of waste into rivers. Over the next 100 years a succession of Acts attempted to deal with water pollution.

Until recently these Acts were ineffectual, paying greater heed to the needs of industry than wildlife or even human health. However, the increasing strength of the later Acts – and in particular the Control of Pollution Act 1974 – coupled with better sewage treatment and the decline of heavy industry, led to an improvement in water quality from the 1960s, followed by the slow return of the salmon. By 2000, salmon had been recorded from almost all the catchments from which they had been lost, albeit often in small numbers. The release of large numbers of bred salmon into these cleaner rivers speeded up this colonization process, as has improved upstream access. For example, a recently completed set of 37 fish passes on the rivers Thames and Kennet (Fig. 9.11) means that for the first time in 150 years salmon can potentially reach high enough up the Thames and its tributaries to spawn.

However, the return has been slow. In the Thames a couple of returning salmon have been recorded annually from the mid-1970s and from 1979 the river and its tributaries have been artificially stocked – in 2005 almost 69 000 smolts and 16 000 fry were released. While hundreds of salmon returned from the sea to the Thames in the 1980s and 90s, and there was a rod catch of 22 Thames salmon in 1996, no fish returned in 2005 and only two in 2006, and it is unclear whether any salmon have spawned recently in the Thames.

Fig. 9.11 *Boulter's Weir, on the River Thames, near Maidenhead, UK, in January 2009.* **Inset**: *Fish ladder added early in the twenty-first century to help salmon reach upstream. A weir has been present in this position for over 600 years. Salmon and eels were trapped at earlier weirs: although 17 kg salmon were caught in the 1790s, catches soon became erratic with increased pollution and the building of locks and weirs downstream, and the last salmon was caught here in 1823/4*

Although the recent decline in numbers of salmon returning to the Thames has been put down to low river flow, salmon stocks elsewhere also continue to decline: in rural British rivers salmon numbers declined by 42% between 1974 and 1998, including rivers least affected by industrial pollution. Wild salmon have recently disappeared from Germany, Switzerland, The Netherlands, Belgium and other European countries. A WWF survey of over 2000 historic salmon rivers in 19 countries during 2000 found that only 43% of these rivers had a healthy wild salmon population, populations in 30% of rivers were endangered or vulnerable, 12% were listed as critical and 15% were extinct (see Fig. 9.1). It is possible that increasing acidification of waters and runoff of effluent and fertilizers from intensive agriculture have had an effect on salmon, but increasing attention is now being paid to the decline in the population of salmon at sea. The number of salmon caught at sea has steadily declined since the 1980s (see Fig. 9.8); and although increasingly restrictive fishing quotas and salmon farming has been implicated recently in this decline, it started earlier and is more widespread to be accounted for by these alone.

It is possible that global climate change could be having an effect. As salmon is a coldwater fish, and inhabits the surface water layers, any increase in sea surface temperature could restrict its range, and some studies have noted a marked effect of increased temperature upon salmon. Similar has been suggested in fresh water: Johnson *et al.* (2009) estimate that climate change will make southern UK rivers a less favourable habitat for salmon in the future.

Another candidate is change observed in global weather patterns linked with the North Atlantic Oscillation Index. This is a measure of sea-level atmospheric pressure distributions between Iceland and the Azores. It is the major influence on storms, rainfall and temperature, and drives North Atlantic climate variability. The start of the steep decline in salmon in the sea corresponds to the North Atlantic regime shift (see Box 3.5 for more details on regime shifts) of the early 1980s, and was followed by a northwards shift of the Gulf Stream a couple of years later. This regime shift, following a shift in climatic variables and increase in sea surface temperature, had a widespread impact on marine organisms, including the prey of the salmon. Warm-water copepods moved north and there was a decrease in the number of coldwater species. Small copepods increased from 1984 and phytoplankton from 1986, while a key zooplankton, *Calanus finmarchicus*, decreased.

In 1988 there was a significant decrease in salmon abundance. Anthropogenic climate change will probably only exacerbate this shift, and it is predicted that the decline in numbers of salmon returning to their home rivers will continue.

It is ironic that, after many decades, when legislation and practical efforts to increase salmon stocks and regulate and reduce our effects upon this species are finally having a significant impact, the salmon is probably under greatest threat from a poorly understood and unpredicted change in global weather patterns.

FURTHER READING

Jones (1959) has still the best account of the salmon life-cycle in fresh water, but is out of date for the sea stage. Netboy (1968, 1980) is also useful, especially for a world-wide view of salmon exploitation, updated by Mills (2003). The ecology of salmon in Britain is covered by Hendry and Cragg-Hine (2003) and Verspoor et al. (2007). Salmon farming is covered by Stead and Laird (2002), the effect of salmon escapes is discussed by Esmark et al. (2005), Fiske et al. (2006), Hindar et al. (2006) and Walker et al. (2006). Wheeler (1979) discusses the loss of salmon from the Thames, and to counter our southern bias see Williamson (1991) and Maitland (2007) for Scotland, and Vickers (1988) for Ireland.

Thorpe (1988) covers aspects of salmon migration, Odling-Smee and Braithwaite (2003) learning and fish orientation, and Lohmann (2010) magnetic field perception in animals.

See Clark (1950) for salmon fishing at Wareham, Jenkins (1974) for salmon trapping on the Severn, and Ayton (1998) for English and Welsh salmon fishing. Tsurushima (2007) reviews records for eleventh century salmon fishing, and McQuade and O'Donnell (2007) describe the Mesolithic fish trap in the Liffey. For possible effects of climate change on salmon see Reist et al. (2006); for regime shifts and the salmon see the references in Chapter 3 and also Lees et al. (2006) and deYoung et al. (2008) for general aspects and Beaugrand and Reid (2003) for specifics related to salmon. The status of salmon rivers is covered in WWF (2001).

For the development of anadromy, see references in Box 9.1 and Ramsden et al. (2003). Smoltification is reviewed by McCormick et al. (1998) and McCormick (2001). Pike and Wadsworth (1999), Boxaspen (2006) and Costello (2006) review salmon sea lice and their control, and McVicar (2004) gives one view on establishing the effect of sea lice on wild salmonids. Freshwater pearl mussels are introduced in Cosgrove et al. (2000), Hastie and Cosgrove (2001) and Skinner et al. (2003). The Atlantic Salmon Trust (www.atlanticsalmontrust.org) is an excellent source of information, and publishes a series of useful 'blue books' on aspects of salmon biology and exploitation.

REFERENCES

Ayton, W. (1998) *Salmon Fisheries in England and Wales*. The Atlantic Salmon Trust, Pitlochry, Scotland.

Beaugrand, G. and Reid, P.C. (2003) Long-term changes in phytoplankton, zoo-plankton and salmon related to climate. *Global Change Biology* **9**, 801–817.

Bevan, J. (ed.) (1983) *Izaak Walton: the compleat angler 1653–1676*. Clarendon Press, Oxford.

Boxaspen, K. (2006) A review of the biology and genetics of sea lice. ICES Journal of Marine Science **63**, 1304–1316.

Clark, H.J.S. (1950) The salmon fishery and weir at Wareham. *Proceedings of the Dorset Natural History and Archaeological Society* **72**, 99–110.

Cosgrove, P., Hastie, L. and Young, M. (2000) Freshwater pearl mussels in peril. British *Wildlife* **11**, 340–347.

Costello, M.J. (2006) Ecology of sea lice parasitic on farmed and wild fish. *Trends in Parasitology* **22**, 475–483.

Esmark, M., Stensland, S. and Lilleeng, M.S. (2005) *On the Run: Escaped farmed fish in Norwegian waters.* WWF-Norway, Oslo (www.wwf.no/core/pdf/wwf_escaped_farmed_fish_2005.pdf).

Fiske, P., Lund, R.A. and Hansen, L.P. (2006) Relationships between the frequency of farmed Atlantic salmon, *Salmo salar* L., in wild salmon populations and fish farming activity in Norway, 1989–2004. *ICES Journal of Marine Science* **63**, 1182–1189.

Gross, M.R., Coleman, R.M. and McDowall, R.M. (1988) Aquatic productivity and the evolution of diadromous fish migration. *Science* **239**, 1291–1293.

Guthrie, R.D. (2005) *The Nature of Paleolithic Art.* University of Chicago Press, Chicago, USA.

Hastie, L. and Cosgrove, P. (2001) The decline of migratory salmonid stocks: a new threat to pearl mussels in Scotland. *Freshwater Forum* **15**, 85–96.

Hendry, K. and Cragg-Hine, D. (2003) *Ecology of the Atlantic Salmon.* Conserving Natura 2000 Rivers Ecology Series no. 7. English Nature, Peterborough, UK.

Hindar, K., Fleming, I.A., McGinnity, P. and Diserud, O. (2006) Genetic and ecological effects of salmon farming on wild salmon: modelling from experimental results. *ICES Journal of Marine Science* **63**, 1234–1247.

Ishiguro, N.B., Miya, M. and Nishida, M. (2003) Basal euteleostean relationships: a mitogenomic perspective on the phylogenetic reality of the 'Protacanthopterygii'. *Molecular Phylogenetics and Evolution* **27**, 476–488.

Jenkins, J.G. (1974) *Nets and Coracles.* David & Charles, Newton Abbot, UK.

Johnson, A.C., Acreman, M.C., Dunbar, M.J. *et al.* (2009) The British river of the future: how climate change and human activity might affect two contrasting river ecosystems in England. *Science of the Total Environment* **407**, 4787–4798.

Jones, J.W. (1959) *The Salmon.* Collins, London.

Keegan, P. (ed.) (2003) *Ted Hughes: collected poems.* Faber & Faber, London.

Lees, K., Pitois, S., Scott, C., Frid, C. and Mackinson, S. (2006) Characterizing regime shifts in the marine environment. *Fish and Fisheries* **7**, 104–127.

Lohmann, K.J. (2010) Magnetic-field perception. *Nature* **464**, 1140–1142.

Magnus, O. (1555) *Historia de Gentibus Septentrionalibus* [transl. P. Fisher and H. Higgens, ed. P. Foote]. The Hakluyt Society, London [3 vols 1996–1998].

Maitland, P.S. (2007) *Scotland's Freshwater Fish, Ecology, Conservation and Folklore.* Trafford Publishing, Victoria, BC, Canada.

McCormick, S.D. (2001) Endocrine control of osmoregulation in teleost fish. *American Zoologist* **41**, 781–794.

McCormick, S.D., Hansen, L.P., Quinn, T.P. and Saunders, R.L. (1998). Movement, migration, and smolting of Atlantic salmon (*Salmo salar*). *Canadian Journal of Fisheries and Aquatic Science* **55** (Supplement 1), 77–92.

McDowall, R.M. (2001) The origin of the salmonid fishes: marine, freshwater . . . or neither ? *Reviews in Fish Biology and Fisheries* **11**, 171–179.

McDowall, R.M. (2008) Why are so many boreal freshwater fishes anadromous? Confronting 'conventional wisdom'. *Fish and Fisheries* **9**, 208–213.

McQuade, M. and O'Donnell, L. (2007) Late Mesolithic fish traps from the Liffey estuary, Dublin, Ireland. *Antiquity* **81**, 569–584.

McVicar, A.H. (2004) Management actions in relation to the controversy about salmon lice infections in fish farms as a hazard to wild salmonid populations. *Aquaculture Research* **35**, 751–758.

Mills, D. (ed.) (2003) *Salmon at the Edge.* Blackwell Science, Oxford.

Netboy, A. (1968) *The Atlantic Salmon: a vanishing species?* Faber & Faber, London.

Netboy, A. (1980) *Salmon: the world's most harassed fish.* André Deutsch, London.

Odling-Smee, L. and Braithwaite, V.A. (2003) The role of learning in fish orientation. *Fish and Fisheries* **4**, 235–246.

Pike, A.W. and Wadsworth, S.L. (1999) Sealice on salmonids: their biology and control. *Advances in Parasitology* **44**, 233–337.

Ramsden, S.D., Brinkmann, H., Hawryshyn, C.W. and Taylor, J.S. (2003) Mitogenomics and the sister of Salmonidae. *Trends in Ecology and Evolution* **18**, 607–610.

Reid, C. (ed.) (2007) *Letters of Ted Hughes.* Faber & Faber, London.

Reist, J.D., Wrona, F.J., Prowse, T.D. *et al.* (2006) General effects of climate change on Arctic fishes and fish populations. *Ambio* **35**, 370–380.

Skinner, A., Young, M. and Hastie L. (2003) *Ecology of the Freshwater Pearl Mussel.* Conserving Natura 2000 Rivers Ecology Series no. 2. English Nature, Peterborough, UK.

Stead, S.M. and Laird, L. (eds) (2002) *Handbook of Salmon Farming.* Springer-Verlag, Berlin, Germany.

Thorpe, J.E. (1988) Salmon migration. *Science Progress, Oxford* **72**, 345–370.

Tsurushima, H. (2007) The eleventh century in England through fish-eyes: salmon, herring, oysters, and 1066. In: *Anglo-Norman Studies XXIX, Proceedings of the Battle Conference 2006* (ed. C.P. Lewis), pp. 193–213. Boydell Press, Woodbridge, UK.

Verspoor, E., Stradmeyer, L. and Nielsen, J. (eds) (2007) *The Atlantic Salmon: genetics, conservation and management.* Blackwell Publishing, Oxford.

Vickers, K. (1988) *A Review of Irish Salmon and Salmon Fisheries.* The Atlantic Salmon Trust, Pitlochry, Scotland.

Walker, A.M., Beveridge, M.C.M., Crozier, W., Maoiléidigh, N.Ó. and Milner, N. (2006) Monitoring the incidence of escaped farmed Atlantic salmon, *Salmo salar* L., in rivers and fisheries of the United Kingdom and Ireland: current progress and recommendations for future programmes. *ICES Journal of Marine Science* **63**, 1201–1210.

Wheeler, A. (1979) *The Tidal Thames: the history of a river and its fishes.* Routledge & Kegan Paul, London.

Williamson, R. (1991) *Salmon Fisheries in Scotland.* The Atlantic Salmon Trust, Pitlochry, Scotland.

WWF (2001) *The Status of Wild Atlantic Salmon: a river by river assessment.* WWF.

deYoung, B., Barange, M., Beaugrand, G. *et al.* (2008) Regime shifts in marine ecosystems: detection, prediction and management. *Trends in Ecology and Evolution* **23**, 402–409.

THE TRUTH ABOUT OAK FORESTATION

Oak

CHAPTER 10

"Hallo, Pooh," said Piglet.

"Hallo, Piglet. This is Tigger."

"Oh, is it?" said Piglet, and he edged round to the other side of the table. "I thought Tiggers were smaller than that."

"Not the big ones," said Tigger.

"They like haycorns," said Pooh, "so that's what we've come for, because poor Tigger hasn't had any breakfast yet."

Piglet pushed the bowl of haycorns towards Tigger, and said, "Help yourself," and then he got close up to Pooh and felt much braver, and said, "So you're Tigger? Well, well!" in a careless sort of voice. But Tigger said nothing because his mouth was full of haycorns . . . .

After a long munching noise he said: "Ee-ers o i a-ors."

And when Pooh and Piglet said "What?" he said "Skoos ee," and went outside for a moment. When he came back he said firmly: "Tiggers don't like haycorns."

A.A. Milne, *The World of Pooh*

UNLIKE TIGGERS, many animals thrive on acorns, the nut of the oak tree. In this chapter we shall first consider this and other uses of oak: as a source of tannin for leather production, wood for charcoal manufacture, timber for building ships and houses, and as a cultural icon. Then we investigate the basis (physiological and evolutionary) for the extraordinary profligacy with which oaks produce acorns in certain 'mast' years, and the consequences of this behaviour for the ecological community centred on oak. We discuss how oak woods composed of the two species considered native to Britain, *Quercus robur* (pedunculate oak) and *Q. petraea* (sessile oak) (Box 10.1) look today, and how they may have looked in prehistoric times. The progress of development of the primeval 'wildwood' down to present times is described, along with recent DNA-based evidence on the origins of present-day oaks. Finally, we look ahead to a time when fossil fuels have been exhausted and the wood economy can be expected to resume something of its former importance, very likely with oak at its heart. Our perspective on oak is inevitably British at root; but we look outwards to the rest of the world wherever possible, acknowledging the importance of the broader picture of the genus *Quercus*.

Wild consumers of the acorn include rabbits, mice, deer, bears, squirrels, jays, rooks and woodpeckers. Domestic pigs were traditionally pastured in oak (and beech) woods in Britain, the quantities of woods in eastern England being

Biological Diversity: Exploiters and Exploited, First Edition. Paul Hatcher and Nick Battey.
© 2011 John Wiley & Sons, Ltd. Published 2011 by John Wiley & Sons, Ltd.

BOX 10.1   THE TAXONOMY OF OAKS

The oaks are mainly trees (some are shrubs) in the genus *Quercus*, family Fagaceae. Other members of the family are *Castanea* (chestnut), *Fagus* (beech), *Castanopsis, Lithocarpus,* and *Trigonobalanus*; they are rich in tannins, have inconspicuous flowers and a fruit which is usually a single-seeded nut held by a distinctive cupule (which is derived from extensions of the pedicel of the flower). The family is mainly from the northern hemisphere; Fagaceae and the southern hemisphere *Nothofagus* (southern beech) are now thought not to be sister taxa.

The greatest diversity at the genus-level in the Fagaceae is found in south-east Asia, but fossil evidence indicates that western North America was equally diverse in the early Tertiary; maximum diversity was probably attained about 25 mya (in the Upper Oligocene). By this time the genus *Quercus* had achieved widespread distribution.

Oaks are very difficult taxonomically, as Charles Darwin commented in his *Origin of Species*. Long-distance gene flow and regular interspecific hybridization create intermediate forms which are a challenge for classification. Hence, today *Quercus* is estimated to have around 500 species, the exact number depending on the authority. This is about half the total number in the Fagaceae. *Quercus* has two subgenera, and the distribution of these and the sections within subgenus *Quercus* are shown in Fig. 1. Subgenus *Quercus* includes the sections *Quercus* (the white oaks, found in Europe, Asia and North America); *Cerris* (Turkey oak and relatives, of Europe and Asia); *Lobatae* (red and black oaks of North America, Central America and northern

South America); and *Protobalanus* (golden-cup oaks of North America). The other subgenus is *Cyclobalanopsis* (the cycle-cup oaks of south-east Asia).

What is striking about most of these groups, each of which is monophyletic, is that they are found on *either* the Eurasian *or* the American continent, not both. A detailed analysis of the fossil record and the current distribution led Axelrod (1983) to conclude that these main oak groups evolved where they are currently found. The exception is the white oak group, which includes *Q. robur, Q. petraea, Q. alba* (the white oak of eastern North America), *Q. lobata* (California oak, whose acorns were eaten by Native North Americans) and *Q. dentata* (Japanese Emperor oak). The wide distribution of this group across both continents was interpreted by Axelrod to suggest initial evolution in North America, then radiation out to Eurasia, via the Bering land bridge, during the Oligocene. Ancestors of the chestnut oaks, which include *Q. prinus* (rock chestnut oak), are the most likely candidates for this migratory role.

Finally, Axelrod noted that most oaks occupy latitudes below 40°N, with the highest concentration of species in Mexico and Central America, and south-east China. *Q. robur* and *Q. petraea* are exceptions in extending as far as 60°N. Intriguingly, this was also true of oak ancestors in the Tertiary, when generally much warmer, wetter conditions over the planet allowed many other species to extend as far north as Spitzbergen. What is it about northerly latitudes that oaks have always disliked? Axelrod suggested long days, but no-one really knows.

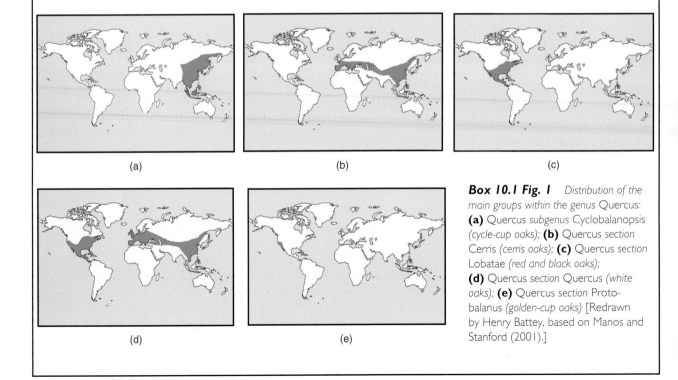

(a)   (b)   (c)

(d)   (e)

Box 10.1 Fig. 1 *Distribution of the main groups within the genus* Quercus: **(a)** Quercus *subgenus* Cyclobalanopsis *(cycle-cup oaks);* **(b)** Quercus *section* Cerris *(cerris oaks);* **(c)** Quercus *section* Lobatae *(red and black oaks);* **(d)** Quercus *section* Quercus *(white oaks);* **(e)** Quercus *section* Protobalanus *(golden-cup oaks)* [Redrawn by Henry Battey, based on Manos and Stanford (2001).]

described in the Domesday Book of 1086 in terms of the number of swine that could be fattened in the autumn, or 'pannage'. Indeed, commoners' rights of pannage still exist in parts of Britain today, for example in the New Forest. Good Spanish chorizo still gains its distinctive quality through the fattening of Iberian pigs on acorns produced by holm oak (*Q. ilex*) and cork oak (*Q. suber*). These oaks extend over an area of some 4 million hectares in the *dehesas*, the traditional arid climate silvopastoral system of central Spain.

Humans have also made use of acorns as food, and they were an important element in the diet of several Native American tribes such as the Miwok. The acorns were cached in raised granaries, safe from rodents, and acorn flour was made using pestles in mortar holes formed in natural limestone slabs; these holes are still visible today in northern California. John Muir, the naturalist and inspiration behind the National Parks of the USA, praised the bread made from acorn flour as a compact and strength-giving food.

So why don't more people eat acorns now? Oak trees are abundant, and in a good mast year a single oak tree can produce 50 000 acorns. Given the high carbohydrate and fat content of acorns, one would expect them to be an important basic food. In some cases, such as the chestnut oaks of North America (see Box 10.1), the raw acorns can be sweet and edible, and in the Far East *Q. cornea* and *Q. cuspidata* are consumed. But often the problem is a high concentration of tannins, phenolic compounds whose main general function is probably to defend plants against predators. These tannins render the acorns indigestible to many animals. Thus the Californian Indians could only exploit the acorn as food after extensive washing of the milled nuts to leach out the tannins. The extent of tannin indigestibility varies with species: pigs seem relatively unaffected, while both acorns and oak leaves can be toxic to horses. The ability to digest acorns may be a contributory factor in the battle for ascendancy between the grey and the red squirrel (Box 10.2).

The high tannin content of the acorn is shared by the rest of the oak tree, and particularly the cambial region[1] immediately below the bark. This is why oak bark

[1]The cambium is the meristematic region giving rise to secondary growth in a tree. In fact there are two cambia – the vascular cambium, which gives rise to secondary phloem and xylem, and allows growth in girth of the trunk; and the cork cambium, responsible for the production of bark. Both cambia are located immediately beneath the bark. As well as being important in the production of new cells, the main sugar transport around the plant occurs in this region; hence the major impact of girdling or 'bark ringing' on the growth and wellbeing of trees. For further information, see Esau (1965).

BOX 10.2   OAKS AND SQUIRRELS

The grey squirrel (*Sciurus carolinensis*) was introduced to Britain from North America during the nineteenth century, the first well-documented release being in Cheshire in 1876. As the grey squirrel dramatically expanded its range across England and Wales during the twentieth century, the native red squirrel (*S. vulgaris*) declined in parallel. The red squirrel has now been replaced across most of its range by the grey, its decline being particularly pronounced in mixed deciduous woodlands. An early hypothesis to account for this decline was that the grey could make better use of acorns. Feeding experiments by Kenward and Holm (1993) indicated that red squirrels were much less efficient at digesting acorns, and this was associated with a failure to metabolize tannins. This observation fits with the general pattern of feeding competition between the two species: young red squirrels show a lower growth rate when grey squirrels are present, and females breed less. Additional factors, such as pilfering of red-squirrel seed caches by the grey, also contribute to the competitive advantage of the grey squirrel.

Although feeding competition may partly explain the success of the grey squirrel in oak-dominated forests, it does not account for its overall success, particularly in conifer forests. Recent explanations for the decline of the red squirrel have focused on its susceptibility to squirrel poxvirus, which the grey squirrel carries. This susceptibility probably combines with the competitive advantages of the grey squirrel in feeding, the result being the overwhelming of the red by the grey.

Current concern centres on the potential spread of the grey squirrel from northern Italy, where it has also been introduced and led to a decline of the red squirrel, across the rest of Europe. If left unchecked, this spread is predicted to be complete within the next 100 years (Bertolino et al., 2008). A trial eradication of the grey squirrel from Italy was stopped as a result of legal action by animal rights groups (Bertolini and Genovese, 2003), since when no further action has been attempted. The problem of how (or even whether) we should try to deal with the spread of the grey squirrel remains unresolved, even though failure to act could lead to the extinction of the red squirrel over much of its native range, as well as the de-barking and consequent damage to trees over a huge area of European forest.

was used for traditional tanning of animal hide into leather. In the agrarian economy of Britain in the sixteenth and seventeenth centuries, leather working was an important manufacturing industry, second only to textiles. Boots, shoes, bags, gloves, saddles and harnesses were all essentials of life. The industrial revolution added new uses for leather; leather buckets, bellows, piston-sleeves and machine belting were required for coal mines, steel furnaces and textile factories. Also, during the Napoleonic wars, the army became a major consumer, requiring leather for boots, saddles, holsters and scabbards. Throughout this period of high demand, British oak bark was the main source of tanning material, and it remained so until the middle of the nineteenth century when substitutes (such as Turkish valonea, the dried acorn cups of the Valonea oak, *Q. macrolepis*) became available from overseas.

Indeed, during the eighteenth and early nineteenth centuries the demand for tanned leather had increased so much that most oak was probably grown principally for its bark. The woodland historian Oliver Rackham has estimated that, in 1810, 500 000 tons of oak would have been felled for its bark, compared with not more than 200 000 tons for timber for shipbuilding. Rackham associates the large areas of oak coppice still remaining in Britain today with this need for bark, which could be obtained from coppiced underwood, in contrast to timber which demands mature trees; the best bark came from about 20 years' growth. Once stripped of its bark the remainder of the coppiced oak was used for other economic purposes, including charcoal production.

Leather tanneries were found in most towns, typically being located on the outskirts because of their unpleasant smell. The traditional tanning process involved soaking the raw animal hides in lime, and scraping off excess flesh and hair. Material for light leathers was then placed in infusions of bird and dog dung, while hides for heavier leather were treated with fermenting barley or rye meal. Lastly, the hides were taken through a series of 20–30 open-air tan pits containing tannin at increasing concentrations, obtained from whole or shredded oak bark. The whole process took from six months to two years. During tanning the oak tannins cross-link the collagen protein, converting the putrescible skin into leather, a stable material that is resistant to microbes and heat.

In England and Wales, oak bark was supplied from local woods and also major deciduous forests, those in Sussex and Hampshire being of esteemed, tannin-rich character. In Scotland in the eighteenth century much of the lowlands was treeless, so bark was often obtained from the highlands and transported by cart to tanneries in centres such as Inverness. The bark harvest, along with charcoal made from the barked stems, was an important source of income, and the cultivation of oak for this purpose probably accounts for the large amount of 'scrub' oak (abandoned coppice) still found in the highlands. Smout (2005) gives a fascinating account of the way this exploitation of oak has created many of the even-aged Atlantic oak woods of western Scotland valued today by conservationists, not least because of their exceptionally rich moss and lichen flora.

Bark could only be 'peeled' from April to June when the cambium becomes active, and was stripped, using a 'barking iron', from the main trunk of the tree before felling (Fig. 10.1). Typically the rest of the bark would be removed from the smaller branches as the tree lay on the ground, although Edlin[2] relates how barkers in the Forest of Dean traditionally peeled the whole tree (including the smaller, most elevated branches) before felling, an activity evidently requiring much skill and bravery. The peeling of bark of *Cinchona* for quinine has some similarities and is discussed in Chapter 12.

[2]Edlin (1973), p. 88.

(a)

(b)

Even though the prosaic product (bark) became for a time quantitatively more important than the well-known one (timber), oak timber has over the centuries been basic to the British construction industries. Its distinctive properties (hardness, durability, strength) made it sought after for both houses and ships. Medieval building appears to have been heavily dependent on oak; at least it is the most common timber in those buildings still standing. The difficulty is that this impression may be a biased one, due to the durability of oak! However, timber frames do seem mainly to have been of oak, with a trend from the use of large numbers of small trees to make the timbers, to larger trees sawn into components of the required size by the early seventeenth century.

But it is for its role in the construction of ships for the British navy that oak is most renowned. This association has, at least since Tudor times, been a curious mixture of fact and fictional rhetoric. Oak was unquestionably vital to the construction of both the naval and the merchant fleets. Shipwrights believed implicitly in the superiority of British oak—preferably sourced from the woods of Sussex, Surrey or Hampshire—and there was continual concern about the availability and sustainable supply of large oaks suitable for construction of ships of the line. As an indication of the quantities involved, it is estimated that 36 acres (14.5 ha) of woodland was required to provide the oak for the *Mary Rose* (built 1509–10). It was against this background that John Evelyn's *Sylva* was written in 1664, and his concern over the vulnerability of the British navy to a shortage of oak explains the priority given to it by Evelyn.

Yet it is striking how the new edition of Evelyn's book was still being used over 100 years later (in 1776) for a similar purpose, this time by Evelyn's disciple Alexander Hunter, to lobby and agitate for the planting of acorns (and other trees). The difficulty was not planting, the various campaigns having had considerable success, but the tension between the need to allow oaks to grow for 100 or 200 years in order to provide (in posterity) for the good of the nation, and the pressure to harvest trees young for a quick profit. This conflict dogged the British landowning class, and its navy, for generations. The requirements for oak were not only increasingly prodigious (construction of a 74-gun ship of the line at the time of the Napoleonic wars would typically require 2000 two-ton oaks) but also quite specific, the shapes offered by oak being unique (Fig. 10.2). For instance, 438 'knees' of oak (single pieces of timber naturally curved through 90 degrees) were required to build *HMS Victory*, Admiral Nelson's flagship at the Battle of Trafalgar in 1805. Here the natural tendency of oak to branch almost horizontally was crucial.

The thought of oak branching reminds us that, alongside the undoubted qualities of oak for shipbuilding, there has always been a more rhetorical dimension to oak. This is summed up by the title of the report of the Liverpool

Fig. 10.1 **(a)** *Bark stripping [From Edlin, H.L. (1973). Woodland Crafts in Britain. David & Charles, Newton Abbot, UK];* **(b)** *Remains of a bark peelers hut, Parrock Wood, Haverthwaite, Cumbria, UK*

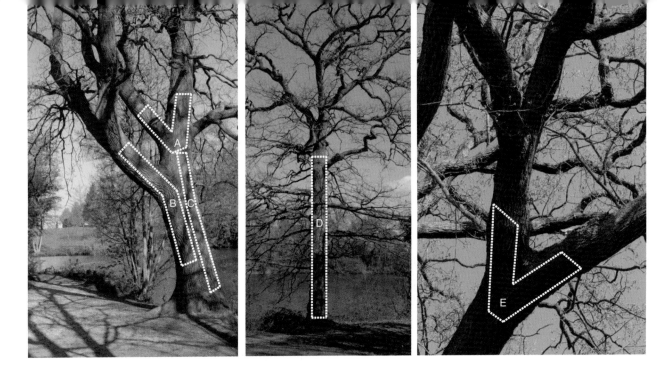

Fig. 10.2 *Particular shipbuilding cuts possible from some oaks*, Quercus robur: *A = crutch; B = cathead or top piece; C = futtock; D = stern post; E = ordinary knee*

shipwright Roger Fisher to a parliamentary committee enquiring into the oak shortage in 1763: *Heart of Oak: the British bulwark*. The oak is tough in its heart, individualistic and dependable, reflective of the (desired) character of the British. This symbolic function may have been most developed in Britain because of its island status, the 'wooden walls' of the navy protecting the kingdom; yet the oak is of national importance in many other countries, including France, Germany, Estonia, Poland, and the USA. Clearly, as well as being a common, well-loved tree (Rackham has estimated that 12% of English town-, village- and hamlet-names mention oak), it has a character that inspires awe and respect. This is best captured by Oliver Wendell Holmes:

> There is a mother-idea in each particular kind of tree, which, if well marked, is probably embodied in the poetry of every language. Take the oak, for instance, and we find it always standing as a type of strength and endurance. I wonder if you ever thought of the single mark of supremacy which distinguishes this tree from all our other forest-trees? All the rest of them shirk the work of resisting gravity; the oak alone defies it. It chooses the horizontal direction for its limbs, so that their whole weight may tell, – and then stretches them out fifty or sixty feet, so that the strain may be mighty enough to be worth resisting. You will find, that, in passing from the extreme downward droop of the branches of the weeping-willow to the extreme upward inclination of those of the poplar, they sweep nearly half a circle. At 90° the oak stops short; to slant upward another degree would mark infirmity of purpose; to bend downwards, weakness of organization.[3]

REPRODUCTIVE BEHAVIOUR (MASTING) IN OAK: CAUSES AND CONSEQUENCES

We now move to a consideration of the spectacular reproductive behaviour of oak trees. We shall focus initially on the oaks native to Britain, but then extend our discussion more widely, to the oak forests of North America.

[3]Holmes (1906), p. 224.

What, then, are the oaks of Britain, and what do they look like? *Q. robur*, pedunculate (English) oak, is typically the massive, pioneering oak of the lowlands of the east and south of England. It is the solitary oak of hedgerow and field. The wide branch angle of the oak is a feature shown very strikingly by *Q. robur* when grown in an open situation (Fig. 10.3a). In closed stands found in woods and forests, both oaks will tend to produce straighter taller trees, but this is particularly clear in the case of the sessile oak (*Q. petraea*) (Fig. 10.3b). Thus, *Q. petraea* is typically more slender, upward-branching; it is common in the oakwoods of the west, particularly in Wales, and is characteristic of highland woods. Yet there is much mixed oakwood: there are woods of *Q. robur* in highland regions on the west, and there are pure stands of *Q. petraea* in the east, for example in Kent. The distributions of the two species are made yet more complex by extensive planting. The species can be most readily distinguished by their acorns, which are stalked (or pedunculate) in *Q. robur* (while the leaves lack a prominent petiole). Conversely, *Q. petraea* has stalkless acorns, and its leaves have petioles. However, the two species hybridize extensively, so intermediate types are frequent. The importance of hybridization to the spread of oaks into Britain is discussed later, in the section 'How the oaks got to Britain'.

The life of an oak tree can be long, and trees can become very large. Under favourable conditions a 20-year-old will be about 8 m tall; a mature tree of 100–200 years may be 30–40 m. With age, oak trees often become 'stag-headed', as the highest branches die back (Fig. 10.4a). Old oaks of 300 years or more are not uncommon and are frequently important in local culture and history (Figs 10.4b and c). Flowering does not generally occur until oak trees are 20–30 years old, although coppicing tends to reduce this to 10–20 years. Therefore, in maiden trees (those produced from seed and not cut back at any time) substantial acorn crops typically occur from an age of about 40 years. Even more striking is the marked yearly variation in acorn crops once fruiting has begun: some years see a bumper crop, only to be followed by an interval of one or several years before another occurs. This erratic behaviour is known as mast seeding, 'mast' being the German word for fattening, recalling the tradition of feeding up pigs and other animals on seeds in the autumn months (see earlier). Such mast seeding is a feature of many oak species, and indeed many tree species in general, and has been the subject of intense research interest. Dipterocarps (e.g. *Shorea* species), which are important rainforest trees over much of the tropics, particularly in Malaysia, are the classic example—in addition to masting, these large trees show synchrony between species: all species in a region flower

Fig. 10.3 *The two oak species native to Britain:* **(a)** Quercus robur *(***inset***: pedunculate acorn, leaf with small petiole);* **(b)** Quercus petraea *(***inset***: sessile acorn, leaf with prominent petiole) [(b) main picture courtesy of Cathy Newell Price.]*

(a)   (b)

(a)

(b)

(c)

Fig. 10.4 **(a)** *Stag-headed oak.*
(b) *Verderers Oak. The trunk is about*
7 m in girth at its base. This tree is
about 320 years old, and lives in the
Forest of Dean, UK. It is named for
the Verderers, who are appointed to
administer Forest Law, to protect the vert
(green—woodland) and the venison (deer).
(c) *Suter's Oak, New Forest, UK, over*
300 years old [(b) Courtesy of Cathy
Newell Price.]

and fruit in one year after a period of ten or more barren years. The question that has intrigued researchers is why trees like this show such irregular seed production.

The earliest explanation for mast seeding was that seed output simply reflects the tree's resource availability: in a good year it will invest heavily in seed; otherwise, if resources are limited, vegetative growth takes priority. However, thinking based on evolutionary concepts raised the question of whether such heavy seed production might be a strategy to reduce losses of seeds to predators. This 'predator-satiation' hypothesis suggests that there is a fitness advantage to the tree species in periodically saturating seed predators so that some seed survives to produce the next generation. Then in 'off-years' predator numbers fall, allowing rapid saturation of the predator population in the next mast year. Note that according to this hypothesis maximum fitness benefit would derive from synchrony between individuals in the occurrence of mast seeding years. For oaks, the situation is complicated because seed predators like jays, squirrels and mice are also seed dispersers.

Quercus petraea is a typical masting species. It produces large crops about every eight years, with very few acorns in the years between. *Q. robur*, on the other hand, produces some acorns every year, with heavy crops roughly every other year. Crawley and Long have studied this alternate bearing in *Q. robur*, and their data are shown in Fig. 10.5. This research indicates that the pattern of seed production was fairly synchronous between individual oaks, although there was no environmental signal which correlated with 'on' (or 'off') years. This led the authors to favour a resource control model, in which heavy fruiting occurs only when resources exceed a critical threshold. They also found evidence for predator satiation in the case of the gall wasp *Andricus quercuscalicis* (Box 10.3) whose galls inflicted lower acorn mortality during 'on' than 'off' years. Vertebrate predators (woodmice, rabbits, grey squirrels, jays and wood pigeons) were a major factor in the low survival of acorns: where rabbits were very prominent they had the potential to eliminate oak recruitment, even in 'on' years. However, where rabbits were kept to low densities, predator satiation seemed to be effective, leading to the overall conclusion that irregular bearing in *Q. robur* facilitates a higher seedling recruitment rate than would a more regular pattern of acorn production.

Victoria Sork's group have studied masting in North American oaks (see Box 10.1) and found that it is effective in satiating insect predators. In addition,

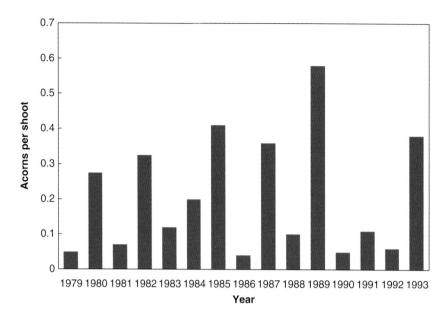

Fig. 10.5 *Acorn production by* Quercus robur *in Silwood Park, Berkshire, between 1979 and 1993. The mean number of acorns (including those with knopper galls) per shoot, averaged over 30 trees, is shown [Redrawn by Henry Battey from Crawley and Long (1995).]*

they emphasize that selection for increased acorn size (bigger acorns produce more vigorous seedlings than small ones, and allow a broader range of habitats to be colonized) may have contributed to the evolution of masting, because the production of larger acorns requires more resource, increasing the length of the intermast period.

The occurrence of a mast year causes a dramatic pulse in resource (acorns) to which the woodland community responds. The impact of this has been illustrated graphically in studies of the oak forests of the eastern USA. Here, a mast year has important consequences for rodents, deer, deer ticks (which carry Lyme disease, a bacterial disease of humans), birds, and the introduced gypsy moth (*Lymantria dispar*) (Fig. 10.6). Rodent populations (the white-footed mouse *Peromyscus leucopus* and the eastern chipmunk *Tamius striatus*) increase in mast years; rodent predators (such as owls) therefore also increase, while increased nest predation means that understory birds such as thrushes are reduced. Most sinister for humans, white-footed mice are hosts for the larval stage of black-legged ticks (*Ixodes scapularis*) and infect these larvae with the bacterium responsible for Lyme disease (*Borrelia burgdorferi*). The tick larvae then moult into nymphs and seek a vertebrate host the following year. Where this host is a human, Lyme disease can be the result. Nymphs moult into adults and seek a deer host; the location of deer therefore determines where larvae are found the following year. All this means that a boom in the mouse population, and a migration of deer in response to masting, is likely to be associated some time later (depending on the life-cycle duration of the tick, prevalence of the bacterium, and location of the deer) with an outbreak of Lyme disease.

A consequence of mast failure can be an outbreak of the gypsy moth (see also Box 5.2) which causes extensive defoliation (and death) of oaks and other trees.

| BOX 10.3 | *ANDRICUS QUERCUSCALICIS* **AND OAK GALLS** |

In Britain the oak has more insect species (at least 500) associated with it than any other plant. This is probably a result of its long period of dominance in the British flora, and the diversity of habitats which its bark, twigs, leaves, buds, catkins, acorns and roots offers. Galls are a common response of plants to the presence of a parasite, and in the case of the oak, insects are their most common cause. In the gall, the number and/or size of the host cells increases, and the gall structure is used by the parasite to enable completion of its life-cycle. The gall wasps (Cynipidae) have a particular liking for oak, which in Britain hosts 30–35 species, more than any other plant. They have two generations in their life-cycles: one is agamic, exclusively female; the other is sexual, with both males and females. The agamic generation typically passes the winter in strong resistant galls, often on the roots or at the base of the tree. Then in spring the females emerge and lay eggs in buds, catkins or leaves; these hatch out as males and females during the summer, mate and give rise to the next agamic generation.

There are several species of gall wasp which are well known through their galls. For example, 'oak-apples' are the highly coloured bud galls which house the sexual generation of *Biorhiza pallida* (Fig. 1a). In England, Oak Apple Day, 29 May, commemorates the restoration of the monarchy, Charles II having returned from exile at this time in 1660. Branches with oak-apple galls were traditionally worn as decoration in the associated festivities. Another common oak gall is the 'oak-marble' gall, which houses the agamic generation of *Andricus kollari*, a species deliberately introduced from the Middle East into Devon in about 1830 because of the value of the galls it causes for dyeing and ink production (due to their high tannin content). This introduction created controversy at the time because of fears that the galls would reduce the acorn harvest and deprive farmers of pannage. The oak-marble gall is a brown, woody sphere (Fig. 1b) which allows the agamic generation to overwinter on the branches of *Q. petraea* or *Q. robur*. Remarkably, the galls housing the sexual generation of *A. kollari* form in the axillary buds of the Turkey Oak (*Q. cerris*), a species from a different taxonomic section of oaks (see Box 10.1).

Another invasive gall wasp which alternates between these different oak species to complete its life-cycle is *A. quercuscalicis*, first recorded in Britain in 1962. Similar to *A. kollari*, the species probably invaded through Devon, though in this case without human assistance. This wasp causes the oak to produce the distinctive 'knopper galls', shiny green (later brown) ridged structures that occupy the acorn cups of *Q. robur* in early July (Fig. 1c). *Andricus quercuscalicis* has been so successful that it is found in Britain almost wherever the Turkey oak is common, and can destroy up to 60% of the *Q. robur* acorn crop.

Eleven species of *Andricus* (including *A. kollari* and *A. quercuscalicis*) have sexual generations that induce galls on oaks in the section *Cerris*. All these species are monophyletic, and the shift to host-alternation is evolutionarily unusual compared with shifts in the location of galls (buds, leaves, or catkins) within an existing host, which seem to happen quite frequently (Cook *et al.*, 2002). Inclusion of the new host species in the life-cycle led to a marked increase in speciation in *Andricus*, implying that it opened up new evolutionary possibilities for the genus. The success of the recent British invasions by host-alternating *Andricus* species suggests that, in addition, artificial extension of the range over which oaks from the two sections co-exist (i.e. co-planting of *Q. robur* and *Q. cerris*), has offered excellent opportunities for gall wasp species with relatively specialized life-cycles.

(a)

(b)

(c)

Box 10.3 Fig. 1 *Oak galls.* **(a)** *oak-apple* [From http://en.wikipedia.org/wiki/File:Oak_apple. jpg]; **(b)** *oak-marble* [Courtesy of Chris Madgwick]; **(c)** *knopper*

This occurs because white-footed mice are important predators of the pupae of the gypsy moth. Poor supply of acorns diminishes the mouse population, allowing the moth population to boom, with devastating consequences for the oaks.

Current interest in this forest ecosystem focuses on its instability; masting events lead to effects that are in a general sense predictable, but the details can vary and the responses take many years to work their way through the system. The relatively recent extinction (during the late nineteenth century) of the passenger pigeon (*Ectopistes migratorius*), which formerly consumed huge quantities of acorns, is thought to have contributed to the instability.

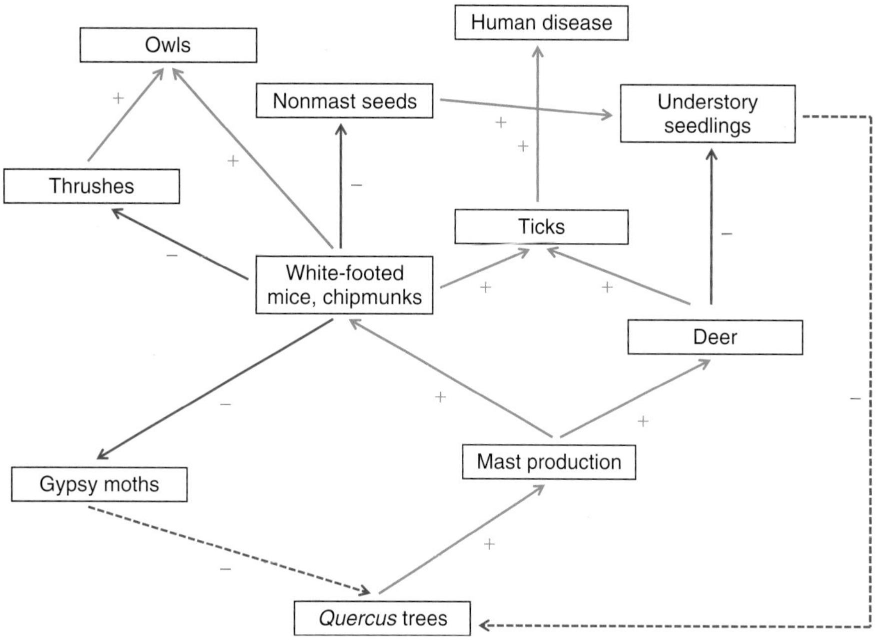

One phenomenon that ought to enhance rather than reduce stability is the anticipation of mast seeding by seed predators. This has been described for red squirrels in conifer forests, whose reproductive effort has been shown to increase in the year leading up to a masting event. An alternative strategy is shown by the edible dormouse (*Glis glis*) in beech forests in Germany: this species skips reproduction in unfavourable years, adapting its life-history strategy in tune with the availability of seed. It is possible that at least some of the inhabitants of oak forests exhibit similar strategies to escape the tactic of reproductive unpredictability pursued by the trees.

Owls — Human disease — Nonmast seeds — Understory seedlings — Thrushes — Ticks — White-footed mice, chipmunks — Deer — Gypsy moths — Mast production — *Quercus* trees

Fig. 10.6 *Interactions in the oak forests of eastern USA. Arrows show the directions of predominant influence. Plus symbols (green) indicate that an increase of the donor level results in an increase in the recipient level; minus (red) indicate a decrease in recipient level. Dashed lines show (relatively minor) feedback loops. [Redrawn by Henry Battey from Kelly et al. (2008).]*

ACORN TO OAK: ESTABLISHING AND REGENERATING OAK WOODS

We have seen that the irregular pattern of acorn production has a major impact on the dynamics of oak ecosystems. An equally important question is what determines where acorns germinate, and whether they persist as seedlings long enough to create new oak woods. This issue of oak regeneration is a hot topic among conservationists because of its apparent failure in woods – elsewhere, in open ground, hedgerows and open areas of wood-pasture, oaks regenerate freely (Fig. 10.7). But in many woodlands there are far fewer young oaks (under 50 years old) than expected. This had already been recognized as a problem at the beginning of the last century, and Watt (1919) carried out the first systematic investigation of it.

Watt cites the opinions of others of that era on the causes of the problem as including: over-felling and lack of replanting; pannage; degeneration of soil humus due to clearing of underwood; soil impoverishment by leaching; and excessive consumption of seeds and seedlings by rabbits (associated with the removal of holly and thorn by humans). While acknowledging the general

(a)   (b)

Fig. 10.7   *Oaks regenerating in heath*

significance of these factors, Watt's own studies led him to conclude that conceal-
ment of acorns was key to their survival, and that this, and successful germination, were both favoured in damp oak woods compared to dry oak woods and
oak–birch–heath woods. But perhaps most significantly from the modern perspective, he states:

> The efficacy of mildew in producing fatal effects on oak seedlings is
> shown to be greater on sandy soils than on clay, and the bracken, by its
> increasing frequency in the Dry Oakwood and Oak–birch-heath associa-
> tions, by greatly diminishing the supply of light and impairing the vitality
> of the seedlings, materially assists the fungus in producing these effects.[4]

At the time Watt was working and writing, oak mildew (*Microsphaera alphi-
toides*) was a relatively new introduction from North America, it first being noted
in 1908. It rapidly became ubiquitous on oaks in Europe, and it seems likely that
this was the fungus noted by Watt. It is now regarded by some as a key factor in
reducing oak regeneration. Its white bloom on the leaf is of little consequence
to oaks in open ground; but under shade, in woods, it seems crucial, effectively
making oak shade-intolerant. Other more recent changes in woodland management, including the decline in coppicing, may have further aggravated the shading
problem for oak, so that regeneration is now often dependent on the (artificial)
creation of cleared areas within existing oak woods.

There is, however, another interpretation of the failure of oak regeneration, one which asserts that oak has always been relatively shade-intolerant;
which, in other words, places less stress on the role of American mildew. This is
a view associated primarily with the Dutch ecologist Frans Vera. Vera's general
argument is that the natural state of much of Europe before human influence
became extensive (i.e. in the pre-Neolithic era, until about 6000 BP) was one of
wood-pasture: clumps of woodland, interspersed with grassland across which
ranged grazing animals. A present-day example with some similarity to this
might be the woodland pasture found in parts of the New Forest in southern
England (Fig. 10.8). This contrasts, particularly for Britain, with the view estab
lished by the twentieth century botanist Sir Arthur Tansley, of a primeval forest
(or wildwood) with trees stretching almost unbroken from Land's End to John
O'Groats. Vera's innovation is to stress the importance of the grazing animals,
which in Britain would mean auroch (*Bos primigenius*), moose (*Alces alces*), red and
roe deer (*Cervus elaphus, Capreolus capreolus*) and, early in the interglacial period, wild
horse (*Equus ferus*), in maintaining a much more open wood-pasture ecosystem.[5]
Crucially for the current context, Vera argues that such grazing would provide
the opportunity for regeneration of the shade-intolerant oak. Thorny plants such

[4]Watt (1919), p. 203.

[5]Of these herbivores, red deer are likely to
have been the major one in Britain. Red
deer are the preferred prey of wolves; it is
therefore of interest that the introduction
of wolves into Yellowstone National Park,
USA (see Box 19.3), had the effect of locally
suppressing the herbivorous elk, which in
turn created a mosaic of woodland mixed
with grassland over the territory occupied
by the wolves. This suggests that predators
may also have had an important impact on
the structure of Britain's wildwood. For
further discussion, see Bullock (2009) and
Box 19.3.

as hawthorn and blackthorn would repel grazers, and provide a protected environment in which oak could germinate, grow in the absence of heavy shade, and establish a new oak grove.

Whichever is the correct interpretation, one of the keys to the failure of oak to regenerate in woods seems to be its shade-intolerance. In the modern situation, and notwithstanding Vera's model, another key factor is overgrazing. Take just one example: Wistman's Wood on Dartmoor (Fig. 10.9). This is upland pedunculate oak at its most characterful, but it is probably only there because the granite boulders provide protection from overgrazing. Here is what Peter Marren has to say:

Fig. 10.8 *Woodland pasture in the New Forest, Hampshire, UK*

> There is much about this wood that remains mysterious, but the deformity of the trees is really no more than one should expect from a wind-blasted situation four hundred metres above the sea in the heart of Dartmoor. What is surprising is that the wood should be here at all. However, woods do not grow in places that are best for trees; they occur where the trees cannot be eaten. All three high-level woods of Dartmoor – the other two, at opposite ends of the moor, are called Black Tor Copse and Piles Copse – are made up of twisted oaks sunk in rocky 'clitter'.[6]

Another peculiarity of Wistman's Wood is that the oak is *Q. robur* when all around Dartmoor are woodlands of sessile oak, *Q. petraea*. Attempts have been made to account for this distribution in terms of soil-type preference, but the explanation does not seem to hold generally. Perhaps the most interesting observation on the question of the complex distribution patterns of the two species is that *Q. petraea* does best in its own company, whereas *Q. robur* is most successful on its own or in competition with other tree or herbaceous species. The gregarious character of *Q. petraea* deserves further study, and may be related to its tendency to swamp out *Q. robur*; the genetic mechanisms for this are discussed in the next section.

HOW THE OAKS GOT TO BRITAIN

Fig. 10.9 *Pedunculate oak (Quercus robur), Wistman's Wood, Dartmoor, UK [Courtesy of Cathy Newell Price.]*

How did Britain come to have its present-day distribution of oak trees? How are British oaks related to those in nearby France and Spain where the oak is widespread; where it is equally iconic, but in different ways to the oaks of Britain? To answer these questions, we must go back to the last glaciation, and trace forward the history of the (re-)occupation of the European continent by trees. Twenty thousand years ago, at glacial maximum, Britain was covered by a thick layer of ice north of a line stretching from South Wales to East Anglia. South of this was permafrost, and steppe tundra with some hardy dwarf birch (*Betula nana*) and tree birch (*Betula* species). Glaciers also extended outwards

[6]Marren (1992), p. 117.

from the Pyrenees, Alps and Carpathians, so that most of the European plain (present-day France and Germany) was a harsh, mainly treeless environment. Only south of the European mountains, in the Iberian peninsula, Italy and the Balkans, were there areas warm enough to allow most trees to survive. Driven south as the glaciers advanced, these trees clung on at the fringes of mountain regions, far enough south to avoid lethal freezing conditions, yet at a high enough altitude to ensure sufficient rainfall.

From 20 000 BP the ice started to recede, as we entered the current interglacial period (also known as the Holocene or Flandrian; see Box 10.4). As temperatures rose the trees gradually returned north, the time at which species returned reflecting in part their relative cold-hardiness, and also the distances they had to cover and their speeds of migration. Thus birch is often an early presence in the pollen record, reflecting its cold-tolerance, light fruits (distributed by air), and its likely survival through the glacial period on the European mainland, not far south of Britain. The next arrivals include pine (*Pinus*), followed by the less hardy broadleaves: oak (*Quercus*), elm (*Ulmus*), lime (*Tilia*), ash (*Fraxinus*), hornbeam (*Carpinus*) and beech (*Fagus*). The times of arrival of key tree species in Britain, and their general relation to cold-hardiness (reflected by present northern limit), are depicted in Fig. 10.10. Hazel (*Corylus*) is an interesting anomaly, because it is an early presence, even though its present northern limit suggests only moderate cold-hardiness. Clearly, other factors than temperature (such as rainfall) must have played a role.

An important, related point is that several taxa present in Europe during the previous interglacial (115 000–70 000 BP) failed to survive into the current interglacial. These include walnut (*Juglans*), *Liquidambar* and *Magnolia*, trees which, however, did survive in North America, where the north–south distribution of the

BOX 10.4 ICE AGES AND PALEOCLIMATE

The Earth has been in a relatively cold phase during the Pleistocene Epoch (Fig. 3.2). It has seen a whole series of glacial ('Ice Age') and interglacial periods, as discussed in more detail by Imbrie and Imbrie (1979), and Cox and Moore (2000). During this cold phase the polar ice caps and glaciers in high regions such as the Alps extend back and forth but are generally well-developed. We are therefore currently in the interglacial period of a cold phase which has lasted, intermittently, for the past 1–2 million years. The most recent glacial period lasted from ~115 000 years ago until ~10 000 years ago. Glacial periods are much longer than interglacials, giving populations of trees in refugia time to accumulate mutations and evolve differences.

During the 4500-million-year history of Earth there have been several cold phases, but they are relatively rare – a specific set of circumstances is needed to cause them. As prerequisite to our most recent cold phase, the Antarctic landmass had moved to its present position over the South Pole by 15 million years ago, and the South Polar ice cap had formed. By about 5 million years ago the continents of the northern hemisphere had moved so as to isolate the Arctic Ocean from warm currents, so it became frozen. The formation of these two polar ice caps increased the reflectivity of the earth and therefore reduced the amount of energy reaching the Earth. The scene was set for a cold phase.

An explanation for the pattern of alternating warmer and colder periods within this overall cold phase was suggested by the Scotsman James Croll in the late 1800s, and developed into a series of calculations of radiation input through the year, at different latitudes, by the Yugoslav mathematician Milutin Milankovitch in 1910–40. This model took account of known variations in the eccentricity of the earth's elliptical orbit (100 000-year periodicity), its tilt (41 000-year periodicity) and its precession around the angle of tilt (21 000-year periodicity).

These cycles influence the yearly pattern of energy input from our Sun to the Earth. The Milankovitch model showed that they combine to yield a periodic cycle in energy distribution across the planet that would cause a glacial–interglacial cycle of about 100 000 years, with less pronounced cycles of warm and cold of 41 000 and 21 000 years superimposed upon it. In the 1960s and 70s, it was realized that the ratio of the two oxygen isotopes (^{18}O and ^{16}O) in the ocean varies according to the extent of polar ice formation. Measurement of this ratio in the skeletons of micro-organisms that collect over hundreds of thousands of years as sediments on the ocean floor therefore provides a record of the past climate. When analysed in detail, this record showed the periodicity predicted by the Milankovitch model.

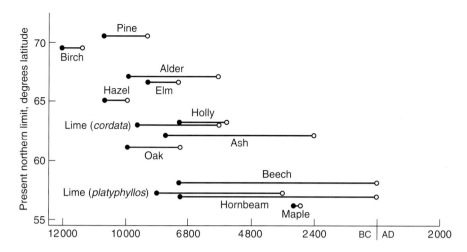

Fig. 10.10 *Progress of colonization of tree species related to their current northern limits of distribution. Closed circles indicate the time at which pollen quantity is sufficient to indicate local presence of the tree in Britain. Open circles indicate the time at which the pollen record indicates that the tree became abundant [Redrawn from Rackham, O. (2003) Ancient Woodland: its History, Vegetation and Uses in England. Castlepoint Press, Dalbeattie.]*

main mountain ranges (Appalachians and Rockies) meant a clearer line of retreat southwards. This accounts for the relative species-richness of present-day North American forests, compared to those in Europe.

By about 6000 BP, British woodlands had achieved their maximum development. This wildwood dominated the landscape. It was varied in composition: birch–pine in the highland region of Scotland; oak–hazel in northern England and much of Wales; and lime-dominated woods further south (Fig. 10.11). Shortly after this time the land bridge between Britain and continental Europe was flooded, limiting the potential for the arrival of new species. About the same time the first signs of farming due to Neolithic people appear in Britain, the techniques of settled agriculture having been transmitted northwards from their area of origin in the Fertile Crescent, between the Tigris and Euphrates rivers. This marked the beginnings of the extensive forest clearance associated with cereal production and cattle farming; elm and then lime leaves were gathered for fodder and these species show a corresponding decrease in pollen abundance. There was also climatic deterioration from about 3000 BP, cooler wetter conditions encouraging the formation and spread of blanket bog in many upland areas. By the end of the Bronze Age (~2750 BP) farming had developed to the extent that it occupied about a sixth of England, the other five-sixths being wildwood. During the Iron Age and period of Roman occupation (~2700 to 1600 BP), agriculture developed quickly: Oliver Rackham has estimated that about half of the lowland zone of England and Wales was cleared of forest in 700 years. Thus, the Anglo-Saxons inherited a fully agricultural landscape, which is essentially the one we have today. At the time of Domesday Book (AD 1086) only about 15% of the land area remained wooded.

Fig. 10.11 *Wildwood at its zenith (6000 BP). The map shows the provinces of the fully developed wildwood just before the Neolithic [Redrawn by Henry Battey from Rackham (2003).]*

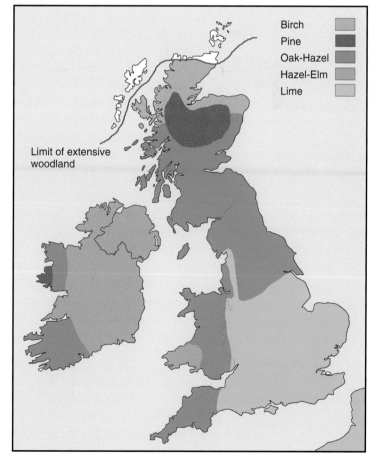

In the Middle Ages there was initially a trend towards further grubbing of woodland (e.g. in the Weald of Kent between AD 1100 and 1300), but then the value of woods for iron smelting (as charcoal) and shipbuilding was (re)realized (the Romans had very extensive ironworks in the Weald, and also in the Forest of Dean). Woodland management by the ancient techniques of coppicing and pollarding[7] was widely practised and wood products were an important part of the economy, as we saw in the first section of this chapter. The decline of these industries took place in the nineteenth century following the development of coal fuel and the industrial revolution.

How does the story of oak fit into this general picture? The first records of oak pollen appear in Cornwall 9500 BP. Progress across Britain was steady and oak was a consistent presence in the pollen record at 6000 BP. Thus oaks invaded Britain from the south, colonization occurring at a rate of 350–500 m per year through England and Wales. The 'bog oaks' preserved in the Fenland of East Anglia are a graphic reminder of the massive stature oaks could attain at this time – a distance of 27 m to the first branch has been reported. With subsequent intensification of human activity, oak was not so rapidly diminished as elm and lime, probably because its tough wood was not readily susceptible to the polished stone axe. However, by the late Bronze Age, with the expansion of the human population, oak began to be extensively employed in trackways, houses and boats, and in consequence declined in abundance. Iron Age timber roundhouses such as the one at Pimperne in Dorset needed 60-year-old oaks for the porch posts. This and the timber requirements of similar buildings have led to the suggestion that by this time oak woodlands must have been carefully managed. The late Roman quay in London also made use of massive oak timbers, which as well as showing the importance of oak have an additional present-day use in dating the successive phases of construction of the quay.[8]

Although the oak species considered native to Britain (Q. robur and Q. petraea) cannot be distinguished by their pollen, chloroplast DNA (cpDNA) variation has proved exceptionally useful in tracing the origin of British oak trees. In the most comprehensive study, over 1000 mature oaks (of either Q. robur or Q. petraea) were sampled from 224 British ancient woodland sites by Joan Cottrell and co-workers. This was carried out as part of a wider analysis of European oaks, and revealed that 98% of British oaks possess one of three cpDNA genotypes (haplotypes 10, 11 or 12) in a common maternal lineage (lineage B).[9] These haplotypes are also found at high frequency in Spain and western France, supporting previous deductions that British oaks survived the last glaciation in the Iberian peninsula and re-colonized Britain via western France. The distribution of these haplotypes, particularly the dominance of haplotype 11 in East Anglia, suggested that there may have been two colonization events – one from the south through Cornwall, and another in the east through Norfolk. One very rare haplotype (haplotype 7), from a different lineage (lineage A, most common in the Balkans), and found only as scattered trees along the north coast of France and in a single population near Rennes, was found to be present in the border region of England and South Wales. Subsequent detailed study of this haplotype in the Forest of Dean provided no evidence that it was introduced there as a result of deliberate planting, leaving open the possibility that its presence in the region is the result of a single, long-distance colonization event from northern France. The Verderers Oak (see Fig. 10.4b) is an example of this haplotype.

One striking feature of this analysis is the clumped distribution of single haplotypes in particular areas, such as Cornwall, North Wales and East Anglia.

[7]Coppicing and pollarding are methods in which the tree is regularly cut back – either to ground level or just above head height – in order to induce sprouting of new shoots. They enable a regular cycle of straight wood of intermediate size to be harvested, typically over an interval of about 10–15 years. Pollarding was traditionally employed to reduce damage by browsing animals, but it is frequently used today in order to contain the growth – for instance of lime and London plane trees in cities. Neglected pollard oaks are a common sight in many English fields (Fig. 10.4a).

[8]Indeed, the oak is fundamental to the tree-ring dating method that has been widely used in the archaeology of northern Europe. A chronology has been constructed for the past 5000 years based on matching patterns of tree-rings in oak samples from Ireland, England, Wales and Germany. This technique depends on the long-lived nature of oak, which allows overlaps between successive samples; and the general availability of oak, which reflects its wide distribution and extensive use in construction. One vivid example of the wider relevance of the method is the marked gap in oak tree-ring records available around 1350. This gap reflects the decimation of the human population of Europe due to the Black Death (see Chapter 16 and Box 16.3) and the concurrent hiatus of building activity. Extrapolating, in the absence of historical records, the (un)availability of tree-ring data can be used to deduce the occurrence of other likely pandemics. Furthermore, the character of tree-rings can indicate environmental events – such as volcanic eruptions – which in turn may lead to population migrations, and so be the indirect cause of disease pandemics. For an accessible account of this fascinating (and perhaps unexpected) example of oak-exploitation, see Baillie (1995).

[9]The word haplotype is a contraction of haploid genotype. Because cpDNA is haploid, heritable sequence variants are referred to as haplotypes, rather than the more longwinded cpDNA genotype. cpDNA is maternally inherited; studies using cpDNA therefore reveal colonization by seed without the mixing of DNA that arises in nuclear DNA due to the paternal influence transmitted through pollen. Another use of cpDNA has been to determine the source of oak used to make barrels for the French wine industry. Deguilloux et al. (2004) used cpDNA haplotype analysis to show that oak from Eastern Europe was being sold as French wood.

This is known as a leptokurtic pattern, and reflects single long-distance seed dispersal events, followed by establishment of a small number of individuals, before the arrival of the main colonization front. It is likely that birds, in particular jays (*Garrulus glandarius*), are responsible for this pattern, carrying acorns long distances and establishing new populations. This also helps to explain the unexpectedly rapid colonization of oaks across Europe, birds (and perhaps rivers) allowing jumps of hundreds of kilometres, rather than the distance of a few hundred metres facilitated by scatter-hoarding animals like squirrels. A further interesting finding of the cpDNA studies is the sharing of haplotypes across oak species: ancient woods containing both Q. *robur* and Q. *petraea* often have a single haplotype. This is believed to reflect frequent interspecific hybridization and unidirectional chloroplast capture. In the case of Q. *robur* and Q. *petraea*, it also suggests a mechanism by which hybridization facilitates invasion. Q. *robur* acorns are dispersed more effectively than Q. *petraea* (there is evidence to suggest that jays prefer them), and so Q. *robur* will tend to act as the pioneer species. However, in crosses Q. *robur* typically acts as the maternal parent, so pollen from Q. *petraea*, transported long distances by wind, can allow this species to 'catch up' with Q. *robur* by hybridization. If (as seems to occur) the resultant hybrid is preferentially pollinated by Q. *petraea*, after several generations Q. *petraea* will have invaded new territory based on the success of initial Q. *robur* dispersal. Clearly, for the mechanism to succeed in the long term, other factors must enable the persistence of Q. *robur*. One recent factor in favour of Q. *robur* may be its preferential planting by humans.

THE FUTURE OF OAK: RETURN TO A WOOD ECONOMY?

In discussing oak we have considered its practical value to humans and its impact on the community of organisms it supports. Of course oak is just one tree among many, and a similar story of interdependence amongst humans, other animals and plants could be told for many trees. We are daily reminded of the importance of trees in the battle against climate change: not chopping them down; planting new ones; locking up their carbon as charcoal. Fake plastic trees have even been proposed, to absorb and process CO_2 in the absence of sunlight. Here the emphasis is all on capturing carbon to reduce global temperature.

But what will happen when fossil fuels have been used up? One example of a positive step emphasizing the potential of wood as fuel is the English Forestry Commission initiative to increase to two million tonnes annually the output of wood fuel from managed woodlands. Wind, tide and nuclear sources are, however, likely to be quantitatively the most important sources of energy in the post-fossil fuel world. But we depend on oil and coal for many other things, most obviously plastic. Trees may therefore turn out to be most valuable as wood, rather than as fuel, and many of the traditional wood-based trades that have been diminished in our modern, oil-based economy will see a renaissance. Already vegetable tanning, substituted post-industrially by chromium and other metals, is now considered more sustainable when based on plant products. Wood use for items currently made of plastic will surely return, which means we need trees like the oak and we will need to remember how to exploit their products. Fake plastic trees will be made of wood, and our fake plastic watering cans, should we insist on them, from distillates of wood gas.

FURTHER READING

The starting point for oak and woodland history in Britain is Oliver Rackham (2003, 2006). The collection edited by Christopher Smout (2003) provides a Scottish perspective on woodland history which has much relevance elsewhere. For woodland ecology, Peterken (1993) and Marren (1992) are recommended. Vera (2000) is a key book which challenges orthodox accounts of the wildwood. A still useful source for many scientific aspects of the oak in Britain is Morris and Perring (1974). General biogeography is well summarized in Cox and Moore (2000) and a detailed treatment of plant vegetation history is given in Tallis (1991). The following sources deal with more specific aspects, information on which can also be found in the above books.

For the history of acorn use by the North American Indians, Saunders (1920) is an interesting source and can be accessed on-line. An entry point to the literature on oaks in the Spanish *dehesa* is provided by Cañellas *et al.* (2007), while the history of the bark trade is described by Clarkson (1974) and woodland crafts in general by Edlin (1973); an excellent article on tanning leather is that by Covington (1997). The stripping of bark from cork oaks is a sustainable industry of great importance in many Mediterranean regions, and is threatened by the rise of the plastic wine stopper. This is leading to a decline with some parallels to that of the Atlantic oak woods; a useful summary is by WWF (2006). For aspects of the historical place of oak in British society, Evelyn (1664), Albion (1926), and Watkins (1998) are good value.

There is a large literature on mast seeding, a good overview of which is provided by Kelly and Sork (2002). Crawley and Long (1995) is specifically focused on *Q. robur* in Britain; while Jones *et al.* (1998) and Kelly *et al.* (2008) provide fascinating accounts of the oak forest ecosystem of eastern USA. The analysis of the post-glacial path of oaks across Europe has a similarly large literature. For the picture from pollen analysis, Godwin (1975) is seminal, while the chapter by Godwin and Deacon in Morris and Perring (1974) is more specific to oak. The work on chloroplast DNA analysis of oak is summarized thoroughly in Lowe *et al.* (2004); key primary papers are Dumolin-Lapègue *et al.* (1997), and Cottrell *et al.* (2002, 2004). The role of hybridization in invasion is reviewed by Petit *et al.* (2003), while the selection processes which act to maintain the two species in the face of extensive hybridization are discussed by Scotti-Saintagne *et al.* (2004). For the general ecology of *Q. robur* and *Q. petraea* in Britain, Jones (1959) is very useful, and for general botanical information about *Quercus*, Mabberley (1987) is invaluable.

The English Forestry Commission strategy to increase wood fuel provision is described in the 'Woodfuel strategy for England' at www.forestry.gov.uk/england-woodfuel. Oak pests and diseases are not covered here, but information on Acute Oak Decline, the most recent threat to oak in Britain, can be found at www.woodlandheritage.org.

REFERENCES

Albion, R.G. (1926) *Forests and Sea Power: the timber problem of the Royal Navy 1652–1862*. Harvard University Press, Cambridge MA, USA.

Axelrod, D.I. (1983) Biogeography of oaks in the Arcto-Tertiary province. *Annals of the Missouri Botanical Garden* **70**, 629–657.

Baillie, M.G.L. (1995) *A Slice Through Time: dendrochronology and precision dating*. Batsford, London.

Bertolino, S. and Genovesi, P. (2003) Spread and attempted eradication of the grey squirrel (*Sciurus carolinensis*) in Italy, and consequences for the red squirrel (*Sciurus vulgaris*) in Eurasia. *Biological Conservation* **109**, 351–358.

Bertolino, S., Lurz, P.W.W., Sanderson, R. and Rushton, S.P. (2008) Predicting the spread of the American grey squirrel (*Sciurus carolinensis*) in Europe: a call for a co-ordinated European approach. *Biological Conservation* **141**, 2564–2575.

Bullock, D.J. (2009) What larger mammals did Britain have and what did they do? *British Wildlife* **20** (Supplement), 16–20.

Cañellas, I., Roig, S., Poblaciones, M.J., Gea-Izquierdo, G. and Olea, L. (2007) An approach to acorn production in Iberian dehesas. *Agroforest Systems* **70**, 3–9.

Clarkson, L.A. (1974) The English bark trade 1660–1830. *Agricultural History Review* **22**, 136–152.

Cook, J.M., Rokas, A., Pagel, M. and Stone, G.N. (2002) Evolutionary shifts between host oak sections and host-plant organs in Andricus gallwasps. *Evolution* **56**, 1821–1830.

Cottrell, J.E., Munro, R.C., Tabbener, H.E. *et al.* (2002) Distribution of chloroplast DNA variation in British oaks (*Quercus robur* and *Q. petraea*): the influence of post-glacial colonisation and human management. *Forest Ecology and Management* **156**, 181–195.

Cottrell, J.E., Samuel, C.J.A. and Sykes, R. (2004) The species and chloroplast DNA haplotype composition of oakwoods in the Forest of Dean planted between 1720 and 1993. *Forestry* **77**, 99–106.

Covington, A.D. (1997) Modern tanning chemistry. *Chemical Society Reviews* **26**, 111–126.

Cox, C.B. and Moore, P.D. (2000) *Biogeography*, 6th edn. Blackwell, Oxford.

Crawley, M. J. and Long, C. R. (1995) Alternate bearing, predator satiation and seedling recruitment in *Quercus robur* L. *Journal of Ecology* **83**, 683–696.

Deguilloux, M-F., Pemonge, M-H. and Petit, R.J. (2004) DNA-based control of oak wood geographic origin in the context of the cooperage industry. *Annals of Forest Science* **61**, 97–104.

Dumolin-Lapègue, S., Demesure, B., Fineschi, S., Le Corre, V. and Petit, R.J. (1997) Phylogeographic structure of white oaks throughout the European continent. *Genetics* **146**, 1475–1487.

Edlin, H.L. (1973) *Woodland Crafts in Britain*. David & Charles, Newton Abbot, UK.

Esau, K. (1965) *Plant Anatomy*. John Wiley & Sons, Chichester, UK.

Evelyn, J. (1664) *Sylva: or a discourse of forest-trees, and the propagation of timber in His Majesties dominions* [1972, Scolar Press Facsimile edition]. Scolar Press, Yorkshire, UK .

Godwin, H. (1975) *The History of the British Flora: a factual basis for phytogeography*. Cambridge University Press, Cambridge.

Holmes, O.W. (1906) *The Autocrat of the Breakfast Table*. J.M. Dent & Sons, London.

Imbrie, J. and Imbrie, K.P. (1979) *Ice ages: solving the mystery*. Macmillan, London.

Jones, C.G., Ostfeld, R.S., Richard, M.P., Schauber, E.M. and Wolff, J.O. (1998) Chain reactions linking acorns to gypsy moth outbreaks and Lyme disease risk. *Science* **279**, 1023–1026.

Jones, E.W. (1959) Biological flora of the British Isles: *Quercus* L. *Journal of Ecology* **47**, 169–222.

Kelly, D., Koening, W.D. and Liebhold, A.M. (2008) An intercontinental comparison of the dynamic behavior of mast seeding communities. *Population Ecology* **50**, 329–342.

Kelly, D. and Sork, V.L. (2002) Mast seeding in perennial plants: why, how, where? *Annual Review of Ecology and Systematics* **33**, 427–447.

Kenward, R.E. and Holm, J.L. (1993) On the replacement of the red squirrel in Britain: a phytotoxic explanation. *Proceedings of the Royal Society of London* B **251**, 187–194.

Lowe, A., Harris, S. and Ashton, P. (2004) *Ecological Genetics: design, analysis and application.* Blackwell, Oxford.

Mabberley, D.J. (1987) *The Plant Book.* Cambridge University Press, Cambridge.

Manos, P. S. and Stanford, A. M. (2001) The historical biogeography of Fagaceae: tracking the tertiary history of temperate and subtropical forests of the northern hemisphere. *International Journal of Plant Sciences* **162** (Supplement), S77–S93.

Marren, P. (1992) *The Wild Woods: a regional guide to Britain's ancient woodland.* David & Charles, Newton Abbot, UK

Morris M.G. and Perring, F.H. (eds) (1974) *The British Oak: its history and natural history.* Classey, Farringdon, UK.

Peterken, G. (1993) *Woodland Conservation and Management.* Chapman & Hall, London.

Petit, R.J., Bodénès, C., Ducousso, A., Roussel, G. and Kremer, A. (2003) Hybridization as a mechanism of invasion in oaks. *New Phytologist* **161**, 151–164.

Rackham, O. (2003) *Ancient Woodland: its history, vegetation and uses in England.* Castlepoint Press, Dalbeattie, UK.

Rackham, O. (2006) *Woodlands.* HarperCollins, London.

Saunders, CF. (1920) *Useful Wild Plants of the United States and Canada.* R.M. McBride & Co, New York. Accessible at www.swsbm.com/ManualsOther/UsefulPlants/Useful_Plants.html.

Scotti-Saintagne, C., Mariette, S., Porth, I. et al. (2004) Genome scanning for interspecific differentiation between two closely related oak species [*Quercus robur* L. and *Q. petraea* (Matt.) Liebl.]. *Genetics* **168**, 1615–1626.

Smout, T.C. (ed.) (2003) *People and Woods in Scotland: a history.* Edinburgh University Press, Edinburgh.

Smout, T.C. (2005) Oak as a commercial crop in the 18th and 19th centuries. *Botanical Journal of Scotland* (Special Issue) **57**, 107–114.

Tallis, J.H. (1991) *Plant Community History: long-term changes in plant distribution and diversity.* Chapman & Hall, London.

Vera, F.W.M. (2000) *Grazing Ecology and Forest History.* CABI, Wallingford, UK.

Watkins, C. (1998) 'A solemn and gloomy umbrage': changing interpretations of the ancient oaks of Sherwood Forest. In: *European Woods and Forests: studies in cultural history* (ed. C. Watkins), pp. 93–113. CABI, Wallingford, UK.

Watt, A.S. (1919) On the causes of failure of natural regeneration in British oakwoods. *Journal of Ecology* **7**, 173–203.

WWF (2006). *Cork screwed? Environmental and economic impacts of the cork stoppers market.* WWF Report. Available at http://assets.panda.org/downloads/cork_rev12_print.pdf.

MR FIBONACCI
COUNTS RABBITS IN 1202

NUMBER OF PAIRS EACH MONTH

1

1

2

3

5

8

AND SO ON...

Fibonacci first introduced the sequence of numbers which bears his name to European mathematics. He described the theoretical growth of a population of rabbits; although biologically unrealistic, the sequence of monthly increase has interesting properties — for instance, the ratio of successive members converges to a limit, known as the golden ratio.

The Rabbit

In some years they were everywhere, burrowing under fences, undermining quarries, and consuming almost all vegetable material in sight: you could shoot them as fast as you could reload, and still more would come. Yet in other years they were strangely absent, their presence only indicated by shadowy half-dead individuals by the side of the road, seemingly intent on committing suicide under the wheels of the next vehicle.

THIS IS NOT the screenplay from some zombie horror film, but rather our recollections of rabbits in southern England in the last quarter of the twentieth century. Few organisms exhibit the evolution from exploited into exploiter better than the rabbit, and thus this species is a fitting bridge between the two sections of this book. To explain this shift, and how the rabbit ended up in the situation described above, we shall discuss the history of the rabbit in Britain and Australia, and subsequent efforts to control it.

The European rabbit, *Oryctolagus cuniculus*, is a member of the mammal order Lagomorpha, family Leporidae. This family includes about 50 species worldwide, and in the UK also includes the brown hare (*Lepus capensis*) and the mountain hare (*L. timidus*). The rabbit is smaller than the hare, with proportionally shorter ears (without black tips) and legs. It originated in south-western Europe, and is now present in much of western Europe north into southern Scandinavia. It has been introduced into much of the New World, including the USA, Australia and New Zealand.

All rabbit life and behaviour is centred on the warren: the rabbit is unusual among lagomorphs in digging a series of underground tunnels and dens for living and breeding. The shape and size of the warren is mainly determined by soil conditions; the female rabbit (doe) carries out most of the work, and can excavate up to 2 m of burrow per night, with the male rabbit (buck) looking on but rarely helping. Up to 30 rabbits can live in one warren and there are distinct adult male and female dominance hierarchies. These are determined by mock fighting among young and are maintained by adult scent marking. The individual's social position determines its mate (rabbits tend to pair with mates of a similar status), length of breeding season, access to food, and nesting position in the burrow. The dominant pair sleep and breed in the centre of the warren, the safest place; more subordinate rabbits have to use the periphery of the warren, and may be excluded from the warren altogether.

Rabbits are mainly nocturnal, emerging at dusk to feed and ranging only about 200 m from the warren. While feeding they have their head close to the ground and with a limited field of view are vulnerable; thus, warning of predators is communicated by rapid thumping of the hind legs of look-outs. The other main communication is by scent-marking, with scenting posts of urine and droppings to demarcate the warren's territory. Rabbits are generalist herbivores, feeding on a wide range of usually herbaceous plants; but they favour cereals, root vegetables, and young shoots of most meadow plants. They will also nibble the bark of trees. To obtain maximum nutrition from their food, the soft faecal pellets produced are immediately eaten and pass through the digestive system again. This often takes place in the burrow; the hard fibrous black pellets which result from this second digestion are always left above ground. Cellulose digestion is aided by bacteria housed in the caecum, a large sac between the small and large intestine.

Rabbits are famed as prolific breeders, and they have a number of adaptations to enable the population to respond quickly to changes in the environment and food availability, and to make optimal use of their resources. Does become sexually mature at 3.5 months, bucks when a couple of weeks older. In the UK the breeding season is mainly between January and August. After a short courtship there is quick repeated mating and, unlike in larger mammals, ovulation is induced by mating. Rabbits are thought to be mainly monogamous, and may pair for life, which may be 8–9 years. Bucks wait for does to emerge from their nest burrow, and thus the doe often conceives again within 24 hours of giving birth. If a pregnant doe is stressed, possibly by predator attack or lack of food, she will resorb her embryos rather than aborting them. This is another adaptation to make best use of scarce resources. The likelihood of resorbtion may be psychological and based on how confident the doe feels, and the more dominant the doe the less likely she is to resorb. In a breeding warren resorbtion can account for up to 60% of pregnancies.

The pregnant doe has a gestation period of about 30 days, and the litter of two to seven blind, deaf, and immobile kittens are born into a nursery den in a nest lined with grass and fur from the mother. The mother returns to this nest only once a night for 5–10 minutes to suckle her young, closing the tunnel mouth with hay and straw as she leaves. At birth the kittens thermoregulate poorly, and in wetter climates many die from hypothermia in wet nests or by drowning in flooded burrows. Survivors double in weight within a week and start to explore the warren after ten days, first leaving it a week later. They graze freely by 21 days old, and are then weaned. Given that a doe in good condition could produce a litter a month over the breeding season, and the earlier of these could also produce offspring that year, it would be possible for one female to be the progenitor of over 60 offspring a year. However, taking mortality into account, 10–20 is probably the maximum.

FROM EXPLOITED TO EXPLOITER IN BRITAIN

The rabbit is not a species native to the British Isles. While it is possible that it was here before the last Ice Age, it then died out. It is unlikely that the rabbit was introduced by the Romans in any number (Box 11.1), but it was introduced by the Normans for meat, sometime after their conquest in the eleventh century, probably from France.

Rabbits were first introduced to offshore islands (one of the first records is from the Scilly Isles in 1176) probably to prevent escape and also to minimize predation and poaching—factors that have been important ever since. The rabbit

BOX 11.1	RABBITS AND ROMANS

The early history of the rabbit is sketchy and confused. They were apparently unknown to the ancient Greeks, and are not mentioned by Aristotle – he mentions hares, but none that burrow. However, by the second century BC rabbits had been introduced from Spain into the Italian peninsular.

The Romans often regarded the rabbit as indistinguishable from the hare, and this causes further confusion, compounded by no clear distinction between rabbits and hares in pictorial representations from this period. For example, Varro in 36 BC describes three types of hare. The third type, found in Spain, is smaller, lower in build than the Italian hare, and is called *cuniculus* because it digs tunnels and mines – this is the rabbit.

Varro further suggests that all three types of hare should be kept in *leporia*. These were walled enclosures (to prevent predators from entering) next to the villa, with sufficient cover to protect the hares from eagles. By late Republican times, *leporarium* became the term for an enclosure or game reserve, often quite extensive, for a variety of animals including deer, cattle and boars.

Several Roman authors note that *cuniculi* come from Spain, where they are very abundant. For example, Strabo writing in the first century AD notes that the burrowing hares in Spain damage plants by eating their roots and describes several methods of hunting, including sending muzzled Libyan ferrets into the burrow to bolt the rabbits into a net. Strabo also mentions that the occupants of the Balearic Islands asked to be moved elsewhere as they were being driven out by rabbits, which they could not control.

Therefore, the Romans associated rabbits with Spain, and not in a positive manner, as illustrated by Catullus in the first century BC. Many of Catullus's poems are about his infatuation

with his mistress, Clodia, who more often than not seems to have become attached to other men, in this case 'Egnatius, son of rabbity Celtiberia [*cuniculosae Celtiberiae fili*], made a gentleman by a bushy beard and teeth brushed with Spanish piss.'[a]

Apparently, rabbit embryos were a Roman delicacy, but Apicius, in his *Art of Cooking*, mentions only recipes for hares.

There has been considerable debate over whether the Romans introduced the rabbit to Britain. It is generally accepted that they did not, and until earlier this century there were no reliable archaeological records. However, rabbit bones have been recently reported from Roman sites in Norfolk and East Sussex and are thought to be reliably Roman, and not a later addition.

There are no written rabbit records from Roman Britain. However, Appian does mention that his soldiers fed upon hares (the brown hare might have been introduced into Britain by the Romans), and representations of hares occur, for example on the mosaic of Chedworth Roman Villa, Gloucestershire.

Some writers have suggested that although evidence is scant, 'logic' suggests that the Romans *should* have introduced the rabbit. This is a dangerous hypothesis, as it can never be falsified; it is far better to stick with the falsifiable hypothesis that they did not, until good evidence emerges.

It should not be surprising that the Romans did not introduce the rabbit to Great Britain, or at least not in great numbers. It was probably a rather tender domesticated animal to the Romans (wild rabbits are still absent from most of Italy), and nowhere near as important a food source as the hare. Faced with the barbaric, cold and wet hinterland of the empire that was Britannia at the time, the introduction of the rabbit would have been a low priority.

[a]Catullus xxxvii, translation from Cornish *et al.* (1988)

was recognized as a sickly animal in the British climate, prone to drowning in its burrows in wet and heavy soils, and thus was usually confined to sandy free-draining soils. On the mainland it was probably introduced first to coastal sand dunes, many of which are still named as such (e.g. Newborough Warren, Anglesey, one of the largest sand dune systems in the UK and a commercial rabbit warren until the 1930s). Rabbit warrens gradually spread inland and in many cases artificial burrows were provided. These ranged from holes dug in the ground or in a pile of loose soil, to elaborate artificial burrows with side passages, stone capping covered with banks of soil, and drainage ditches (Fig. 11.1). These artificial burrows not only helped the rabbit become established in an area, a process which could take several years, but by having few exits, made trapping by ferrets (*Mustela furo*) and netting much easier. In some areas toads, crabs and lobsters, with a lighted candle on their backs, were used at least until the nineteenth century to drive rabbits out; the latter two used in South Devon were called 'sea ferrets'.

Many early rabbit warrens were owned by monasteries (it was believed that rabbit embryos were aquatic, and thus could be eaten on Fridays) and by the

Fig. 11.1 *Characteristic remains of a rabbit warren, a pillow mound on Minchinhampton Common, Gloucestershire. This is a post-medieval warren. Note the drainage ditch around the mound. Ordnance survey map on mound for scale*

king. For example, by 1235 Henry III had rabbits in his royal coneygarth (enclosure of coneys) at Guildford, later warrens in Hampshire, Surrey and Kent had to supply rabbits for his Christmas feast, and in 1244 his new park at Windsor was stocked with rabbits from Sussex. The right to kill any beast, including rabbits, was the gift of the king, but he was prepared to sell the rights to hunt certain animals outside the Royal Forests and grant a charter of free-warren. This gave the recipient the sole right to hunt beasts of the warren (pheasant, *Phasianus colchicus*, partridge, *Perdix perdix*, hare and rabbit – animals that it was possible for a large hawk to catch), and this was the original meaning of 'warren'; only later did it come to exclusively signify a managed series of rabbit burrows. Free-warren was a sought-after right, and was quickly obtained if possible – free-warren charters were granted to most East Anglian villages by about 1280, and rabbit warrens soon established.

The original little enclosures containing small warrens next to manors and monasteries were maintained in some areas up to the nineteenth century. The once-common presence of these enclosures is reflected in 'conygar' and other corruptions of coneygarth, which can still be found as field and farm names in Britain: until the seventeenth century adult rabbits were called coneys, and 'rabbit' only referred to the young. From the thirteenth century larger warrens appeared, particularly in areas such as the East Anglian Breckland which had the ideal characteristics of dry sandy soil in undulating heathland. However, these warrens were still small, and the rabbit remained a costly luxury item both for meat and fur, and made the warreners little profit.

After the Black Death in the middle of the fourteenth century (Chapter 16), the first expansion of rabbit warrens started in Britain. Historians still argue over the cause of this change, which affected other former luxury goods as well, but it was probably related to the collapse in the grain market, which made warrening a profitable use of the land, coupled with increased purchasing power, which turned rabbits into an affordable foodstuff for more of the population. These factors led to an increase in the size and stocking density of rabbit warrens in East Anglia, and warrens were set up over much of southern England. The increasing value of these warrens led to them being physically enclosed by fences, banks or walls, with watchtowers to spot poachers and often an imposing lodge for the warrener (Fig. 11.2). As numbers of rabbits being sent to market increased during the early fifteenth century there was oversupply, the price of rabbits fell, and remained at a stable, low level throughout the fifteenth century.

The second increase in rabbit warrens occurred in the sixteenth and seventeenth centuries. This was due partly to the creation of a new class of merchants, lawyers and officials who were able to purchase land and buildings cheaply after the dissolution of the monasteries in 1536–9. These new gentry were interested in increasing the profitability of their land and rabbit warrens were considered the best and most profitable use of otherwise barren waste and upland areas. Williamson (1997) uses the term 'intermediate exploitation' for the warrening of rabbits at this time. They, along with deer, fish in ponds, and pigeons, were neither wholly domesticated nor wholly wild, and this mode of exploitation fitted with the land-owning gentry's philosophy. Traces of many of the artificial warrens constructed during this period remain as 'pillow mounds' (Fig. 11.1 and Box 11.2), or vermin traps (Fig. 11.3).

(a)

Fig. 11.2 **(a)** The warrener's lodge at the centre of Minchinhampton Common, Gloucestershire. The tall central section is probably the original lodge, dating from the early seventeenth century. The early fifteenth century warren lodge at Thetford, Norfolk **(b)** is a fortified structure with small windows, deeply splayed inside to allow a good angle of fire **(c)**, and with a hole in the arched ceiling of the only entrance to enable objects to be dropped on to the unwelcome **(d)**

(b)

(c)

(d)

Fig. 11.3 Vermin trap in the rabbit warren on Minchinhamton Common, Gloucestershire. Low walls (dashed lines) were constructed, funnelling the vermin into a trap (T) placed at the constriction of the walls

BOX 11.2   PILSDON PEN: PILLOW MOUNDS AND RABBIT WARRENS

Pilsdon Pen, some 10 km NW of Bridport, is one of Dorset's best-known landmarks, and one of the highest points in the county (Fig. 1) Like several of the hills in the area it contains a large Iron Age hillfort, with multiple-ditched defences. When surveyed in the 1940s it was noted that inside these defences were several pillow mounds, and the earthworks from a much larger rectangular structure.

    Pillow mound is a term coined in the 1920s by the pioneering aerial archaeologist O.G.S. Crawford to describe the many narrow banks with flanking ditches he found on aerial surveys of Wessex. Many are now marked on the larger scale

Ordnance Survey maps. He suggested that pillow mounds were the remains of medieval and later rabbit warrens. This idea has gradually gained acceptance as in many cases documentary details of site use have corresponded to the remaining features on the ground.

    However, interpreting the banks of the large rectangular structure on Pilsdon Pen was problematic. Between 1964 and 1971 this area of the hill fort was excavated, and parallel flat-bottomed straight-sided trenches with lateral connections and projections were found under the earth bank. These parallel trenches were interpreted as the foundations, or 'sleeper

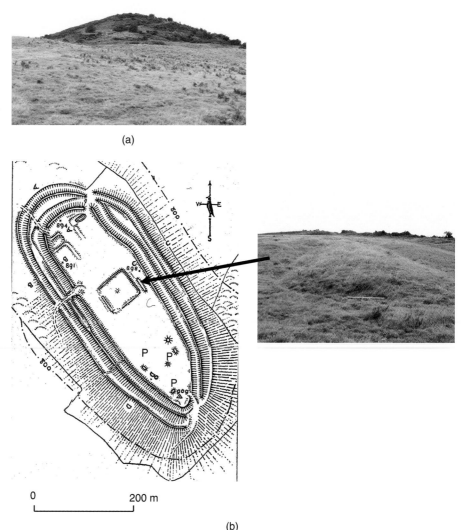

(a)

Box 11.2 Fig. 1 *Pilsdon Pen Hill Fort.* **(a)** *General view from the south.* **(b)** *Plan of 1940s survey: P = pillow mounds, with photo of large pillow mound and flanking ditches (facing north-west, rule = 1 m).* **(c)** *Plan of 1964–71 excavation site of rectangular structure (brown, trenches from Iron-Age huts; grey, trenches from rabbit warren).* **(d)** *Visible banks from central rectangular warren (facing NNE, rule = 1 m)* [(b) from RCHME (1952); (c) redrawn and slightly simplified from Gelling (1977).]

0   200 m

(b)

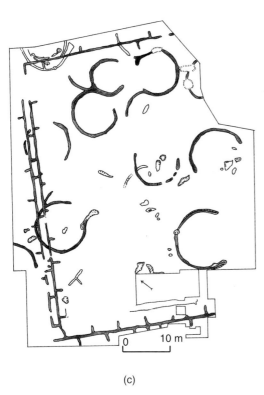

(c)

(d)

Box 11.2 Fig. 1   *(Continued)*

trenches', in which the base timbers of an Iron Age wood-framed building were laid, built around a central courtyard. The lateral projections were thought to represent trenches for buttresses. The presence of closely spaced parallel walls envisaged by this interpretation was hard to explain, but the final report (Gelling, 1977) suggests that store rooms could be contained therein. This building was hailed as one of the most important British prehistoric buildings.

Subsequently, small-scale excavations of this area were carried out by the owners, the National Trust, and by the mid-1980s it was thought most likely that this enigmatic rectangular structure actually represented the remains of a post-medieval rabbit warren. In hindsight, this seems fairly obvious – the size

and shape of the ditches, their infill with better-quality soil and the later bank, all point towards burrows and nests constructed artificially into the hard clay and chert of the site (even the original report notes the considerable disturbance done to the trenches by rabbits).

We discuss this not to belittle or denigrate the original excavators, who were working within the accepted paradigms of their discipline, but to suggest this as an example of the result of a lack of cross-fertilization between disciplines. More worryingly, this and other examples given by Williamson (2007) suggests the lack of a folk-memory of these structures, which would have been a common part of the rural landscape and economy only a few generations ago.

However, from now on the commercial rearing of rabbits was under increasing economic competition from other land uses. The rapid increase in warrens during the seventeenth century (by the eighteenth century over 5000 ha of Breckland was devoted to warrens), with the rapid expansion of the population and growth of towns, and the need for affordable meat, gave way to a decline during the eighteenth century. This was due to the increased profitability of wheat and barley and the requirement for land either to grow these crops, or on thin soils (otherwise suitable for rabbit warrens) for increased grassland to rear the flocks of sheep needed to fertilize these crops. In some counties, for example

Wiltshire, commercial warrening had almost ceased by the end of the eighteenth century, and surviving warrens were concentrated into the more marginal land of the East Anglian heaths and upland areas such as Dartmoor. Some of these commercial warrens survived into the twentieth century.

The last increase in warrening took place from the middle of the nineteenth century. This time warrens were used to produce game animals for shooting, and some commercial warrens in Breckland were almost completely given over to game rabbit production. Few records of this survive, but King and Sheail (1970) have collected oral records of the operation of one such warren on Fyfield Down, near Avebury in Wiltshire. This warren ran for at least 30 years from about 1880, and was situated on thin chalk downland littered with sarsen boulders—very difficult to plough or to cut hay from. Between five and six thousand rabbits were shot here annually.

The rabbit was also a cause of great social division in the countryside. During the earliest period of warren expansion rabbits were seen as a symbol of the feudal order—the common countryperson could not kill one—and thus poaching became a form of social protest as well as supplying much-needed meat. For example, the Peasant's Revolt of 1381 expressly demanded that all men should have the right to hunt 'hares' in the field: this was not granted. During the second expansion of warrening, matters got steadily worse. Seventeenth and eighteenth century law made a fetish of property; the property of the landed classes. Thus, the 1692 Game Act excluded rabbits and they became the property of the landowner. Consequently, while a poacher of game birds at night was treated under the relatively lenient game laws and could receive a still substantial £5 fine, or three months in gaol if he could not pay, a poacher of rabbits under the same circumstances could receive, from 1765, seven years' transportation. In 1723, the infamous 'Black Act' was passed and an armed hunter of rabbits, or one in disguise, could receive a summary trial without jury, and execution.

Furthermore, the ancestral 'free-warren' right of landowners to establish rabbit warrens often clashed with commoner's rights to graze animals on the same land, and warrens were deliberately overstocked or extended to drive down the value of the common land and make subsequent sale and enclosure easier (see also Chapter 17). The commoners could not legally kill or remove the rabbits, but could take the case to court claiming that the rabbits were a 'private nuisance' against their commoner's rights. These cases could take years to pass their expensive way through court, and judges would usually find in favour of the landowner. Thus, it is not surprising that many commoners took direct action. For example, after waiting three years for their case to be heard in court, 200–300 commoners of Cannock Chase worked for two weeks after Christmas 1753 to kill up to 15 000 rabbits and to destroy five warrens. The court case was finally heard three years later, and found in favour of the landowner.

Rabbits soon escaped from their warrens, especially if the stocking density became too great or the food supply was poor, and damage to crops was reported from the middle of the fourteenth century. The rabbit became an increasing and more widespread nuisance during the eighteenth and nineteenth centuries as agricultural improvement led to a greater area and range of winter crops being grown (most feral rabbit populations were previously probably limited by winter food availability), and new hedges and hedgebanks resulting from enclosure increased availability of cover and burrowing sites. A warming of the climate from the early nineteenth century would also have helped rabbit survival. To this must be added the rise in gamekeeping from the middle of the eighteenth century—actively protecting feral rabbits as game and exterminating many of their predators. It has been repeatedly stated that the rabbit also evolved to survive in the wild in Britain. It is quite probable that it did, and there would have been selective pressure for it to do so, but we have no evidence for this.

By the 1840s rabbits were regarded as a serious agricultural problem in much of Britain. Yet before the Ground Game Act 1880 it was illegal for lease-hold farmers to control rabbits on their land without the landlord's permission. Subsequently, rabbits killed on farms provided a nice supplementary income and undercut those sold from commercial warrens, hastening the end of commercial warrening in Britain. But the number of rabbits in the British countryside was increasing. Commercial rabbit trapping became very common in Britain, using mass-produced steel gin traps: by 1934 over 50 million skins were sold for hatting in Britain and 30 million exported. This did not halt the increase in the rabbit population: in an ironic reversal of the thought-of effects of illegal rabbit killing earlier, now-legal trapping actually increased rabbit numbers. As Vesey-Fitzgerald (1946) noted: 'The gin trap can clear an area of rabbits, but in the hands of a skilful man it can also maintain a steady population.'[1] The people dependent on this were keen to preserve the rabbit crop: leaving a viable breeding colony; collecting more bucks than does; and continuing to destroy predatory foxes, stoats and weasels.

In another reversal of fortunes, during the twentieth century Acts were passed to try to control the rabbit, including the Corn Protection Order 1917 and the Forestry Act 1919. These emergency acts were unpopular with landowners and soon repealed. Although during the 1920s almost yearly rabbit control bills were presented to Parliament, there was little effective rabbit control during the interwar years. By the Second World War the British rabbit population was estimated at 50 million and the rabbit was declared second only to the rat as a pest. During the war the wild rabbit was reviled as a pest of agriculture, but schools were encouraged to form rabbit clubs to breed them for food. By the early 1950s the rabbit population had increased to 60–100 million. Progress was made only with the Pests Act 1954, whereby the Ministry of Agriculture could declare parts of the country 'rabbit clearance areas' in which all rabbits could be killed, and there was a significant increase in organized rabbit culling. But by this time another, more lethal, rabbit foe had entered the country.

FROM EXPLOITED TO EXPLOITER IN AUSTRALIA

The convicted rabbit thieves transported to Australia may soon have been reacquainted with their nemesis, for five rabbits were listed in the inventory of the First Fleet to Australia in 1787–8, with further consignments arriving subsequently. As early as 1806 the Reverend Marsden tried to establish a warren, unsuccessfully, at Parramatta, near Sydney. Rabbits were more successfully reared in hutches around houses, and by the 1830s there were caged rabbits in many places on the east and south coasts of Australia and offshore islands. On some of these islands the rabbits had been placed deliberately as food for shipwrecked sailors.

From the 1840s further attempts were made to introduce rabbits into the wild on the mainland, and although potential predators such as native cats

[1]Vesey-Fitzgerald (1946), p. 172.

(*Dasyurus* species), dingos (*Canis lupus dingo*) and eagles were poisoned, these introductions were not successful. Introduced rabbits, however, thrived in the wild in Tasmania and became a problem by the 1860s, also needing the attention of the poisoners.

All the rabbits introduced into Australia up to the 1850s had been domesticated European stock, bred for food and not for survival in the wild. However, on the night of 25 December 1859 the sailing ship *Lightning* berthed at Victoria. On board were 24 wild English rabbits imported by Thomas Austin, the wealthy owner of Barwon Park at Winchelsea near Geelong. He wished to create an English-style sporting estate, and the rabbits were duly released into a grassy paddock. They survived, and thousands were present two years later. By 1865, Austin estimated that he had shot 20 000 rabbits on his estate, and the following year alone 14 000 were shot in Barwon Park.

The rabbit spread outside Barwon Park, and given by Austin to his friends, and imported by other landowners, spread quickly throughout suitable habitat in Australia, for example travelling 700 km in eight years in the 1880s (Fig. 11.4).

Fig. 11.4 *The major rabbit barriers built in Australia between 1880 and 1910, and the spread of the rabbit through Australia [Data on rabbit barriers from McKnight (1969), and on spread of rabbits from Myers et al. (1994); after Stodart and Parer (1988).]*

From the 1860s the rabbit had a dual personality in Australia, as both exploited and exploiter. The rabbit not only became a significant source of sport for the wealthy, but enabled the creation of several new industries. In 1869 the canning of rabbits for export to England began, and export of frozen rabbits started from Victoria in 1894. Around 1874 it was discovered that rabbit skin could be felted and used for hats, and this became a major industry in London, later using Australian rabbits. Throughout the 1920s an estimated 9 million rabbits a year were

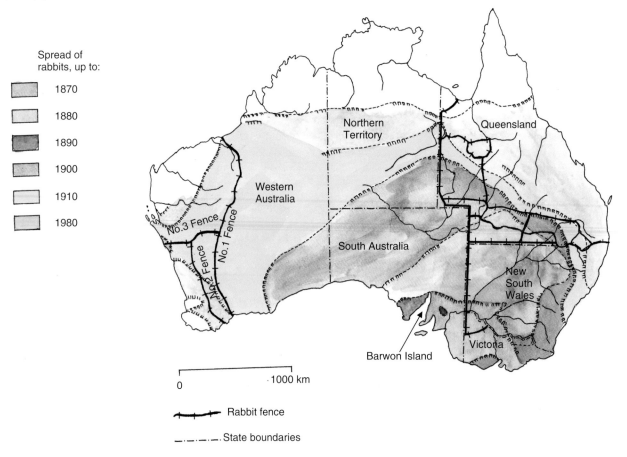

Spread of rabbits, up to:

1870
1880
1890
1900
1910
1980

Rabbit fence

State boundaries

shipped from Australia. These industries relied on rabbit trappers. Commercial rabbit trapping started in the early 1870s, for meat alone. Meat trappers worked mainly with spring traps, working up to 100 traps a night, and 40 pairs of rabbits was considered a good night's catch. Skinners had a possibly easier job—they could use poison—and a good pair could skin 600 rabbits an hour.

However, the numbers of rabbits available to trappers underline the problem that the rabbit had become in Australia by the last quarter of the nineteenth century. As the rabbits spread they stripped crops and, by grazing much closer to the ground than sheep or cattle could, significantly reduced the stock-carrying capacity of much already marginal land, especially in drought years. From the 1870s there were petitions from stock-holders for government rabbit control, but a succession of Acts had no effect on rabbit numbers. As long as the rabbit also remained a commercially valuable creature, effective control was unlikely. Several control measures were tried and many predators suggested: stoats (*Mustela ermina*) introduced for this purpose became a serious problem in New Zealand, devastating the populations of native ground-nesting birds and now have a control programme of their own.

Government-sponsored trapping schemes were attempted. An 1885 law in New South Wales promised a bonus paid for all rabbit skins with scalps attached, but this had to be repealed three years later when the cost of these bonuses far outstripped the amount of state money available (a scenario we will encounter again with wolf bounties, in Chapter 19). These changes halted commercial rabbiting in New South Wales for a couple of years, until the frozen carcass trade started.

Barrier fences were a useful method of excluding rabbits from small plots, if they used the correct wire mesh (for many years the Australian government recommended a mesh with too large a diameter to contain rabbits), with 15 cm dug into the ground and, most importantly, constant maintenance. Continent-scale fencing against the spread of the rabbit was another matter. Yet from the 1880s thousands of kilometres of rabbit fence was built across Australia (Fig. 11.4). By the time that government approval and funding for a fence had been obtained the rabbits had often already passed, yet the fences were still built. For example, No. 1 rabbit fence in Western Australia ran for over 1800 km and took five years to build, being completed in 1907. The fences were poorly maintained and largely ineffective: wombats (*Vombatus* species) would burrow underneath the fence and push the wire out of the way, emus (*Dromaius novaehollandiae*) and kangaroos (*Macropus* species) would charge and break the fence, and sandstorms could cover a fence in a couple of days.

Poisoning became recognized as the best method of rabbit control, and is still used. Strychnine (see Box 17.3) was first used, added to grain bait, or to lures such as quince jam. However, strychnine works so fast that other rabbits are frightened away from the bait. Phosphorous and arsenic were also used, and millions of kilometres of bait trails were laid. Latterly, sodium fluoroacetate (Compound 1080) has been used, chiefly in baited carrots. All of these poisons are non-selective, and by also killing native wildlife have only added to the effect of the rabbit on the ecosystem. Ultimately, to obtain rabbit-free land the warrens have to be physically destroyed. Tractors and ripping ploughs are often used, but Rolls (1984) suggested that it was best to use a massive single-furrow plough digging to almost 1 m depth, pulled by 18 bullocks. These animals, unlike horses, are not spooked by holes in the ground, and the pounding of 72 large hooves is a useful addition to the plough.

By the late 1940s there could have been 500 million rabbits in Australia. Why was the rabbit such a good colonizer? Myers *et al.* (1994) suggest that this was because the rabbit is an excellent generalist, physiologically preadapted to much of the climate of southern Australia and with few diseases and parasites when introduced, and was able to utilize the burrows abandoned by native herbivores, such as wombats. It was also able to capitalize on the effects of the recent human immigrants: habitat disturbance caused by cattle and sheep—which led to the replacement of the native perennial shrub, tree and tall grass vegetation with introduced annual grasses and unpalatable shrubs—was probably the single most important factor, along with the depletion of native predators by the settlers.

MYXOMATOSIS AND BEYOND

As the biology of infectious diseases of humans and animals became better understood during the nineteenth century (see Chapters 5, 12 and 16), it was not long before diseases were utilized as control agents for the exploiters of mankind, and scientific biological control was born.

Pasteur had been working during the late nineteenth century on the chicken cholera bacterium, *Pasteurella multocida*, and had some success inoculating rabbits with this in France. Thus, when the New South Wales government established a Royal Commission of Enquiry into 'Schemes for the Extermination of Rabbits' in April 1888, with a prize of £25 000 for a rabbit control method that was not injurious to other creatures, Pasteur's team entered this bacterium. After problems due to the possibility of this bacterium infecting domestic fowl, Pasteur's team was allowed to experiment in a specially built laboratory on Rodd Island in Sydney Harbour. Experiments demonstrated that rabbits that ate feed contaminated with the bacteria rapidly died, and that it was harmless to domestic animals caged with the rabbits. Unfortunately, it was also harmless to rabbits fed on uncontaminated feed in the cages. Pasteur seemed to have considered contagiousness of a control agent unimportant, concentrating on the kill rate after infection, and his team had to be ordered to carry out the experiments which demonstrated that chicken cholera was not contagious in rabbits.

Pasteur's proposal was rejected in April 1889 by the Commission, as were all the almost 1500 proposals received, and the prize was not awarded. The contentious interaction between Pasteur, his team and the Commission caused bad feeling for many years.

Not long afterwards, in 1896, the scientist Guiseppe Sanarelli imported some domestic European rabbits from Brazil to his laboratory in Montevideo, Uruguay. They died from an unusual disease, which he named myxomatosis from its main symptom, a 'mucous tumour' (Box 11.3). In 1911 the infective agent was isolated from Brazilian rabbits by the scientist Moses. This 'infective particle' was able to pass through the filters used to trap bacteria and was almost impossible to see under the light microscope; thus it was termed a 'filterable virus' or 'invisible microbe'. These were the early days of virus research.

In 1919, the Brazilian researcher Aragão suggested that the myxoma virus could be used against rabbits in Australia, and in 1926 Dr Seddon of the NSW Department of Agriculture obtained some of this virus from Aragão and carried out direct inoculation laboratory tests on rabbits. They were not promising—it was difficult to get the virus to spread from rabbit to rabbit. In the 1930s, Dr Jean Macnamara, an Australian paediatrician, met Dr Richard Shope in New York, who was also working on the myxoma virus, and sent a vial of the virus back to Australia. It was not allowed into the country and was destroyed.

| BOX 11.3 | **THE MYXOMA AND RHD VIRUSES** |

The myxoma virus is a large DNA virus of the Poxviridae family (which contains viruses such as smallpox), genus Leporipoxvirus, with a characteristic brick-shaped virion. Unlike many viruses that replicate in the cell nucleus and utilize the host's DNA and protein replication machinery, the pox viruses replicate in the host cytoplasm and their DNA encodes almost 200 genes (many originally acquired from their hosts) necessary for replication.

The myxoma virus originates from South America. It first infected domestic European rabbits via a mosquito vector from the disease reservoir present in the native Brazilian rabbit, *Sylvilagus brasiliensis*. In this species, one of its normal hosts, the myxoma virus produces only localized and benign skin tumours. It is only in the European rabbit that a generalized and lethal disease is produced.

The cause of death is various; it is not due to the growth of the virus in any one vital organ. After inoculation with the Moses strain, a lump is seen at the inoculation site after three days. After replicating here, the virus passes to the local lymph nodes and causes a generalized viral infection manifested as secondary skin lesions. These appear after 6–7 days and become widely distributed over the body. The virus also infects many organs, including the testes. Thickening of the eyelids starts at this time, and by day nine the eyes are fully closed. The rabbit will continue to eat and drink until just before its death.

Infertility is common in male rabbits recovering from the virus: after recovery rabbits show a high degree of resistance to new infection, and resistance can be transferred to embryos. This passive immunity, however, is only effective for a couple of weeks after birth.

The RHD virus is rather different. It is a member of the Caliciviridae family, genus Lagovirus, of small round RNA viruses, which include the noroviruses, one of which caused 'winter vomiting disease' in humans during recent winters. The RHD virus is shed in faeces and nasal secretions, and can be ingested and inhaled, and transmitted in infected meat. It is stable in the environment, and like myxoma virus can be passively transported by insects. Rabbits less than four weeks old do not show symptoms from infection and acquire lifetime immunity, but do excrete infectious virus particles.

However, RHD is less efficient in controlling rabbits in more temperate areas of Australia. This is now thought to be due to infection by a non-pathogenic lagovirus – RCV-A1 – that gives cross immunization to RHD.

Currently, scientists are using recombinant technology, whereby genes from other organisms are added to the virus genome, to develop viruses for two applications. One involves developing a recombinant myxoma virus that can be used for immunocontraception of rabbits. It is hoped that this virus will spread naturally and control rabbit populations without the unpleasant effects of myxomatosis. Laboratory trials have demonstrated that this can work, but much more information about the autoimmune response of rabbits will be needed before field trials can proceed. Although immunocontraception may dampen population fluctuations, alone it is unlikely to control rabbits and will have to interact synergistically with the two virus diseases.

On the other hand, there is a need for an effective vaccine for these viruses; one that can be used in the wild. In particular, wild rabbit populations need to be safeguarded in their native range in Spain where they are a key part of the ecosystem and hunting economy (see Box 11.6). While effective vaccines to both the myxoma and RHD viruses are already available, they have to be given to each rabbit, and booster vaccinations are needed. It is not feasible to catch a significant proportion of wild rabbits to administer a vaccine, and thus from the mid-1990s work has been carried out in Spain to develop a recombinant attenuated myxoma virus that will provide protection against both myxomatosis and RHD. It is hoped that such a virus would spread naturally, via insect vectors, once released, and field trials on a small uninhabited Mediterranean island have demonstrated that limited spread can occur.

Macnamara was convinced that the myxoma virus was the answer to the rabbit problem in Australia and she embarked on a publicity campaign to get this virus into the country. She was put in contact with Sir Charles Martin, who in 1933 began working on the virus for the Australian government, in Cambridge, England. He used both 'Strain A' supplied by Aragão, and the more infectious one originating from Moses (the 'Moses strain'), and carried out experiments in a large netted compound in Cambridge. Although he was able to get a good infection rate in rabbits, he doubted that myxomatosis would spread in the wild.

From 1936, tests on the virus in Australia demonstrated that it was very host-specific and from 1937 trials were carried out on Wardang Island in the Spencer Gulf, South Australia. Inoculated rabbits were introduced into the rabbit population on the island and the progress of the disease monitored. In experiments over the next two years the results were the same: the inoculated rabbits soon died, and although there was transmission between rabbits, the disease died out within a year. Further field experiments on the mainland in 1941 and 1942 gave the same results.

Trials of the myxoma virus were suspended during the Second World War, but after further arguments between Macnamara and government scientists

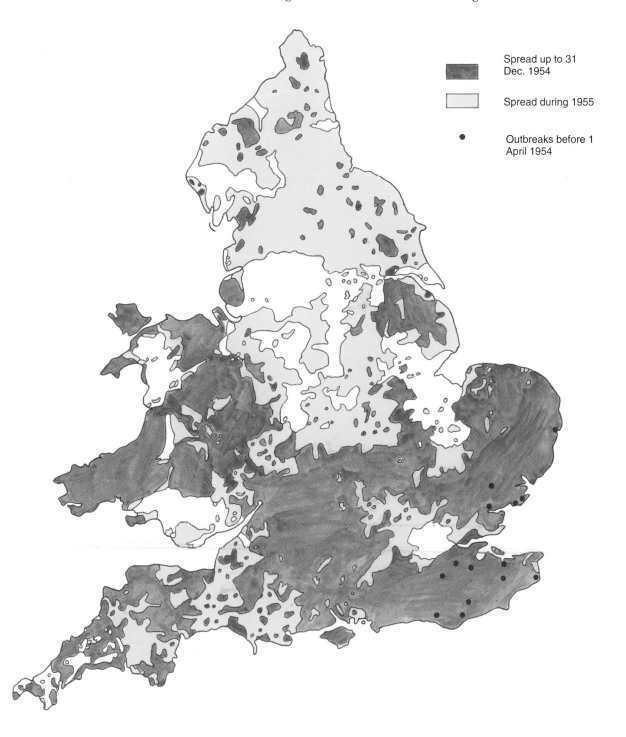

Fig. 11.5   *The early spread of myxomatosis in Britain [Redrawn from Fenner and Ratcliffe (1965), data originally from Thompson (1956).]*

another field trial took place in 1950 at three sites in New South Wales. Infected rabbits were introduced into burrows, and again, although there was initial transmission of the virus, the infection died out within a few months. This was thought to be final proof that the virus could not work in the field.

However, in late December 1950 the scientists began receiving calls that sick rabbits were in abundance near one of the release sites at Balldale, with up to 90% mortality reported within a month of the appearance of the first symptoms. Nine weeks after the initial report, a 1600 by 1750 km area was infected – the virus had entered the wild rabbit population. Although there were initial quarantine restrictions on the spread of the virus and sick rabbits, these were soon removed as they were useless, and in 1951 a state rabbit myxoma virus inoculation scheme started. Landowners brought their rabbits to central depots to be inoculated by the virus: one favoured method was by dipping an office stapler into the virus and then stapling the rabbit's ears. Inoculation continued into the 1960s.

The effects of myxomatosis were significant: some areas reported a 25–50% increase in stock-carrying capacity after myxomatosis appeared (in 1953 after three seasons of myxomatosis it was estimated that the reduction in rabbits had contributed £24 million in increased wool and £10 million in increased meat from sheep alone), and landowners who were spending $2000 on 4–6 weeks' rabbit control before myxomatosis were by 1960 spending a tenth of that.

Why did the virus work in 1950 but not in the previous experiments? A major factor was having suitable conditions for the vectors of myxoma virus to operate. Unlike malaria (Chapter 12) and plague (Chapter 16) in which the infectious organism is vectored inside an insect, the myxoma virus is transmitted on the outside of the mouthparts of a number of sucking insects, in Australia mainly mosquitoes. The outbreaks of myxomatosis in Australia are seasonal, and are linked to the breeding season of the mosquitoes and weather conditions which favour different mosquito species with differing infection capabilities.

The first deliberate release of the myxoma virus into the wild, however, was in Britain. The virus was released in rabbits inoculated with the Moses strain on Skokholm Island, off the Pembrokeshire coast between 1936 and 1938. The results were as disappointing as those in Australia at that time. Similar also were the results of eight separate introductions of the virus into the Heisker Islands, Outer Hebrides, between July 1952 and May 1953.

During this time the retired physician Dr Armand Delille wished to clear his estate near Paris of rabbits, and on 14 June 1952 inoculated two rabbits with the myxoma virus obtained from a laboratory in Lausanne, Switzerland. This strain had been isolated from wild rabbits in Brazil only in 1949, and probably because it had not passed through many generations in the European rabbit (unlike the Moses strain used in Australia, isolated 40 years previously), was extremely virulent to its new host. This disease spread throughout France quickly, travelling up to 45 km a month, and it was initially unclear what the cause was: by October 1952 it had been identified as myxomatosis. Delille, who had initially kept quiet about this introduction, confessed in June 1953. He was rewarded with a gold medal from

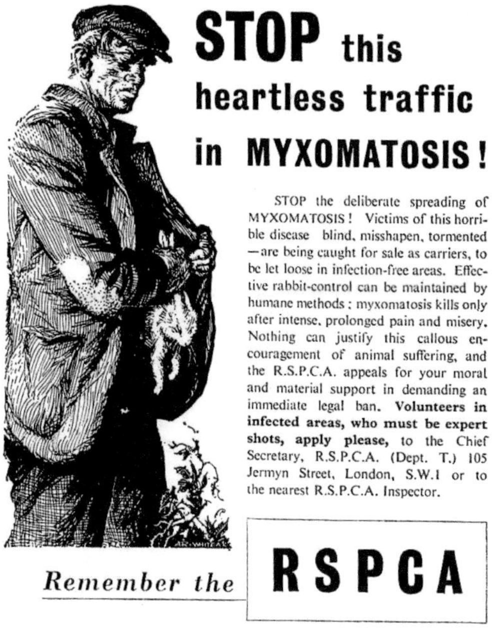

STOP this heartless traffic in MYXOMATOSIS!

STOP the deliberate spreading of MYXOMATOSIS! Victims of this horrible disease blind, misshapen, tormented – are being caught for sale as carriers, to be let loose in infection-free areas. Effective rabbit-control can be maintained by humane methods ; myxomatosis kills only after intense, prolonged pain and misery. Nothing can justify this callous encouragement of animal suffering, and the R.S.P.C.A. appeals for your moral and material support in demanding an immediate legal ban. **Volunteers in infected areas, who must be expert shots, apply please,** to the Chief Secretary, R.S.P.C.A. (Dept. T.) 105 Jermyn Street, London, S.W.1 or to the nearest R.S.P.C.A. Inspector.

Remember the **RSPCA**

Fig. 11.6 *Advertisement placed in British newspapers during 1954 by the Royal Society for the Prevention of Cruelty to Animals [From Fenner and Ratcliffe (1965).]*

BOX 11.4   RABBITS IN VERSE: HARDY AND LARKIN

Few poets and novelists have managed to write about rabbits without sinking into a mawkish sentimentality. However, Thomas Hardy achieved it, for example in his 1922 poem 'The Milestone by the Rabbit-Burrow (On Yell'ham Hill)':[a]

In my loamy nook
As I dig my hole
I observe men look
At a stone, and sigh
As they pass it by
To some far goal.

Something it says
To their glancing eyes
That must distress
The frail and lame,
And the strong of frame
Gladden or surprise.

Do signs on its face
Declare how far
Feet have to trace
Before they gain
Some blest champaign
Where no gins are?

Philip Larkin also managed it in 'Myxomatosis', written in 1954.[b] He describes coming across a rabbit suffering from the disease (very soon after it had been introduced into the UK–did

Larkin actually have such an encounter, or was he working from second-hand knowledge?). The rabbit seems to him to be asking what new sort of trap is this disease, as if a trap-image is engrained into the rabbit's psyche. Larkin cannot explain and answers by killing the rabbit with his stick.

Although at first seemingly different–from different viewpoints and generations, and in different metres–these two poems share many similarities.

By the time Larkin wrote 'Myxomatosis' he had left behind the Yeats-inspired lyrical mysticism of his earlier poetry and had embraced Hardy's pessimistic naturalism and use of commonplace language. Yet Larkin goes further than Hardy. Hardy is conditionally pessimistic–there *could* be good times ahead: the 'blest champaign' (the best quality agricultural land), both for rabbits and humans, with an absence of traps–while for Larkin the pessimism is absolute, there is no escape from myxomatosis and hope of recovery is futile. One cannot help but think that this reflects Larkin's attitude to life in general.

The rabbit is juxtaposed with the human in both these poems, and the difference between rabbit traps and myxomatosis is similar to that between conventional warfare familiar to Hardy, and the cold-war threat of nuclear conflict during the 1950s to Larkin. Larkin suggests that a conventional response to such a new threat is of no use.

[a]First published in Thomas Hardy, *Late Lyrics and Earlier* (1922); from Gibson (1976).

[b]Larkin's poem, first published in *The Spectator*, 26 November 1954, can be read in Thwaite (2003).

---

agriculturalists and foresters, and a lawsuit from hunting organizations. The number of rabbits hunted fell from 55 million in 1952–3 to less than 7 million the following year and 1 million the year after. It was also estimated that 35% of domestic rabbits had died from myxomatosis in France by the end of 1953.

Myxomatosis reached Britain in August or September 1953, with the first confirmed cases in October 1953. It was the Lausanne strain, from France, almost certainly introduced on a diseased rabbit by a private individual. The first reaction of the government was to try to eradicate the local outbreaks (Fig. 11.5) by extermination of infectious rabbits, quarantine and rabbit fencing. This was soon found to be useless, and after becoming quiescent during the winter of 1953–4 myxomatosis rapidly spread the following year, reaching Wales by May and Scotland by July 1954.

As with the widespread use of gin traps to control rabbits, there was considerable opposition to the spread of myxomatosis in Britain on humanitarian grounds (Fig. 11.6 and Box 11.4), and the Ministry of Agriculture rejected calls for myxomatosis to be deliberately spread. The Pests Act 1954 dealt almost entirely with rabbits, and along with banning the gin trap, also made it illegal to deliberately spread myxomatosis. Although much of the early spread of the virus was by deliberate introduction of diseased rabbits, no one was convicted of this until 1958, but Bartrip (2008) suggests that by preventing the active dissemination of virulent strains of the virus to rabbits that survived the initial outbreak the chance of exterminating the rabbit in Britain was lost.

In Britain, myxomatosis is less seasonal than in Australia, and this is thought to be because its main vector in Britain is the rabbit flea (Box 11.5) which can persist in rabbit burrows during the winter. The probable reason for the lack of spread of myxomatosis on Skokholm Island earlier was the lack of rabbit fleas there.

BOX 11.5 THE RABBIT FLEA

The European rabbit flea, *Spilopsyllus cuniculi*, is a member of the insect order Siphonaptera. This order contains about 2500 species, all of which are laterally flattened ectoparasites which feed on the blood of mammals and occasionally birds. Although they are wingless and are usually distributed by their hosts, they have large powerful hind legs for jumping. The spring of the jump is provided by a pair of structures near the bases of the hind limbs containing pads of the highly elastic protein resilin.

Even among such a specialized group as the fleas, the rabbit flea has an unusually close relationship with its host. These fleas usually remain on the head and especially the ears of the rabbit, fixed firmly to the skin. The fleas regulate their breeding to coincide with the production of rabbit offspring by detecting host hormones. Thus, ovarian maturation only occurs in fleas feeding on rabbits during their last 7–10 days of pregnancy, or on a newborn rabbit during its first week of life. Male fleas also require contact with rabbits in the same condition before they can mate. At parturition, most fleas leave the female rabbit and enter the nest, feeding upon the young, copulating and laying eggs mainly in the nest. The larval fleas feed on faecal remains, skin cells, or dried blood excreted by the adult fleas. They pupate in a silken cocoon, and a new adult may remain in this for up to a year until stimulated to emerge by the presence of a rabbit.

This close association is probably necessary, as rabbits, unlike many small mammals, do not have a permanent nest. By maturing eggs only just before parturition the flea ensures that resources are used efficiently, and that it is likely to be carried to a nest.

The rabbit flea is the principal vector of the myxoma virus in Britain, and an important vector in continental Europe. The virus particles are not ingested (unlike the plague bacterium, see Chapter 16), but are carried on the outside of the mouthparts, particularly the serrated cutting plates of the laciniae (Fig. 1), from probing virus-rich lesions. These mouthparts, although shorter than those of mosquitoes, seem to be pre-adapted for maximal ease of contamination by the virus.

In Australia, mosquitoes are responsible for most myxoma virus transmission. This is thought to be limiting: the mosquitoes have a limited distribution; are most prevalent during the hotter summer months when rabbits could best recover from the virus; and confer a selective advantage on attenuated strains of the virus. Therefore, as indigenous fleas are poor vectors, the European rabbit flea was released in Australia in 1968 as a year-round vector, and soon spread amongst the rabbit population. The fleas, however, are limited to the cooler and damper parts of the country. In hot and dry conditions there is insufficient nest humidity for breeding, and if rabbits do not breed for prolonged periods during a drought, the adult fleas may die before reproducing.

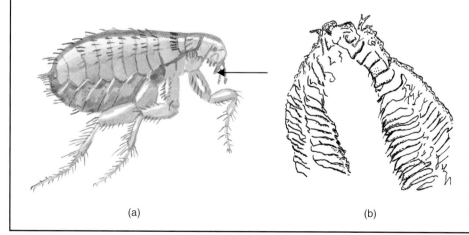

(a)   (b)

Box 11.5 Fig. 1 *European rabbit flea* Spilopsyllus cuniculi, *adult, length 1 mm.* **(b)** *The outer surface of the left lacinia of the rabbit flea, arrowed in (a)* [(b) *drawn from a photo in Mead-Briggs (1977).*]

BOX II.6   THE WIDER EFFECTS OF MYXOMATOSIS

We have concentrated on the effects of myxomatosis and RHD to the rabbit population, but due to their abundance rabbits form an important part of many ecosystems. Thus, the sudden reduction in their numbers was likely to have important effects throughout the community (Fig. 1).

In Britain, the first community effects of myxomatosis were positive; better tree regeneration and a greater variety of wild flowers in grassland, both common ones such as cowslip (*Primula veris*) and rockrose (*Helianthemum* species) and rarer orchids and pasqueflowers (*Pulsatilla vulgaris*). However, within a couple of years these plants had been out-competed by tall weedy grasses and scrub, dominated by shrubs such as hawthorn (*Crataegus monogyna*), blackthorn (*Prunus spinosa*), gorse (*Ulex europaeus*) and bramble (*Rubus fruticosus*). This effect was particularly noticeable on southern British chalk and limestone grassland. The characteristic short grazed, species-rich vegetation of these areas had been maintained by sheep grazing, but with the virtual disappearance of sheep from lowland chalk downland from the 1940s, this habitat had been maintained solely by rabbit grazing. Without this there was nothing to stop it reverting, via scrub, to woodland (Fig. 2).

This change in grassland had positive, if short-lived, effects for some species such as the Duke of Burgundy butterfly (*Hamearis lucina*), which started to colonize downland

after myxomatosis, laying on the now more abundant cowslips. However, some birds were negatively affected, including lapwings (*Vanellus vanellus*), stone curlews (*Burhinus oedicnemus*) and wheatear (*Oenanthe oenanthe*) which need bare ground on which to nest.

These changes in grassland habitat probably hastened the extinction of the large blue butterfly (*Maculinea arion*), which was already rare before myxomatosis. This has an unusual life-cycle: the young larvae feed on wild thyme (*Thymus polytrichus*) growing in chalk grassland, but after a couple of weeks drop to the ground and are carried into nests of the red ant *Myrmica sabuleti* where it feeds upon ant larvae until it is fully grown. This butterfly can only utilize effectively this one ant species, and this ant needs longer periods of warmth to survive than other *Myrmica* species. Increasing cover by tall grasses, herbs and shrubs after myxomatosis caused the microclimate to become colder over the nests, and this ant species was gradually replaced by others better adapted to survive these conditions. By 1979 the large blue had become extinct in Britain.

The lack of rabbits after myxomatosis had a more direct effect on predators. Breeding of buzzards in the year after myxomatosis was significantly decreased, but recovered after a couple of years, and other birds of prey, such as red kites (Chapter 17), may have been affected for a short time. Other

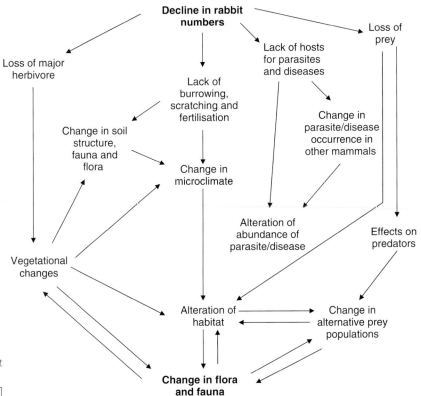

Box II.6 Fig. I   *Summary of the ecological effects of the reduction in rabbit numbers after myxomatosis* [Redrawn from Sumption and Flowerdew (1985).]

Box 11.6 Fig. 2   *Sheep grazing was reintroduced in the 1970s on this National Nature Reserve at Aston Rowant, Oxfordshire, to control the scrub that started to colonize the chalk grassland after myxomatosis wiped out the rabbit population. The sheep do a good job of keeping the scrub at bay, but they cannot recreate the very closely grazed sward that the rabbits created*

predators such as foxes and weasels may have benefited from the great increase in field voles (*Microtus agrestis*) that occurred due to the increase in height of vegetation after myxomatosis.

In Spain, where the rabbit is native, the decrease in rabbit numbers after myxomatosis and RHD has had a serious effect on the red kite but has been critical for two endangered carnivores: the Spanish imperial eagle (*Aquila adalberti*) and the Iberian lynx (*Lynx pardinus*). These two species are restricted to the Iberian peninsular, and are thought to have evolved there recently as rabbit prey specialists from their close relatives the Eurasian lynx (*L. lynx*) and Eastern imperial eagle

(*A. heliaca*) respectively. Both Spanish species feed almost entirely on rabbits and are thus directly affected by decreases in rabbit numbers. The geographic range of the lynx declined by 80% after myxomatosis, and again significantly after RHD. Only two populations of the lynx remain in the wild, containing fewer than 150 adults, and only 200 breeding pairs of the eagle remain, confined to south-western Spain. Supplementary feeding has been attempted for both, but the only long-term solution to the near-extinction of these species will be to increase rabbit numbers in the wild by effective vaccination (see Box 11.3).

In both Australia and Britain, the initial effect of the myxoma virus was to kill at least 70–95% of the rabbits in the infected areas, with considerable effects on the native wildlife (Box 11.6). However, within a couple of years attenuated virus strains were found in the wild. These strains take longer to kill rabbits (and a greater proportion recover) than the original strains, and this means that the rabbit is infectious for longer. Thus, it is not surprising that these attenuated strains rapidly became the dominant form of the virus. Resistance to the virus in rabbits also increased rapidly from the late 1950s in Australia and later in Britain. There is now a balance between host resistance and virus virulence: as resistance increased so did the virulence of the virus and 40–60% of rabbits are now killed by a myxomatosis outbreak. There is also anecdotal evidence that British rabbits changed their behaviour, becoming more solitary and nesting above ground, which would reduce infection from the rabbit flea.

Thus, by the end of the 1950s the mortality rate from myxomatosis in Australia had fallen to 25% in some places, and by the mid-1960s it could not be relied upon to reduce rabbit populations by more than 75%. The rabbit population soon started to increase again, even though new, more virulent strains of the virus were inoculated into rabbits and more efficient vectors were introduced (Box 11.4).

By the mid-1980s rabbits remained under reasonable control in much of the better-quality agricultural land of Australia, but continued to be a problem in the arid interior. In Britain the rabbit population had remained at only 5% of its pre-1953 total until the early 1970s, then increased rapidly, reaching 20% of its pre-1953 total by 1979 and 33% by the 1990s (Fig. 11.7). By this time, the rabbit population in parts of Britain was reported to be reaching or exceeding its pre-myxomatosis levels, and was again becoming a major issue for farming and nature conservation (Box 11.6).

Fig. 11.7 *Distribution of rabbits in England and Wales:* **(a)** *before myxomatosis and* **(b)** *in 1980. The mean abundance index is a measure of the relative number of rabbits in the county. Although the density measurements are not directly comparable, note the change in density distribution of the rabbit after myxomatosis. The very high abundance for rabbits in Berkshire in (b) is thought to be an error [(a) redrawn from Thompson and Worden (1956), (b) from MAFF (1982).]*

In 1984, a new highly contagious and usually fatal disease was discovered in China in a batch of commercially bred Angora rabbits imported from Germany. This disease, termed rabbit haemorrhagic disease (RHD), was characterized by a very rapid development of severe hepatitis and death usually within 48 hours, and was later found to be caused by another virus (Box 11.3). The disease spread rapidly through China, killing over 14 million domestic animals in nine months. It was first found in Europe in 1986, and in 1992 RHD was first found in Britain in domestic rabbits, and a couple of years later in the wild population. The disease spread throughout Europe, killing up to 75% of wild rabbits in Spain, for example, and rabbit populations have been slow to recover subsequently.

It was thought that the virus had one origin, in China, and then subsequently spread throughout the world through trade in domestic rabbits. However, recent work has shown that the story is more complicated: RHD virus has been present in European rabbits since the 1950s yet was symptomless. Furthermore, virulent forms of the virus differ significantly throughout the world, suggesting that these arose on more than one occasion, and that different lineages of the virus started to produce epidemics at the same time in Europe and Asia. We do not know how this happened, or how this virulent form was able to infect rabbits which already had antibodies to the non-virulent form. The virus seems to circulate mainly as a non-virulent form and the considerable genetic recombination that occurs naturally may have created the virulent forms.

Noticing the great mortality caused by this disease, the Australian and New Zealand governments investigated its use for rabbit control – this and losses caused by rabbits were still costing more than US$310 million annually in Australia in the late 1980s. A Czech strain of the virus was imported into quarantine in Australia during 1991, and after host-range tests was allowed to be used in a controlled field test on Wardang Island in 1995, the island previously used for myxomatosis tests. Considerable efforts were made to prevent the escape of rabbits, and multiple rabbit-proof fences were erected, but little was known about the transmission of the virus. Although all rabbit fleas and wild rabbits were removed before the trials began, fly traps were used and surrounding swamps treated to kill mosquitoes, there were no barriers to prevent insects entering or leaving the RHD-infected areas.

The virus crossed the quarantine barriers on Wardang Island in late September 1995 (it is thought to have been vectored by a swarm of biting flies) and spread to the mainland, being found over 350 km from the island within weeks. By May 1996 it had spread to all states and territories of mainland Australia, travelling up to 400 km a month. There was little that could be done and in October 1996 the Australian government officially approved the release of the virus. Since then the virus has been effective particularly in arid and semi-arid areas (unlike myxomatosis), killing up to 95% of rabbits in some areas, and has led to the collapse of rabbit commerce. Rabbit populations have remained at 15% of pre-RHD levels since, and in the 13 years since RHD has been released A$6 billion has been saved.

It is tempting to see this as a repeat of the myxomatosis story: it is not clear whether an intentional release of the RHD virus into the wild would have been allowed in Australia, but by carrying out a trial, was it hoped that the virus would become established by itself, and thus present a fait accompli to the regulatory authorities and the public? Certainly, once the virus had escaped scientists wanted to release RHD quickly to control rabbits before they became immune, and the government attempt to rename RHD as rabbit calicivirus disease could be seen as a means to make it sound innocuous. The New Zealand government decided not to recommend release of the virus (they had managed rabbit populations without introducing myxomatosis), but with rising rabbit numbers and a perceived lack of government action over rabbit control, a group of New Zealand farmers introduced and spread RHD illegally in 1997. The disease spread over a large area of the South Island before being detected by government officials, and reduced the rabbit population by 50% within six weeks of release. The government decided that it was not feasible to control RHD and accepted that it had become endemic in New Zealand. The virus was soon legally on sale for rabbit control.

As long as the rabbit is able to be both exploiter and exploited it is unlikely ever to be controlled. The only way forward is 'decommercialization'. New Zealand, which had a rabbit problem similar to that in Australia, did decommercialize the rabbit in 1947, making trade in rabbits illegal and sacrificing a lucrative export industry in rabbit skins and carcasses. After this, the rabbit was controlled in New Zealand without the introduction of myxomatosis, while in Australia several million rabbits are killed each year for meat and fur, and millions further die of the effects of myxomatosis and RHD. As noted by Rolls (1984):

> [T]he history of the rabbit the world over is a history of the incomprehensible lethargy of landowners and deficiency of governments'.[2]

[2]Rolls (1984), p. 260.

FURTHER READING

Sheail (1971) remains the best history of the rabbit, supplemented by Sheail (1978). A good introduction to the rabbit is given by Leach (1989), with more detailed information available in Thompson and King (1994). For the archaeology of rabbit warrens, Williamson (2007) is essential. Tittensor and Tittensor (1985) give a particularly useful account of the rise and fall of one rabbit warren, and rabbit warrens in other districts are covered by Bailey (1988) for East Anglia and Bettey (2004) for Wiltshire.

The labyrinthine game laws are covered by Munsche (1981), and Hay (1975) gives a useful description of the legal and physical battle between commoners and warreners in one area. The standard work on the history of the rabbit in Australia is Rolls (1984), Kerr (2008) is a useful summary, and see McKnight (1969) for rabbit fences in Australia.

Fenner and Fantini (1999) give an authoritative account of myxomatosis, concentrating on it in Australia, and also cover the beginning of RHD. For later work see Cooke and Fenner (2002), Forrester et al. (2006), Lawrence (2010) and Strive et al. (2010), and Drollette (1996) and Hayes and Richardson (2001) for the controversy surrounding the release of the virus in Australia. Bartrip (2008) discusses the spread of myxomatosis in Europe.

For the rabbit in Roman times see Toynbee (1973) and Alcock (2001). Williams et al. (2007) summarize the effects of imposed sterility on rabbit populations, and Angulo and Bárcena (2007) review attempts to develop a myxoma and RHD virus vaccine. Mead-Briggs (1977) gives a comprehensive review of the rabbit flea. The effects of the decline in rabbit numbers due to myxomatosis on other organisms is reviewed by Sumption and Flowerdew (1985) and Ferrer and Negro (2004).

REFERENCES

Alcock, J.P. (2001) *Food in Roman Britain*. Tempus, Stroud, UK.
Angulo, E. and Bárcena, J. (2007) Towards a unique and transmissible vaccine against myxomatosis and rabbit haemorrhagic disease for rabbit populations. *Wildlife Research* **34**, 567–577.
Bailey, M. (1988) The rabbit and the medieval East Anglian economy. *Agricultural History Review* **36**, 1–20.
Bartrip, P.W.J. (2008) *Myxomatosis: a history of pest control and the rabbit*. Tauris Academic Studies, London.
Bettey, J. (2004) The production of rabbits in Wiltshire during the seventeenth century. *The Antiquaries Journal* **84**, 380–393.
Cooke, B.D. and Fenner, F. (2002) Rabbit haemorrhagic disease and the biological control of wild rabbits, *Oryctolagus cuniculus*, in Australia and New Zealand. *Wildlife Research* **29**, 689–706.
Cornish, F.W., Postgate, J.P. and Mackail, J.W. (transl.) (1988) *Catullus, Tibullus and Pervigilium Veneris*. 2nd edn. Loeb Classical Library, Harvard University Press, Cambridge, MA, USA.
Drollette, D. (1996) Australia fends off critic of plan to eradicate rabbits. *Science* **272**, 191–192.

Fenner, F. and Fantini, B. (1999) *Biological Control of Vertebrate Pests: the history of myxomatosis, an experiment in evolution.* CABI Publishing, Wallingford, UK.

Fenner, F. and Ratcliffe, F.N. (1965) *Myxomatosis.* Cambridge University Press, Cambridge.

Ferrer, M. and Negro, J.J. (2004) The near extinction of two large European predators: super specialists pay a price. *Conservation Biology* **18**, 344–349.

Forrester, N.L., Trout, R.C., Turner, S.L. *et al.* (2006) Unravelling the paradox of rabbit haemorrhagic disease virus emergence, using phylogenetic analysis: possible implications for rabbit conservation strategies. *Biological Conservation* **131**, 296–306.

Gelling, P.S. (1977) Excavations on Pilsdon Pen, Dorset, 1964–71. *Proceedings of the Prehistoric Society* **43**, 263–286.

Gibson, J. (ed.) (1976) *The Complete Poems of Thomas Hardy.* Macmillan, London.

Hay, D. (1975) Poaching and the game laws on Cannock Chase. In: *Albion's Fatal Tree: crime and society in eighteenth-century England* (eds D. Hay, P. Linebaugh, J.G. Rule, E.P. Thompson and C. Winslow), pp. 189–253. Allen Lane, London.

Hayes, R.A. and Richardson, B.J. (2001) Biological control of the rabbit in Australia: lessons not learned? *Trends in Microbiology* **9**, 459–460.

Kerr, P. (2008) Biocontrol of rabbits in Australia. *Outlooks on Pest Management*, August, 184–188.

King, N.E. and Sheail, J. (1970) The old rabbit warren on Fyfield Down, near Marlborough. *Wiltshire Archaeological and Natural History Magazine* **65**, 1–6.

Leach, M. (1989) *The Rabbit.* Shire, Princes Risborough, UK.

Lawrence, L. (2010) Australian rabbits are doing what comes naturally – again. *Outlooks on Pest Management*, February, 19–21.

MAFF (1982) *Mammal and Bird Pests 1981.* Agricultural Science Service Research and Development Reports. HMSO, London.

Mead-Briggs, A.R. (1977) The European rabbit, the European rabbit flea and myxomatosis. *Applied Biology* **2**, 183–261.

McKnight, T.L. (1969) Barrier fencing for vermin control in Australia. *Geographical Review* **59**, 330–347.

Munsche, P.B. (1981) *Gentlemen and Poachers: the English game laws 1671–1831.* Cambridge University Press, Cambridge.

Myers, K., Parer, I., Wood, D. and Cooke, B.D. (1994) The rabbit in Australia. In: *The European Rabbit: the history and biology of a successful colonizer* (eds H.V. Thompson and C.M. King), pp. 108–157. Oxford University Press, Oxford.

RCHME (1952) *An Inventory of the Historical Monuments in Dorset. Volume 1: West.* HMSO, London.

Rolls, E.C. (1984) *They All Ran Wild: the animals and plants that plague Australia*, revised edn. Angus & Robertson, London.

Sheail, J. (1971) *Rabbits and Their History.* David & Charles, Newton Abbot, UK.

Sheail, J. (1978) Rabbits and agriculture in post-medieval England. *Journal of Historical Geography* **4**, 343–355.

Stodart, E. and Parer, I. (1988) Colonisation of Australia by the rabbit *Oryctolagus cuniculus* (L.). Project Report no. 6, CSIRO Division of Wildlife and Ecology, Canberra, Australia.

Strive, T., Wright, J., Kovaliski, J., Botti, G. and Capucci, L. (2010) The non-pathogenic Australian lagovirus RCV-A1 causes a prolonged infection and elicits partial cross-protection to rabbit haemorrhagic disease virus. *Virology* **398**, 125–134.

Sumption, K.J. and Flowerdew, J.R. (1985) The ecological effects of the decline in rabbits (*Oryctolagus cuniculus* L.) due to myxomatosis. *Mammal Review* **15**, 151–186.

Thompson, H.V. (1956) Myxomatosis: a survey. *Agriculture* **63**, 51–57.

Thompson, H.V. and King, C.M. (eds) (1994) *The European Rabbit: the history and biology of a successful colonizer.* Oxford University Press, Oxford.

Thompson, H.V. and Worden, A.N. (1956) *The Rabbit.* Collins, London.

Tittensor, A.M. and Tittensor, R.M. (1985) The rabbit warren at West Dean near Chichester. *Sussex Archaeological Collections* **123**, 151–185.

Toynbee, J.M.C. (1973) *Animals in Roman Life and Art.* Thames and Hudson, London.

Thwaite, A. (ed.) (2003) *Philip Larkin: collected poems.* Marvell Press and Faber & Faber, Victoria and London.

Vesey-Fitzgerald, B. (1946) *British Game.* Collins, London.

Williams, C.K., Davey, C.C., Moore, R.J. *et al.* (2007) Population responses to sterility imposed on female European rabbits. *Journal of Applied Ecology* **44**, 291–301.

Williamson, T. (1997) Fish, fur and feather: man and nature in the post-medieval landscape. In: *Making English Landscapes: changing perspectives* (eds K. Barker and T. Darvill), pp. 92–117. Bournemouth University School of Conservation Sciences, Occasional Paper 3. Oxbow Monograph 93, Oxbow, Oxford.

Williamson, T. (2007) *Rabbit Warrens & Archaeology.* Tempus, Stroud, UK.

PART 2
EXPLOITERS

SOME SYMPTOMS OF
MOSQUITO FEAR!

i - SWATTING.

ii - SPRAYING.

COUGH
PSST!
COUGH
PSST!

iii - PUTTING UP FLY PAPER.

iv - BUYING CITRONELLA CANDLES...

ATISHOO!!
COUGH!

v - MOSQUITO NETS...

vi - AND A FAKE, DECOY PERSON TO LURE MOSQUITOS AWAY.

OTHER SYMPTOMS TO WATCH OUT FOR INCLUDE...

FEVER OF ACTIVITY, CAUSING SWEATING

SHIVERING WITH ANXIETY ABOUT BEING BITTEN

CULMINATING IN UTTER **EXHAUSTION**

I've been up all night swatting mosquitos!

Malaria

MALARIA IS A world-wide human disease of the first importance. Forty per cent of the world's population live in areas where there is a risk of malaria, and it annually infects around 250 million people and claims about one million lives (Fig. 12.1). Yet it is a disease that can be treated, its vectors controlled, and because it has no non-human animal reservoir it should be possible to eradicate it. Thus, any study of malaria raises a number of questions. For instance, how can it be controlled? Why did previous eradication attempts fail? Will current eradication attempts be more successful? Why is there still no malaria vaccine? Is malaria a cause of poverty or caused by poverty? In this chapter we will address these questions, starting with the biology of malaria.

WHAT CAUSES MALARIA?

Malaria is caused by single-celled eukaryotic organisms of the genus *Plasmodium*, in the Kingdom Chromalveolata, Phylum Apicomplexa. Members of this phylum are exclusively obligate intracellular parasites of animals, and include genera such as

Fig. 12.1 *Estimated numbers of* (**a**) *malaria cases and* (**b**) *deaths in 2008 per WHO region. Cases are estimated by different methods for different countries, and then deaths estimated by different methods from cases. The estimated value is given by the horizontal bold line, the bar represents the 5% to 95% range of the estimate. WHO regions: AFR, Africa; AMR, Americas; EMR, Eastern Mediterranean; SEAR, South-East Asia; WPR, Western Pacific* [Data from WHO (2009).]

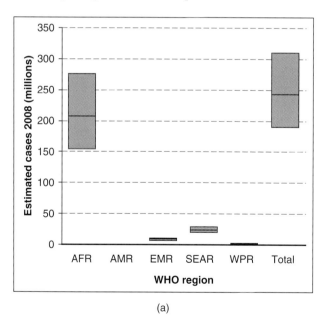

(a)

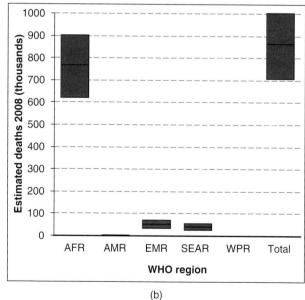

(b)

Biological Diversity: Exploiters and Exploited, First Edition.   Paul Hatcher and Nick Battey.
© 2011 John Wiley & Sons, Ltd. Published 2011 by John Wiley & Sons, Ltd.

Toxoplasma and *Cryptosporidium*. Plasmodia are vectored by mosquitoes of the genus *Anopheles*. The *Plasmodium* life-cycle is complex: the parasite is a shape-shifter, changing its morphology ten times during its life-cycle, for much of its time remaining hidden inside host cells. Although we have simplified it below, this complexity is important: to understand this is to start to understand why malaria control is so difficult.

There are over 170 species of *Plasmodium*, causing malaria in most major groups of land vertebrates, 25 infect primates and four infect humans: *P. malariae*, *P. ovale*, *P. vivax* and *P. falciparum*. Much of the work below has been carried out on *P. falciparum*, which currently causes over 90% of malaria mortality, but the other species should not be forgotten: *P. vivax* is the most widespread species outside of Africa.

During its life-cycle *Plasmodium* goes through three asexual replication phases and one sexual non-replicating phase. After landing on a human, the infected female mosquito probes the skin with her mouthparts, piercing the epidermis, and releasing saliva into the skin. In this saliva are *Plasmodium* sporozoites (it is not now thought that she injects them directly into the blood). Over the next few hours these 100 or so sporozoites move into the bloodstream or lymph system and thence towards the liver. Most of the sporozoites do not make it, and roughly five 'bites' from an infected mosquito are needed to ensure an infection.

In the liver, the sporozoite passes through Kupffer cells (the phagocytes that line the liver capillaries) and several liver cells (the hepatocytes) before finally coming to rest in one of them (Fig. 12.2). Here it changes rapidly, loosing its invasion structures and becoming a rounded hepatic trophozoite. Within 2–10 days this has divided (shizogony) by mitosis many times, producing 10–30 000 merozoites all contained within the original cell membrane: this cell is now called the hepatic schizont. This then dies and bursts, releasing the merozoites into the

Fig. 12.2 *Human stages of the malaria life-cycle. Liver cells shown in green, blood cells in red and Plasmodium in blue. Hypnozoites are only formed in some Plasmodium species (see text), and only P. falciparum produces knobs (arrowed) on erythrocytes in the trophozoite stage [Based on many sources, especially Bannister and Mitchell (2003).]*

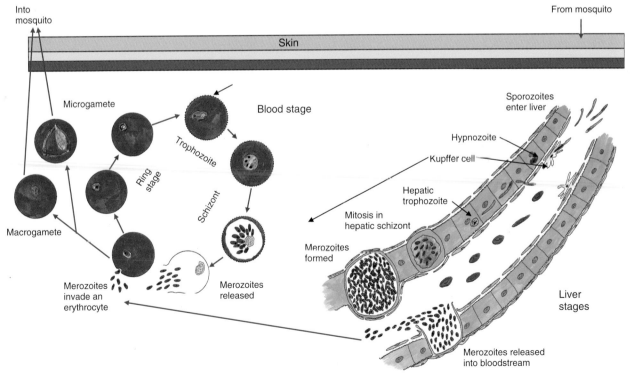

bloodstream. Here the human malarias differ. In P. falciparum and P. malariae all the sporozoites pass through this route, but in P. vivax and P. ovale some sporozoites do not develop immediately but enter a resting stage in the liver cells—the very small hypnozoite. These can remain dormant for weeks, months or years and once activated start producing merozoites. We still know very little about this—P. vivax hypnozoites were detected only in 1982, and we do not know what breaks dormancy, but P. vivax relapsing fever has been known for millennia.

The merozoites are the smallest Plasmodium stage, only about 1.2 μm long, and they seek out red blood cells. Each recognizes and attaches to one, engaging with binding receptors (Fig. 12.3), reorientates itself and within 30 seconds invades it, shedding its outer surface coat and entering the vacuole it has induced. It now rapidly develops into the trophozoite, and feeds on haemoglobin. During the next 1–2 days, up to 20 merozoites are formed by shizogony—the limited space in the erythrocyte does not allow more to be formed—and the red blood cell bursts, releasing more merozoites (see Fig. 12.2). These re-invade red blood cells: in a severe infection over 20% may be inhabited by P. falciparum merozoites.

The process of merozoite production and re-infection of red blood cells continues in cycles, and after a while becomes synchronized. At some point (P. vivax does this immediately, P. falciparum waits up to two weeks), some merozoites develop not into trophozoites but into micro- and macro-gametocytes. These stay within the blood cells, and tend to move to the peripheral circulation.

When an Anopheles mosquito takes a blood meal from an infected human, she may ingest many infected red blood cells. Any trophozoites and merozoites are killed and digested, but the gametocytes are not. In the mosquito's gut they rapidly develop, within minutes they swell and burst, releasing gametes in a sudden violent action. Male sperm seek out and enter the female gamete within the hour, and their nuclei fuse (Fig. 12.4). At this point a diploid zygote is formed. This is the only diploid stage in the Plasmodium life-cycle; it is haploid for the rest of the time. Next, by meiosis four haploid genomes are formed within the one cell. The zygote now differentiates over the next 5–10 hours into a cigar-shaped mobile ookinete. This glides through the blood in the mosquito's gut and, once it reaches the stomach wall, goes between or through the epithelial cells, resting between these and the basement membrane.

Here the ookinete becomes an oocyst, and grows and divides to produce sporozoites, a process that can take several weeks. A mature oocyst is readily visible in a dissected mosquito and the midgut can be completely covered by them,

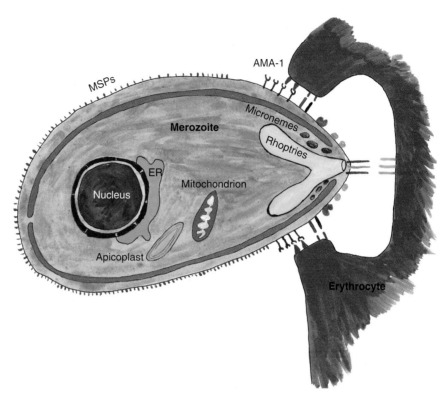

Fig. 12.3 Invasion of an erythrocyte by a Plasmodium falciparum merozoite. The parasite interacts with the erythrocyte via a number of parasite ligands that interact with host receptors. Some of these are being investigated for vaccine development [Redrawn from Kappe et al. (2010).]

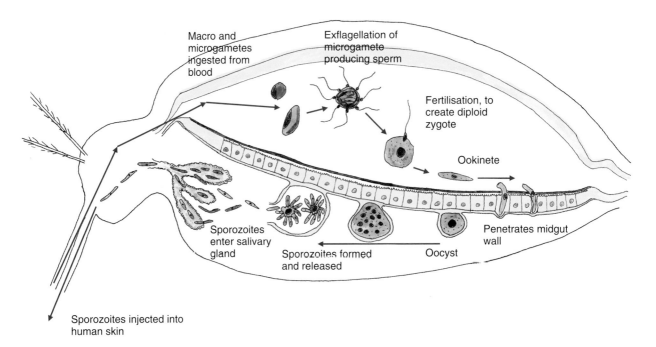

Macro and microgametes ingested from blood

Exflagellation of microgamete producing sperm

Fertilisation, to create diploid zygote

Ookinete

Sporozoites enter salivary gland

Sporozoites formed and released

Oocyst

Penetrates midgut wall

Sporozoites injected into human skin

Fig. 12.4 *Malaria life-cycle in the mosquito [Based on several sources, especially Bannister and Mitchell (2003) and Baton and Ranford-Cartwright (2005).]*

looking like bunches of grapes. The oocyst then bursts, releasing sporozoites into the haemolymph, and they make their way to the salivary gland ducts from where they can be injected into the next human.

SIGNS AND SYMPTOMS

Until merozoites are released from the erythrocytes, *Plasmodium* has little effect, and is unlikely to be noticed. But the synchronized bursting of the red blood cells is noticed and causes the classic recurring fever of malaria. Each attack is abrupt and severe; the patient starts to feel unwell, has a headache, and then has the feeling of extreme cold, with violent uncontrollable shivering. Within an hour the body temperature rises to 40–41°C, now giving the feeling of unbearable heat. This lasts 1–2 hours, with profuse sweating, and 6–8 hours after the attack started the body temperature returns to normal and the patient is left exhausted (Fig. 12.5). *Plasmodium vivax* and P. *ovale* produce a tertian fever where the patient has one day free of fever in three days, while P. *malariae* produces a quartan fever, with fever every other day.

Plasmodium *falciparum* also produces a cyclical tertian fever, but matters quickly become much more serious with deterioration into stupor, fits and a coma, often leading to death. Partly this is due to the greater proportion of erythrocytes that become infected, but P. *falciparum* differs from other malarias in an important respect—sequestration. Knobs develop on the P. *falciparum*-infected erythrocyte (see Fig. 12.2), containing a protein secreted by the merozoite. This interacts with the endothelial cells lining the blood vessels, causing the red blood cells to adhere. This is probably a parasite adaptation to prevent the cells being transported to the spleen where they are destroyed, but leads to a reduction in blood flow and blocking of capillaries. All the major organs can be affected by this.

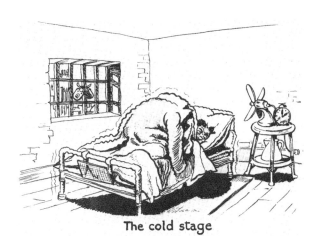

The cold stage

The hot stage

The three other human plasmodia can also cause serious injury. *Plasmodium malariae*, which occurs in all malaria endemic areas of the world, can cause severe kidney disease: if this becomes established it is untreatable and fatal. Although without a liver resting stage, it can remain in the blood for years, and if not completely eradicated can recur decades later. *Plasmodium ovale* has a more limited distribution and is mainly found in the parts of West Africa where P. *vivax* is absent. This is also a relapsing malaria and can lie dormant for years.

Plasmodium vivax is increasingly being recognized as a serious disease that can cause severe anaemia, lung injury and the rare and little understood vivax-associated coma. It has an infectious life of over two years in humans and is also the species most likely to colonize areas inhospitable or marginal to malarias. This is due to its fast production of sporozoites in the mosquito at low temperatures, rapid production of gametes in the blood, and its relapsing nature. These features allow P. *vivax* to maintain transmission in variable conditions, in which survival of the vector is less certain.

The sweating stage

Fig. 12.5 *Cartoons by an anonymous Second World War RAF artist showing the three clinical stages of a malaria attack [From Bruce-Chwatt and de Zulueta (1980), from Wellcome Museum of Medical Science.]*

Across sub-Saharan Africa, where malaria is endemic and most people are continuously infected by P. *falciparum*, the majority of adults experience no overt disease. This is due to natural acquired immunity, about which we still know little. Infants are also reasonably resistant to severe malaria until about six months old. This is possibly due to maternal antibodies, and also inhibitory factors in breast milk. However, from about 3–4 months infants start to become susceptible to malaria, and the risk of severe malaria and death increases at between two and four years old. As the child gets older the risk of severe malaria falls, and immunity is built up, although this is not completed until puberty. This immunity is very slow to develop—it can take 10–15 years of being infected five times a year. From then, as long as the individual is regularly infected, he or she will maintain immunity. However, pregnant women have a markedly increased susceptibility to malaria, probably due to reduced immunosuppression during pregnancy and the accumulation of infected erythrocytes cytoadhering to the placenta.

The spatial distribution of *Plasmodium falciparum* malaria endemicity in the World

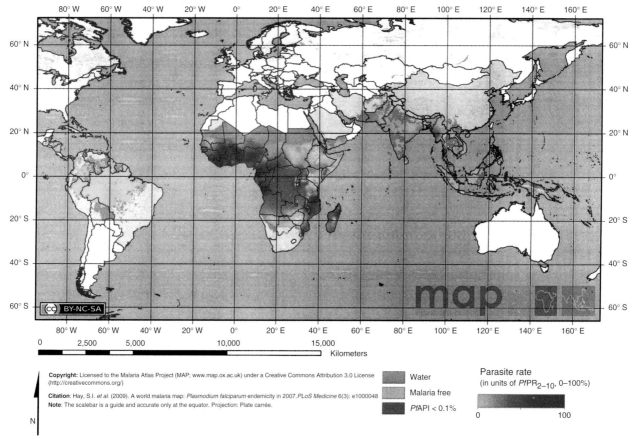

Fig. 12.6 *Estimated spatial distribution of Plasmodium falciparum in 2007, produced by the Malaria Atlas Project. The transmissibility of P. falciparum increases from dark blue (very low) through yellow (low), orange (medium) and red (high) [From Hay, S.I., Guerra, C.A., Gething, P.W. et al. (2009) A world malaria map: Plasmodium falciparum endemicity in 2007. PLoS Medicine **6**(3): e1000048.]*

Although the distribution of P. *falciparum* has been fairly accurately estimated (Fig. 12.6), the distribution of the other three species is far less certain, with considerable under-reporting and misidentification. Areas in which most of the adult population are resistant owing to high continuous malaria transmission are called 'areas of stable malaria'; these include much of sub-Saharan Africa. In other areas, malaria can be seasonal or more occasional, populations do not build up resistance and malaria can affect all ages. These areas, including most of Europe and the Mediterranean until recently, Western Pacific, Central and South America and Asia are termed 'unstable malaria areas'. Sometimes these areas may be subjected to greater than usual amounts of malaria transmission, perhaps due to influx of a malarious population or unusual weather conditions. This can lead to epidemic malaria, which spreads rapidly through a non-resistant population causing very high mortality.

THE MOSQUITO

Plasmodia are vectored by mosquitoes, members of the insect order Diptera, family Culicidae. These insects have complete metamorphosis, like the silkworm (Chapter 5). Only mosquitoes of the genus *Anopheles* can transmit *Plasmodium*, and only the female adults are blood feeders, needing a blood meal to mature each batch of eggs. She forages during the evening and into the night, locating humans by odours from sweat and breath. She rests during the day in a cool place to digest the blood meal and to allow egg development; this may take two days. She then flies to suitable water and lays 50–150 eggs, after which she forages for another blood meal. The legless larvae hatch in 2–3 days and usually moult three times in three days before moulting to the pupa. The larvae lie parallel to the water surface feeding on algae and bacteria and breathing through a tail spiracle. Anopheline larvae are distinguishable from those of other mosquitoes which lie at an angle to the water surface (Fig. 12.7). After another 2–3 days the adults emerge, and mate immediately. *Anopheles* mosquitoes are very variable in their feeding preferences, and the type of water body in which they lay their eggs. Throughout the world different mosquito species are the main malaria vectors: in sub-Saharan Africa the *A. gambiae* complex of species is most important, while in England the most efficient malaria transmitter was *A. atroparvus* (Box 12.1).

(a)

(b)

(c)

Fig. 12.7 *Man-made mosquito breeding sites.* (**a**) *Water containers in a SE Asian cocoa plantation.* (**b**) *Water butts and stagnant water in the garden of one of the authors in southern England. The latter is suitable breeding ground for* Anopheles claviger, *although in June 2010 was inhabited only by culicine mosquito larvae* (**c**)*, anopheline mosquito larvae do not have a long respiratory siphon on the abdomen (arrowed)*

The *Anopheles* mosquito is far more than just a mechanical vector (e.g. fleas vectoring myxomatosis, see Chapter 11), as the *Plasmodium* undertakes an essential part of its life-cycle, including mating, within it. Yet, as with other vector-borne organisms, this is the weak part of the parasite's life-cycle. The *Plasmodium* gametocytes have to be picked up by an *Anopheles* mosquito that preferentially infects humans (many *Anopheles* species will feed from humans but prefer to feed on animals) and sporozoites have to form and get into another human before the mosquito dies. This is a race against time, and the speed of development of *Plasmodium* in mosquitoes is strongly related to temperature: in cold conditions most mosquitoes will have died before *Plasmodium* has developed sporozoites within it.

There is increasing evidence that infection alters mosquito behaviour to the advantage of the *Plasmodium*. For example, the presence of sporozoites in the salivary gland can induce increased feeding persistence and probing, thus increasing opportunities for depositing sporozoites.

ERADICATING MALARIA: FALSE HOPE NO. I

After the early evolution of malaria (Box 12.2), *Plasmodium* species spread with human migrations throughout the world, and early sources record this spread: Chinese and Greek observations record malaria in the first millennium BC, and it has recently been discovered in the mummy of the Egyptian 'boy king' Tutankhamen, who died in 1324 BC. It may have been present in Greece as early as the eighth century BC, and by the third century BC malaria was present in Italy. It gradually spread north and west, reaching Spain by the eleventh century AD, and reaching southern England by at least the sixteenth century (see Box 12.1). The prevalence of malaria was linked to suitable ecological conditions, and in some cases modifications to the environment by mankind assisted the spread and severity of malaria. For example, Sardinia was one of the first places in Europe to

BOX 12.1 MALARIA IN BRITAIN

For at least several centuries malaria was endemic to parts of Britain, and died out only relatively recently.

It is hard to determine the early history of malaria in Britain as the terminology used in the past to describe it was imprecise. Malaria-like diseases were called 'agues', 'intermittent fevers', 'tertian fevers' and 'quartian fevers' – terms also used for many other infections. This has led to a range of views, from malaria epidemics sweeping the country during the Middle Ages to malaria being not endemic at all.

However, in some instances we can be reasonably sure that malaria was involved. In particular, 'marsh ague' prevalent in many of the wetlands of southern Britain has symptoms – alternating cold and hot fevers, periodicity, enlargement of the spleen, relapsing, and anaemia – which characterize malaria, and which later were treatable with cinchona bark (see Box 12.3). In the early twentieth century it was confirmed that marsh ague was caused by *Plasmodium vivax* in England.

Of the six species of *Anopheles* mosquito present in the UK, only *A. atroparvus* breeds in sufficient numbers close to humans, and bites them sufficiently frequently to be a good transmitter of *Plasmodium*. This mosquito breeds mainly in brackish water along estuarine marshes, and malaria was certainly present in these areas in the sixteenth century. From this time serious efforts began to reclaim this land, long recognized as both very rich agriculturally and extremely unhealthy. William Cobbett sums this up nicely in 1823, referring to Sandwich, Kent:[a]

> Rottenness, putridity is excellent for land, but bad for Boroughs. This place, which is as villainous a hole as one would wish to see, is surrounded by some of the finest land in the world. Along on one side of it, lies a marsh. On the other side of it is land which they tell me bears *seven quarters* of wheat to an acre.

Early drainage efforts probably made the malaria problem worse. Whereas the fens would have been washed by tidal waters, reclamation schemes prevented tidal flow, but were unable to effectively drain the land and stagnant pools were formed. Furthermore, as the peatland was drained it shrank as it dried, and also wasted away due to bacterial action, lowering the level of the land progressively by several metres. This made drainage even more difficult. The introduction of wind pumps in the early eighteenth century may have helped for a while, but with increasing peat shrinkage they became less effective, and it was only with the introduction of steam-powered pumps from the 1820s that the fens became progressively drained.

Malaria probably reached a peak in the UK during the seventeenth century, then gradually declining through the eighteenth and nineteenth centuries. This decline is likely to have been due to a number of factors; drainage of the marshes certainly helped, but *A. atroparvus* is still frequent in the areas where malaria was endemic in England. However, the transmission rate of malaria between humans could have been lowered significantly by the increasing number of cattle grazed on the new root crops planted in the reclaimed land, the separation of animals and humans into different buildings (giving mosquitoes attracted to animals less of an opportunity also to bite humans), and the construction of better ventilated and lit human habitations. The effect of malaria could have been lessened by improved health, water supply and diet, and the increasing affordability of quinine: between 1840 and 1890 its price fell 100-fold.

Between the sixteenth and nineteenth centuries, the marsh and fen lands were recognized as very unhealthy, especially to outsiders. In 1722, Daniel Defoe records in relation to the Essex marshlands:[b]

> [A]ll along this country it was very frequent to meet with men that had had from five or six, to fourteen or fifteen wives .... The reason, as a merry fellow told me ... was this; That they [the husbands] being bred in the marshes themselves, and season'd to the place, did pretty well with it; but that they always went up into the hilly country ... for a wife: That when they took the young lasses out of the wholesome and fresh air, they were healthy, fresh and clear, and well; but when they came out of their native air into the marshes among the fogs and damps, there they presently chang'd their complexion, got an ague or two, and seldom held it above half a year, or a year at most; and then, said he, we go to the uplands again, and fetch another ....

[a]William Cobbett, diary 3.9.1823, p. 319; Cobbett (1893).
[b]Daniel Defoe, Letter 1 (1722), p. 13; Defoe (1928).

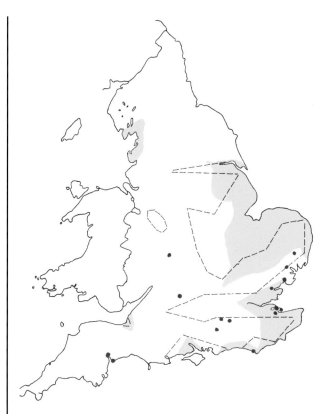

Box 12.1 Fig. 1   *Malaria in the UK. Yellow, distribution of indigenous malaria about 1860; dashed line, distribution of indigenous malaria (to the east and south of the line) 1917–21; red spots, cases of indigenous malaria 1941–48 [Based on Shute and Maryon (1974).]*

The mortality in these areas was commonly twice that found in more upland rural parishes, and during the seventeenth and eighteenth centuries it was not uncommon for 10–20% of the marshland population to die per year. Populations were only sustained by immigration to work for landlords willing to pay high wages on the rich land, and by copious amounts of alcohol and opium to ward off the effects of the ague.

Plasmodium vivax malaria is often seen as benign, yet between 1840 and 1910 there were 8200 deaths largely attributable to malaria recorded in the UK. Thus, was the strain of *vivax* present in the UK particularly virulent, or was another *Plasmodium* species present? Certainly by the twentieth century the UK *vivax* strain was of low virulence compared to those found elsewhere, and it is uncertain whether mosquitoes found in Britain can transmit *P. falciparum*. It is more likely that malaria

was one of many exacerbating mortality factors, including other infections, poor health, and malnutrition. Hutchinson and Lindsay (2006) go further and suggest that malaria was not even an important contributory mortality factor, but that diarrhoeal diseases from poor water and sanitation during the summer and acute respiratory infection during the winter were responsible.

By the early twentieth century malaria had become rare in the UK, although the last indigenous case did not occur until 1953. However, the conditions for malaria transmission still occurred. From 1915, 160 000 British troops were stationed in Salonica, Greece, during the First World War. Many contracted malaria, hospitals became choked and thus soldiers with acute or persistent malaria were sent back to hospital camps in the UK. One was foolishly set up on the Isle of Sheppey, Kent, known for indigenous malaria. From the spring of 1917 many civilians near this camp started to suffer from *P. vivax*, spread by the local *A. atroparvus* mosquitoes, rising to almost 500 cases by 1919. The outbreak was able to gain a hold owing to slow diagnosis of the disease as malaria. Subsequently, quinine and bed nets were distributed and the epidemic subsided by the end of 1919 without fatalities.

During and after the Second World War, better surveillance, precautionary DDT spraying and early therapy ensured that, although many thousands of returning service personnel had malaria, there were fewer than 50 cases of indigenous malaria recorded in the UK between 1940 and 1948 (Fig. 2). Since the 1970s the increasing affordability of overseas air travel has led to an increase in cases of malaria in returning travellers (Fig. 2). In 1983 the first cases of 'airport malaria' were recorded in the UK, when two people who had not been to a malarious country contracted *P. falciparum* in the UK. Both lived close to Gatwick Airport and had been in contact with airport personnel, and it is likely that an infected mosquito stowaway on a flight was responsible.

Will projected climate change increase the risk of malaria returning to the UK? At present the most sensible answer is that of Kuhn *et al.* (2003) who conclude that the projected rise in temperature in the UK could increase the risk of local malaria transmission by 8–14%, but as the transmission rate in the UK is extremely small (no transmission recorded from any of the thousands of malaria cases imported into the UK since 1953) this is unlikely to appreciably increase the risk. However, it is much harder to predict the possibility that other species of mosquito, possibly from southern Europe, better able to transmit *Plasmodium* (including *P. falciparum*) could become established in the UK with altering climatic conditions.

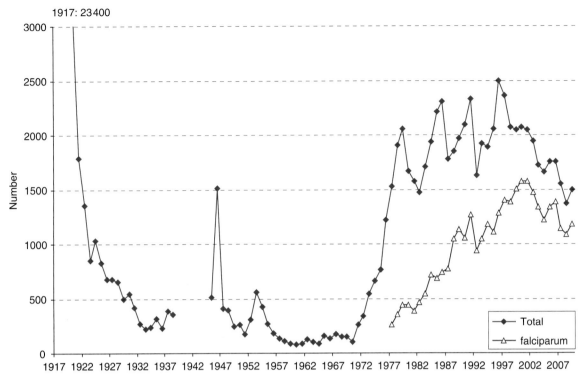

Box 12.1 Fig. 2 *Total number of imported cases of malaria recorded in the UK from 1917, and number of* P. falciparum *cases from 1977* [Data from Bruce-Chwatt and de Zulueta (1980), Phillips-Howard *et al.* (1988), Smith *et al.* (2008) and the Health Protection Agency (www.hpa.org.uk).]

get malaria, and this was linked with the Carthaginian invasion of sixth century BC which introduced extensive grain cultivation to the coastal lowlands, removed forests, and increased flooding and swamps. This increased breeding grounds for the efficient local vector *Anopheles labranchiae*, which transmitted the P. falciparum which may have been brought into Sardinia by slaves from North Africa. Early schemes to drain swamps and fens also probably increased malaria.

From Europe, malaria was introduced into the hitherto malaria-free New World by colonists. In North America it gained a foothold first in the Jamestown Colony (Chapter 5) probably due to P. vivax carried by settlers from the English malarial hotspots of East Anglia, Kent and Essex (see Box 12.1). Later, P. falciparum was introduced in slaves from Africa, brought over especially to work in sugarcane plantations (Chapter 6). Early agriculture in the southern American states involving rice production in inland swamps was also ideal for vector development.

Malaria probably reached its greatest geographical extent during the nineteenth century, and by 1940 had retreated slightly (Fig. 12.8). In part this was due to improvements in land management and agriculture as seen in England which meant that, without any specific control, communities grew out of malaria and conditions were eliminated in which the malodorous air (mal-aria) thought to lead to the disease resided. In other, poorer, areas such as southern Europe and the southern states of North America, agricultural development stagnated and malaria became entrenched.

BOX 12.2	EVOLUTION OF MALARIA

Plasmodium is thought to have evolved from a free-living protozoan that became adapted to live in the gut of aquatic insects, including the Diptera (appearing 150–200 mya). Subsequently the proto-plasmodium acquired the ability to undergo intracellular shizogony, greatly increasing its ability to proliferate.

The evolution of the four human *Plasmodium* species is still uncertain. *Plasmodium malariae* is thought to be an early evolved species: it can be transmitted in temperate climates and in sparse populations and is likely to have travelled with humans as they migrated out of Africa. *Plasmodium ovale* is also thought to have arisen in Africa, where it is still most prevalent.

The evolution of the other two *Plasmodium* species is more controversial. *Plasmodium falciparum* is closely related to, and probably evolved from, *P. reichenowi*, a parasite of chimps, but is only distantly related to other mammalian malarias. It is possible that a mutation altering the surface of the red blood cell may have protected people from *P. reichenowi* but made them more susceptible to *P. falciparum*, with final mutations to increase susceptibility occurring in the last 5000–10 000 years. There are two views on when *P. falciparum* evolved: the first that it evolved 2–3 million years ago in human ancestors and persisted; and the other that it evolved much more recently, persisting at very low levels until undergoing a significant expansion about 6000 years ago. This coincides with the development of agriculture: *P. falciparum* needs a large human population in which to survive, as it has no latent phase and is often fatal. The occurrence of agriculture allowed the development of larger settled communities which could have come into contact with *P. falciparum*'s most efficient vector: *Anopheles gambiae*. This mosquito prefers to breed in small temporary sunlit freshwater pools, which would have been created by the clearing of forests by the early African agriculturalists, allowing the mosquitoes and humans to interact.

Plasmodium vivax is generally thought to have evolved in Africa. Evidence for when it evolved is provided by a human mutation that gives almost complete protection against *P. vivax*: the Duffy negative genotype. This is a partially recessive trait in which the Duffy antigen, an essential receptor for *P. vivax* merozoites to enter host erythrocytes, is missing. In West Africa, the allele for Duffy negative has about a 97% frequency, and thus can be said to be fixed in the population, and *P. vivax* no longer occurs there. This fixation is estimated to have taken over 10 000 years, and selection for this allele began during the last 100 000 years. It also suggests that *P. vivax* is a sufficient selective force, and is or was not as benign as sometimes supposed.

The Duffy negative condition carries no disadvantage to humans, unlike some anti-malaria mutations. For example, there are high frequencies of thalassemias, a class of anaemias due to abnormalities in genes coding for haemoglobin, around the Mediterranean, Africa, Middle East and Asia. These confer about a 50% protection from malaria. Glucose-6-phosphate dehydrogenase (G6PD) deficiency has a similar distribution, and gives a similar protection against *P. falciparum*. This was probably selected for in African populations about 4–12 000 years ago, but causes 'favism', a haemolytic crisis caused by consumption of foods such as fava beans (Chapter 7).

Probably the most famous anti-malarial mutation is sickle cell trait or haemoglobin S. This is a single point mutation in the beta chain of haemoglobin; the heterozygote gives 90% protection against *P. falciparum*, but the homozygote is harmful—without medical care infants with this will die within two years. That such a deleterious mutation has become fixed in Africa is testimony to the selective force exerted by *P. falciparum*. Indeed, our past contact with malaria has left us with a considerable legacy—the burden of the mutations discussed above is thought comparable with the burden caused by malaria.

Our interaction with malaria is still evolving; there are several non-human malaria species that can infect humans. One is *P. knowlesi* which, although first recorded in humans in 1965, has only recently attracted attention. This species primarily infects macaques in Asia and has an early blood stage that looks identical to *P. falciparum* and other stages similar to *P. malariae*. Often misdiagnosed, *P. knowlesi* is much more harmful than *P. malariae*, producing often-fatal 24-hour relapsing fevers. It is now understood that *P. knowlesi* often infects humans in Sarawak, Malaysian Borneo, and elsewhere, with hundreds of cases reported annually. It is not thought to be a newly emerged species, rather that it has been occasionally infecting humans since they arrived in Asia 70 000 years ago. Its vectors are forest-dwelling mosquitoes, and as forests were not settled in, human contact would have been rare. At present it primarily infects people entering the forest or working on its edges, and as there is no recorded human–human transmission it is best not thought of at the moment as the fifth human malaria. This could change, however, if deforestation and increased contact between macaques and humans shift relationships and this species begins the complete adaptive shift into humans.

Fig. 12.8 *Distribution of malaria in 1940 [Redrawn from WHO (2008).]*

Towards the end of the nineteenth century European colonial expansion into areas of high malaria transmission in Africa and Asia led to increased research into malaria and its life-cycle. First, in 1880 Alphonse Laveran (a French military doctor based in Algeria) discovered the malaria parasite in human blood, followed by Ronald Ross (a medical doctor in British India) discovering the role of anopheline mosquitoes in the transmission of malaria in birds in 1898 and Giovanni Grassi discovering *Anopheles* transmission in humans soon after.

Subsequently, two competing malaria control approaches developed: either the malaria parasite could be controlled (favoured by continental workers such as Koch); or the vector could be targeted, as favoured by US and UK researchers. Early successes were achieved in vector control, for example during the building of the Panama Canal, by targeting the breeding sites and applying oil to the water surface to prevent

---

BOX 12.3   QUININE

Indigenous peoples throughout the world have adopted thousands of plant remedies against malaria. For example, even in the UK, Nicholas Culpeper lists, in his 1653 *The Complete Herbal and English Physician Enlarged*, 59 species of herb that could be used against agues. However, quinine was the first drug that not only alleviated the symptoms of malaria but actually killed the blood stage of *Plasmodium*. Quinine comes from the bark of *Cinchona* species (Rubiaceae) (Fig. 1). These evergreen trees grow native only on the west side of the Andes usually 1000–2000 m above sea level, in high humidity on well-drained soils.

The early history of quinine is lost in myth and falsehood, and so is difficult to piece together. For example, the generic name *Cinchona* was given to the trees by Linnaeus in 1742, as a misspelling of Chinchon. This came from the legend that the first European cured of malaria by the bark was the Second Countess of Chinchon in Peru, 1638. However, it is interesting that although *Cinchona* is a South American native, malaria is not: it was introduced by the European explorers to the region. Thus, it is unlikely that *Cinchona* was a folk remedy for malaria, rather one of many barks used in traditional medicine to alleviate fevers. At some point it came to the notice of the early explorers and became known as a 'cure' for malaria. In the 1630s, the Society of Jesus (the Jesuits) sent bark samples back to Spain and Rome, it became used in Europe and the dried and ground bark became known as 'Jesuit's powder'. The Jesuits kept the source of this bark secret and attempted to develop a monopoly in its supply, and it was not until 1735 that the trees were found in Peru by others.

Box 12.3 Fig. 1 Cinchona [From Freeman and Chandler (c. 1908).]

There was considerable scepticism over the powers of this bark. Not only from physicians, as its use was contrary to accepted medical practice still based on classical writers such as Galen, but its provenance was doubted, and all this was tied up with extreme anti-Catholic sentiment against the Jesuits. It was not until the successful treatment of Charles II's malaria by a quack doctor using a mysterious powder, later revealed as *Cinchona* bark, that it started to become accepted in the UK.

During the seventeenth and eighteenth centuries, supply of the bark was restricted and very expensive, and its use was confined to the wealthy. Large-scale trials of the bark in the UK were undertaken during the great epidemic of agues (some of which may have been malaria; see Box 12.2) in the UK during 1780–6. During this time, the household of James Woodforde, Parson of Weston Longville in Norfolk, was constantly suffering from malaria, and in this extract from his diary for 13 March 1784 he gives an excellent account of the use of cinchona bark at the time, and its side effects:[a]

Nancy [James Woodforde's niece] brave to day (tho' this Day is the Day for the intermitting Fever to visit her) but the Bark has prevented its return … Dr. Thorne's Method of treating the Ague and Fever or intermitting Fever is thus – To take a Vomit in the Evening not to drink more than 3 half Pints of Warm Water after it as it operates. The Morn' following a Rhubarb Draught – and then as soon as the Fever has left the Patient about an Hour or more, begin with the Bark taking it every two Hours till you have taken 12 Papers which contains one Ounce. The next oz. &c. you take it 6. Powders the ensuing Day, 5 Powders the day after, 4 Ditto the Day after, then 3 Powders the Day after that till the 3rd oz. is all taken, then 2 Powders the Day till the 4th oz: is all taken and then leave of. If at the beginning of taking the Bark it should happen to purge, put ten Dropps of Laudanum [a tincture of opium in alcohol] into the Bark you take next, if that dont stop it put 10. drops more of Do. in the next Bark you take – then 5 Drops in the next, then 4, then 3, then 2, then 1 and so leave of by degrees. Nancy continued brave but seemed Light in her head. The Bark at first taking it, rather purged her, and she took 10 Drops of Laudanum which stopped it.

The published diaries do not mention the cost of this treatment, but it would have been high, and Parson Woodforde, although not a parsimonious man, would often try to treat Nancy's agues with rhubarb and other generally available medicines for several days, without effect, before calling for Dr Thorne and his bark. The total amount of bark taken by Nancy seems quite large, but as there were no standards, the active ingredient was unknown, and many ineffective counterfeit barks were being sold, she could have had anything from a homeopathic to an industrial dose of quinine.

[a]Beresford (1926), pp. 121–122.

It was not until 1820 that the main active ingredient, the alkaloid quinine, was extracted from the bark, and during the nineteenth century powdered *Cinchona* bark was gradually replaced by quinine solution. The structure of quinine was confirmed in the 1930s but, as it is too expensive to synthesize commercially, quinine production still relies on cultivation of *Cinchona* trees.

The extraction of quinine led to greater dose standardization and demand for the drug. Demand was further fuelled by European colonial expansion into malarious areas of the world. Thus, by the 1850s there was serious concern that the supply of bark would run out, as *Cinchona* trees were killed by bark stripping. Therefore, the British and Dutch undertook expeditions to collect *Cinchona* samples to grow in their new colonies. Most of these attempts failed, as *Cinchona* is not easy to propagate, and even the successful plantations produced plants too low in quinine for economical extraction. By the end of the nineteenth century millions of *Cinchona* trees in plantations were being destroyed in India and Sri Lanka as useless.

However, some had more success. In the middle of the nineteenth century the British bark dealer Charles Ledger heard of some high-yielding trees in Bolivia from his servant, and sent him to collect seed. This took several years, but eventually in 1865 Ledger had several kilograms of seeds. He first tried to sell them to the British government but, mindful of the problems they were already having in growing *Cinchona*, they refused. Later he sold about 500 g of seeds to the Dutch government. They established seedlings in Java, and the bark was found to contain over 8% quinine – some had over 13%. This was the tree they were after, and to prevent outcrossing with *Cinchona* varieties with lower quinine concentrations, they established remote plantations in the Javan forests (Fig. 2). From these seeds arose, through much breeding, the species now called *C. ledgeriana*. Although high yielding it is difficult to grow, and thus is usually grafted on a more vigorous one-year old *C. succirubra* rootstock.

Quinine is mainly found in the outer epidermis and periderm of the bark of stems and branches. Although the tree will recover if the bark is removed carefully, this procedure is too labour-intensive today and the trees are either uprooted when 6–20 years old or coppiced when seven years old. The bark is stripped by hand and allowed to dry in the shade (see Fig. 2) before being transported to the processing plant.

The mode of action of quinine is still unclear, but its effects are confined to the trophozoites, starting to work within 24 hours and clearing the infection within five days. The most common side-effect is hypoglycaemia, and high concentrations of quinine are likely to give rise to 'cinchonism' – high-tone deafness, tinnitus, nausea, vomiting, dizziness, blurring of vision – and an overdose can lead to blindness, coma and death. No wonder that, coupled with the extremely bitter taste, it was very difficult to get patients to finish a course of quinine treatment – including at times Nancy Woodforde mentioned above.

The Dutch established a virtual monopoly in quinine production, which was maintained until the Japanese seized the Java plantations during the Second World War. This cut Allied supplies of quinine and for a while malaria rose in Allied troops. A massive screening programme for synthetic anti-malarial drugs led to one, atebrine, being selected. This, although toxic in large doses and with a tendency to turn the skin of users yellow, was very effective. Later in the war the much safer chloroquine was developed, and this became the mainstay of malaria medication until resistance developed to it through the 1970s. With increasing resistance to synthetic drugs, interest returned to quinine, and it is still a drug of choice for the treatment of severe *falciparum* malaria. However, almost half of the quinine produced today goes into the food and drink industry, as it is still a very useful bittering agent.

Box 12.3 Fig. 2 Cinchona *cultivation at the beginning of the twentieth century.* **(a)** *Netherlands government* Cinchona *plantation, Java.* **(b)** *Establishing a* Cinchona *plantation.* **(c)** *Drying* Cinchona [From Freeman and Chandler (c. 1908).]

the mosquito larvae from breathing, However, this needed careful identification and targeting of the *Anopheles* species involved and was very costly, but if successful could lead to the long-term control of malaria. It was cheaper to focus on the parasite, and this approach also had some early successes. For example, distribution of free quinine (Box 12.3) in southern Italy in the first decades of the twentieth century reduced deaths significantly. However, it did not reduce transmission, and when the quinine distribution system broke down during the First World War, malaria mortality rates rose again.

The conflict between the two control methodologies reached a head during the 1920s and led to the League of Nations establishing a Malaria Commission in 1924 to adjudicate. Their report supported parasite control and rural development, stating that mosquito control had often promised much but delivered little. Thus, during the 1920s and 30s a number of schemes were set up to alleviate rural poverty and malaria. For example, in Italy during the 1930s an expensive scheme of generalized rural uplift, landscape engineering, job creation, better agricultural methods and housing—called *bonifica integrale* or bonification—took place and contributed to significant reductions in malaria. Likewise, rural development schemes in the USA—including the Tennessee Valley Authority scheme (from 1933) of large-scale public works, flooding swamps, moving people to higher ground and draining wetlands—almost eradicated malaria from large areas of country. This was due to social and economic developments, not specific malaria control methods.

However, vector control became the dominant post-war methodology. Improved mosquito identification allowed control to be targeted to the species that bit humans, and a widely publicized success against malaria in Brazil helped. In 1930, *A. gambiae*, the most efficient vector of P. *falciparum*, was introduced into Brazil from Africa and led to major malaria epidemics. A vigorous policy during 1938–9 of attacking its breeding and resting sites led to the cessation of the epidemic.

The development of the organochloride insecticide DDT during the Second World War—considered both very effective (it is a long lasting contact insecticide, and thus ideal for indoor residual spraying of walls, killing resting mosquitoes) and very safe—confirmed the targeting of vectors in malaria control. For the first time there was talk of vector elimination, and this was trialled in Sardinia in the 1940s. Sardinia had the reputation as being one of the worst areas for malaria (especially P. *falciparum*) in Europe, and in 1946 a UN-funded programme to eradicate the malaria vector began. It turned out to be nearly impossible to eradicate the mosquito, and after four years, the application of 267 tonnes of DDT to the island and a three-fold budget overrun, the project was halted. The vector was not eradicated, but reduced to very low numbers, and malaria transmission was stopped. This was hailed as a victory for malaria eradication (note that the goals had changed).

The Sardinia experiment directly influenced the World Health Assembly's resolution in 1955 to abandon long-term malaria control and instead undertake short-term malaria eradication, by eliminating the vector and interrupting transmission, after which local flare-ups could be treated with the improved and less toxic drugs such as chloroquine that were becoming available.

In 1957 the WHO started coordinating the global campaign for malaria eradication, and this was successful in many countries, including Italy, Greece, Cyprus during the 1950s and 1960s, and by 1970 all of Europe, the USSR, most of North America and parts of the Middle East were free from malaria. These were all in temperate areas with unstable malaria and well-developed infrastructures. Little attempt was made to control malaria in Africa, and thus malaria became seen as a tropical disease.

Why was malaria not eradicated? Compared to some diseases, such as plague (Chapter 16), there are many points in favour of malaria eradication: the parasite is host-specific and thus does not have any non-human reservoirs; it has a limited number of vectors; vector behaviour of resting immobile for many hours after ingesting a blood meal means control is easier; the long time needed for sporozoites to develop in the mosquito gives many control opportunities before she becomes a vector; and new cheap insecticides such as DDT and inexpensive drugs such as chloroquine to treat patients became available. Yet over-reliance on DDT led to resistance developing within five years, and reliance on chloroquine led to resistance spreading through the 1970s. Replacement pesticides and drugs were much more expensive. Tactical problems, including inflexible top-down organization and lack

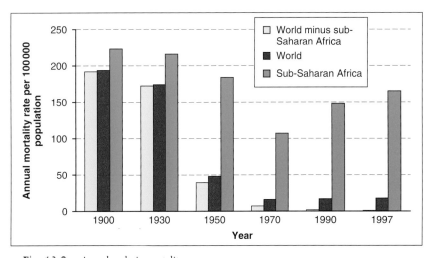

Fig. 12.9   *Annual malaria mortality rates since 1900 [Data from WHO (1999).]*

of local knowledge, hindered progress, and insufficient progress during the 1960s caused funding donors to turn to more readily controllable diseases.

In 1969 the World Health Assembly declared that eradication was probably not feasible in the near future in many parts of the world and that there should be a return to malaria control. Unfortunately this led to a further significant reduction of funds, and during the 1970s the number of malaria cases rose 2.5-fold. In 1990, malaria was affecting more people than in 1960. The situation in Africa and parts of Asia deteriorated markedly (Fig. 12.9) and major malaria epidemics erupted in India (1974–7), Turkey (1976–8) and Brazil (1985–9).

ROLLING BACK TO ERADICATE MALARIA: FALSE HOPE NO. 2?

After many years of stagnation, lack of leadership and funding, in the 1990s world-wide malaria control was again suggested. In 1992 a revised global malaria control strategy was approved by WHO members, malaria control was recognized to be an essential part of overall health development, and sustainable progress towards control was emphasized. Although through the 1990s some successes were recorded, for example in Brazil, world-wide progress was very slow. In 1998 the Roll Back Malaria (RBM) initiative was launched by the WHO in partnership with UNICEF, UNDP, and the World Bank, to activate the global malaria control strategy with the aim of halving malaria-associated mortality by 2010 and halving it again by 2015.

RBM focused on the use of a limited number of cost-effective interventions: artemisinin-based drugs (Box 12.4) and insecticide-treated bed nets (ITNs; see Box 12.5). Early progress of RBM was slow: during its first five years malaria deaths actually increased world-wide and the use of artemisinins and ITNs was limited by distribution problems and lack of funding.

One of the major problems in determining the effectiveness of any anti-malaria campaign is in obtaining accurate data on the extent of the problem and the efficacy of controls. Many global disease eradication programmes routinely spend around 10% of their budget on surveying and data gathering, yet RBM was criticized for spending much less, and the WHO has been accused of over-reliance on nationally supplied data, which may overestimate control, rather than those from independent surveys. This is still evident in the latest WHO *World Malaria Report* (2009) where extent of malaria (see Fig. 12.1) and effectiveness of control is estimated from limited data by some fairly complex modelling.

Furthermore, the WHO was criticized at this time for inertia in replacing older drugs (to which malaria often had resistance) with artemisinin-based therapies and for focusing solely on bed-nets for mosquito control, and not supporting residual spraying. The use of persistent insecticides such as DDT for the indoor

BOX 12.4   ARTEMISININS

During the 1960s, the North Vietnamese premier requested urgent assistance from China to counter the high incidence of malaria in his troops fighting in the Vietnam War. Lacking access to western synthetic anti-malarial drugs, the Chinese turned to their traditional medicine for remedies and after a systematic examination one was found to have useful properties. This was *qing hao*, or sweet wormwood, *Artemisia annua* (a relative of absinthe and tarragon), juice extracts of which were mentioned from AD 340 as a cure of intermittent fevers.

In 1972 the active compound *qinghaosu* or artemisinin was purified, and the drug entered clinical trials in China. Artemisinin is an unusual sesquiterpene lactone that is rather unstable and soluble neither in water nor oil. To improve its properties the molecule was modified, producing artemether, which is more oil-soluble, as well as artesunate which is water-soluble and five times as potent as artemisinin.

We still do not know how artemisinin works against *Plasmodium*, but it is potent and remarkably well tolerated in the human body, unlike quinine and many synthetic anti-malarial drugs. It is active against a wide range of parasite ages, from those that have just entered the red blood cell to the mature stages, and has some effect against gametocytes. Along with quinine, it is very effective against severe malaria, and it inhibits cytoadherence of *P. falciparum*-infected erythrocytes to blood vessels more effectively than other anti-malarials. It does not work against the hepatocytes, however.

Although used for many years in China and SE Asia without any reported side-effects, the WHO and Western governments were not prepared to grant approval for these drugs as they had not been produced in factories that complied with international standards. Thus, in 1984 the WHO embarked on a programme to develop arteether, a drug already abandoned by the Chinese in favour of artemether. After 10 years, when arteether had just entered clinical trials, the WHO was able to recognize the other artemisinins, and subsequent development of arteether was abandoned. However, it took years to *prove* the safety of artesunate, which only received US regulatory approval in 2004.

Artemisinins are eliminated from the blood within hours, meaning that they must be taken for at least five days to prevent a relapse, and even after the end of a course of treatment some parasites may still exist. Thus, for uncomplicated malaria it is best combined with another drug that has longer-lasting effects (ACT: artemisinin combined treatment), and this will probably also slow down the development of artemisinin resistance. To ensure that both drugs are taken, fixed-dose ACTs are needed, but these have been slow in deployment, with the WHO recommending them only in 2004.

Like quinine, artemisinin can be synthesized, but not economically. Thus, artemisinin supply depends on growing *Artemisia annua*. High-yielding hybrids have been produced containing 1–1.5% artemisinin concentration, and 5–10 thousand hectares

of *Artemisia* are estimated to be planted world-wide. Yields of artemisinin are still very variable, but in 2010 a genetic map of *A. annua* was produced and key loci that could improve yield have been identified, which should aid breeding efforts.

The tiny and expensive hybrid seeds are planted in nurseries, and transplanted to the field. They are harvested after 150–200 days when the artemisinin concentration is at its maximum, just before flowering. The plants are cut at the base, air dried and then threshed to remove the leaves. As the artemisinin content soon decreases the dried leaves must be passed to the extraction factories within a couple of months. The fluctuating demand for artemisinin and the decentralized and commercial production of *Artemisia* means that its production follows boom and bust economics. For example, increased funding for ACTs in 2005 led to an acute shortage of artemisinin and a trebling in its price. This led to many unregulated growers and manufacturers entering the market, and a subsequent oversupply of the drug. Thus, by 2008 the price of artemisinin had fallen to a fifth of its peak, uneconomic for many growers and processors who stopped production: the WHO predicted a shortage of artemisinin in 2010.

The inability to economically synthesize artemisinin is seen as one reason why its price is high and supply erratic. Steps have been taken to address this. One route has been to genetically engineer yeast *Saccharomyces cerevisiae* to produce an artemisinin precursor by inserting genes from *A. annua*. Commercial production is likely to start in a couple of years. Alternatively, drugs can be synthesized that mimic the effects of artemisinins. This is hampered by lack of knowledge of the mode of action of artemisinins, and few of the thousands of molecules synthesized have shown efficacy against *Plasmodium*. The most promising, OZ277, entered clinical trials in 2004, but these had to be stopped owing to its instability in blood. Others are currently in development.

At present ACTs and insecticide-treated bed nets (see Box 12.5) are the main weapons against malaria, and just as there are no current alternatives to the use of pyrethroids on bed nets, there are no current alternatives to the use of artemisinins for the treatment of most types of malaria. Thus, the possibility of resistance developing is being taken very seriously. Although the short half-life of artemisinins in the blood minimizes the period available for the selection of resistance, artemisinin monotherapies have been used in some countries for 30 years, are still being widely sold, and the deployment of ACTs has been slow. Resistance to artemisinins, manifested as a delay in clearing the parasite from the blood, is currently developing on the Thai–Cambodian border. With replacement drugs more than a decade away, efforts are being taken to stem this resistance, including reducing drug pressure and an ambitious plan to quickly eradicate malaria from these areas before resistance takes hold.

BOX 12.5   BED-NETS

Bed-nets have been used for centuries against biting insects, being first recorded in the sixth century BC in the Middle East. Marco Polo noted their use in southern India in the thirteenth century against mosquitoes—hundreds of years before the link between mosquitoes and malaria was established. However, bed-nets were rarely effective: they could offer protection if free of holes, were tucked in securely and the sleeper did not touch the side of the net during the night, but these conditions were seldom met.

During the 1970s a new class of insecticide became available: the pyrethroids. These are contact acting, persistent, and combine high insect with low mammalian toxicity. They also have a useful 'knock-down' action—insects coming into contact with a pyrethroid-treated surface are rapidly paralysed and fall to the ground. These features make pyrethroids ideal for mosquito control, and from the 1980s successful large-scale trials were carried out on pyrethroid-treated bed-nets. These reported an average 18% drop in mortality of under five-year-olds. By the early 1990s insecticide treated bed-nets (ITNs) were hailed as the most powerful malaria control tool to be developed since DDT and chloroquine in the 1940s.

However, ITNs need re-treating with insecticide after three washes, or at least every year, to maintain efficacy, which can be problematic. Thus, from the early 1990s long-lasting insecticide-treated bed-nets (LLINs) have been developed. The problem is analogous to that of maintaining the biofouling ability of marine coatings discussed in Chapter 13, and has been solved in two ways. In one product the pyrethroid is incorporated into the polyester net fabric. Only 2–5% of the insecticide is available on the surface at any one time and after washing it is replaced by diffusion from within the polymer. The other product has

the pyrethroid bound to the surface of the net within a resin that slowly releases it. Both types can give effective control for 3–5 years and resist at least 20 washes, limited mainly by the life of the net, not the insecticide.

By 2007 the WHO recommended the use of LLINs only, and production is rapidly increasing. However, distribution of bed-nets has been slow. In part this is due to understandable logistical, supply and financial problems, but until recently this has also been due to policy differences. With inadequate funds for mass deployment of bed-nets, WHO targeted their use to pregnant women and children under five years, and tried to develop a system of subsidized government intervention while encouraging commercial sector involvement.

However, there is a powerful argument that bed-nets should be seen as a public good. In areas with high bed-net usage not only are the bed-net users protected, but everyone has a measure of protection. This effect comes from a reduced risk of being bitten combined with a lower probability of being bitten by an infectious mosquito: by a process of attrition the proportion of mosquitoes in the population old enough to contain infective sporozoites decreases. In 2007 the WHO recognized this, and recommended full coverage of all people at risk in an area with LLINs.

At present, pyrethroids are the only suitable insecticides for bed-net treatment, and the spectre of mosquito resistance is looming. Resistance currently does not reduce the protection given by bed-nets, but resistance based on mosquitoes producing modified or greater quantities of detoxifying enzymes has been reported recently, and this is likely, in time, to reduce the efficacy of bed-nets.

spraying of houses had largely ceased, based on health concerns first voiced in the 1960s. Yet these justified concerns were based on heavy usage in agriculture, rather than the much lower concentrations used in residual spraying (a similar situation arose over the use of organochloride insecticides in locust control in the 1970s; see Chapter 15). For many years the WHO, although not banning the use of residual spraying, would not fund it. It was not until September 2006 that the WHO reversed its policy, supporting indoor spraying, and the use of DDT in particular, admitting that they stopped supporting it despite the evidence that it worked well.

Given that, half-way through the time period allocated by the RBM programme for halving deaths from malaria, little progress had been made, it was perhaps surprising that in 2007 the ante was upped. In October 2007, Bill and Melinda Gates called for the eradication of malaria. The e-word had not been spoken in relation to malaria for 40 years and many were incredulous at this statement, which the WHO hurriedly endorsed, with *Science* headlining a report 'did they really say … eradication?' To date, only smallpox has been eradicated, in that it no longer exists in the wild anywhere, and to eradicate malaria transmission-blocking vaccines and new drugs to replace those to which *Plasmodium* is developing resistance will be needed.

BOX 12.6	MALARIA ELIMINATION

Renewed interest in global malaria eradication has meant a greater focus on the process of country-wide malaria elimination. For example, the Global Malaria Action Plan sets a target of at least 8–10 countries in the elimination stage to have achieved no incidence of locally transmitted infection by 2015. The current process recognized by the WHO is very similar to that used in the 1950s, by which countries pass through a number of stages on the way towards certified elimination. However, there is no longer an indicative timescale to this process, recognizing that it will be different in each country, and could be lengthy.

Figure 1 gives the current stage reached by countries. Most malarious countries are still in the malaria control stage, including all of sub-Saharan Africa, most of East Asia and South America. In this phase the aim is to reduce the incidence of malaria to a marginal level, for example, not more than 5% of febrile illnesses caused by malaria during its high-transmission season. At this point, malaria transmission is close to being inter-rupted; although the WHO recognizes that currently available tools do not allow all countries to reach this stage.

Once malaria has been controlled, the country may enter the pre-elimination phase. This involves a re-organization of resources from control, to greater emphasis on surveillance, de-tection and treatment of any remaining cases, and elimination of remaining malaria foci. In 2008–9, six countries moved from the pre-elimination phase to join four countries in the elimination phase. Six of the countries had already eliminated malaria once before, but malaria returned owing to state breakdown, war, and movement of refugees, and in Turkey due to agricultural de-velopment and insecticide resistance. However, three countries went from the elimination phase to the pre-elimination phase during this time, owing to an upsurge of malaria.

When a country has had no locally acquired malaria cases for at least three consecutive years its government can ask the WHO to certify that malaria has been eliminated. These countries should then strive to prevent malaria re-introduction. Currently nine countries are in this category, and join the 24 countries that were certified free of malaria by the WHO between 1961 and 1987.

Currently no sub-Saharan African country has entered the pre-elimination stage. Although it is likely that many other countries, with renewed efforts, will be able to progress through the elimination process, it will be a real milestone when the first sub-Saharan African country enters the pre-elimination stage.

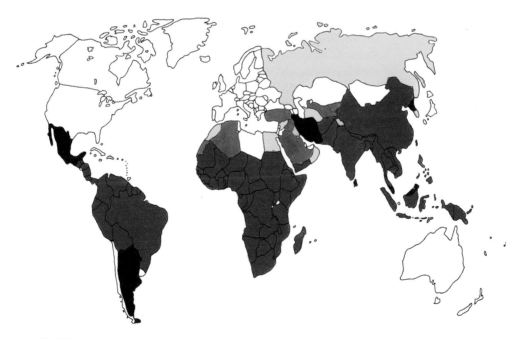

Box 12.6 Fig. 1   *Progress in malaria elimination in 2009. Countries in control phase (red), pre-elimination (brown), elimination (orange) and prevention of re-introduction (yellow)* [Redrawn from WHO (2009).]

Still, a Global Malaria Action Plan (GMAP) was assembled by RBM, and published its goals in September 2008: to reduce malaria deaths to near zero by 2015 and then progressively eliminate malaria from countries and regions until it is eradicated (Box 12.6). This has been called 'totally unrealistic' by some scientists. At the time of its launch, greater funding had been pledged to malaria control than ever before; US\$ 1.1 billion in 2008 compared to US\$ 51 million five years before (but still far short of the annual US\$ 5–6 billion GMAP says it needs), yet only months afterwards the world was plunged into financial crisis and recession. Will funding be maintained in the near future? (The US government has requested a significant reduction in its aid contributions for 2011.) And how well will it be spent? A *Nature* editorial in February 2008 stated:

> [T]he WHO-led Roll Back Malaria initiative is mired in bureaucracy and anything but effective .... The international malaria effort is still geared towards maintaining donor support instead of getting teams into the field gathering data and delivering basic items such as bed nets.[1]

The latest *World Malaria Report* from the WHO (2009) states that its findings 'are cause for cautious optimism', with estimated 2009 global funding of US\$ 1.7 billion, yet targets are still a long way from being met. Four targets, with 2008 African figures given in parenthesis, are: getting ITNs to 80% of the population (31%); treating 80% of acute malaria cases with artemisinin combination treatments (<15%); routine use of other drugs to prevent malaria in 80% of pregnant women to protect unborn children (unknown); and indoor spraying of residual insecticides to protect 80% of households (unknown – 19 African countries reported to be implementing this).

Some progress is, however, being made: the number of courses of ACTs and supply of ITNs has increased markedly in the last few years, and these have made a real difference in some countries. For example, in Eritrea 1.1 million bed-nets were distributed to a population of 3.8 million between 2001 and 2008. Combined with almost a quarter of a million people annually protected by indoor residual spraying and all reported cases of P. *falciparum* malaria being treated with ACTs between 2006 and 2008, this led to a fall of over 90% in malaria outpatients and 86% fewer deaths from malaria in 2008, compared to 2001. In Zambia, enough bed-nets were distributed between 2002 and 2006 to protect three-quarters of the 12.6 million population, indoor residual spraying was covering around half the population by 2008, and over 3 million courses of ACTs were distributed. In 2008 the number of inpatients and deaths from malaria had decreased by over 50% compared to 2001.

These and similar success stories from Rwanda, São Tomé and Principe, and Zanzibar show that with sufficient effort and money the burden of malaria can be significantly reduced. Yet, these are all relatively small countries that do not have the highest malaria transmission rates. To control malaria throughout Africa will require even greater effort and the WHO recognizes that 'global eradication cannot, however, be expected with existing tools'.[2] One of the new tools they are going to have to rely on is a malaria vaccine.

[1]Anon (2008).
[2]WHO (2009), p. 46.

WOT NO VACCINE?

During the Second World War a street graffito became common in the UK, the 'Chad': a long-nosed head and pair of hands peering over a wall and complaining about some shortage, for example 'wot no bread?' A similar sentiment has appeared throughout the scientific and world development literature: why do we not yet have a malaria vaccine?

It is not, certainly in recent years, for want of trying. Malaria vaccine research is a century old, but early studies based on inadequate knowledge of the malaria parasite had little success. The 1955 WHO eradication programme did not mention vaccines, and by the early 1960s investigation into the biology, biochemistry and immunology of human malaria had been neglected for decades as workers were convinced that malaria would be soon eradicated. By the 1970s the fading of this dream, and advances in molecular biology, led to increased funding for malaria vaccine work.

There have been many potential malaria vaccine candidates over the years, and the reader is referred to Sherman (2009) for a full coverage of the story and the personalities involved. Here we can cover only some of the major strands pursued. One line of investigation started with the goal to develop a vaccine that would accelerate the acquired immunity to P. falciparum that develops in adults with repeated infection. In 1981 the first and most abundant merozoite surface protein (MSP-1) of P. falciparum and P. vivax was discovered. This is essential for parasite survival and binding of the merozoite to the erythrocyte (see Fig. 12.3). Antibodies against MSP-1 block invasion and thus this was suggested as a blood stage vaccine candidate. By 2005, 26 out of 93 candidate malaria vaccines were based principally on MSP-1, yet phase I trials were unsuccessful,[3] and the limited published data have not shown protection: by 2007 only two MSP-1-based vaccine projects were still running.

Another membrane protein critical for parasite survival, AMA-1 (apical membrane antigen 1), was characterized from P. falciparum in the late 1980s. Antibodies to this do not inhibit merozoites from attaching to erythrocytes but inhibit orientation and successful penetration (Fig. 12.3). High levels of these antibodies are expressed in humans and thus this was also thought to be a promising vaccine candidate. However, shortly afterwards it was found that the AMA-1 gene is highly polymorphic: would a vaccine work against all of the antigens, and would it induce the parasite to switch antigens, rendering the vaccine useless? Regardless, several AMA-1 vaccines have entered preliminary trials.

One promising area of the Plasmodium life-cycle to target is the pre-erythrocyte stage. In the late 1970s a sporozoite antigen was found, named the circumsporozoite protein (CSP). After the gene responsible was cloned and sequenced in P. falciparum in 1984, it became possible to develop a vaccine. The patent was licensed to GlaxoSmithKline (GSK) and they tried to make CSP in recombinant E. coli. They could not make the full-length protein, but could make four parts of it, and one was tested. The results published in 1987 showed that one out of six volunteers inoculated became protected—the first time an individual had been protected by a subunit malaria vaccine. By 1992 the vaccine RTS,S had been

[3]*Clinical trials.* Any prospective drug or vaccine has to pass through a series of trials before it can be registered. The process is not unlike undertaking an undergraduate degree, with testing becoming more rigorous at each stage, failure at any stage possible, and candidates possibly being asked to retake certain stages, start again or drop out. Clinical trials are carried out in phases, each asks some new questions, and repeat others in an iterative process. After research and preclinical development the product enters phase I testing. This aims to find a safe dosage, evaluates its effect on the body and ability to produce a human immune response. If the product is found to be safe it proceeds to phase II where safety, potential side-effects and preliminary efficacy against infection are tested and the optimum dose is determined. These tests can last two or more years and involve a few thousand volunteers. Phase III then evaluates safety and efficacy on a scale large enough to ensure that the vaccine works under varied conditions within endemic countries. These trials can last for 3–5 years. If these trials are successful the manufacturer submits the vaccine to the regulatory authorities for approval, and if granted the vaccine is made available. Subsequent phase IV trials over the next 4–6 years monitor safety and effectiveness, rare side-effects and also the duration of protection once the vaccine has been released. A full set of clinical trials can take more than 10 years to complete and can cost over $US 500 million.

constructed. This combines the repeat sequence parts of the CSP protein that induce formation of antibodies (R) with portions recognized by T cells in the blood (T) fused to the hepatitis B surface antigen (S, which helps trigger a better immune response). This is self-assembled with the unfused S antigen (one part RTS to four parts S) to make the complete vaccine. Early clinical trials showed that formulation of the vaccine was critical. Vaccines are combined with adjuvants to help them work; this is a black art, as often there is no clear reason why they are effective. Crucial to the effectiveness of the RTS,S vaccine is the extract from the Chilean soap-bark tree *Quillaja saponaria*.

Subsequent trials have shown that RTS,S is safe in both adults and infants and that it can give some protection. 1998 trials in The Gambia on 250 men demonstrated a 34% reduction in the first appearance of *P. falciparum* in the blood, over 16 weeks. In January 2001, GSK entered into partnership with the PATH-Malaria Vaccine Initiative (MVI, the major private–public partnership for vaccine development), with money from the Bill and Melinda Gates Foundation, to develop the vaccine for use in infants and young children in Africa. Subsequent phase II trials of over 2000 children 1–4 years old in Mozambique in 2003 demonstrated a 35% efficacy against appearance of parasites in the blood and 49% efficacy against severe malaria, which lasted for 18 months after vaccination. This was sufficiently encouraging for a phase III trial – the first for a malaria vaccine. This started during May 2009 in seven African countries and will involve over 16 000 children under two years old over the next three years. It is planned to submit RTS,S to regulatory authorities in 2012 and then the vaccine could be available from 2013.

By the end of these trials GSK will have spent over $500 million on this vaccine and the rationale behind this is expressed well by the president of GSK Biologicals in 2008, Jean Stéphenne: 'we are convinced that if we develop the malaria vaccine it will hard for society to refuse to pay for it',[+] although it is unlikely to make the company significant amounts of money. Note the use of the phase 'the malaria vaccine': GSK has been criticized for going it alone in malaria vaccine research and some would like to see RTS,S trialled with other candidates in a combination vaccine.

Malaria vaccine research is finally coming of age, and in November 2009 the MVI announced a major review. Along with only supporting combined trials in the future they announced that they would now be investing in several under-funded, but promising, areas of research. This includes two different approaches to vaccination: live attenuated vaccines and vaccines that block transmission in the mosquito.

In the early 1940s, workers found that injection of inactivated sporozoites into birds rendered them partially immune to later malaria infection: this is a live attenuated vaccine. In the late 1960s, trials took place using human volunteers, and direct infection from x-irradiated mosquitoes. The irradiated sporozoites penetrate the hepatocytes and begin normal development, but then stop growing, undergo little or no nuclear division and persist in a dormant state in the liver. Trials of several hundred volunteers in the early 1970s gave variable results: when it worked over 90% protection was given and it was effective against multiple *P. falciparum* strains for up to 10 months. However, it was difficult to give the correct dose of radiation to the mosquitoes: too little and the patient was likely to be given malaria. These trials ended in 1974 in a morass of ethical and legal issues.

[+]Maher (2008), p. 1042.

Trials of γ-irradiated sporozoites and volunteers started in 1989, with over 90% immunization being recorded after about 1000 mosquito bites. A private company was formed in 2002 to develop this into a vaccine, and a patent was filed in 2005. Work has shown that a vaccine can be made and that it is safe, so phase I trials started in April 2009. Yet there is still criticism of this approach: is it safe, and if so, can enough vaccine be produced, given that manufacture involves dissecting the irradiated sporozoites from mosquito salivary glands? It is also unclear how this vaccine works. An alternative approach is to produce a genetically weakened sporozoite. In 2005, Mueller and colleagues published the first studies on a *Plasmodium* that cannot divide in mouse liver because a gene essential for early liver stage development had been knocked out. This protected against sporozoites injected later.

As studies on bed-nets have demonstrated (see Box 12.5), blocking the transmission of *Plasmodium* by mosquitoes is essential for the control of malaria. The vaccines considered so far, as well as most anti-malarial drugs, do not do this. Yet a mosquito transmission-blocking vaccine may be possible. During their 24-hour development in the mosquito gut the *Plasmodium* gametes are bathed in the blood meal, and are thus accessible to immune factors from the vertebrate host. Recently, work has started into investigating vaccines that target the mosquito antigens instead. *Plasmodium* targets the most abundant ligands on the midgut lumen surface (see Fig. 12.4) to which the gametes bind, and when antibodies against possible ligands are included in the blood meal the vectors are immunized. Any vaccine produced will have to be complex, given the number of ligands involved, but as all *Plasmodium* species use the same mosquito gut ligands there is the real possibility of a multi-species malaria vaccine here.

The first-generation malaria vaccines currently being tested give protection far short of that usually considered essential from a vaccine, and the MVI has given a realistic deadline of 2025 to have a second-generation vaccine available that has over 80% efficacy and a life of at least four years. Workers have finally realized just how complex *Plasmodium* is and how difficult it will be to create an effective vaccine against it. However, work is still hampered on many fronts including a lack of a good laboratory immunological test for malaria or a suitable non-human malaria model.

Furthermore, not only is *Plasmodium* orders of magnitude more complex than bacteria and viruses (the usual subject for vaccines), but humans never develop a complete immunity to it. After several years' infection a degree of immunity develops, reducing morbidity and reducing parasite burden, but this is soon lost without continual re-infection. Yet, this is the very immunity that many vaccines seek to emulate. The future of vaccine research must lie in combination vaccines targeting the sequence of *Plasmodium* development in the liver and blood, and also in blocking its transmission.

HAVEN'T WE FORGOTTEN SOMETHING?

So far we have considered malaria and its control predominantly from the scientific and technological viewpoint. However, this ignores a vital component: humans, and more particularly their ecology. Indeed, an alternative history of malaria could be written from this perspective. Such a history would have to consider how

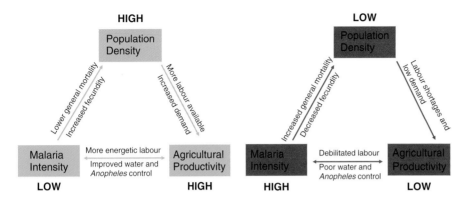

Fig. 12.10 *The Malaria Blocks Development model [Redrawn from Brown (1997).]*

malaria control came to embody one particular cultural model, that of 'malaria blocks development', or MBD (Fig. 12.10). Malaria is certainly linked with poverty, but which came first? In other words, does malaria cause poverty (the MBD model), or does poverty cause malaria? This is not mere semantics as adherence to the MBD model in malaria control has profound implications.

As we have seen, the MBD model has not always dominated: early efforts to control malaria, for example in Italy and the southern states of the US, were based on rural development. Yet even during the early twentieth century the MBD model was developing. This is a cultural model, a type of socially learned cognitive tool used to explain the world. It is never explicitly stated as it is based on shared assumptions, and has been used by elites to promote malaria eradication. Its use by colonial powers to increase the productivity of the local workforce is clear—malaria was seen as the 'great debilitator', and if it could be eradicated then the natural process of economic development would take over, largely to the benefit of the landowners and ruling classes. Furthermore, adherence to the MBD model allows the *status quo* to be maintained: there is no need to tackle exploitation, land reform, social justice and other factors leading to inequality and poverty, as these problems can be attributed to malaria, not to the rulers.

Use of the MBD model was cemented after the Second World War both by the development of techniques such as DDT residual spraying that meant for the first time that malaria eradication might be a possibility, and by the failure of the international community to re-establish rural development programmes. The reasons for this failure are complex, but previous schemes that could have formed benchmarks here, such as bonification in Italy, or the Tennessee Valley Authority scheme in the US, were written off as fascist or proto-communist respectively, and large integrated schemes generally seen as 'statist'. Indeed, post-war rural development schemes became mired within cold war rhetoric and anti-communist development programmes, seen as a threat to dominant economic interests, and were demoted within the nascent WHO and de-prioritized. Thus, by the mid-1950s it became inevitable that the only programme for malaria control that was acceptable was eradication, by vector control.

This programme followed the MBD model, in its faith in technology, privileging the skills of outside experts, marginalizing or ignoring the skills and experience of the local population, and in the absence of research, especially into malaria and local conditions. In part, the failure of this programme is due to the failure to recognize the role of rural poverty in causing the conditions that lead to malaria outbreaks—poor housing, proximity to animals, living near wetland crops, and so on—and failing to treat the underlying causes of lack of agricultural development.

Has anything changed? Certainly the Roll Back Malaria programme recognizes the importance of developing local infrastructure, but it still views malaria as a cause of underdevelopment: 'Tackling malaria is thus a major battle in the war against poverty.'[5] This was given support by an influential paper by Gallup and Sachs (2001) which restated MBD, albeit only with anecdotal evidence, concluding: 'The location and severity of malaria are mostly determined by climate and ecology, not poverty per se.'[6]

The views from Sachs' team have become more nuanced: by 2002 'we tend to favour the explanation that causation runs in both directions, with the causal link from malaria to underdevelopment much more powerful than is generally appreciated,' although 'it is certainly true that poverty itself can be held accountable for some of the intense transmission recorded in the poorest countries.'[7] But they link this to inability to afford technological solutions, such as bed-nets and medicines, rather than underlying conditions *per se*. Finally in 2004 with further work, Malaney *et al.* (2004) conclude: 'the causal effect of malaria on poverty cannot readily be isolated from the effect of poverty on malaria';[8] and in 2010, Bonds *et al.* conclude:

> [W]hat may be more important in these debates is therefore not whether the effect of health on poverty is more significant than that of poverty on health, but whether the combined effect is powerful enough to generate self-perpetuating patterns of development or the persistence of poverty'.[9]

We wait to see whether poverty alleviation is given due attention in the Malaria Eradication Research Agenda (malERA), the new road map due for publication late in 2010, or whether another magic bullet, such as DDT was in the 1940s, overrides all. In the meantime, the conclusions of the eminent malariologist, the late Leonard Bruce-Chwatt in an oration to the Royal Society of Tropical Medicine and Hygiene in 1979 are still apt:

> We must not expect to conquer malaria, a disease rooted in the physical environment of tropical areas and in their unsatisfactory socio-economic conditions, by a miraculous vaccine, drug or insecticide. Each of such scientific tools will be of great value but their proper use will depend on other factors closely related to human ecology in its broadest sense and especially on the national will to fight the disease and on the international determination to co-operate with the developing countries in this endeavour.[10]

FURTHER READING

Of all the subjects discussed in this book, malaria is the fastest developing; we have to significantly revise our malaria lectures each year and it can be difficult to keep up with new developments. Starting points are the comprehensive websites produced by the World Health Organization (www.who.int) and Roll Back Malaria (www.rollbackmalaria.org). Finding informed comment and discussion of developments and the science behind them can be more difficult, but *Science, Nature* and *Trends in Parasitology* are useful; for example a series of articles on malaria appeared in *Science* on 14 May 2010 (vol. 328, no. 5980).

Two recent books are very useful overviews: Packard (2007) and Webb (2009). Warrell and Gilles (2002) give a medical approach to the subject. Collins

[5]WHO (1999), p. 61.

[6]Gallup and Sachs (2001), p. 95.

[7]Sachs and Malaney (2002), p. 681.

[8]Malaney *et al.* (2004), p. 143.

[9]Bonds *et al.* (2010), p. 1191.

[10]Bruce-Chwatt (1979), p. 614.

and Jeffery (2005, 2007) give useful summaries of *Plasmodium ovale* and *P. malariae* respectively, and Bannister and Mitchell (2003), Frevert (2004) and Baton and Ranford-Cartwright (2005) expand on the *Plasmodium* life-cycle. For history in Europe see Bruce-Chwatt and de Zulueta (1980).

Sherman (2009) gives an excellent and fair assessment of the development of malaria vaccines. Recent developments are covered by a series of articles in *Human Vaccines* (2010, vol. 6, no. 1). The MVI website is also a good source of up-to-date information (www.malariavaccine.org). Knockout vaccines are discussed by Ménard (2005) and Mueller *et al.* (2005), and transmission-blocking vaccines by Dinglasan and Jacobs-Lorena (2008). For the human ecology of malaria, see Packard (2007) and the papers in *Medical Anthropology* (1997, vol. 17, no. 3). The first evidence for dual causation between malaria and socioeconomic status was given by Somi *et al.* (2007).

For a good overview of the evolution of malaria, see Carter and Mendis (2002). For more recent developments, see Carter (2003), Hume *et al.* (2003), Hayakawa *et al.* (2008) and Rich *et al.* (2009). For *Plasmodium knowlesi*, see Cox-Singh and Singh (2008) and Galinski and Barnwell (2009). Dobson (1980, 1997) covers the history of malaria in the UK, especially in Essex and Kent; and see Shute and Maryon (1974) for twentieth-century UK malaria, and Snow (1990) for mosquitoes in Britain. The best account of the draining of the fens is still Darby (1956), and for recent discussions see Kuhn *et al.* (2003) and Hutchinson and Lindsay (2006).

Good accounts of quinine are given by McHale (1986) and Rocco (2003), and for artemisinins see Woodrow *et al.* (2005) and White (2008). For *Artemisia* production see Ellman (2010), its genetic map Graham *et al.* (2010) and Milhous and Weina (2010). Ro *et al.* (2006) describe the genetic modification of yeasts to produce artemisinic acid, and Dondorp *et al.* (2010) give a good review of resistance to artemisinins. Bed-nets are reviewed by Hill *et al.* (2006) and Enayati and Hemingway (2010), the latter also a useful review of all aspects of malaria control. See Roberts (2007) for discussion of policy battles.

REFERENCES

Anon (2008) Time to take control. *Nature* **451**, 1030.

Bannister, L. and Mitchell, G. (2003) The ins, outs and roundabouts of malaria. *Trends in Parasitology* **19**, 209–213.

Baton, L.A. and Ranford-Cartwright, L.C. (2005) Spreading the seeds of million-murdering death: metamorphoses of malaria in the mosquito. *Trends in Parasitology* **21**, 573–580.

Beresford, J. (ed.) (1926) *The Diary of a Country Parson: the Reverend James Woodforde. Volume II: 1782–1787.* Oxford University Press, Oxford.

Bonds, M.H., Keenan, D.C., Rohani, P. and Sachs, J.D. (2010) Poverty trap formed by the ecology of infectious diseases. *Proceedings of the Royal Society B* **277**, 1185–1192.

Brown, P.J. (1997) Malaria, *Miseria*, and underpopulation in Sardinia: the 'malaria blocks development' cultural model. *Medical Anthropology* **17**, 239–254.

Bruce-Chwatt, L.J. (1979) Man against malaria: conquest or defeat. *Transactions of the Royal Society of Tropical Medicine and Hygiene* **73**, 605–617.

Bruce-Chwatt, L.J. and de Zulueta, J. (1980) *The Rise and Fall of Malaria in Europe: a historico-epidemiological study.* Oxford University Press, Oxford.

Carter, R. (2003) Speculations on the origins of *Plasmodium vivax* malaria. *Trends in Parasitology* **19**, 214–219.

Carter, R. and Mendis, K.N. (2002) Evolutionary and historical aspects of the burden of malaria. *Clinical Microbiology Reviews* **15**, 564–594.

Cobbett, W. (1893) *Rural Rides. Volume 1.* Reeves and Turner, London.

Collins, W.E. and Jeffery, G.M. (2005) *Plasmodium ovale*: parasite and disease. *Clinical Microbiology Reviews* **18**, 570–581.

Collins, W.E. and Jeffery, G.M. (2007) *Plasmodium malariae*: parasite and disease. *Clinical Microbiology Reviews* **20**, 579–592.

Cox-Singh, J. and Singh, B. (2008) Knowlesi malaria: newly emergent and of public health importance? *Trends in Parasitology* **24**, 406–410.

Darby, H.C. (1956) *The Draining of the Fens*, 2nd edn. Cambridge University Press, Cambridge.

Defoe, D. (1928) *A Tour Through England and Wales.* Dent, London.

Dinglasan, R.R. and Jacobs-Lorena, M. (2008) Flipping the paradigm on malaria transmission-blocking vaccines. *Trends in Parasitology* **24**, 364–370.

Dobson, M. (1980) 'Marsh fever': the geography of malaria in England. *Journal of Historical Geography* **6**, 357–389.

Dobson, M.J. (1997) *Contours of Death and Disease in Early Modern England.* Cambridge University Press, Cambridge.

Dondorp, A.M., Yeung, S., White, L. *et al.* (2010) Artemisinin resistance: current status and scenarios for containment. *Nature Reviews Microbiology* **8**, 272–280.

Ellman, A. (2010) Cultivation of *Artemisia annua* in Africa and Asia. *Outlooks on Pest Management*, April, 84–88.

Enayati, A. and Hemingway, J. (2010) Malaria management: past, present, and future. *Annual Review of Entomology* **55**, 569–591.

Freeman, W.G. and Chandler, S.E. (no date, c. 1908) *The World's Commercial Products.* Sir Isaac Pitman & Sons, London.

Frevert, U. (2004) Sneaking in through the back entrance: the biology of malaria liver stages. *Trends in Parasitology* **20**, 417–424.

Galinski, M.R. and Barnwell, J.W. (2009) Monkey malaria kills four humans. *Trends in Parasitology* **25**, 200–204.

Gallup, J.L. and Sachs, J.D. (2001) The economic burden of malaria. *American Journal of Tropical Medicine and Hygiene* **64**, 85–96.

Graham, I.A., Besser, K., Blumer, S. *et al.* (2010) The genetic map of *Artemisia annua* L. identifies loci affecting yield of the antimalarial drug artemisinin. *Science* **327**, 328–331.

Hay, S.I., Guerra, C.A., Gething, P.W. *et al.* (2009) A world malaria map: *Plasmodium falciparum* endemicity in 2007. *PLoS Medicine* **6**(3): e1000048.

Hayakawa, T., Culleton, R., Otani, H., Horii, T. and Tanabe, K. (2008) Big bang in the evolution of extant malaria parasites. *Molecular Biology and Evolution* **25**, 2233–2239.

Hill, J., Lines, J. and Rowland, M. (2006) Insecticide-treated nets. *Advances in Parasitology* **61**, 78–128.

Hume, J.C.C., Lyons, E.J. and Day, K.P. (2003) Human migration, mosquitoes and the evolution of *Plasmodium falciparum*. *Trends in Parasitology* **19**, 144–149.

Hutchinson, R.A. and Lindsay, S.W. (2006) Malaria and deaths in the English marshes. *The Lancet* **367**, 1947–1951.

Kappe, S.H.I., Vaughan, A.M., Boddey, J.A. and Cowman, A.F. (2010) That was then but this is now: malaria research in the time of an eradication agenda. *Science* **328**, 862–866.

Kuhn, K.G., Campbell-Lendrum, D.H., Armstrong, B. and Davies, C.R. (2003) Malaria in Britain: past, present, and future. *PNAS* **100**, 9997–10001.

Maher, B. (2008) The end of the beginning. *Nature* **451**, 1042–1046.

Malaney, P., Spielman, A. and Sachs, A. (2004) The malaria gap. *American Journal of Tropical Medicine and Hygiene* **71** (Supplement 2), 141–146.

McHale, D. (1986) The cinchona tree. *Biologist* **33**, 45–53.

Ménard, R. (2005) Knockout malaria vaccine? *Nature* **433**, 113–114.

Milhous, W.K. and Weina, P.J. (2010) The botanical solution for malaria. *Science* **327**, 279–280.

Mueller, A.-K., Labaied, M., Kappe, S.H.I. and Matuschewski, K. (2005) Genetically modified *Plasmodium* parasites as a protective experimental malaria vaccine. *Nature* **433**, 164–167.

Packard, R.M. (2007) *The Making of a Tropical Disease: a short history of malaria.* Johns Hopkins University Press, Baltimore, USA.

Phillips-Howard, P.A., Bradley, D.J., Blaze, M. and Hurn, M. (1988) Malaria in Britain 1977–86. *British Medical Journal* **296**, 245–248.

Rich, S.M., Leendertz, F.H., Xu, G. *et al.* (2009) The origin of malignant malaria. *PNAS* **106**, 14902–14907.

Ro, D.-K., Paradise, E.M., Ouellet, M. *et al.* (2006) Production of the antimalarial drug precursor artemisinic acid in engineered yeast. *Nature* **440**, 940–943.

Roberts, L. (2007) Battling over bed nets. *Science* **318**, 556–559.

Rocco, F. (2003) *Quinine: malaria and the quest for a cure that changed the world.* HarperCollins, London.

Sachs, J. and Malaney, P. (2002) The economic and social burden of malaria. *Nature* **415**, 680–685.

Sherman, I.W. (2009) *The Elusive Malaria Vaccine: miracle or mirage?* ASM Press, Washington, DC, USA.

Shute, P.G. and Maryon, M. (1974) Malaria in England past, present and future. *Royal Society of Health Journal* **1**, 23–29.

Smith, A.D., Bradley, D.J., Smith, V. *et al.* (2008) Imported malaria and high risk groups: observational study using UK surveillance data 1987–2006. *British Medical Journal* **337**, 103–106.

Snow, K.R. (1990) *Mosquitoes.* Naturalists' Handbooks no. 14. Richmond, Slough, UK.

Somi, M.F., Butler, J.R.G., Vahid, F., Njau, J., Kachur, S.P. and Abdulla, S. (2007). Is there evidence for dual causation between malaria and socioeconomic status? Findings from rural Tanzania. *American Journal of Tropical Medicine and Hygiene* **77**, 1020–1027.

Warrell, D.A. and Gilles, H.M. (eds) (2002) *Essential Malariology,* 4th edn. Arnold, London.

Webb, J.L.A. Jr. (2009) *Humanity's Burden: a global history of malaria.* Cambridge University Press, Cambridge.

White, N.J. (2008) Qinghaosu (artemisinin): the price of success. *Science* **320**, 330–334.

WHO (1999) *The World Health Report* 1999. World Health Organization, Geneva, Switzerland.

WHO (2008) *Global Malaria Control and Elimination: report of a technical review.* World Health Organization, Geneva, Switzerland.

WHO (2009) *World Malaria Report* 2009. World Health Organization, Geneva, Switzerland.

Woodrow, C.J., Haynes, R.K. and Krishna, S. (2005) Artemisinins. *Postgraduate Medical Journal* **81**, 71–78.

Biofouling and the Barnacle

I hate a Barnacle as no man ever did before, not even a Sailor in a slow-sailing ship.

Charles Darwin, 1852[1]

There was a Barnacle Junior ... leaving the Tonnage of the country, which he was somehow supposed to take under his protection, to look after itself, and, sooth to say, not at all impairing the efficiency of his protection by leaving it alone.

Charles Dickens, 1856[2]

PLACE ALMOST ANY non-organic surface (and many an organic one) into water and the process of accumulating organisms begins. In many situations this is not a problem and leads to the richly diverse marine and intertidal communities that can be seen by diving or the more leisurely pursuit of examining rock pools and wandering along rocky coasts (Fig. 13.1). In some cases, for example aquaculture production of algae and shellfish, this process of bioaccumulation is actively encouraged. However, in others, particularly when occurring on built structures, biofouling leads to increased corrosion, reduced movement of fluids through pipes and inhibited movement of vessels through water. Just as weeds are only plants that are in the wrong place according to humans, so biofoulers are just aquatic organisms that are inhibiting mankind's activities. In this chapter we will consider the marine biofouling process and then examine the barnacle—one of the most troublesome marine biofouling organisms—before discussing techniques

Fig. 13.1 *The same organisms accumulate on natural and built structures in this Devon, UK harbour. Only on the bridge could these be considered as biofoulers*

Biological Diversity: Exploiters and Exploited, First Edition. Paul Hatcher and Nick Battey.
© 2011 John Wiley & Sons, Ltd. Published 2011 by John Wiley & Sons, Ltd.

[1]Letter to W.D. Fox, October 1852; Burkhardt and Smith (1989), p. 100.
[2]Dickens (1857, first published 1856), p. 394.

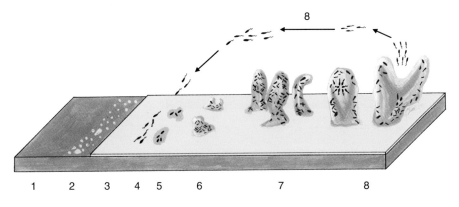

Fig. 13.2 *Bacterial biofilm development.* **1**, *clean surface;* **2**, *the conditioning biofilm (yellow) of organic molecules starts to form;* **3**, *conditioning biofilm complete;* **4**, *initial attachment of bacterial cells to the surface;* **5**, *production of extracellular polymeric substances (pink) anchor cells to the surface;* **6**, *early development of biofilm architecture;* **7**, *maturation of biofilm architecture;* **8**, *dispersal of cells from the surface [Partially based on Stoodley et al. (2002).]*

Fig. 13.3 *Schematized time-line of the biofouling sequence [Based on Wahl (1989).]*

(and their effects on the wider environment) for combating biofouling.

THE FOULING PROCESS

The biofouling process begins as dissolved chemicals (mainly macromolecules such as glycoproteins, proteoglucans and polysaccharides) adsorb to the surface, and in a couple of hours a thin (less than 100 nm thick) conditioning film appears (Fig. 13.2). The process is similar in quite different environments−if you have just brushed your teeth it is going on now in your mouth−in all cases the medium is a dilute salty aqueous solution of proteins and sugars.

About one hour after the immersion of the surface bacterial colonization starts. The first bacteria may be able to approach the conditioning film and attach to it, attracted largely by physical forces, but many will be repulsed by electrostatic forces and will only be able to approach so far. These bacteria form bridges to the surface by producing polysaccharide fibrils which attach to the macromolecular film. Subsequent enzymatic shortening of these fibrils pulls the bacteria towards the surface. Then the establishment of covalent bonds between the bacteria and the film sticks it irreversibly to the surface.

The bacterial colony now develops and there can be a succession of bacterial species. Rod-shaped ones that are capable of growing in high nutrient conditions are often the first colonizers, followed by coccoid and later stalked and filamentous species that can cope with lower nutrient levels. This growing colony forms a biofilm by the secretion of a slime-like matrix of extracellular polymeric substances (mainly polysaccharides, proteins and nucleic acids) (Fig. 13.2). This protects the colony against environmental stress, enables the exchange of nutrients and can increase resistance to biocides. The bacterial biofilm can take over 10 days to reach a structural maturity and by then may contain hundreds of proteins. This biofilm was thought to be a homogenous

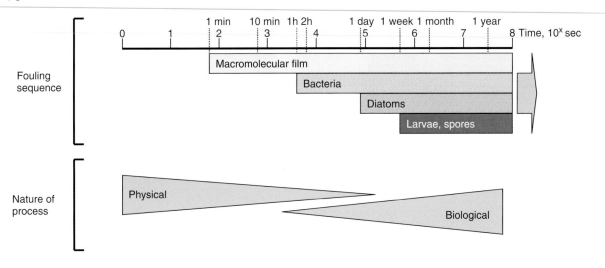

layer, but recent studies have shown it to be highly structured, with open water-channels so that nutrients can be delivered into the community. There can also be a degree of cellular specialization and, once the community is mature, motile forms of the bacteria are produced to repeat the colonization process. If it has been a couple of hours since you brushed your teeth, a plaque which may contain hundreds of species of bacteria has formed.

Several days later unicellular eukaryotes start to attach to the biofilm. These include yeasts, protozoa (some are sessile filter feeders, other are mobile predators of other microbes), marine fungi and diatoms. The diatoms are especially important and attach by mucous secretion; they generally stick better to biofilms than to clean surfaces, and contribute significantly to the evolution of the substrate.

Fig. 13.4 *A complex biofouling community on a breakwater. The limpet (Patella species) is covered with biofoulers, which probably help camouflage it. Limpets are herbivores, feeding especially on green algae, they also inhibit the settlement and growth of young barnacles by barging into them. Likewise, barnacles inhibit grazing by limpets, and in this community the only green algal growth is on the limpets and barnacles*

After at least one week the first multicellular eukaryotes start to attach to the fouled surface (Fig. 13.3). The initial species are attracted by the chemicals released from the surface; later species may be attracted by the presence of particular species (e.g. colonizers of the algae, Box 13.1), may be inhibited by the presence of other species (and may have to wait for these species to be disturbed), or many be quite tolerant of what is there already.

The biofouling of a surface is an excellent example of the ecological principle of succession, whereby a new surface is colonized by a largely predictable series of organisms, with gradually increasing individual and community complexity. Slowly, a complex community builds up, many constituents of which can be easily studied on a rocky shore or seaside structures (Fig. 13.4). These communities may contain green algae such as *Ulva* and *Cladophora* species, brown algae including *Fucus* and *Laminaria* species, and red algae such as *Corralina*, *Chondrus crispus* (carragheen) and *Palmaria palmata* (dulse) (Box 13.1). This is analogous to the *Sargassum* community (Chapter 2). Animals are represented by coelenterates

BOX 13.1   USES FOR SEAWEED

It is worth remembering that many biofoulers, especially the seaweeds, have been exploited by mankind. Many of the more delicate red and green seaweeds have been used as food around the coasts of Britain and Ireland. For example the red seaweed *Palmaria palmata* 'dulse' (Fig. 1a) can be eaten raw, cooked like a vegetable or dried and chewed like gum; *Porphyra* species 'laver' can be washed and boiled and made into cakes with oatmeal and fried; and carragheen (*Chondrus crispus*) is used in the preparation of jellies and as an emulsifier. Agar agar is extracted from a variety of red seaweeds and is used for the culture of bacteria, being water-soluble and setting to a jelly. The green seaweed *Ulva* species 'gutweed' (Fig. 1b) is excellent when stir-fried until crisp.

The tough brown seaweeds, the wracks (Fig. 1c), are less suitable for human food, but have formed the food of coastal livestock: there is still a breed of sheep on North Ronaldsay, Orkney, that has seaweed as its principal food. Seaweed was used until the late twentieth century as a commercial fertilizer, especially after being fermented in heaps; it has more nitrogen and potassium than animal manure, but less phosphorus, and also conserves moisture. Liquid seaweed extracts are still available as garden fertilizers.

The greatest use of brown seaweeds was in the production of soda ash as an alkaline flux to be added to lower the melting point of silica in glass-making and also used in soap-making (see Chapter 14). The preferred source of this was from the saltmarsh plant *Salicornia*, imported from Spain as 'barilla'. When this became unavailable or too expensive, interest turned to locally available seaweeds and an important coastal industry developed from the late seventeenth century producing soda ash. The kelp (the word can refer both to the raw material and the product), generally *Fucus* and *Laminaria* wrack species, were collected, heaped and dried and then burnt in shallow pits on the shore until after 6–8 hours it began to liquefy. This was then stirred vigorously and allowed to set, which could take weeks, before being removed and stored. There was considerable resistance to this industry to start with, because people were concerned that the foul fumes from kelp burning would taint crops and poison livestock; but as the price of soda ash rose it became the main source of income for many coastal communities during the eighteenth century. These communities often neglected farming and fishing during part of the year to burn kelp. At the peak of the industry 20000 tonnes of soda ash a year were produced from the Hebrides alone, requiring 20 times as much seaweed.

After the Napoleonic war, barilla became cheaper to import again and kelp burning declined. However, with the discovery of iodine in 1811 from kelp there was a resurgence in kelp processing as for a time this was the only source of this element. This was ended by the discovery later in the nineteenth century of Chilean saltpetre from which cheaper iodine could be extracted, and by the 1920s the British kelp industry had become extinct.

Seaweed is increasingly cultivated, and industry consumes about 8 million tonnes wet weight a year: the production of seaweed meal and alginates from wracks is increasing; the latter are used as thickeners and gelling agents.

(a)

(b)

(c)

Box 13.1 Fig. 1 *Common seaweeds:* **(a)** Palmaria palmata *(red alga, Rhodophyta);* **(b)** Ulva *species (green alga, Chlorophyta, Protista);* **(c)** Fucus vesiculosus, *bladderwrack (brown alga, Stramenopiles, Chromalveolata). All are potential biofoulers, and all have also been exploited by humans. Although often free of colonizers themselves, a number of species do attach to seaweeds including bryozoans, minute colonial animals – arrowed in (a)*

(hydroids), bryozoans, porifera (sponges, see Fig. 13.11), polychaete worms, annelids (tubeworms), ascidians (sea-squirts), tunicates, molluscs (mussels, limpets, clams, oysters), and crustaceans (barnacles).

THE BARNACLE

Barnacles are among the commonest biofouling organisms, and are probably of the greatest economic importance, certainly to shipping. Although some barnacles were thought to be the eggs and young of geese for a long time (Box 13.2), in 1758 Linnaeus placed then in the Mollusca, based on their adult forms which have hard shells attached to a substrate and with a soft animal inside. The free-living

BOX 13.2   BARNACLES AND GEESE

There is an old story which connects the barnacle and the barnacle goose (modern Latin name *Branta leucopsis*) as stages in the life-cycle of a single creature originating from rotting timbers floating in the seas. Gerald of Wales was a twelfth-century cleric commissioned to report on the state of Ireland, which had at the time recently been overrun by Henry II, the Norman King of England. In his report Gerald is responsible for one of the earliest recorded statements of the barnacle goose story:[a]

> There are many birds here that are called barnacles, which nature, acting against her own laws, produces in a wonderful way. They are like marsh geese, but smaller. At first they appear as excrescences on fir-logs carried down upon the waters. Then they hang by their beaks from what seems like sea-weed clinging to the log, while their bodies, to allow for their more unimpeded development, are enclosed in shells. And so in the course of time, having put on a stout covering of feathers, they either slip into the water, or take themselves in flight to the freedom of the air. They take their food and nourishment from the juice of wood and water during their mysterious and remarkable generation. I myself have seen many times and with my own eyes more than a thousand of these small bird-like creatures hanging from a single log upon the sea-shore. They were in their shells and already formed. No eggs are laid as is usual as a result of mating. No bird ever sits upon eggs to hatch them and in no corner of the land will you see them breeding or building nests. Accordingly in some parts of Ireland bishops and religious men eat them without sin during a fasting time, regarding them as not being flesh, since they were not born of flesh.

Medieval bestiaries often present an elaborated version of the story, in which the geese are produced like fruits from a tree. This elaboration may be a consequence of the lack of observation of geese actually developing from logs in the water: the barnacles would then be interpreted as arrested remains of barnacle geese, whose full development was confined to the living tree, somewhere out-of-sight.

This version, in its turn, finds its way into John Gerard's *Herball*, a popular sixteenth-century account of the uses and variety of plants (Fig. 1).

The barnacle associated in this way with the origin of the goose is the ship's barnacle, *Lepas anatifera*, rather than the more familiar acorn barnacle of the seashore. *Lepas* is a stalked barnacle, and hangs by its stalk in a manner which looks quite goose-like.

[a]Gerald of Wales: 'The History and Topography of Ireland', read at Oxford about 1188 (Giraldus Cambrensis, 1982), pp. 41–2.

As well as a passing visual similarity, migration probably contributed to the idea that the goose originates from the barnacle: the barnacle geese fly north to Svalbard (in the Arctic waters of the North Atlantic) to breed, so in Britain they are never seen sitting on nests with eggs. The yearly appearance of the geese (in the autumn) and disappearance (in the spring) would naturally raise a question over their origin, and the answer might appear to be supplied by the humble barnacle. Given the common currency of the idea of spontaneous generation in medieval times (see Box 2.3), the life-cycle can be completed with the barnacles being generated from the log/tree.

However, there may, in addition, be a more sinister explanation. Gerald of Wales was one of the original sources of the story, and his report is full of lies about the Irish people, with tales of bestiality, hermaphroditism (used to mean that which partakes of both genders, such as a woman with a beard; and the progeny of different species, such as a man who was half-ox), and other behaviours both strange and sordid. Nature in Ireland is given the same curious treatment, the report on the timber–barnacle–barnacle goose metamorphosis being only one example. Why would an upstanding member of the clergy sink to such depths of deception and misrepresentation? Gerald could at times be quite an astute and careful observer of nature, so he was not simply a gullible foreigner reporting local tales as if they were fact. His real motives seem to have been at once highly intellectual; and vain, parochial and petty. The medieval scholar David Rollo interprets Gerald's fabrication as deliberately contrived, in a bizarre double-bluff, to create an incredulous response from the reader. By complaining about the nonsense perpetrated by Gerald, these readers showed themselves to be fools, incapable of seeing his deeper message. And what was this message? That the invading Anglo-Norman Plantagenet dynasty had no right to be in Ireland; that right, according to Gerald, lay with his Welsh-Norman family, who had first occupied Ireland, only to be rapidly displaced by the Plantagenets. Why did his message involve the distortion of Ireland and its inhabitants? His language is metaphorical: nature should represent reason; when it is perverted, it stands for empty rhetoric. This was a style and culture which Gerald associated with his Anglo-Norman rivals; his biggest target in this literary snare was his Plantagenet patron, Henry II, whom he simultaneously flattered and despised.

Gerald may, therefore, have been playing an elaborate game, understandable only to the lettered elite of the time, getting back at the King through a courtly sniggering. Nowhere in Gerald's strategy was importance attached to an accurate portrayal of the natural world, of course. His 'nature' is a cultural construct serving very definite, if subtle, literary and political purposes. Incidentally, by depicting the Irish as sub-human beasts, his portrayal justified intervention in the affairs of the Irish by outside powers.

Incidentally again, he perpetuated the idea of the barnacular origin of a goose, which persisted for around 500 years. His ideological word-games fuelled a belief about nature which ensnared popes, professors and even the first president of the Royal Society. On enquiry in 1435 the future Pope Pius II, visiting King James I of Scotland, was disappointed to be told that the goose-bearing trees were to be seen only further north, in Orkney; 'miracles flee further and further', he complained. The later (1526) account of Hector Boece, Professor in the University of Paris, in his 'History of Scotland' was widely influential and denied that the geese grow on trees by their beaks, instead affirming that:

> [T]hey are bred always only by nature of the seas: for all trees that are cast in the seas by process of time appear first worm-eaten, and in the small bores or holes thereof grow small worms: first they show their head and feet, and last of all they show their plumage and wings; finally, when they are coming to the just measure and quantity of geese, they fly in the air as other fowls do: as was notably proven in the year of God 1490, in sight of many people, beside the castle of Petsligo.

The first President of the Royal Society, Sir Robert Moray, presented a paper in 1661 on the topic of the barnacle goose and its origins, in which he confirmed the essentials of Gerald of Wales' twelfth-century account. There were throughout the period of ascendancy of the story those who denied it, including Albertus Magnus and Frederick II (a keen thirteenth-century Sicilian experimenter-king). Such doubters also (presumably) included Pope Innocent III, who in 1215 issued a Bull forbidding the eating of barnacle geese during Lent; the idea of their egg-free origin had spread through the clergy and led to the consumption of the geese as 'fish'. But this should not obscure the extraordinary tenacity of the story of the barnacle goose. It shows the ability of the human mind to create an explanation where one is wanting; and the (less charming) capacity of mankind to create a Nature which suits his own purpose. It is doubtful that Gerald of Wales would have imagined the impact of his fishy tale, which may perhaps have given rise to the expression 'a *canard*'.

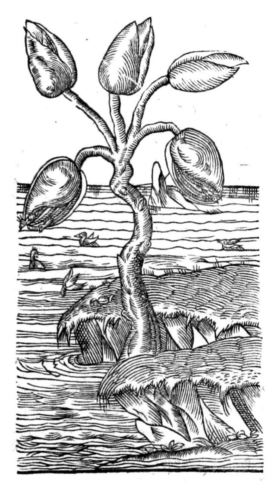

Britannica Conchæ anatiferæ.
The breede of Barnakles.

Box 13.2 Fig. 1 *The breede of Barnakles, from John Gerard's* The Herball or Generall Historie of Plants *(John Norton, London, 1597). Young geese are shown escaping from the lower barnacle 'shells', and others are swimming in the water* [source www.BioLib.de.]

larval stages of the barnacle were described by Thompson only in 1830 and, as they have a characteristic crustacean morphology, in 1834 barnacles joined this subphylum of the arthropods as the class Cirripedia (meaning curled foot). This classification was controversial for years, until Darwin's monographs in the 1850s confirmed barnacles as crustaceans (Box 13.3).

The Crustacea is the only large group of arthropods that are primarily aquatic and are unique among arthropods in having two pairs of antennae. The more than 31 000 crustacean species extant (an impressive species richness, although some insect *families* have more species alive today than this) include crabs, shrimps, lobsters, woodlice, crayfish, and many that form the plankton — including the salmon sea-louse (see Box 9.4).

BOX 13.3   TWO CDS AND MR A AND B

Bibliographers and biographers often refer to their subject by their initials. This can save space but can cause momentary confusion when flipping between subjects. For example, a biologist may automatically assume that CD refers to Charles Darwin, but an English scholar might equally assume that this means Charles Dickens. Both wrote about barnacles.

By 1844, Charles Darwin was completing the write up of his results from the *Beagle* voyage and had that summer finished a 230-page account of his species theory. He sealed this and did not return to it for ten years. In the meantime he studied barnacles—why?

During a discussion in September 1845 on the problem of species with Darwin, Joseph Hooker (1817–1911, botanist and director of the Royal Botanic Gardens Kew 1865–85), one of his best friends, remarked that he had little time for theory that came from people who had not studied deeply many species. Although not an *ad hominem* attack on Darwin, he took it to heart: 'How painfully (to me) true is your remark'.[a] Hooker encouraged Darwin to start on some species work after he had finished the last of the Galapagos expedition volumes, and this he did in October 1846. Darwin had been struck by a specimen he had found on the Chilean coast in 1835: a conch shell with small holes. Inside these holes he found a tiny creature, which certainly had barnacle-like features, yet was digging a home and not secreting it. Darwin called this species Mr Arthobalanus, and initially started work on writing a paper on this species, which he thought would take perhaps a year. This paper was never completed; instead Darwin found it necessary to describe the common species of barnacles first, and this led him to gradually take on the whole group in order to understand and place Mr Arthobalanus. In May 1848 he remarked:[b]

The Barnacles will put off my species book for a rather a long period … my work retrogrades, i.e., as I keep on finding out new points, I have to hark back to genera, which I thought I had completed.

Hooker later came to regret asking Darwin to start on a group of species, repeatedly asking him when he will return to his species theory. In October 1849 he wrote to Darwin that he cared more for the species work than barnacles, eliciting this sharp reply from Darwin:[c]

[T]his is too bad of you, for I declare your decided approval of my plain Barnacle work over theoretic species work, had very great influence in deciding me to go on with former & defer my species-paper.

Darwin appeared to regret this also, as he was drawn into describing all the barnacles his correspondents could send him, both fossil and living. By December 1849 he wrote 'I begin to think I shall spend my whole life on Cirripedia, so slow is my progress,'[d] and by 1852 he was 'wonderfully tired of my job'[e] and 'I hate a Barnacle as no man ever did before, not even a Sailor in a slow-sailing ship.'[f]

However, Darwin did persevere, through bouts of severe illness, the birth of several children and the death of one, and by the autumn of 1854 was packing up his barnacles (and restarting work on his species theory) after producing two volumes totalling over 1000 pages on the living barnacles and two more on the fossil species. These are still regarded as standard works (although they are not without problems, such as confusing cement glands with ovaries), and Crisp (1983) argues that this was Darwin's greatest work, for if he had not written *Origin of Species* someone else would have written something similar soon, but had he not completed the barnacle work, biological knowledge of this group would have been delayed well into the twentieth century.

It is generally suggested that the eight years that Darwin worked on barnacles bought him time and enabled him to perfect his use of taxonomic nomenclature and the use of embryology to study the species question. Yet Darwin himself was not always so sure of its use, writing to Hooker in 1850: 'You ask what effect studying species has had on my variation theories; I do not think much,'[g] and in 1853 writing 'I have spent an almost ridiculous amount of labour on this subject & certainly w[d] never have undertaken it, had I forseen what a job it was.'[h] Although in the 1870s he recognized that it had been of considerable use when he had to discuss principles of natural classification in *Origin of Species*, he still states 'Nevertheless, I doubt whether the work was worth the consumption of so much time.'[i]

All this probably led to Darwin's hatred of barnacles, but a final reason was likely to have been that he found them so variable that it was hard to decide the boundaries between species reliably and this possibly threatened to derail his species theory altogether. Half-way through his barnacle work he states 'Systematic work w[d] be easy were it not for this confounded variation.'[g]

[a]Letter to J.D. Hooker, 10 September 1845; Burkhardt and Smith (1987), p. 253.
[b]Letter to E. Cresy, May 1848; Burkhardt and Smith (1988), pp. 135–6.
[c]Letter to J.D. Hooker, 12 October 1849; Burkhardt and Smith (1988), p. 270.
[d]Letter to A. Hancock, 25 December 1849; Burkhardt and Smith (1988), p. 292.
[e]Letter to W.D. Fox, 7 March 1852; Burkhardt and Smith (1989), p. 83.
[f]Letter to W.D. Fox, October 1852; Burkhardt and Smith (1989), p. 100.
[g]Letter to J.D. Hooker, 13 June 1850; Burkhardt and Smith (1988), p. 344.
[h]Letter to W.D. Fox, 17 July 1853; Burkhardt and Smith (1989), p. 148.
[i]Autobiography, written c.1876; Barlow (1958), p. 118.

About 18 months after Darwin had finished with Mr Arthobalanus, another Mr Barnacle appeared from another CD, Charles Dickens. In February 1856 the third monthly instalment of his best-selling *Little Dorrit* introduced Mr Tite Barnacle, the head of the Circumlocution Office, and a whole shoal of other Barnacles. This monstrous government department, worthy predecessor of Kafka's bureaucratic nightmares and Terry Gilliam's *Brazil*, was populated by Barnacles and 'its finger was in the largest public pie and in the smallest public tart'.[j] Whereas Darwin was worried that his barnacle work would remain 'wholly unapplied' (in fact it has been of great use in antifouling research), biofouling allusions abounded in Dickens' *Little Dorrit*. The role of the Circumlocution Office was to ensure that the solution chosen to any problem was 'not the way to do it'

and would slow down the ship of state as long as possible ('that what the Barnacles had to do, was to stick on to the national ship as long as they could'[k]): 'to arrange with complexity for the stoppage of a good deal of important business otherwise in peril of being done'.[l]

Dickens' satire was directed at the British Government's handling of the country's intervention in the Crimean War, and the subsequent calls for enquiries and government reform, but over 150 years later it would seem that this shoal of Barnacles and their Circumlocution Office is still very active in British politics.

[j]Dickens (1857), p. 103.
[k]Dickens (1857), p. 115.
[l]Dickens (1857), p. 398.

The barnacles are, apart from some parasitic species, the only sessile (i.e. attached to a substrate) group of crustaceans. They are exclusively marine and about two-thirds of the 900-odd extant species are free-living, attached to rocks and other surfaces. The rest are parasitic or commensal on other animals including whales and turtles. The free-living, surface-dwelling barnacles (there are also some boring species) form the order Thoracica. This is divided into three suborders: Lepadomorpha, the stalked barnacles once confused with the barnacle goose (see Box 13.2); Verrucomorpha, asymmetric stalkless barnacles; and Balanomorpha, the symmetric stalkless acorn barnacles. Members of the last suborder, which includes *Balanus*, *Semibalanus* and *Chthamalus* species, are the most important biofouling barnacles on ships today and form the basis of the discussion of life-cycles and adaptations that follow. The information is also relevant to the other stalkless barnacles and to some extent to the stalked ones.

Fig. 13.5 *Life-cycle of an acorn barnacle, Semibalanus balanoides (not to scale). M = moult.* **Insets**: *left shows lateral view of 5th-stage nauplius larva; right shows lateral view of the cypris antennule (ii, iii, iv, segments of the antennule)* [Redrawn from Stubbings (1975).]

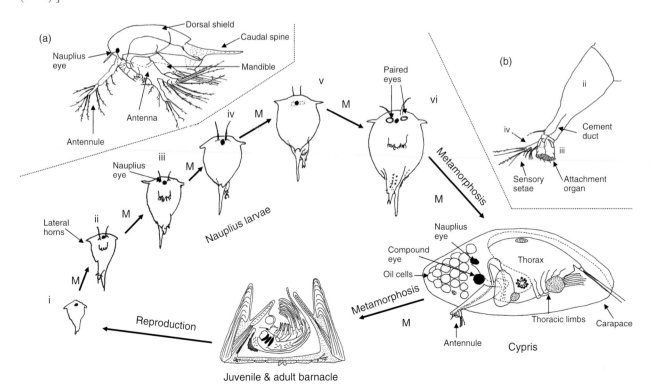

Barnacles have an unusual life-cycle, even among Crustacea, although there are certain parallels with the salmon sea-louse discussed earlier. The adult barnacle broods its eggs all winter in colder climates (we will use the term 'it' here, as most thoracican barnacles are hermaphrodite and thus functionally male and female). The young larvae hatch in the spring and are released into the spring bloom of plankton – up to 13 000 may be released from one adult. These are termed nauplius larvae (Fig. 13.5). They moult progressively through six stages, finally becoming almost 1 mm long in *Semibalanus*. This larva is almost all head and has three pairs of limbs arising from it: antennules; biramous (i.e. branched towards the base into two limbs) antennae; and biramous mandibles behind the mouth. The body is covered by a single-piece dorsal shield. This larva has only one eye, the median nauplius eye, which it uses to orient towards the light.

The nauplius uses all its limbs for swimming, although the antennae are the main source of power – beating at 4–12 times a second for short bursts, alternated with periods of drifting. As the limbs cannot feather their hairs on the backstroke, the nauplius has an inefficient jerky back and forwards movement through the water. It feeds on small phytoflagellates and diatoms, largely caught by the antennae and passed to the mandibles for transfer to its mouth.

Subsequently, the sixth-stage nauplius larva moults into a cypris (Fig. 13.5). This has a quite different structure, and is a specialized non-feeding search and attachment stage. During metamorphosis to the cyprid, all the naupliar feeding and swimming structures are lost: antennae are lost, mandibles are much reduced and antennules modified into specialized attachment structures. The thorax now develops and six pairs of thoracic swimming limbs appear, all within a bivalve carapace. The nauplius eye is still present, but is joined by two compound eyes.

The cypris must find a suitable site and attach itself before it runs out of energy in days to a couple of weeks. The precise cues it uses in site selection are still uncertain, but include a complex of surface proteins. When it comes into contact with a solid surface it attaches to it with the attachment organ on its antennules (Fig. 13.5b) using removable glue. It now walks over the surface on its antennules, first quickly searching and then, when a suitable spot is located, making a more detailed examination. If the site is unsuitable it detaches itself and swims away: factors such as light levels, flow velocity, degree of disturbance, surface texture and topography are important in site suitability. Chemical cues given off by bacterial biofilms and attached barnacles of the same species (attractants, for example 'arthropodins' released from the epicuticle of adult barnacles) or predators such as dog whelks (a repellent; Box 13.4) can also affect settlement. The blobs of glue the searching cypris leaves behind may also induce settlement of other barnacle larvae. If the site is suitable it starts to secrete liquid cement from specialized glands; this hardens over 1–3 hours into a hard tanned protein and the barnacle is now permanently fixed to the spot.

Now a second metamorphosis takes place into the familiar adult barnacle (Fig. 13.5). This process is complex. The body, which is attached to the substrate by the antennules on its head, undergoes flexion: imagine standing on your head and then bending backwards at the neck by 90 degrees so that your arms and legs stick up vertically. The barnacle achieves this and its six pairs of thoracic limbs elongate into feathery cirri for food collection. Primordial plates appear in the juvenile barnacle, and grow gradually by deposition of calcium carbonate (taking calcium from the sea water) throughout its life. The cuticle lining the plates is moulted periodically, but the plates remain.

The adult barnacle is attached to the substrate by cement. This barnacle cement is the toughest and most durable found in aquatic organisms. It is very resistant to chemical breakdown and thus is hard to study, so we have an imperfect understanding of it, but it appears to be made up of five or six proteins with different properties. These proteins become cured and hard in a process little understood but probably involving the formation of disulphide bonds. The strength of this cement and its ability to glue under water has led to investigation of potential uses in dentistry, and repair of blood vessels and nerves. This is helped by the fact that barnacle cement does not induce the production of antibodies. The advent of a self-repairing dental filling without the need for drilling will certainly be something to look forward to!

As the barnacle grows in diameter it applies cement at its growing edge, and may need to reapply cement if it has been damaged, or contact with the substrate has been broken. The cement ducts are kept open by being filled with a flushing fluid after the cement passes through them. If contact with the substrate is lost, the flushing fluid leaks out and more cement can pass down the ducts to seal the gap – the barnacle does not suffer from the DIYer's problem of used tubes of glue with blocked nozzles.

The acorn barnacle is surrounded by a number of immoveable plates, ranging from several whorls of them in the most primitive forms to only four or six in the most advanced species such as *Semibalanus* (Figs 13.6 and 13.7). At the top is an operculum of four plates that can open to enable excretion and reproduction, and for the cirri to emerge for feeding. This fixed life-style gives the adult a number of problems: feeding, competition and predation, and reproduction. These barnacles are limited to feeding by 'fishing' for small organisms using their thoracic limbs. Early in their evolution, barnacles already

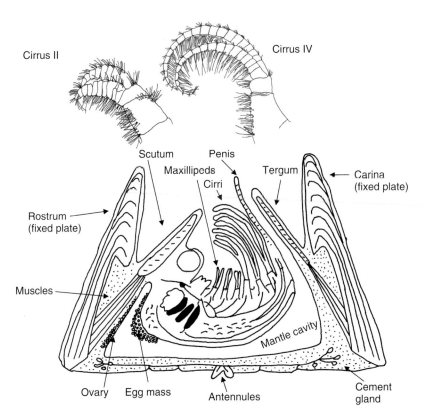

Fig. 13.6 *Cross-section through an adult Semibalanus balanoides acorn barnacle. The scutum and tergum are moveable plates of the operculum. Cirri II and IV are magnified. Note that the cirri are biramous, and differ considerably in shape. Cirrus II is one of three pairs of maxillipeds used to transfer food collected by the other three pairs of cirri (including cirrus IV) to the mouth [Redrawn from Stubbings (1975) and Rainbow (1984).]*

had a division of labour among these limbs, with two pairs developed into maxillipeds used for transferring the food caught by the four pairs of cirri to the mouth. Subsequent species have evolved faster reaction times, and faster beating of the cirri – the more advanced Balanoidea have three pairs of cirri forming a basket, moving towards each other and downwards at up to 2.5 times a second, sweeping the water for small zooplankton such as copepods and larvae and moving them towards the maxillipeds (Fig. 13.6).

The acorn barnacles are hermaphrodite, each with functioning male and female sex organs, but they do not mate with themselves. As they cannot move to find a mate, the male sex organ, the penis, must do the searching – as Anderson (1994) puts it:

> Imagine a mating system involving acting at times as a male, with a penis extendible several times the length of the body, allowing copulation with attached neighbours, also firmly rooted to the spot; and at other times as a female, being probed by persistent penises extended by male-acting neighbours.[3]

Fig. 13.7 *The acorn barnacle* Semibalanus balanoides. *The six fixed plates enclosing the animal and the four opercular plates can be seen. Barnacles of a variety of ages, including the very youngest adults, are present in this young colony. Already some are becoming crowded out and a couple have begun growing over others*

This penis must have a guidance system to find females, and be capable of quick retraction. It searches through 360 degrees, extending up to five times its resting length, and once a female is found copulation may take 90 seconds. Although the penis then withdraws, it returns to inseminate the female repeatedly over the next hour or so, transferring all of the stored seminal fluid. In species with an annual breeding cycle the penis is lost after mating, and regrows again the following year. The eggs are brooded in the barnacle's mantle cavity for several months before release; they obtain shelter but no food from the parent.

Cyprid larvae usually do not settle directly next to another barnacle, but maintain enough space for the young adult to develop. However, as the adult grows and more barnacles settle, the substrate may become crowded, and barnacles may have to compete for space. Here the barnacle is a perfectly passive–aggressive organism: it will continue to grow and a smaller individual caught between several larger ones is liable to be crushed, or one will try to force its side plates under the other, dislodging it (see Fig. 13.7). In some cases the dislodged barnacle fixes itself to the other barnacle and in time may grow over it, smothering its operculum and killing it.

ANTI-FOULING AND THE RISE AND FALL OF TBT

Ever since they first went down to the sea in ships humans have had to contend with biofouling. And there is a lot of it. Untreated vessel bottoms can collect 150 kg per square metre of biofouling in less than six months: on a large oil tanker with 40 000 m² of underwater area this would amount to 6000 tonnes of fouling. Railkin (2004) estimated that in 1978 there was a total of about 5000 km² area of marine structures introduced by mankind, which could support 6 million tonnes of biofouling organisms. These figures may have doubled by now. The costs of this biofouling are great also: even a small amount of fouling can increase fuel consumption by between a quarter and a half, and is reputed to cost the US Navy alone over a billion dollars a year, and European aquaculture annually about 260 million euros.

[3]Anderson (1994), p. xi.

Fig. 13.8 *Oyster and slipper limpet shells; exploiter and exploited washed up together on a British south coast beach*

Biofouling of ships is also an important means of introduction of species into new areas. For example, the slipper limpet *Crepidula fornicata*, so commonly washed up on British beaches, is not a native species but was introduced to Britain from North America in the 1880s possibly originally with oysters, but certainly later moved around Europe on the hulls of ships. This species competes with oysters for space and food and has been a serious problem in oyster culture (Fig 13.8). The Australasian acorn barnacle *Elminius modestus* is also a problem in oyster production. This was introduced into British ports during the Second World War on a ship travelling from Australia, and within 15 years became the most common barnacle in British oyster beds. Unlike the native barnacles, it settles at the same time as the oyster spats (Box 13.5) and competes with them for space, sometimes smothering them. Recently, Williams and Smith (2007) report that hull fouling is the most significant source of seaweed introductions, and that these introductions can affect established communities.

Shipbuilding technology has to some extent been driven by the need to reduce biofouling, but anti-fouling technology has also had to react and adapt to changing materials used in shipbuilding. Furthermore, the need to also protect wooden vessels from boring invertebrates and iron and steel vessels from saltwater corrosion has been an important consideration. Anti-fouling technology culminated in the use of TBT (tributyltin) based self-polishing paints towards the end of the twentieth century, but which by the end of 2008 had been completely banned world-wide.

Many attempts have been made to prevent fouling of wooden vessels. Ancient Phoenicians and Carthaginians in the eighth century BC are said to have used pitch as a protectant, and waxes and tars were used from an early time. Various poisons were also tried, including arsenic and sulphur mixed with oil, recorded from 412 BC. Repeated attempts were made from at least the third century BC to use lead sheathing. This technology was forgotten for several centuries until adopted by Spain early in the sixteenth century, followed by England later. However, lead sheathing is poor at preventing fouling and causes corrosion of any iron in contact with it, and thus was officially abandoned by the Admiralty in the late seventeenth century. At this time shipbuilders reverted to wood sheathing: sandwiching a layer of animal hair and tar between two layers of wood. This may have prevented borers from penetrating, but it was expensive, and would have had little effect against biofoulers.

During the eighteenth century experiments with copper sheathing were undertaken. The first authentic record is of HMS *Alarm*, a 32-gun frigate sheathed with copper in 1758, originally against ship-worm. It was noticed that she did not foul, nor was the wood damaged by the copper, and by 1780 copper sheathing was in general use by the Royal Navy: it was hoped that this would enable British ships to outrun the unsheathed French ships (Fig. 13.9a).

Although copper sheathing was the first effective anti-fouling technology developed, it was not without problems. It was expensive, and the copper corroded quickly in sea water. This corrosion was originally thought to be due to impurities in the copper, but in a series of experiments during the mid-1820s Humphrey Davy demonstrated that this was due to electrolytic corrosion of the copper by sea water. He found that this corrosion could be halted by attaching small pieces of iron to the copper. Although Davy hailed this as a success it was soon found that these hulls rapidly became covered with fouling organisms. Later it was discovered that the anti-fouling action of this copper sheathing arises from the corrosion of the copper, which produces a solution of poisonous copper ions near the hull.

During the nineteenth century there was a revolution in shipbuilding with the adoption of first iron and then steel for hulls. At first copper sheathing was applied direct to iron hulls: the iron hulls soon corroded because of the electrolytic interaction between the metals (as predicted by Davy) and several ships are reported to have been lost due to this–the commander of HMS Triton said in 1862 that the plates of his ship were so thin due to this corrosion that it was only kept afloat by the barnacles! This problem became so acute that in 1847 the Admiralty started to sell all the iron ships in the navy, but this was soon halted owing to the unavailability of suitable wood for building replacements.

This was a quandary. Iron was needed for larger ships, especially as good wood was scarce, and wooden ships could not cope with powerful steam engines or newly invented explosive shells, nor could they be effectively subdivided by watertight bulkheads. Several solutions were tried to enable copper sheathing–still seen as the best anti-fouling solution–to be used. Ships built from iron or steel frames with wooden hull planking (so-called composite construction) could be copper-sheathed relatively easily (Fig. 13.9b), and while there were several other advantages to such construction, this ability played a part in the construction of composite warships until the early twentieth century. Merchant vessels that needed to maintain a fast speed over a long voyage, such as the tea clipper Cutty Sark, were also built of composite construction: an unsheathed vessel could easily gain enough biofoulers during a voyage from the tropics to reduce speed significantly.

Vessels with steel and iron hull plating were also eventually copper-sheathed, with the copper separated from the hull by layers of wood sheathing (Fig. 13.9c). This was complicated: there could be no electrical connection between the hull and the copper, so several layers of wood were needed to ensure that no nails fastening the copper to the wood were in contact with fastenings between the wood and the hull. It was also expensive, heavy, slowed the ship and only lasted a year or two before needing replacement. The problem of fouling of warships based in the tropics, away from dockyards for long periods, was so acute that copper sheathing of steel hulls was used into the twentieth century.

The other alternative was to develop more effective anti-fouling paints. The first practical composition was introduced into Britain in 1860, a 'hot plastic' paint; a metallic soap of copper sulphate, applied hot over a quick-drying primer coat of varnish and iron oxide. Various other formulations using copper, arsenic and mercury were marketed and, by the early twentieth century, these paints were providing fouling protection for up to nine months. However, they were difficult to apply, and caused untold injury to shipyard workers. During the early twentieth century 'cold plastic' anti-fouling paints were developed. These could be applied by brush or spray and hardened by evaporation of solvents: by the 1940s some were lasting up to 18 months.

During the 1950s anti-fouling paint technology developed rapidly and the first organometallic anti-fouling paints were developed, using tin, arsenic and mercury biocides. Owing to acute toxicity, the use of the latter two elements along with pesticides such as DDT in anti-fouling paints was soon prohibited.

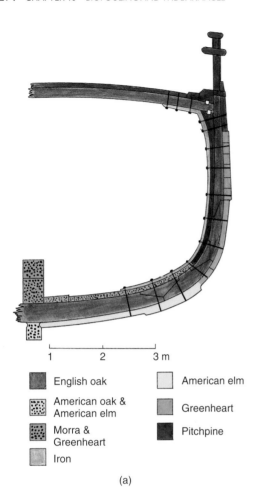

1   2   3 m

■ English oak

▨ American oak & American elm

▨ Morra & Greenheart

▨ Iron

□ American elm

■ Greenheart

■ Pitchpine

(a)

Fig. 13.9 *Structural solutions to biofouling of ships.* **(a)** *Half amidships section through the wooden hull of the 35 m three-masted schooner William Ashburner, launched at Barrow-in-Furness, UK, in 1876. When she entered the transatlantic trade in 1878 it was a relatively easy job to add copper sheathing (actually a cheaper brass substitute was used) as the ship had been built using copper nails. The sheathing was applied over a layer of felt to protect it from rubbing on the hull. She was massively built with frames spaced 5 cm apart on the floor, and the hull at the bows and stern was up to 48 cm thick. She traded until running aground in the Bristol Channel, UK, in 1950 and was abandoned; parts were still visible in the 1970s.* **(b)** *Vertical section through the hull of a sheathed composite (iron frame, wood planking) vessel. Two layers of wood were necessary to prevent contact between the iron fastenings of the hull frames and the copper nails from the sheathing.* **(c)** *Vertical section through the hull of HMS Inconstant, a large sail and steam iron frigate built 1866–9. The first layer of oak was laid down vertically (i.e. the grain ran from the keel to the deck) and was bolted directly to the iron hull, the second layer was laid down horizontally (i.e. the grain ran from bow to stern) and was fastened to the first layer with brass screws. The copper sheathing was nailed to this, making sure that the nails did not touch the iron bolts. 15 cm thickness of oak was used, partially to try to deflect shot which could shatter the iron hull, but also so that the hull could be copper-sheathed [(a) based on Latham (1991), (b,c) on Brown (1997).]*

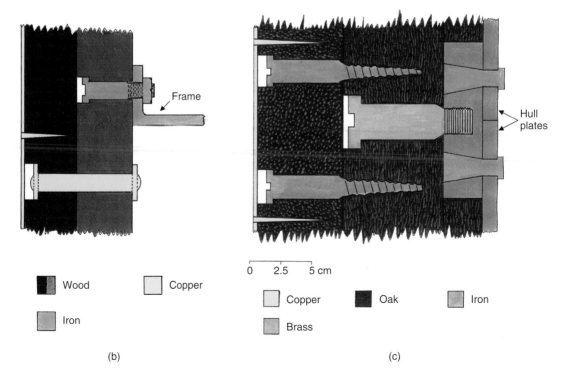

■ Wood

■ Iron

□ Copper

Frame

(b)

0   2.5   5 cm

□ Copper

■ Oak

■ Iron

■ Brass

Hull plates

(c)

| BOX 13.4 | **DOG WHELKS** |

One of the first effects of TBT (tributyltin) on the wider environment was noticed in the dog whelk, *Nucella lapillus*. In the late 1960s it was found that some dog whelks in Plymouth Sound, UK, had become strangely distorted; the female whelks had started to develop a penis and vas deferens. As these developed they grew over the female genital opening and prevented the release of egg capsules, rendering the whelk sterile. The accumulation of aborted capsules in the oviduct could eventually lead to its rupture and the death of the whelk. This condition was termed 'imposex' and was subsequently found in over 70 species of marine gastropod, but it was not until the early 1980s that it was demonstrated that this condition was caused by minute concentrations of TBT, which is concentrated by the whelk.

The proportion of imposex occurring in dog whelk populations was used as a measure of TBT, and by the early 1980s dog whelk populations had declined dramatically along the British coast, disappearing completely from some areas, especially near boatyards and marinas. Dog whelk populations have recovered in many areas since the ban on the use of TBT on small vessels in the later 1980s; but the ban came too late for some populations in the south west which went extinct. Near many commercial ports, however, which remain hotspots of TBT pollution, populations have not recovered.

The dog whelk is a gastropod mollusc (Fig. 1). It is widely distributed on rocky shores, largely in the intertidal zone, bordering the North Atlantic from Portugal to the White Sea and in North America south to Long Island. Dog whelks are carnivorous, preying on other sedentary creatures such as barnacles and mussels. Indeed, they are one of the most important rocky shore predators.

The dog whelk penetrates the shell of its prey using secretions from an accessory boring organ at the front of the foot (which chemically softens the shell) combined with rasps of its radula, the multi-toothed feeding organ important in many molluscs (see Chapter 3). Once the shell has been penetrated the dog whelk inserts its proboscis and injects a choline-based neuromuscular toxin to paralyse the prey, followed by the secretion of several digestive enzymes. The 'soup' produced is sucked up by the dog whelk. Feeding is a slow process—it may take up to a week to bore into a large mussel—but dog whelks can often manage to inject their neurotoxin between the opercular plates of barnacles and thus dispense with the boring. This allows a dog whelk to feed on about one barnacle a day.

Dog whelks have separate sexes and aggregate to spawn, having repeated copulation in between which the female will release a couple of egg capsules, and attach them to rock. Each capsule contains around 600 eggs, but only 6% will develop into embryos. The rest are fertilized but the male and female nuclei do not fuse and development is arrested. These are 'nurse' eggs, and serve as food for the developing whelks during the 4–7

months it takes for them to develop. The larvae hatch as 'crawlaways' which already have their initial shell (the protoconch, which persists as the apex of the adult shell) and seek tiny mussels and barnacles to feed upon. They do not reach sexual maturity for at least a year, and can live for up to six years.

The Phoenicians in about 1500 BC found that extracts from the hypobranchial glands (which produce the neurotoxins mentioned above) of several Mediterranean whelks could produce a vivid dye, Tyrian purple, and an important industry developed around this until the fall of Constantinople in AD 1453. However, there is considerable doubt as to what colour Tyrian purple was, as a range of colours can be obtained from Mediterranean whelks, depending on the species and method of extraction.

The dog whelk (which does not occur in the Mediterranean) was apparently the basis of a much smaller dye industry in Britain, Ireland and Norway until about the seventeenth century. Tyrian purple had been almost forgotten when, in October 1684, a Bristol physician, William Cole, was speaking with two women at Minehead, Somerset (Cole, 1685). They told him that someone in an Irish port was making a good living dying cloth a crimson colour with a liquid taken out of a shellfish. They did not know which shellfish, or which part of it was used, and so Cole experimented with the species he could find, and found one, probably the dog whelk, from which he could extract a 'vein' containing the dye. When placed on linen this gave a light green colour, in the sun this changed over a couple of hours to deep green, then blue, to deep purple. Cole's work started a small resurgence in interest in this dye.

Box 13.4 Fig. 1 *Dog whelks,* Nucella lapillus, *feeding on barnacles on a Devon beach. Dog whelks can be found in many different colours and patterns. Note the many empty barnacle 'shells', their plates still remaining stuck to the rock*

BOX 13.5   OYSTERS

Arcachon Bay is a triangular enclosed estuary on the French Atlantic coast between the Gironde estuary and the Spanish border. In the central portion of the bay 1000 hectares of oyster beds are situated, producing 10–15 000 tonnes of oysters a year (10% of French production). These beds are surrounded by ten marinas and other moorings with accommodation for almost 8000 pleasure craft. It is therefore not surprising that after TBT became the accepted anti-fouling paint for these small craft, if it was going to have an effect on other marine organisms this would soon be seen on oysters in Arcachon Bay.

And it was, in 1974. The oyster shells became wafer-like, bubbled and distorted with the formation of an interlamellar gel in chambers, and recruitment of oysters was also affected. The effect of TBT was demonstrated experimentally by hanging plates coated with TBT in tanks of oysters. Even the lowest concentration tested (0.8 ng of TBT-Sn per litre) affected shell calcification. There was no oyster recruitment to the beds in Arcachon Bay between 1977 and 1981, and by 1982 annual production was down to 3000 tonnes. It was estimated that between 1977 and 1983 losses at Arcachon Bay alone amounted to US$ 147 million. This led to the first ban on TBT in the world, when the French government prohibited from January 1982 the use of TBT-based antifouling paints along their Atlantic coast on craft below 25 m length. This ban was later extended and made permanent; oyster reproduction gradually recovered from 1983.

TBT led to the virtual collapse of the oyster industry in France and Britain, but unfortunately was only one of many recent disasters to have befallen this industry. Traditional oyster culture in Europe used the native flat oyster, Ostrea edulis, a bivalve mollusc, which although it had been collected since the Iron Age had not been farmed significantly until the seventeenth century. Individuals of this species can alternate their sex, for example being male one season and female the next. The larval oyster has a short planktonic stage before a foot develops and a suitable substrate is sought. The oyster produces a thread, the byssus, to stick to the substrate, and then metamorphoses into a spat. This attaches its left valve to the substrate as the shell is secreted and remains there, filter feeding on suspended particles in the water.

Culturing of oysters from the middle of the nineteenth century led to greater overfishing and finally a collapse in production. To overcome this, the Portuguese oyster, Crassostrea angulata, was imported into France from 1868. These rapidly spread naturally and replaced the native oyster; by 1920 completely replacing it in Arcachon Bay. Ostrea edulis suffered from epidemics of viral and protozoal diseases during the twentieth century, which further decreased its numbers. During the 1950s and early 1960s, over 100 000 tonnes of C. angulata were being harvested in some years in France; however, an epidemic of gill disease in 1970–3 led to this species disappearing from the French coast. Thus, from 1972 Crassostrea gigas, the Pacific oyster, was imported, and French oyster production reached over 100 000 tonnes per year again before TBT had its effect.

The Pacific oyster is the most widely cultivated oyster in the world, with Japan and South Korea being the major producers. It has a higher tolerance of fresh water than Ostrea edulis and thus is more suited to estuaries. The spats are not fussy about where they settle, and a wide range of substrates is used. In Arcachon Bay the traditional clay roof tiles coated with lime are used to rear the spats, stacked in cages (in the late 1980s there were 20 million of these in Arcachon Bay alone). After 6–18 months the seed oysters are detached and placed on the sea bed or culture tables, and reared until three years old.

Across the Channel, the oyster fishery in Poole Harbour, UK, was also hit by TBT. Similar in many ways to Arcachon Bay, Poole Harbour is one of Europe's largest lowland basins, with a shoreline of over 100 km, 3600 ha of water at high tide, most in the intertidal zone, and an entrance only 370 m wide. A busy commercial harbour, ten boat yards and many marinas are enclosed in this area.

Oyster fishing has taken place in Poole Harbour since at least Saxon times (a huge late Saxon midden of 4–8 million oyster shells from under present-day Poole suggests that there was an important oyster industry then), and there was a lucrative oyster fishery until the early nineteenth century. During the early 1980s an attempt was made to restart an oyster fishery using C. gigas. Although initial British trials of this species in the 1970s had been successful, it failed in Poole Harbour: oysters had abnormal and extreme shell thickening and were unmarketable. It was originally thought that this oyster was not suited to Poole Harbour; only later was it found that these problems were caused by TBT, and that Poole Harbour was one of the most heavily TBT-polluted areas in Britain. By 1989 the TBT concentration had fallen and subsequently commercial beds have been established in the centre of the harbour, to the north of Brownsea Island. They now extend for over 150 ha, and two million C. gigas spats are placed here every year.

Furthermore, the paint formulations were still far from ideal. In one type (soluble matrix paints) the whole paint slowly dissolved in sea water. Thus, this paint had a low strength and was easily damaged, could only contain a low concentration of biocide, and did not work well on stationary ships. The other type (insoluble matrix paints) had a hard insoluble matrix out of which the biocide leached slowly. However, over time the rate of leaching slowed (as the biocide had further to travel to reach the water) and the honeycomb structure left made the hull rougher. The remaining matrix had to be removed before repainting. Both types of paint provided effective anti-fouling for little more than 18 months.

A revolution in anti-fouling technology occurred in the late 1960s with the development of self-polishing paints and the use of tributyltin (TBT) as a biocide. These paints used an acrylic copolymer with TBT groups (an organometallic compound to 'deliver' the tin, TBT was first used in conventional anti-fouling paints during the 1960s) bound to it. There was a slow and continuous hydrolysis of the top layer of the paint in sea water, releasing the tin biocide. As the surface of the paint was continually dissolving, but leaving a hard surface beneath it, the paint effectively polished itself. These paints had an effective life of five years.

Self-polishing anti-fouling paints containing TBT were introduced in the mid-1970s and within 10–15 years were being used on over 80% of commercial vessels. They were also very popular anti-foulants for fish-farm cages and pleasure craft.

However, TBT is almost universally toxic: it has been used as a fungicide, biocide, insecticide and wood preservative. Its effects on non-target organisms were soon noticed, particularly in dog whelks and oysters (see Boxes 13.4 and 13.5). Although TBT degrades in sea water to harmless compounds in a few weeks, before then it is one of the most poisonous organometallic compounds introduced into the marine environment, and if it becomes incorporated into the sediment can remain active for decades (much longer than originally predicted). Experiments demonstrated that it had effects on marine organisms in concentrations at the limit of detection, for example 1–2 ng per litre. Particularly high levels of TBT were found in coastal waters, around marinas and fish-farms. TBT is lipid-soluble and is bioaccumulated in animals ascending the food chain; thus, herrings in the Baltic were recorded with up to 4600 ng of TBT per gram wet weight in their livers. As the tolerable daily intake of TBT was set at only 15 µg TBT per 60 kg person per day, it was thought possible that people on a seafood-rich diet could be exceeding this level.

Largely because of the almost complete collapse of the oyster industry due to TBT (see Box 13.5), its use on small craft and fish-farms was first banned in France in 1982, followed by Britain in 1987 and other countries by the end of the 1980s.

TBT was described by Goldberg (1986) as the most toxic substance ever deliberately introduced into the aquatic environment and he likened its potential effects through bioaccumulation to that of DDT (see Chapter 12). This statement was taken as a rallying call by the environmental movement who lobbied for a *complete* ban on the marine use of TBT. Although the use of TBT-based anti-fouling paints on pleasure craft (which spend most of their time idle in marinas and can easily be removed from the water for cleaning) can hardly be condoned, its use on commercial shipping is more equivocal. It could be argued that as they spend little time in port the level of coastal TBT pollution generated by commercial ships will be small (TBT hotspots around major ports seldom extend more than a couple of kilometres from the port), and the most polluting activity – cleaning and repainting hulls – can be regulated (Fig. 13.10). Several cogent arguments for the retention of TBT for commercial ships (based on more biofouling = slower ships + more ships needed = more fuel used = increased greenhouse gas emissions) were made in the late 1990s as the International Maritime Organisation (IMO) debated the complete banning of TBT. However, in October 2001 the IMO did adopt an 'International Convention on the control of harmful anti-fouling systems on ships' which mandated signatory countries to stop applying or reapplying organotin compounds which act as biocides from 1 January 2003.

Fig. 13.10 *Even small vessels such as this trawler being repainted in a Norwegian shipyard have a large underwater area needing anti-fouling treatment. Good shipyard control of waste anti-fouling products is essential to avoid hotspots of pollution in these harbours*

This convention was to come into force 12 months after 25 states representing at least 25% of the world's merchant shipping tonnage had ratified it. In September 2007, Panama was the twenty-fifth country to ratify the treaty, and with countries representing over 38% of the world's merchant shipping having ratified it, it came into force in October 2008. By this date organotin biocides were to be removed from all ships, or to be covered by a barrier to prevent them leaching into the sea. It remains to be seen how quickly this will be achieved.

ANTI-FOULING AFTER TBT

The all-round success of TBT-based self-polishing anti-fouling paints inhibited the development of alternatives, and thus with the prohibition of these paints there was no replacement available that was nearly as effective. Anti-fouling technology has developed since along two main lines: the development of effective and environmentally safe biocides for use in conventional paints; and the development of biocide-free surfaces that prevent fouling. There has been increased interest in learning from nature. Can we discover why many marine creatures, slow-moving or stationary, which we would expect to be fouled remain clean, and can we incorporate these features in our anti-fouling coatings (Fig. 13.11)?

With the phasing out of TBT, most anti-fouling paint manufacturers fell back to using copper-based compounds. Copper is an effective biocide against many organisms and has the advantage that it is not accumulated in the environment as rapidly as TBT (although copper accumulation in port sediments has been reported and it has been banned as an anti-foulant in some areas). However, many biofouling organisms, including *Ulva* species, have become tolerant or resistant to copper.[4] Therefore, these paints have to include 'booster biocides' effective against these organisms. Many compounds have been tried; most were already in use in agriculture as herbicides or fungicides. Unfortunately, there was often little data on the toxicity of these compounds to marine species or their fate in the marine environment, and many products turned out either to be too species specific or alternatively too broad, affecting non-target species. Irgarol 1051 (a triazine herbicide) and diuron were the most commonly used booster biocides, but both are persistent: irgarol has a 350-day half-life in sea water and unfortunately degrades to a compound that is even more toxic to some marine organisms. By the early years of the twenty-first century diuron had been banned in the UK, and irgarol restricted to vessels over

[4]Piola *et al.* (2009) suggest that there is an increasing evolution of copper-tolerant biofouling organisms which outcompete less tolerant species, and via shipping invade new habitats, in which communities are less able to resist invasion due to copper pollution. This 'double whammy' may turn out to be a very important deleterious and unintended effect of banning TBT.

(a)

(b)

(c)

Fig. 13.11 *The smoothest inorganic surface soon becomes covered with fouling organisms, yet while in this marine aquarium the bottle in* **(a)** *has become encrusted with sponges (Porifera), the sea anemone in* **(b)** *and sea urchin in* **(c)** *in the same tank remain free of foulers. Can we harness the anti-fouling mechanisms that these organisms use to keep our marine structures free of fouling?*

25 m length. Some biocides, such as isothiazolinones, appear to be more suitable, combining toxicity to a wide range of biofouling organisms with rapid breakdown in sea water to much less toxic chemicals.

Although around three-quarters of the world's shipping uses copper-based anti-fouling products, these have not proved as effective as coatings containing TBT. The effective life is shorter, often only 2–3 years, and not withstanding pollution problems there have been problems in formulating reliable coatings.

More environmentally friendly biocides have been sought. The Australian red alga *Delisea pulchra* produces halogenated furanones, which are released from vesicles within specialized gland cells and strongly inhibit the settling of fouling organisms; these chemicals are being commercially developed. Many other potential anti-fouling chemicals have been detected in marine organisms, including some that deter the settlement of barnacles at extremely low concentrations, but there is a need to develop low-cost artificial analogues to these.

Plutarch (c. AD 45–125) experimented with scrapings of algae and slime mixed with pitch as an anti-fouling coating, and recently researchers have returned to this: can living paints be produced using bacteria in biofilms that inhibit fouling organisms? This research is at an early stage, but some bacteria incorporated into resin have inhibited biofilm production in sea water for several months. Alternative research is attempting to incorporate enzymes into anti-fouling paints. These enzymes, such as chitinases active against barnacles, could degrade the biofouler's adhesives. Many patents for these have been issued, but often with little research into efficacy. It is difficult to evaluate the effectiveness of these products, and their stability over several years must be questioned. Regardless of this, an enzyme-based anti-fouling coating has recently been introduced to the Danish yacht market.

All of these alternative products are still biocides. However, there are some anti-fouling preparations that do not use biocides at all: the fouling-release surfaces.

Since the 1970s attempts have been made to use non-stick coatings on ship hulls. Early efforts, often using Teflon coatings, failed as it was not possible to reliably attach the coating to the hull. However, since 1987 silicone-based polymers have been used and have become quite popular for certain types of vessel, with the coating lasting up to five years. These surfaces still get fouled, but only weakly: larger organisms still attach in port but are generally removed when the vessel moves, and any slime and biofilm can be easily removed in dock by high-pressure water hosing. These coatings are most suitable for faster ships such as ferries and some naval vessels that spend little time in ports, as considerable speed is needed to dislodge some fouling organisms, and fuel consumption is increased until they are released. However, this later release of fouling organisms could lead to their introduction to new areas. These coatings are also sensitive to damage, and are several times more expensive than conventional anti-fouling coatings.

Although the aim is to make these surfaces as smooth as possible, a roughened surface may inhibit biofouling. Microfibres (hairs 50–100 μm long and 2–10 μm diameter) have been incorporated, perpendicular to the surface of coatings; these move with the current and inhibit settling of foulers. Some products containing this technology have recently been marketed; although they can last for 3–5 years, the increase in hull roughness may lead to increased fuel consumption. Researchers have investigated the topography of surfaces favourable to biofouling organisms, and also looked at the surface structure of organisms which remain free of fouling. One product resulting from this is Sharklet™, which has a surface of billions of tiny raised microscopic features arranged in diamond shapes 25 μm across and made from a silicone elastomer. This is based on the topography of the skin surface of slow-moving sharks which remain free of biofouling. This surface can be modified to inhibit the settlement of different organisms; for example a surface with 2 μm features inhibits the settlement of *Ulva* algal spores, while one with 40 μm feature height inhibits the settlement of barnacle cyprids. Unfortunately, this latter surface also encourages the settlement of algal spores. The challenge is to develop a topography that inhibits the settlement of a range of biofouling organisms.

Biofouling demonstrates how little we know about the marine environment, and that technological progress cannot be divorced from wider environmental concerns. When these concerns are taken into consideration it is not at all clear at the moment that biofouling products superior to those used in the past will result: those that do will have to incorporate as much biological as chemical knowledge.

FURTHER READING

Dürr and Thomason (2010) give a good overview of biofouling, and Railkin (2004) and Wahl (1989) introduce the biofouling process. Progress in elucidating the formation and structure of biofilms has been rapid in the last 20 years, rendering earlier accounts inaccurate: O'Toole *et al.* (2000), Stoodley *et al.* (2002) and Qian *et al.* (2007) give useful reviews of this area.

Rainbow (1984), Southward (1987) and Anderson (1994) give good introductions to the barnacle. Khandeparker and Anil (2007) is an up-to-date review of barnacle adhesion, and Winsor (1969) discusses the nineteenth century arguments over the classification of the barnacle.

Woods Hole Oceanographic Institution (1952) has still the best account of the early development of anti-fouling research; Costlow and Tipper (1984) and Almeida *et al.* (2007) bring the story up to date. De Mora (1996) is a useful source on TBT and its effects, and Linley-Adams (1999) discusses the bioaccumulation of TBT. Abel (2000), Abbott *et al.* (2000) and Evans *et al.* (2000) discuss the case for not banning the use of TBT on commercial vessels. The International Maritime Organisation website (www.imo.org) gives much useful information on biofouling as well as the legislation to control it.

Hellio and Yebra (2009) review current anti-fouling technologies, as do Chambers *et al.* (2006), Almeida *et al.* (2007) and Quian *et al.* (2010). Olsen *et al.* (2007) review enzyme-based anti-fouling, Genzer and Efimenko (2006) and Forbes (2008) hydrophobic self-cleaning surfaces, and Schumacher *et al.* (2007) discuss recent work on engineered anti-fouling topographies. The journal *Biofouling* should be consulted for advances in barnacle biology and anti-fouling technology.

Fenton (1978) gives details of the uses of seaweed in some coastal communities, and McHugh (2003) discusses current seaweed exploitation. The barnacle–barnacle goose story is covered by Lankester (1915), Heron-Allen (1929) and Rollo (1998, 2000).

The best way to follow Darwin's progress with his barnacles is through his letters, edited by Burkhardt and Smith (1987–9), although Stott (2003) provides an accessible introduction and useful background to this. Crisp (1983) and Southward (1983) give a useful discussion of the value of Darwin's barnacle work 100 years after his death.

Crothers (1985) gives a good introduction to the dog whelk; Bryan *et al.* (1986) and Gibbs and Bryan (1986) describe imposex in dog whelks. Evans and Nicholson (2000) discuss the use of imposex as a bioindicator and Garaventa *et al.* (2006) give an alternative view. Tyrian purple is reviewed by Baker (1974). Yonge (1960) is still a useful source for the history of oyster cultivation in Europe, and Heral (1990) describes the culture of oysters in France. Alzieu (2000) discusses the effect of TBT on oysters in Arcachon Bay and Dyrynda (1992) in Poole Harbour.

REFERENCES

Abbott, A., Abel, P.D., Arnold, D.W. and Milne, A. (2000) Cost–benefit analysis of the use of TBT: the case for a treatment approach. *The Science of the Total Environment* **258**, 5–19.

Abel, P.D. (2000) TBT: towards a better way to regulate pollutants. *The Science of the Total Environment* **258**, 1–4.

Almeida, E., Diamantino, T.C. and de Sousa, O. (2007) Marine paints: the particular case of antifouling paints. *Progress in Organic Coatings* **59**, 2–20.

Alzieu, C. (2000) Environmental impact of TBT: the French experience. *The Science of the Total Environment* **258**, 99–102.

Anderson, D.T. (1994) *Barnacles: structure, function, development and evolution.* Chapman & Hall, London.

Baker, J.T. (1974) Tyrian purple: an ancient dye, a modern problem. *Endeavour* **33**, 11–17.

Barlow, N. (ed.) (1958) *The Autobiography of Charles Darwin 1809–1882.* Collins, London.

Brown, D.K. (1997) *Warrior to Dreadnought: warship development 1860–1905.* Chatham, London.

Bryan, G.W., Gibbs, P.E., Burt, G.R. and Hummerstone, L.G. (1986) The decline of the gastropod *Nucella lapillus* around south-west England: evidence for the effects of tributyltin from anti-fouling paints. *Journal of the Marine Biological Association* UK **66**, 611–640.

Burkhardt, F.H. and Smith, S. (eds) (1987) *The Correspondence of Charles Darwin. Volume 3: 1844–1846.* Cambridge University Press, Cambridge.

Burkhardt, F.H. and Smith, S. (eds) (1988) *The Correspondence of Charles Darwin. Volume 4: 1847–1850.* Cambridge University Press, Cambridge.

Burkhardt, F.H. and Smith, S. (eds) (1989) *The Correspondence of Charles Darwin. Volume 5: 1851–1855.* Cambridge University Press, Cambridge.

Chambers, L.D., Stokes, K.R., Walsh, F.C. and Wood, R.J.K. (2006) Modern approaches to marine antifouling coatings. *Surface and Coatings Technology* **201**, 3642–3652.

Cole, W. (1685) A letter from Mr William Cole of Bristol, to the Phil. Society of Oxford; containing his observations on the purple fish. *Philosophical Transactions of the Royal Society* **15**, 1278–1286.

Costlow, J.D. and Tipper, R.C. (eds) (1984) *Marine Biodeterioration: an interdisciplinary study.* Spon, London.

Crisp, D.J. (1983) Extending Darwin's investigations on the barnacle life-history. *Biological Journal of the Linnean Society* **20**, 73–83.

Crothers, J.H. (1985) Dog-whelks: an introduction to the biology of *Nucella lapillus* (L.). *Field Studies* **6**, 291–360.

Dickens, C. (1857) *Little Dorrit.* [The Clarendon Dickens, ed. H.P. Sucksmith, Clarendon Press, Oxford, 1979.]

Dürr, S. and Thomason, J.C. (eds) (2010) *Biofouling.* Wiley–Blackwell, Chichester, UK.

Dyrynda, E.A. (1992) Incidence of abnormal shell thickening in the Pacific oyster *Crassostrea gigas* in Poole Harbour (UK), subsequent to the 1987 TBT restrictions. *Marine Pollution Bulletin* **24**, 156–163.

Evans, S.M., Birchenough, A.C. and Brancato, M.S. (2000) The TBT ban: out of the frying pan into the fire? *Marine Pollution Bulletin* **40**, 204–211.

Evans, S.M. and Nicholson, G.J. (2000) The use of imposex to assess tributyltin contamination in coastal waters and open seas. *The Science of the Total Environment* **258**, 73–80.

Fenton, A. (1978) *The Northern Isles: Orkney and Shetland.* John Donald, Edinburgh.

Forbes, P. (2008) Self-cleaning materials. *Scientific American*, August, 68–75.

Garaventa, F., Faimali, M. and Terlizzi, A. (2006) Imposex in pre-pollution times. Is TBT to blame? *Marine Pollution Bulletin* **52**, 696–718.

Genzer, J. and Efimenko, K. (2006) Recent developments in superhydrophobic surfaces and their relevance to marine fouling: a review. *Biofouling* **22**, 339–360.

Gibbs, P.E. and Bryan, G.W. (1986) Reproductive failure in populations of the dog-whelk, *Nucella lapillus*, caused by imposex induced by tributyltin from antifouling paints. *Journal of the Marine Biological Association* UK **66**, 767–777.

Giraldus Cambrensis (Gerald of Wales) (1982) *The History and Topography of Ireland* [transl. J.J. O'Meara]. Penguin, Harmondsworth, UK.

Goldberg, E.D. (1986) TBT an environmental dilemma. *Environment* **28**(8), 17–20, 42–44.

Hellio, C. and Yebra, D. (eds) (2009) *Advances in Marine Anti-Fouling Coatings and Technologies.* Woodhead Publishing, Cambridge.

Heral, M. (1990) Traditional oyster culture in France. In: *Aquaculture, Volume 1* (ed. G. Barnabé), pp. 342–387. Ellis Horwood, Chichester, UK.

Heron-Allen, E. (1929) *Barnacles in Nature and Myth.* Oxford University Press, Oxford.

Khandeparker, L. and Anil, A.C. (2007) Underwater adhesion: the barnacle way. *International Journal of Adhesion and Adhesives* **27**, 165–172.

Lankester, R. (1915) *Diversions of a Naturalist.* Methuen, London.

Latham, T. (1991) *The Ashburner Schooners: the story of the first shipbuilders of Barrow-in-Furness.* Ready Rhino, Manchester.

Linley-Adams, G. (1999) *The Accumulation and Impact of Organotins on Marine Mammals, Seabirds and Fish for Human Consumption.* WWF-UK project no. 98054 report. WWF, UK.

McHugh, D.J. (2003) *A Guide to the Seaweed Industry.* FAO Fisheries Technical Paper no. 441. FAO, Rome, Italy.

De Mora, S.J. (ed.) (1996) *Tributyltin: case study of an environmental contaminant.* Cambridge University Press, Cambridge.

Olsen, S.M., Pedersen, L.T., Laursen, M.H., Kiil, S. and Dam-Johansen, K. (2007) Enzyme-based antifouling coatings: a review. *Biofouling* **23**, 369–383.

O'Toole, G., Kaplan, H.B. and Kolter, R. (2000) Biofilm formation as microbial development. *Annual Review of Microbiology* **54**, 49–79.

Piola, R.F., Dafforn, K.A. and Johnston, E.L. (2009) The influence of antifouling practices on marine invasions. *Biofouling* **25**, 633–644.

Qian, P.-Y., Lau, S.C.K., Dahms, H.-U., Dobretsov, S. and Harder, T. (2007) Marine biofilms as mediators of colonization by marine macroorganisms: implications for antifouling and aquaculture. *Marine Biotechnology* **9**, 399–410.

Qian, P.-Y., Xu, Y. and Fusetani, N. (2010) Natural products as antifouling compounds: recent progress and future perspectives. *Biofouling* **26**, 223–234.

Railkin, A.I. (2004) *Marine Biofouling: colonization processes and defenses.* CRC Press, Boca Raton, FL, USA.

Rainbow, P.S. (1984) An introduction to the biology of British littoral barnacles. *Field Studies* **6**, 1–51.

Rollo, D. (1998) *Historical Fabrication: ethnic fable and French romance in twelfth-century England.* French Forum, Lexington, USA.

Rollo, D. (2000) *Glamorous Sorcery: magic and literacy in the High Middle Ages.* University of Minnesota Press, Minneapolis, MN, USA.

Schumacher, J.F., Aldred, N., Callow, M.E. *et al.* (2007) Species-specific engineered antifouling topographies: correlations between the settlement of algal zoospores and barnacle cyprids. *Biofouling* **23**, 307–317.

Southward, A.J. (1983) A new look at variation in Darwin's species of acorn barnacles. *Biological Journal of the Linnean Society* **20**, 59–72.

Southward, A.J. (ed.) (1987) *Barnacle Biology: crustacean issues 5.* A.A. Balkema, Rotterdam, The Netherlands.

Stoodley, P., Sauer, K., Davies, D.G. and Costerton, J.W. (2002) Biofilms as complex differentiated communities. *Annual Review of Microbiology* **56**, 187–209.

Stott, R. (2003) *Darwin and the Barnacle.* Faber & Faber, London.

Stubbings, H.G. (1975) *Balanus balanoides: LMBC Memoirs on Typical British Marine Plants and Animals 37.* Liverpool University Press, Liverpool.

Wahl, M. (1989) Marine epibiosis. I: Fouling and antifouling: some basic aspects. *Marine Ecology Progress Series* **58**, 175–189.

Williams, S.L. and Smith, J.E. (2007) A global review of the distribution, taxonomy, and impacts of introduced seaweeds. *Annual Review of Ecology, Evolution, and Systematics* **38**, 327–359.

Winsor, M.P. (1969) Barnacle larvae in the nineteenth century, a case study in taxonomic theory. *Journal of the History of Medicine and Allied Sciences* **24**, 294–309.

Woods Hole Oceanographic Institution (1952) *Marine Fouling and its Prevention.* US Naval Institute, Annapolis, Maryland, USA.

Yonge, C.M. (1960) *Oysters.* Collins, London.

BRACKEN OVERWHELMS HEATHER

Bracken

BRACKEN IS A FERN of nearly global distribution, being absent only from Antarctica. It is almost as comprehensively regarded as a pest. Yet this has not always been the case; its rhizome (underground stem) was valued as food by Australian Aborigines, New Zealand Maoris and North American Indians, and its fronds have been used for a variety of purposes in many cultures and communities. Here we explore four faces, or sides, of bracken: it can be conceived as provider, competitor, colonizer and protector. We show how its characteristics (biochemistry, morphology and physiology, and life-cycle) contribute to these different aspects of the plant. Overall, bracken is rarely now exploited, but is instead an aggressive exploiter; from being a useful resource, it has become a serious weed.

BRACKEN AS PROVIDER

> We find John Birkett hath a bracken dalt in Swinside pasture beginning at the steps in Holgill Sike and so from thence under a footgate along to the oak stump near the end of the intake wall.[1]

The landscape historian Angus Winchester has unearthed this statement from the records of the manor court at Braithwaite in Cumbria for 1678. Manor courts managed the way in which common rights over 'waste' land of the manor were allocated. John Birkett was a commoner who was given the right to harvest the bracken 'dalt' (a stand of bracken) whose extent was defined in the court record. Such rights were jealously guarded, and reflect the local importance of bracken in the economy of pre-industrial Britain. Bracken was harvested for use as thatch, fuel and, as we shall see, ash. It was also valued as bedding, for both animals and humans. In the Roman settlement at Vindolanda, close to Hadrian's Wall in Northumberland, the inhabitants appear to have performed all their bodily functions in rooms whose floors were strewn with bracken, an absorbent which was periodically replenished, if not replaced.

In medieval and early modern Britain, bracken ashes were important in the manufacture of soap. Soap was used, amongst other things, in linen production, a process requiring specialist trades still reflected in surnames common today: Weaver, Webster, Bowker (Boucher), Fuller and Dyer are some of them. Another is Ashburner, a name that reflects the preparation of plant ash to provide the key element in soap production, alkali – or 'lye'. Alkali whitens and cleanses wool and cloth of grease and dirt. The English Lake District is a region which according

[1]Winchester (2006), p. 6.

Biological Diversity: Exploiters and Exploited, First Edition.   Paul Hatcher and Nick Battey.
© 2011 John Wiley & Sons, Ltd. Published 2011 by John Wiley & Sons, Ltd.

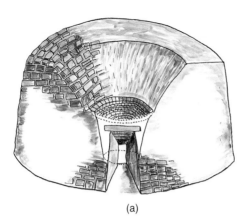

(a)

(b)

Fig. 14.1 **(a)** *Example of pit, deduced by Davies-Shiel to have been used for burning bracken in the English Lake District in the eighteenth century and earlier. A cut-away section is shown [After Davies-Shiel (1972)].*
(b) *Remains of a putative bracken burning pit in 2010, Parrock Wood, Haverthwaite, Cumbria, UK*

to the researches of Davies-Shiel still (in the late twentieth century) had a good number of Ashburners, and was dotted with the remains of 'lyekilns' (Fig. 14.1). The latter are approximately circular stone-built structures with a vent tunnel, and would have held a large pot or cauldron in which bracken was burnt slowly and the ashes collected. James Dunbar's 'Smegmatologia' (1736) provides an early description of this process:

> Take a piece of burning *Peat-Coal*, about the Bigness of your Fist, and lay on it a handful or Two of dry Breckens by Degrees, and you will soon have a Fire; continue thus till you lay on a Cart-Load or Two, but you must take care they do not Flame … and, being put all in a Heap, they will burn themselves for some Days, till all turn white, and be sure they will be turned twice or thrice a Day, from the Bottom to the Top, and they will be whiter; then, when they are cold, sift and barrel them.[2]

Presumably the early Ashburners tended these lyekilns; they would certainly have coveted bracken because of its high potassium content, which makes its ash a rich source of alkali.

Potash can also act as a flux which allows silica to be fused in glass manufacture, and bracken is therefore linked with medieval glass production. This association is again reflected in names, this time of places. In France, 'Verre Fougère' described glass from the region of Fougères, which means bracken/fern; and in the Weald of south-east England, Fernfold near Wisborough was the site of an early glass industry. The importance of local supply of bracken is later shown by that fact that, in 1701, James Montgomery applied successfully to build a glass factory in Glasgow partly because ferns were in plentiful supply in the area.

BRACKEN AS COMPETITOR

One of the observations made by Davies-Shiel, in his study of the Lake District potash industry, is that there is an association of areas infested with bracken today and disused lyekilns. The suggestion is, therefore, that the present distribution of bracken at least partly reflects the decline in its use by humans. This widespread trend has been compounded by other changes that have offered opportunity for bracken invasion. The encroachment of bracken into many upland areas of Britain is one example (Fig. 14.2). It was associated in Scotland with the change from cattle to sheep grazing, and the forced abandonment of arable land during the Highland clearances of the eighteenth and nineteenth centuries. Here reduced

[2]Dunbar, 1736; summary from passage quoted in Rymer (1976), p. 161.

cultivation, and less trampling, both gave opportunity to bracken. In Ireland, bracken is well known as a frequent invader of disused farmland; but one study of islands in the lakes of Connemara suggested an alternative sequence. Apparently, desolation of the native tree cover by colonies of breeding birds (cormorants) was followed by a bracken invasion, which appeared permanently to prevent the re-establishment of trees. In Wales, invasion of common land (land over which commoners have historically had use rights) by bracken is notable and is associated with reduction in cattle grazing and the traditional harvest of bracken. But bracken invasion is a much more general problem; in 1986 Taylor commented that 1–3% of land in Wales was being lost every year to bracken encroachment.

A particularly clear example of the problems created by bracken is shown by its effects on heather moors. These are dominated by ling heather (*Calluna vulgaris*) in combination with the low-growing species *Vaccinium myrtillus* (bilberry). Other species can become prominent according to soil type and climate, including cross-leaved heath (*Erica tetralix*), crowberry (*Empetrum nigrum*), bearberry (*Arctostaphylos uva-ursi*) and hare's-tail cottongrass (*Eriphorum vaginatum*). Such

Fig. 14.2 *Bracken growing near Grudie, Parish of Rogart, Sutherland, Scotland [Courtesy of Chris Madgwick.]*

moors are located most famously in the highlands of Scotland, but are also treasured landscapes of the Welsh uplands, and the North York Moors and Peak District of England.

Lowland heaths, which include gorse (*Ulex* species) and bell heather (*Erica cinerea*) as well as ling heather, used to be very extensive but have been badly hit by development. For example the Dorset heaths declined from an area of 30 000 ha in 1811 to 10 000 ha in 1960 and less than 6000 ha today.

These heather moors and lowland heaths have long been valued as natural environments, but are not really 'natural'. Humans have maintained them by careful burning and managed grazing over the centuries. This keeps in check their tendency to give way to woodland. They are threatened by reclamation (either for agriculture or for building); by fragmentation associated with development; by overgrazing; and by bracken (Fig. 14.3a). Bracken can invade where burning is too severe, so that it kills the heather. Established bracken can survive fire, due to

Fig. 14.3 **(a)** *Bracken and heather.* **(b)** *Bracken invading where conifer plantation has been cleared*

(a)

(b)

(a)

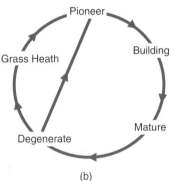

(b)

Fig. 14.4 **(a)** *Bracken as a competitor.* **(b)** *The bracken regeneration cycle as proposed by Watt (1945, 1947) for his study site at Lakenheath Warren in Suffolk, UK [Based on Watt (1945).]*

the presence of the rhizome; and if not already present, it can invade, probably because its spores are well-adapted to colonize fire-damaged sites (see later). Bracken can also invade where heather is overgrazed by sheep and deer. Indeed, because heather establishes from seed, and has a limited life-span (about 40 years), even-aged patches are almost inevitably vulnerable to invasion by a rhizome-based plant like bracken. Maintenance of heather therefore requires continuous management to create a mosaic of plant ages that can withstand bracken and other challenges.

All these examples of the tendency of bracken to take over can be readily understood in the light of its biological properties, which make it both a strong competitor and a successful invader (Fig. 14.3b). In his analysis of plants and their survival strategies, Philip Grime describes three extreme types: the competitor, the stress tolerator, and the ruderal. Bracken is a fine example of a competitor:

> Competitive ability is a function of the area, the activity, and the distribution in space and time of the plant surfaces through which resources are absorbed and as such it depends upon a *combination* of plant characteristics.[3]

Bracken exhibits many of what Grime considers these crucial competitor characteristics (Fig. 14.4a). Its deep, perennating underground stem (rhizome) allows energy for growth early in the season to be rapidly mobilized. This means that bracken fronds emerge and expand quickly and thereby exclude many other plants through shading. A related feature is the size of bracken: its fronds are large (typically around 1 m long, but can be up to 4 m) and quickly create a dense canopy. The rhizome also gives bracken the ability to spread laterally, so that as well as dominating the vertical environment (in the absence of trees) it also controls the horizontal one. Bracken is particularly effective in temperate regions, where it achieves greatest leaf area during mid-summer, when light, temperature and day length allow photosynthesis to be maximized.

A further aspect of the competitive nature of bracken is the large amount of leaf litter it generates. This restricts the germination and establishment of seeds of other species, including trees. Some studies, particularly the detailed work of A.S. Watt at Lakenheath Warren in Suffolk, suggest that bracken's deep litter may, however, contribute to its eventual decline. The rhizome begins to grow in bracken litter, where it can be damaged by frost and drought. Other species can then have success; where these include trees, bracken may eventually give way to woodland, but this will be less likely if grazing is significant. Grazing may thus contribute to a cyclical recurrence of bracken infestation (Fig. 14.4b), emphasizing a further aspect of bracken's competitive ability: it is itself little troubled by grazing, and is not seriously damaged by insects or other pathogens. The toxic properties of bracken and the causes of its robustness are considered in Box 14.1.

[3]Grime (1979) *Plant Strategies and Vegetation Processes*, 1st edn, p. 9.

| BOX 14.1 | THE TOXIC PROPERTIES OF BRACKEN; ALLELOPATHY |

Bracken produces a wide range of secondary compounds that are presumed to deter herbivores and pathogens. Phytoecdysteroids (insect moulting hormones) have been found to be active when injected into locusts; phenolic compounds, tannins, lignins and silicates may provide more general protection. Cyanogenesis (the production of cyanide on tissue damage) is a variable, polymorphic feature in bracken. It is more common in young fronds than old, and depresses larval growth in bracken-associated insects. Cyanogenic fronds also deter sheep and deer grazing. This information indicates that bracken is well adapted to minimize herbivory, something which is also suggested by bracken having, for a plant so widespread, a very depauperate herbivorous insect fauna of only 27 species. The lack of significant damage caused by these insects on bracken may be because of their natural enemies, which keep them rare relative to the amount of bracken. This hypothesis underpinned the suggested introduction of alien insects to control bracken (see Box 14.2).

Bracken is toxic to many mammals. An indirect effect arises because it harbours the sheep tick, *Ixodes ricinus*, which can act as vector for Louping ill in sheep, and Lyme disease in humans. More directly, bracken contains a thiaminase (a vitamin B_1-degrading enzyme) which causes thiamine deficiency in single-stomached animals like horses and pigs. Bright blindness in sheep is caused by degeneration of the retina and is associated with bracken consumption, as is haemorrhagic disease in cattle and sheep. Bracken also contains ptaquiloside, a carcinogenic and mutagenic norsesquiterpene. Cancer is associated with prolonged

consumption of bracken in cattle and sheep. Ptaquiloside can be transferred to humans though milk, indicating that the grazing patterns of cattle near bracken need to be carefully monitored. Spores have been reported to be carcinogenic in mice, but the effects of airborne spores on animals have not been studied. Where bracken is eaten regularly, as it is traditionally in Japan where the croziers are considered a delicacy, an associated increased risk of oesophageal cancer has been found. In general there are reasons to be wary of bracken, and its consumption is probably not a good idea. But as Wilson *et al.* (1998) conclude in their review of the topic: 'Bracken may be a hazard to humans, but, as yet, there is no convincing evidence that people are at serious risk from it.' This statement draws usefully on the distinction between hazard (circumstances that may have harmful consequences) and risk (the probability of harmful consequences occurring from a hazard). It also emphasizes the difficulty of separating out real effects where there is a large number of potential risk factors.

Finally, bracken fronds, litter, rhizomes and roots can release potential toxins to the soil, leading to the suggestion that it inhibits the growth of other (competitor) plants by allelopathy. Thus, leaching by heavy rainfall from green fronds of bracken growing in tropical Costa Rica was suggested to have an allelopathic effect on other plants. However, the variable experimental results obtained when this topic is surveyed more extensively indicate that an allelopathic effect of bracken cannot be regarded as definitely established (see Marrs and Watt, 2006).

Most striking of all is the generally agreed view of the basic ecological place of bracken. This is as a woodland plant. Atkinson (1989) describes it as 'primarily a non-aggressive understorey species of open-canopied woodland'. Yet its response to removal of woodland is to become a highly aggressive competitor. What is behind this ability to respond to deforestation (typically as a result of human activities) in this way? One crucial feature seems to be bracken's photosynthetic capacity: its maximum photosynthetic rate (i.e. in unshaded environments) greatly exceeds that of other ferns, and it can respond to shade removal by strong growth. Further, it is able to modify the morphology of its frond and the associated structure of its stands according to environment, reflecting a developmental plasticity which may be generally important to its success. A further critical feature is the rhizome: for upland areas, a study on Ilkley Moor in Yorkshire concluded that, while changing land use associated with woodland clearance offered the potential for bracken dominance, to take this opportunity it had to be able effectively to mine subsoil moisture and possibly nutrients. The rhizome is key to this ability. The overall conclusion is that a combination of characteristics – physiological, morphological and biochemical – combine in bracken to enable it to exploit the opportunities created by woodland clearance, especially when this is followed by reduced agricultural use, and to come to dominate in a range of environments.

<hr>

| BOX 14.2 | CONTROL OF BRACKEN |

In the UK the current agricultural and conservation policy in relation to bracken is to reduce infestation; in addition, restoration of *Calluna*-dominated heathlands is a common objective. It is difficult to eradicate bracken, and persistent control over many years is required; this is both costly and time-consuming. Where heather restoration is not the objective, the ultimate conversion of bracken-infested land to woodland is therefore a good long-term strategy, and one that can attract grant support.

Initial control can be achieved by mechanical methods (ploughing in, cutting, rolling/bruising, or burning), or by herbicide application. Of the mechanical methods, a recent (Stewart *et al.*, 2008) country-wide analysis over a 10-year period indicated that culling twice per year was most effective. The first cut needs to be done during early summer (June), the aim being to harvest fronds in such a way as to drain to a maximum rhizome reserves. The second cut was done in August in the above study. Herbicide control using asulam [methyl (4-aminobenzenesulphonyl) carbamate] is also effective; it can be applied by air and the optimum time is when the fronds are fully expanded (mid-July to late-September). Asulam is translocated to the rhizome, where it kills buds and can reduce frond production by 95% the following season. However, repeat applications are essential to maintain suppression of the bracken. It should be noted that asulam is preferable to its alternative, glyphosate, because it is selective; but it can, nevertheless, damage other ferns and also some bryophytes. Herbicide application can be used in combination with cutting.

After control it is essential to manage vegetation succession. This requires removal of bracken litter, and addition of propagules of the required speces – for example, heather. Grazing must then be minimized. Importantly, the large-scale analysis by Stewart *et al.* (2008) demonstrated an increase in plant species richness following bracken control.

There was a strong and persistent push towards development and implementation of biological control for bracken in the 1980s. The proposal was to introduce from South Africa two species of moth (*Conservula cinisigma* and *Panotima* sp. near *angularis*) whose caterpillars feed on bracken fronds. The potential advantages were that herbicide use could be reduced, and cost would be low. However, disadvantages were that effects would be irreversible and uncontrollable. The uncertainty, and the potential for indiscriminate elimination of bracken, meant that the idea was not adopted.

So what is the extent of the current bracken problem? For the UK, Pakeman, Marrs and colleagues in 2000 indicated, first, that the present extent of bracken is probably not exceptional in terms of vegetation patterns since the last glaciation. Second, their analysis of data from the Countryside Survey in 1990 revealed that bracken is concentrated in lightly grazed areas at altitudes of 200–400 m on the west of Britain, where it not economic or practical to cultivate the land or to graze it more intensively. Additionally, quite large areas contain non-dominant bracken that could become a problem with further changes in land use, or climate. Increased temperature, a longer growing season, and higher CO_2 levels could all conspire to make bracken more invasive in northern parts of Britain, where frost, waterlogging and the length of the growing season variously act to limit its development at present. Probably the most important contribution of bracken research has been to provide an understanding of how practically to control bracken (Box 14.2). This means that where funding and time are available, effective steps can be taken to limit its spread.

BRACKEN AS COLONIZER

We have seen that bracken can be a highly competitive plant, shading out other species and creating dominant stands that persist and spread through the action of the rhizome system. But another aspect of bracken is its spore and the tiny, weak and vulnerable plant that grows from it. This is the gametophyte phase, and leads to sexual reproduction which reinstates the dominant life phase, the sporophyte. The sporophyte is the bracken plant we all recognize; the gametophyte, most of us have probably never seen. The life-cycle of bracken is summarized in Fig. 14.5, and it is discussed in relation to the life-cycles of the other major plant groups in Box 14.3.

The spore is the long-distance dispersal agent of bracken, and spore production in bracken is immense: up to 300 million can be produced by a single frond, according to one estimate. These spores are small (about 30 μm across) and very light, so they move long distances on air currents. This means that bracken, along with other ferns, is an early colonizer of new terrain, particularly where this has been sterilized by extreme heat or fire. When the volcano on Krakatoa blew itself to bits in 1883, bracken was one of the earliest colonizers of the cooled lava on the remains of the island. It established within three years, presumably from spores borne on the wind from the nearest land mass (Java, about 40 km away), or from other still more distant places. Similarly, in the rubble of bombed-out London in 1944, J.E. Lousley reported that bracken was an early presence, again establishing from spores. Its apparent affinity for lime-rich mortar is an interesting conundrum, since bracken is generally considered to be an acid-loving plant. It has been suggested that the spore is adapted to germinate in a wide range of pH conditions, including the relatively alkaline ashes that result from fire; and that the sporophyte becomes more acid-loving with age.

The gametophyte phase lasts only a few weeks and even today its general importance in establishing bracken stands (relative to vegetative expansion of the sporophyte) is not well understood. It is regarded as the long-distance dispersal phase because there is very little evidence that gametophytes establish in pre-existing vegetation. One expectation, however, is that because the gametophyte is very small, being only a few millimetres across, and the spermatozoids rely on a water film to swim to the archegonia and have no other means of dispersal, it will self-fertilize. This would mean that the resultant sporophyte becomes homozygous at all loci. Yet incompatability mechanisms are known to operate, and bracken is demonstrably heterozygous. Clearly gametophytes do cross-fertilize, presumably because the sperm can swim a very long way, across soil and other detritus, between little green gametophyte islands, seeking out the attractant chemical which archegonia produce to guide them.

The association of bracken colonization with fire is so strong that the ages of bracken stands in Finland have been estimated by Eino Oinonen (based essentially on a diameter–age relationship), and correlated with known battle sites (Fig. 14.6). Some of these stands are very extensive; one with a diameter of 217 m near Sulkava is suggested to have originated following a battle there around 1330. Even larger stands exist (almost 500 m diameter) which would be expected to be around 1450 years old. Although this is very interesting and

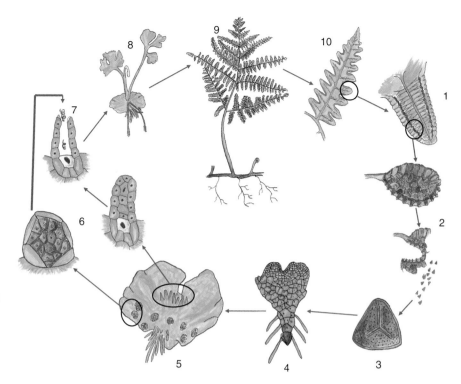

Fig. 14.5 *Life-cycle of bracken.* **1**, *lower surface of fertile frond showing the continuous marginal sorus producing sporangia;* **2**, *mature sporangium (above) and after release of spores (below);* **3**, *spore;* **4**, *germinated spore showing prothallus of gametophyte phase, with rhizoids;* **5**, *prothallus with antheridia and archegonia (these are usually produced sequentially, antheridia first);* **6**, *archegonium containing egg cell (right); antheridium containing developing spermatozoids (left);* **7**, *spermatozoids released by antheridium move towards egg cell;* **8**, *sporophyte developing on gametophyte prothallus;* **9**, *frond of sporophyte;* **10**, *frond becomes fertile when sporophyte is about 3–4 years old* [Based on an original by Rumsey and Sheffield, in Thomson (1990).]

BOX 14.3	ALTERNATION OF GENERATIONS IN FERNS AND THE OTHER MAJOR PLANT GROUPS

In flowering plants (angiosperms), the products of meiosis (the spermatozoid and the ovule) are generated within the flower; and the pollen, carrying the sperm, is transmitted by wind or insects to the female, ovule-bearing part of the flower. In the angiosperms the gametophyte phase is highly reduced (typically to three cells in the male gametophyte, and to seven cells in the female gametophyte), and is completely dependent on the sporophyte. In contrast, meiosis in bracken produces the spore, which though haploid is neither male nor female (but is potentially both) (see Fig. 14.5). The spore germinates to create the gametophyte plant which is free-living, and eventually forms on its surface antheridia and archegonia – the sperm and ovule-producing cells. Fusion of the sperm and egg then creates the zygote, which is diploid and grows into the bracken sporophyte. Initially the sporophyte is dependent on the gametophyte for its survival. This cycle between haploid and diploid phases is known as alternation of generations.

The relative importance of the gametophyte phase compared with the sporophyte phase varies systematically in the different plant groups. Thus, in many of the algae (note that some taxonomists now classify these in the Kingdom Protista, rather than the Kingdom Plantae), alternation of generations is isomorphic: the diploid and haploid phases look the same. In mosses, liverworts and hornworts, the haploid (gametophyte) phase is dominant; the fruiting body is the sporophyte and is dependent on the gametophyte for its survival. In ferns and horsetails, on the other hand, the sporophyte is dominant but the haploid gametophyte retains its independence. In higher plants (angiosperms and gymnosperms) the gametophyte has lost independence and is reduced. Thus in groups other than the algae, alternation of generations is heteromorphic. There is also a clear evolutionary trend associated with the increasing dominance of the diploid phase, presumably because of its ability to include more genetic information during the course of development.

suggestive research, it should be noted that bracken might have been present before the fire events. It is also not clear that the stands measured were single clones. However, studies elsewhere show that clones of bracken arising from a single colonization event can be very ancient: in the Appalachian mountains of Virginia, some clones were up to 1015 m across and had a minimum estimated age of 1180 years. This makes them very old individuals; the analogy has been made between a bracken clone, sprouting leaves from its underground rhizome, and a tree laid on its side. In contrast to a tree, though, bracken has not been forced by natural selection to develop a hefty woody structure to support a height struggle against shade from other plants; rather, it has evolved to branch laterally and very extensively underground, throwing up leaves to harvest light and CO_2 to fund its underground explorations. Probably the most typical situation is for bracken stands to consist of a number of individuals, each of which is quite extensive.

The key to the transformation from the tiny gametophyte to the invasive monster that is the sporophyte occurs early in the development of the latter. After making about ten leaves, the sporophyte shoot apex bifurcates, generating two shoots that turn away from the light, burying themselves in the ground. Here they branch extensively, throwing out primordia which push to the surface to

Fig. 14.6 *Correlation of sizes/ages of bracken stands with historical sites of battles (and presumed fires) in Finland* [*Redrawn from Oinonen (1967b).*]

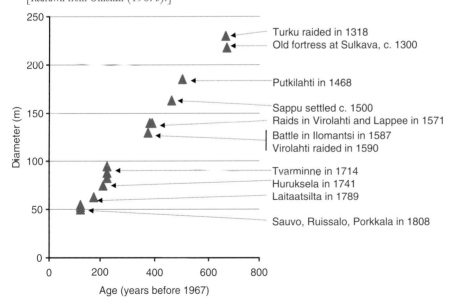

- Turku raided in 1318
- Old fortress at Sulkava, c. 1300
- Putkilahti in 1468
- Sappu settled c. 1500
- Raids in Virolahti and Lappee in 1571
- Battle in Ilomantsi in 1587
- Virolahti raided in 1590
- Tvarminne in 1714
- Huruksela in 1741
- Laitaatsilta in 1789
- Sauvo, Ruissalo, Porkkala in 1808

Diameter (m) vs Age (years before 1967)

emerge as croziers that unfurl into the bracken frond. The rhizome can plunge to depths of 2.5 m, or remain very superficial, for example in waterlogged soils. Most impressive is the extent of the rhizome: for every frond there may be as much as 3 m of rhizome below ground. As the rhizome grows the older sections die back behind it, so that what was originally a single, interconnected plant becomes a number of separated, genetically identical clones. It can remain dormant for many years before producing new fronds, so that eradication of bracken requires repeated destruction of fronds until the rhizome below is so weakened that it dies away (see Box 14.2).

As a result of its successful colonization, a single bracken spore may therefore create, via the gametophyte, a community of clones over a vast area. The ecologist A.S. Watt provided a description of the bracken community from Lakenheath, Suffolk, which he found to be strikingly structured. At the front, pioneering into the region to be invaded, were shorter fronds, arising from the rhizome which itself extended well ahead of the first fronds. Behind this vanguard was the region of tallest fronds; these were followed by an extensive zone which represented the mature part of the bracken community. Behind the mature zone came the 'hinterland', a patchwork of variously sized fronds. Here succession to other vegetation types could occur, or new pioneering bracken might once more regenerate itself.

BRACKEN AS PROTECTOR

In the entry for *Hyacinthoides non-scripta* (bluebell) (Fig. 14.7) in Clapham, Tutin and Warburg's *Excursion Flora of the British Isles*, the description of the plant is followed by:

> Common in woods, hedge-banks, etc., rarely in pastures; throughout the British Is, except Orkney and Shetland, and often dominant in coppiced woods on light soils.

Blackman and Rutter (1950) wanted to understand the 'rarely in pastures' part of this summary: that is, the situations where bluebells *are* found in association with grasslands, which in many cases include bracken. The explanation usually offered, that such bluebells occur where there have in the past been woodlands, seemed at variance with instances of coastal cliff communities of bluebells, grass and bracken, where trees are not likely to have existed in the past. Blackman and Rutter noted that the general absence of bluebells from open ground was associated with the occurrence of grazing and trampling by animals; and their experiments indicated that, along with a requirement for good light levels between March and June, a lack of animal grazing was indeed a key factor for their survival.

In this situation, the presence of bracken is compatible with bluebells because the fronds do not cause heavy shading until after the early spring growth of bluebells. Bracken,

Fig. 14.7 **(a)** *Bluebells (Hyacinthoides non-scripta) carpeting a woodland floor in spring, Oxfordshire, UK.* **(b)** *Bluebells and bracken in early summer, Dartmoor, Devon, UK [Courtesy of Cathy Newell Price.]*

(a)

(b)

nevertheless, excludes other plant species (such as tall grasses like oat-grass, *Arrhenatherum elatius*) which would tend to out-compete bluebell; and, particularly, bracken excludes grazing animals. According to this work, the explanation for the typical association of bluebells with woodlands is that grazing animals are less dominant there. The obvious idea, that bluebells thrive in shade, is not true: they respond to light like most other plants, and survive in woodlands through early growth before the trees leaf-out later in the spring. Bracken acts therefore as surrogate woodland, and protector of the bluebell. The bluebell, primrose and lesser celandine communities on the sea cliffs of Cornwall and south Devon can similarly benefit from bracken protection.

This positive dimension to bracken applies to a few other species, too. In the 'ffriddland' of Wales, which lies between the open moor and the fertile valley bottom, and on the North York Moors, bracken provides cover to the whinchat (*Saxicola rubetra*), a bird now rare in lowland Britain. Bracken is also important as a surrogate woodland offering cover to understorey plants such as violets which are food for two threatened butterfly species – the high brown fritillary (*Argynnis adippe*), and the heath fritillary (*Mellicta athalia*). This only remains true so long as the bracken is not allowed to become too dense, and careful management is therefore required. Lastly, the vibrant orange-browns of bracken create a beautiful spectacle on the hills and valley sides during autumn, giving pleasure to many countryside users. All this is not to say that bracken is not a problem; in fact, it has an overall negative impact on biodiversity. But it is one of a large group of plants and animals which, usually through the actions of humans, find themselves in a new environment and there become a nuisance. Understanding the dynamics of these ecological situations is the central problem of the modern field of invasion biology.

THE BIGGER PICTURE: BRACKEN, A NATIVE INVASIVE

Bracken is a leptosporangiate fern, a group described as modern ferns, to contrast them with the evolutionarily ancient ferns that dominated the Earth during the Carboniferous period (300 mya). Bracken therefore evolved alongside the angiosperms, and fossil evidence of it can be found from the Tertiary period. It probably achieved a wide distribution before the separation of the continents and associated barriers to species movement (see later). It is perhaps a surprise that it has not subsequently differentiated into a wider range of morphological types. It may be, however, that the relative similarity in appearance of bracken the world over is deceptive, since distinctions can be made based on DNA and isozyme analysis which are not obvious morphologically. There is a northern hemisphere species, *Pteridium aquilinum*, which includes 11 subspecies; and a mainly southern hemisphere species *Pteridium esculentum*, which has two subspecies. Other generalists of wide distribution (like bees and wolves, for example; see Chapters 4 and 19) have a large number of subspecies that differ strikingly from each other.

So it can generally be said that bracken is (or appears to be) 'native' wherever it is found. And it invades where humans have created opportunities: by abandoning land, by ceasing to harvest bracken following enclosure in England, and following the Highland clearances in Scotland. Elsewhere in the world the key role of humans is equally pronounced; we have seen how fires associated with battle allowed new bracken stands to establish in Finland; and from the Yucatán peninsula in Mexico to the forests of Normandy in France, disturbance and interference create opportunities for bracken invasions. Yet bracken is just one of many invasive species. The characterization of the behaviour of invading populations, and their impact on other species, has formed the backbone of invasion biology.

In 1833, Charles Darwin provided the first account of the impact of invasive plants on the landscape. He was struck by the invasion of European species during his journey on *HMS Beagle*, as he passed along the coast of South America, opting for a land-based journey wherever possible owing to persistent seasickness. Here he is describing the cardoon thistle in present-day Uruguay:

> Near the Guardia we find the southern limit of two European plants, now become extraordinarily common. The fennel in great profusion covers the ditch-banks in the neighbourhood of Buenos Ayres, Monte Video, and other towns. But the cardoon (*Cynara cardunculus*) has a far wider range: it occurs in these latitudes on both sides of the Cordillera, across the continent. I saw it in unfrequented spots in Chile, Entre Rios, and Banda Oriental. In the latter country alone, very many (probably several hundred) square miles are covered by one mass of these prickly plants, and are impenetrable by man or beast. Over the undulating plains, where these great beds occur, nothing else can now live. Before their introduction, however, the surface must have supported, as in other parts, a rank herbage. I doubt whether any case is on record of an invasion on so grand a scale of one plant over the aborigines. As I have already said, I nowhere saw the cardoon south of the Salado; but it is probable that in proportion as that country becomes inhabited, the cardoon will extend its limits.[4]

The points Darwin brings out are the displacement of other species by the introduced cardoon thistle, and the dependence on human habitation for its spread. With the increase in global traffic over the next century the examples multiplied, but it was not until 1958 that they were gathered together in Charles Elton's *The Ecology of Invasions by Animals and Plants*.

Elton set a particular tone for invasion biology, using language derived from the warlike activities of humans. He began with case histories of seven invasions (from African mosquitoes in Brazil, to Chinese mitten crab in Europe), emphasizing how humans have, world-wide, broken down in a few hundred years the barriers to species movement established around 50 mya, during the Tertiary period, as the continents separated and impassable barriers such as the Himalayas were created. It is a brilliant read, passionate and dynamic, with foreign species bombarding native populations, and ecological explosions tyrannizing the world. It is a wartime book written in the shadow of the nuclear threat.

Fifty years on, and invasion biology has become a massive research field. Richard Mack and co-workers have done incisive work on temperate grasslands, analogous to those described by Darwin, and the characteristics that make some of them vulnerable to invasions associated with humans. Mark Williamson, in 1996, summarized the work of a generation on patterns of invasion. The key features of invasive plants have been identified; these often include the presence of a rhizome, and a prolific seed/spore production rate, characteristics we have seen to be important to the success of bracken. In 2009, Mark Davis produced an insightful book, *Invasion Biology*, which emphasizes the anachronisms of invasion biology and the need

[4]Darwin (1959), p. 113.

Fig. 14.8 *Origin of the name 'eagle-fern' (*Pteridium aquilinum*) for bracken? The double-headed eagle (right) is suggested by the pattern of vasculature in transverse section; an alternative orientation suggests an oak tree [Based on Thomson (2004).]*

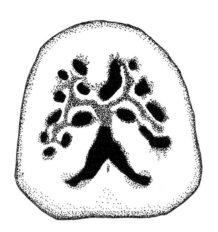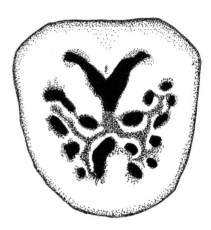

to integrate with the rest of ecology. One issue is the 'alien' (or foreign) character of many invasive species. A familiar example in Britain is the Himalayan Balsam (*Impatiens glandulifera*), which has attracted attention as a problem invasive. But is it really more of a problem than our familiar, native, smothering ivy? This brings us back to bracken: the crucial distinction is not whether any organism is 'native' or 'alien', but whether it is invasive, what its impact is on the biodiversity characteristic of a region, and what can be done about it. Bracken is a native wherever it is found, but the problems arising from its invasive behaviour, and its ability to exploit the areas abandoned by humans, are just as real as if it were an alien.

POSTSCRIPT: THE FIFTH FACE OF BRACKEN

The specific Latin name of European bracken, *aquilinum* (eagle), was given by Linnaeus and may relate to the pattern of vascular bundles visible in sections of the bracken rhizome and stipe. Medieval scholars and herbalists considered that this pattern symbolized a double-headed eagle or, when inverted, an oak tree (Fig. 14.8). Bower (1928) did not detect such patterns, but ruminated on the complexity of the vasculature of bracken, considering that it indicated great evolutionary advancement, in contrast to other features of the plant which he considered primitive. Just as genome-based analysis reveals diversity among brackens that are apparently similar morphologically, the vasculature may turn out to be not only complex but also variable, reflecting the plant's true evolutionary diversity.

FURTHER READING

The most comprehensive, up-to-date and easily available source of information on bracken is Marrs and Watt (2006). For all aspects of bracken management the Marrs Applied Vegetation Dynamics website (www.appliedvegetationdynamics. co.uk) is invaluable. For historical aspects of bracken use, consult Winchester (2006), Rymer (1976), Davies-Shiel (1972, 1974), and Smedley and Jackson (2002). For successive discussions of bracken expansion, see Taylor (1986) and Pakeman *et al.* (1996, 2000). Heathland ecology and the management of bracken are discussed in Gimingham (1972), Sutherland and Hill (1995) and Thompson *et al.* (1995).

Grime (2001) discusses the features of bracken that make it a competitor, and Harper (1977) also has useful insights. Its high photosynthetic capacity relative to other ferns is discussed by Caporn *et al.* (1999). The natural history of bracken as a colonizer is discussed by Page (1976, 1982), and Wolf *et al.* (1987) provides a starting point to gametophyte/sporophyte genetics. The study of Finnish fire-sites and bracken stands was carried out by Oinonen (1967a, 1967b), and studies of stand genetic composition and age by Parks and Werth (1993). Early development of the sporophyte and its morphology are described by Dasanayake (1960) and discussed by Thomson (1990). Watt (1947) gives an influential view of the community structure of bracken. The bluebell/bracken analysis was by Blackman and Rutter (1950) and more on the maritime community can be found in Malloch (1971) and Rodwell (2000). The conservation value of bracken in the context of its status as a problem species is discussed by Pakeman and Marrs (1992).

Invasion biology in general is discussed by Elton (1958), Williamson (1996) and Davis (2009); Mack (1989) provides an overview of the invasion of temperate grasslands. Kendle and Rose (2000) provide a useful review of the issues around the categories 'native' and 'alien'. Bracken toxins are discussed in Cooper-Driver (1990) and Wilson *et al.* (1998), and additional detail on bracken control can be found in Stewart *et al.* (2008) and Lawton (1988). For alternation of generations, see Raven *et al.* (2005).

REFERENCES

Atkinson, T.P. (1989) Seasonal and altitudinal variation in *Pteridium aquilinum* (L.) Kuhn: frond and stand types. *New Phytologist* **113**, 359–365.

Blackman, G.E. and Rutter, A.J. (1950) Physiological and ecological studies in the analysis of plant environment. V: An assessment of the factors controlling the distribution of the bluebell (*Scilla non-scripta*) in different communities. *Annals of Botany* **14**, 487–520.

Bower, F.O. (1928) *The Ferns (Filicales). Volume III: The Leptosporangiate Ferns.* Cambridge University Press, Cambridge.

Cooper-Driver, G.A. (1990) Defense strategies in bracken, *Pteridium aquilinum* (L.) Kuhn. *Annals of the Missouri Botanical Garden* **77**, 281–286.

Caporn, S.J.M., Brooks, A.L., Press, M.C. and Lee, J.A. (1999). Effects of long-term exposure to elevated CO_2 and increased nutrient supply on bracken (*Pteridium aquilinum*). *Functional Ecology* **13** (Supplement 1), 107–115.

Darwin, C. (1959) *The Voyage of the Beagle.* J.M. Dent & Sons, London.

Dasanayake, M.D. (1960) Aspects of morphogenesis in a dorsiventral fern, *Pteridium aquilinum* (L.) Kuhn. *Annals of Botany* **24**, 317–329.

Davies-Shiel, M. (1972) A little known late medieval industry. I: The making of potash for soap in Lakeland. *Transactions of the Cumberland and Westmorland Antiquarian and Archaeological Society* **72**, 85–111.

Davies-Shiel, M. (1974) A little known late medieval industry. II: The ash burners. *Transactions of the Cumberland and Westmorland Antiquarian and Archaeological Society* **74**, 33–64.

Davis, M. (2009) *Invasion Biology.* Oxford University Press, Oxford.

Elton, C.S. (1958) *The Ecology of Invasions by Animals and Plants.* Methuen, London.

Gimingham, C.H. (1972) *Ecology of Heathlands.* Chapman & Hall, London.

Grime, J.P. (2001). *Plant Strategies and Vegetation Processes*, 2nd edn. John Wiley & Sons, Chichester, UK.

Harper, J.L. (1977) *Population Biology of Plants*. Academic Press, London.

Kendle, A.D. and Rose, J.E. (2000) The aliens have landed! What are the justifications for 'native only' policies in landscape plantings? *Landscape and Urban Planning* **47**, 19–31.

Lawton, J.H. (1988). Biological control of bracken in Britain: constraints and opportunities. *Philosophical Transactions of the Royal Society of London B* **318**, 335–355.

Mack, R.N. (1989) Temperate grasslands vulnerable to plant invasions: characteristics and consequences. In: *Biological Invasions: a global perspective* (eds F. di Castri, R.H. Groves, F.J. Kruger *et al.*), pp. 155–179. John Wiley & Sons, New York.

Malloch, A.J.C. (1971) Vegetation of the maritime cliff-tops of the Lizard and Land's End peninsulas, West Cornwall. *New Phytologist* **70**, 1155–1197.

Marrs, R.H. and Watt, A.S. (2006) Biological flora of the British Isles: *Pteridium aquilinum* (L.) Kuhn. *Journal of Ecology* **94**, 1272–1321.

Oinonen, E. (1967a) Sporal regeneration of bracken *Pteridium aquilinum* (L.) Kuhn in Finland in the light of the dimensions and the age of its clones. *Acta Forestalia Fennica* **83**(1), 1–96.

Oinonen, E. (1967b) The correlation between the size of Finnish bracken (*Pteridium aquilinum* (L.) Kuhn) clones and certain periods of site history. *Acta Forestalia Fennica* **83**(2), 1–51.

Page, C.N. (1976) The taxonomy and phytogeography of bracken. *Botanical Journal of the Linnean Society* **73**, 1–34.

Page, C.N. (1982). The history and spread of bracken in Britain. *Proceedings of the Royal Society of Edinburgh* **81B**, 3–10.

Pakeman, R.J. and Marrs, R.H. (1992) The conservation value of bracken *Pteridium aquilinum* (L.) Kuhn-dominated communities in the UK, and an assessment of the ecological impact of bracken expansion or its removal. *Biological Conservation* **62**, 101–114.

Pakeman, R.J., Marrs, R.H., Howard, D.C., Barr, C.J. and Fuller, R.M. (1996) The bracken problem in Great Britain: its present extent and future changes. *Applied Geography* **16**, 65–86.

Pakeman, R.J., Le Duc, M.G. and Marrs, R.H. (2000) Bracken distribution in Great Britain: strategies for its control and the sustainable management of marginal land. *Annals of Botany* **85** (Supplement B): 37–46.

Parks, J.C. and Werth, C.R. (1993) A study of spatial features of clones in a population of bracken fern, *Pteridium aquilinum* (Dennstaedtiaceae). *American Journal of Botany* **80**, 537–544.

Raven, P.H., Evert, R.F. and Eichhorn, S.E. (2005) *Biology of Plants*. W.H. Freeman, New York.

Rodwell, J.S. (2000) *British Plant Communities. Volume 5: Maritime Communities and Vegetation of Open Habitats*. Cambridge University Press, Cambridge.

Rymer, L. (1976) The history and ethnobotany of bracken. *Botanical Journal of the Linnean Society* **73**, 151–176.

Smedley, J.W. and Jackson, C.M. (2002) Medieval and post-medieval glass technology: a review of bracken in glassmaking. *Glass Technology* **43**, 221–224.

Stewart, G., Cox, E., Le Duc, M., Pakeman, R., Pullin, A.W. and Marrs, R. (2008) Control of *Pteridium aquilinum*: meta-analysis of a multi-site study in the UK. *Annals of Botany* **101**, 957–970.

Sutherland, W.J. and Hill, D.A. (eds) (1995) *Managing Habitats for Conservation*. Cambridge University Press, Cambridge.

Taylor, J.A. (1986) The bracken problem: a local hazard and global issue. In: *Bracken: ecology, land use and control technology* (eds R.T. Smith and J.A. Taylor), pp. 21–42. Parthenon Publishing, Carnforth, UK.

Thompson, D.B.A., Hester, A.J. and Usher, M.B. (eds) (1995) *Heaths and Moorland: cultural landscapes*. HMSO, Edinburgh.

Thomson, J.A. (1990) Bracken morphology and life cycle: preferred terminology. In: *Bracken Biology and Management* (eds J.A. Thomson and R.T. Smith), pp. 333–339. Australian Institute of Agricultural Science Occasional Publication no. 40. AIAS, Sydney, Australia.

Thomson, J.A. (2004) Towards a taxonomic revision of Pteridium (Dennstaedtiaceae). *Telopea* **10**, 793–803.

Watt, A.S. (1945) Contributions to the ecology of bracken (Pteridium aquilinum). III: Frond types and the make-up of the population. *New Phytologist* **44**, 156–178.

Watt, A.S. (1947) Pattern and process in the plant community. *Journal of Ecology* **35**, 1–22.

Williamson, M. (1996) *Biological Invasions*. Chapman & Hall, London.

Wilson, D., Donaldson, L.J. and Sepai, O. (1998) Should we be frightened of bracken? A review of the evidence. *Journal of Epidemiology and Community Health* **52**, 812–817.

Winchester, A. (2006) Village byelaws and the management of a contested common resource: bracken (Pteridium aquilinum) in highland Britain, 1500–1800. IASCP Europe Regional Meeting, Brescia, Italy. Available at www.lancs.ac.uk/fass/history/profiles/Angus-Winchester/.

Wolf, P.G., Haufler, C.H. and Sheffield, E. (1987) Electrophoretic evidence for genetic diploidy in the bracken fern (Pteridium aquilinum). *Science* **236**, 947–949.

JEKYLL & HYDE
LOCUST BEHAVIOUR

SOLITARIOUS

GREGARIOUS

The Locust

CHAPTER 15

When all the birds are faint with the hot sun
And hide in cooling trees, a voice will run
From hedge to hedge about the new-mown mead—
That is the grasshopper's. He takes the lead
In summer luxury; he has never done
With his delights, for when tired out with fun
He rests at ease beneath some pleasant weed.

From On the Grasshopper and Cricket, John Keats[1]

For a nation is come up upon my land, strong, and without number, whose teeth are the
teeth of a lion, and he hath the cheek teeth of a great lion.
He hath laid my vine waste, and barked my fig tree: he hath made it clean bare, and cast
it away; the branches thereof are made white.

Joel 1:6–7.[2]

W HAT LINKS THE peaceful chirping grasshopper of Keats with the ravenous locust plague in the Old Testament book of Joel? In this chapter we will find out, and discuss the locust's life-cycle, what features enable it to become such a serious pest, and the continuing development of control methods.

WHAT ARE LOCUSTS?

Locusts and grasshoppers are both members of the insect order Orthoptera, part of a disparate group of insects (Polyneoptera) that also includes earwigs, cockroaches and stick insects. These insects have the ability to flex their wings over their abdomen while at rest, thus enabling them to be stored and also to protect the abdomen. This is in contrast to the dragonflies which cannot do this. The Orthoptera, with about 22 500 described species, is by far the most diverse polyneopteran order and probably arose about 320–310 mya. The key characteristics of this order, and which are well illustrated in the locust, are the development of a shield (the pronotum) that extends over the whole of the thorax and down its sides, jumping hind legs that have thick femurs packed with muscles, and wings inclined over the abdomen at rest (Fig. 15.1).

[1]Written by John Keats on 30 December 1816. Keats and Leigh Hunt challenged each other to write a sonnet on the grasshopper and cricket in 15 minutes. Both succeeded. See M. Allott (ed.) (1970) *The Poems of John Keats*, pp. 97–98. Longman, London.

[2]The Old Testament's *Book of Joel* gives one of the best accounts of a locust plague (probably the desert locust) in antiquity. It was probably written in the fifth century BC. For this and all Biblical quotes the Authorised Version is used.

Biological Diversity: Exploiters and Exploited, First Edition. Paul Hatcher and Nick Battey.
© 2011 John Wiley & Sons, Ltd. Published 2011 by John Wiley & Sons, Ltd.

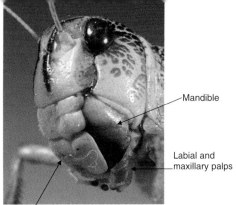

Mandible

Labial and
maxillary palps

Labrum

Pronotum

Fig. 15.1 *The adult desert locust, Schistocerca gregaria. Note the short antennae and the shield-like pronotum covering the thorax. The large mandibles are clearly visible. In front is the sensory labrum, and behind are two pairs of jointed palps which help process the food*

The modern Orthoptera suborders Ensifera and Caelifera are known from the Permian (250 mya) and Triassic (235 mya) respectively. Roughly half of known Orthoptera occur in the Ensifera, and its most diverse grouping is the superfamily Tettigonioidea, the bush crickets, with about 6000 species. These have four-segmented tarsi (the outermost leg segment) and produce sound by rubbing specialized areas of the wings against one another. The long-distance sounds they produce are detected by tympana on their forelimbs: these face forward, and when their legs are spread apart they can determine the direction of the sound (Fig. 15.2). The main superfamily within the Caelifera is the Acridoidea, with 8000 species. This includes the grasshoppers and locusts. These have three-segmented tarsi, and make sounds by scraping their legs against the stiff edges of their wings. The sound is amplified by membranes in the wing and is detected by tympana on each side of the abdomen.

Life history

Fig. 15.2 *A well-camouflaged adult speckled bush-cricket, Leptophyes punctatissima, a common member of the Tettigoniidae in Britain. Note the very long antennae and the forward-facing 'ear' on each forelimb (arrowed)*

Compared with many of the organisms discussed in this book, the life-cycle of the locust is relatively simple. The adult female desert locust, *Schistocerca gregaria* (which has a life-cycle typical of many locust species), lays slightly curved sausage-shaped eggs of about 6 mm length in groups of about 70, up to 15 cm below the surface of bare ground. She digs the hole with her abdomen, which trebles in length during the process, by opening and closing the valves of the ovipositor—the egg-laying structure at the end of the abdomen. The eggs are held together in the hole by a coating of froth—a tanned protein from the oviducts—which fills the hole, forming a solid plug to the surface. This reduces water loss from the eggs and makes an easy route for the hatchlings to reach the surface. The eggs have to take up water from the surroundings in order to mature, and thus the incubation period can be as short as 10 days in warm humid conditions or up to 70 days in cold and dry weather. During maturation the egg doubles in weight and hatches below ground into a worm-like protonymph that wriggles through the plug to the surface. Here it moults into a first instar nymph (Fig. 15.3).

Unlike honey bees (Chapter 4) and silkworms (Chapter 5), orthopterans do not have distinct and different larval and adult forms separated by metamorphosis through a pupal stage. Rather, the young hatched nymph is similar to, albeit much smaller than, the adult (Fig. 15.3). There are some differences; the nymphs do not have wings, and the sexual organs have not developed, but the mouthparts are identical and both adults and nymphs can feed on the same food: plant material. The desert locust nymphs pass through five instars, each separated by a moult, before reaching adulthood. As they develop so do their wings, as non-functional buds on the abdomen which are clearly visible by the third instar.

The newly moulted adult locust is termed a fledgling; it is sexually immature and can only manage a weak descending flight from bushes to the ground. If conditions are favourable the adult matures over about a week: it puts on weight, flight muscles grow and the wing structures harden. If weather conditions are poor (i.e. too cold or too dry), or food is in short supply, this process of maturation can take six months.

Mating does not involve any special courtship: a mature male simply jumps on to the back of a female and holds on with his front legs. If she is receptive he is tolerated, if not he is kicked off. Sperm is transmitted from the male in a spermatophore internally to the female. One spermatophore contains sufficient sperm to fertilize all of the female's eggs (which can be up to ten clutches, but probably only two or three are laid in the field) and fertilization may take place days or weeks after insemination. However, multiple mating is common in locusts, and the last sperm received is most likely to fertilize the eggs.

Phase change

We have yet to define what a locust is, apart from being a member of the insect order Orthoptera. Locusts cannot be defined taxonomically much further than this, as the twelve or so species of Orthoptera recognized as locusts are members of several subfamilies of the Acrididae (Table 15.1). Rather, locusts are defined by their behaviour: all locusts can undergo a transformation from solitary individuals (termed the solitarious phase) to gregarious, swarming populations. It is this phase change that can lead to the formation of locust plagues. In contrast, although many grasshoppers are agricultural pests and can collect in large numbers, they do not swarm in the manner of locusts. However, this distinction between grasshoppers and locusts is not absolute; rather there is a gradation from grasshoppers which are always solitary through to locusts proper, such as the desert locust and those in Table 15.1, in which the phase change takes place readily.

The phase change between the solitarious and gregarious forms of the locust is complex, and involves behavioural, morphological and physiological changes under the control of over 500 genes. Indeed, the differences between the forms are so great that until B.P. Uvarov published his phase-change theory in 1921 the different phases of the migratory locust (Locusta migratoria) and others were thought to be different species; for example, the solitarious phase of the migratory locust was called L. danica. The solitarious forms are usually inconspicuously coloured to blend in with the background, and exist in low-density patches. They can migrate, but fly only at night, and alone; the nymphs move slowly and avoid one another. The gregarious form is the reverse: it is brightly coloured and both nymphs and adults actively seek one another out to form bands or swarms respectively.

The process of phase change has been the subject of considerable research as it is crucial for predicting locust outbreaks: not all gregarious locust populations

Fig. 15.3 *The life-cycle of the desert locust,* Schistocerca gregaria. *Wing buds gradually develop and become easily visible from the third instar (arrowed), but wings are not fully formed until the moult into the adult. Note the uneven growth of parts of the body; the head and thorax are proportionally larger in the nymphs than adult*

Table 15.1 Species of locust, based on Chapman (1976)

Scientific name	Common name	Sub-family	Wingspan of female (mm)	Distribution
Schistocerca gregaria	Desert locust	Cyrtacanthacridinae	125	North Africa, Asia (see Fig. 15.4)
Schistocerca americana	American desert locust	Cyrtacanthacridinae	115	See note[a]
Anacridium melanorhodon	Sahelian tree locust	Cyrtacanthacridinae	143	Across Africa south of Sahara in a belt from about 10° to 20° N
Anacridium wernerellum	Sudanese tree locust	Cyrtacanthacridinae	128	In a belt to the south of *A. melanorhodon*
Nomadacris septemfasciata	Red locust	Cyrtacanthacridinae	131	Africa, mainly south of equator
Patanga succincta	Bombay locust	Cyrtacanthacridinae	138	SW Asia
Melanoplus spretus	Rocky Mountain locust	Catantopinae	43	North America, extinct (see Box 15.3)
Chortoicetes terminifera	Australian plague locust	Oedipodinae	58	Australia (see Box 15.4)
Locusta migratoria	Migratory locust	Oedipodinae	101	See note[b]
Locustana pardalina	Brown locust	Oedipodinae	93	South Africa
Dociostaurus maroccanus	Moroccan locust	Gomphocerinae	67	Middle East and Mediterranean

[a]Subspecies *americana* occurs in Central America, subspecies *paranensis* in South America.
[b]There are many subspecies, including: *migratorioides* in Africa (Fig. 15.8), *migratoria* from France to Japan, *cinerascens* in the Mediterranean area; and other Indian, Australian and Arabian subspecies.

go on to produce a plague, but solitarious locusts never do so. The phase change is triggered by an increasing population density of solitarious locusts. For example, solitarious desert locusts generally inhabit dry semi-desert areas with annual rainfall less than 200 mm, and they move between seasonal breeding grounds determined by rainfall and vegetation. It is thought that their cryptic habits, colouration and low population density enable them to hide from predators, and they do not exhaust the poor food supplies. However, other factors can all lead to dense locust populations building up: weather conditions, such as winds driving many locusts into the same area or successive heavy rains allowing breeding; vegetation structure (small dense patches in sparse vegetation, concentrating locusts); or good quality vegetation that allows successful breeding over several generations. This is the trigger for phase change.

When a solitarious desert locust is placed in a crowd of locusts its behaviour changes into that of a gregarious one within two hours. There are two distinct sensory pathways for this: a thoracic pathway driven by the mechanical stimulation of the hind femur, and a cephalic pathway with sight and sound of other locusts as the stimulus. In 2009, Steve Simpson's team at Oxford found that the concentration of the neurochemical serotonin in the thoracic ganglia increases with either of the stimuli, and this chemical mediates the change in behaviour during the early phase change.

The immediate behavioural change noticeable in gregarious locusts is that they actively seek each other out. Morphological changes take longer and can occur only after a moult, and include not only colour changes but also changes in some body proportions. These changes may take several generations to appear. Several generations are also required for a swarm of locusts to form, and gregariousness has to be maintained. To aid this, the phase state is maternally transmitted to the offspring: the gregarious female secretes a recently identified compound into the egg foam that causes the eggs to hatch gregarious—the nymphs emerge black not green and are larger and immediately congregate. The gregarization process is reversible: isolated gregarious locusts lose gregariousness as rapidly as they gain it.

The process by which gregarization occurs is becoming better understood, but what evolutionary benefit does it give to the locust? There is much argument over this. The greater mobility of gregarious locusts enables them to move rapidly from patch to patch, overwhelming predators and possibly plant defenses; but—as we will see—all locust plagues end up failing in mass starvation, with only a transient increase in numbers.

HOPPER BANDS AND SWARMS

The most impressive feature of an outbreak of locusts is the numbers that aggregate together. Both nymphs and adults do this, the former into hopper bands and the latter into swarms.

The hopper band has a fairly well defined daily schedule. The nymphs spend the night and early morning roosting, resting on bushes and low vegetation. As the temperature rises, groups on the ground begin to form, until they start to move. Here they stimulate each other to begin and continue moving by short hops or by walking, and unless it gets too hot (above about 36°C) they will spend the greater part of the day moving, covering up to 1 km or more. They may stop briefly to feed on low vegetation during the day, but most feeding takes place in the evening at the roosting site. The hopper bands vary greatly in size, but desert locust bands of several square kilometres have been reported.

> They shall run like mighty men; they shall climb the wall like men of war; and they shall march every one on his ways, and they shall not break their ranks:
>
> Joel 2:7

The movement undertaken by the hopper bands cannot really be called migration—they are just moving from food source to food source—but the movement undertaken by swarms of adult locusts can be migratory, and can cover long distances. A desert locust can flap its wings continually for over 17 hours, and swarms regularly cross the Red Sea, a distance of over 300 km. In 1988, a large swarm crossed the Atlantic Ocean to the West Indies, a distance of 5000 km and requiring at least 40 hours of continuous flying.

In their flight, locusts are aided by the wind speed (in still air their speed is 10–25 km/h) and most long-distance flight is in the direction of the wind. This downwind displacement eventually takes locusts into zones of wind convergence, providing rains for successful breeding (Fig. 15.4).

> The earth shall quake before them; the heavens shall tremble: the sun and the moon shall be dark, and the stars shall withdraw their shining:
>
> Joel 2:10

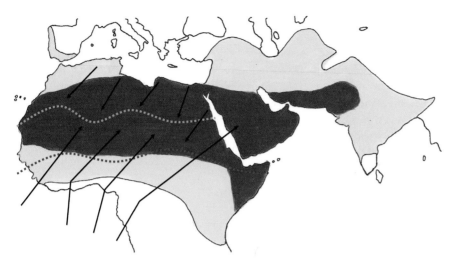

Fig. 15.4 *In contrast to the migratory locust (Fig. 15.8), the desert locust Schistocerca gregaria has no one outbreak area, but has a potential breeding area (red) encompassing much of Africa and Asia. Its invasion area (yellow) is even larger, affecting some 57 countries and 29 million square kilometres — more than 20% of the total land surface for the world. Superimposed are the approximate intertropical convergence zones for January (blue) and July (green), where opposing air masses (from the SW and NE, arrows) meet, rise and may produce rain, allowing the locust to breed [Based on Steedman (1990).]*

The swarms of gregarious locusts mainly fly by day. Flight is limited by body temperature; their flight muscles function efficiently only when they are above 30°C, but as flight itself raises the muscle temperature by about 8°C flight can be maintained in an air temperature above about 23°C. Therefore, adults need some time in the morning to get going: after roosting overnight on vegetation, around sunrise the locusts start to crawl slowly over the vegetation and for the next two hours or so orientate their bodies to the sun to warm up. By 9 am they start to fly, and by 10 am most are airborne, again stimulating each other into action, and there is a mass departure and flight until dusk (unless it is too hot — they cannot sustain flight above about 40°C).

> Thy crowned are as the locusts, and thy captains as the great grasshoppers, which camp in the hedges in the cold day, but when the sun ariseth they flee away, and their place is not known where they are.
>
> *Nahum 3:17*

When they settle there is heavy feeding. Locust flight endurance depends on fat reserves in the body, so when at rest they consume considerable amounts of food. As an adult desert locust may eat 1.5 grams of plant material a day, and the density of a settled swarm may reach 30–150 per square metre, a 10 km² swarm could thus eat 2000 tonnes of vegetation a day. This can not only quickly destroy crops, but pastures also: a swarm of locusts may eat 100–1000 times as much vegetation as the livestock present would eat.

A swarm of locusts is not random. Each locust must react to the others around it or the swarm will disperse, and they have to keep distance (usually 2–4 m apart) to prevent touching. Although at the top of the swarm the locusts are flying in the direction of the swarm, those lower down may be flying in any direction, and those at the edge of the swarm are heading back into it, descending and landing on the ground (Fig. 15.5). This gives a characteristic rolling motion to the swarm and this movement of individuals is essential for group cohesion. A swarm can stay together for weeks, can reach 1.5 km above the ground and may cover several hundred square kilometres.

> [A]nd the sound of their wings *was* as the sound of chariots of many horses running to battle.
>
> *Revelation 9:9*

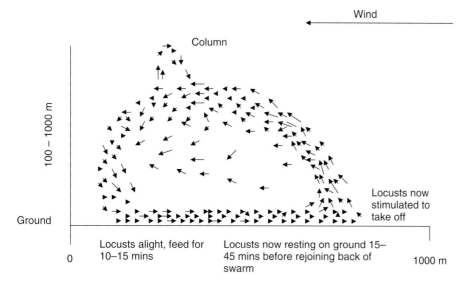

Fig. 15.5 *Structure of a high-flying rolling locust swarm [Based on Farrow (1990).]*

A PLAGUE OF LOCUSTS

> For they covered the face of the whole earth, so that the land was darkened; and they did eat every herb of the land, and all the fruit of the trees which the hail had left: and there remained not any green thing in the trees, or in the herbs of the field, through all the land of Egypt.
>
> Exodus 10:15

Successful breeding over several generations due to favourable weather conditions can lead to an outbreak of locusts, and eventually a plague may result (Box 15.1). Plagues of both the desert and migratory locusts have lasted for

---

BOX 15.1   LOCUST OUTBREAK DEFINITIONS

Over the years a set of definitions for the stages of a locust outbreak has developed and are now in general use for the desert locust. The definitions and quotations below come from van Huis *et al.* (2007).

o *Recessions*: 'periods without widespread and heavy swarm infestations'. Most of the locusts will be in the solitarious phase, and any swarms will only be small and transitory.

o *Outbreaks*: 'when concentration and multiplication cause a marked increase in locust numbers and densities so that individuals gregarize'. This is the stage at which control operations should usually start, and unless controlled, outbreaks lead to the formation of hopper bands and swarms.

o *Upsurges*: 'periods in which a widespread and very large increase in locust numbers initiates contemporaneous outbreaks'. This is followed by two or more successive seasons of gregarious breeding in the same or neighbouring regions.

o *Plagues*: 'occur when widespread infestations of swarms and hopper bands affect extensive areas and generate large numbers of reports during the same year and in each of successive years'.

It can be seen that the definitions of these terms are not precise, and in particular whether the post-1965 outbreaks of the desert locust led to plagues or just upsurges is arguable. This is exemplified in Fig. 15.7 where desert locust outbreaks are labelled as onset–peak–decline, with no attempt to define a plague.

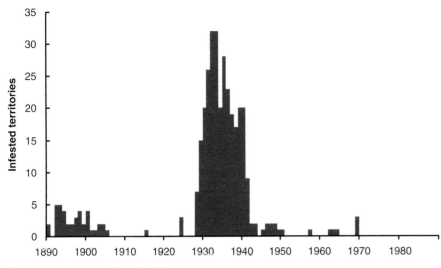

Fig. 15.6 *Territories infested with the African migratory locust, 1890–1990. The migratory locust control organization was established in 1948 [Data from Steedman (1990).]*

decades and covered a vast area of Africa and Asia (Figs 15.6 and 15.7). A plague would contain hundreds of locust swarms and hopper bands, and although the overall cost and loss of crops due to these plagues is unknown, some figures for more local areas indicate their effect. For example, desert locusts consumed 19% of Sudanese vine cultivation in 1944 and ten years later ate 55 000 tonnes of grain, and in Ethiopia consumed 167 000 tonnes of grain in 1958 – enough to feed a million people for a year.

The locust has been seen for millennia as a pestilential animal, a visitation of divine wrath and bringer of disease and death:

> He spake, and the locusts came, and caterpillers, and that without number. And did eat up all the herbs in their land, and devoured the fruit of their ground.
>
> *Psalm* 105:34–35

> And I will restore to you the years that the locust hath eaten, the cankerworm, and the caterpillar, and the palmerworm, my great army which I sent among you.
>
> *Joel* 2:25

And the reek of decomposing locusts has been long associated with human disease. St Augustine reports a swarm of locusts in the second century BC that were hurled into the Mediterranean, and when their bodies were washed up on the North African coast and decomposed a pestilence arose that killed over 800 000 people. Hence the term 'locust plague' that is still used to refer to a notable long-lasting outbreak of numerous swarms and hopper bands affecting large areas (see Box 15.1). Some locust plagues have even affected Europe. In AD 591 a migration of locusts from Africa to Italy is said to have caused a pestilence that killed a million people and animals. A disastrous plague of locusts was reported in 1613 in Europe, beginning in the Camargue of southern France, passing up the Rhône valley, destroying crops and grass as it went, before perishing during the winter. In 1720 another plague of locusts started in the spring in France, and the Archbishop of Avignon ordered prayers to be said for their cessation, and mounted an exorcism of the locusts. An outbreak of the black death (see Chapter 16) started soon after, and the locusts were thought to have been a harbinger of it.

Exorcism of insects was seen as a bit harsh by many in the Church, and so during the Middle Ages locusts in Europe tended to be excommunicated. However, only a communicant can be excommunicated and insects are not, so the only legiti-

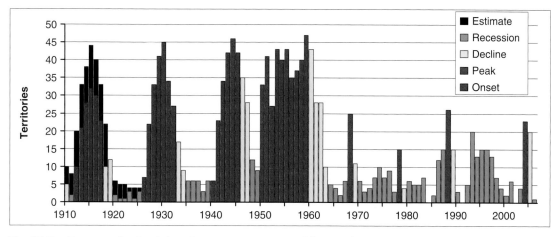

Fig. 15.7   *Territories infested with desert locusts, 1910–2006. For definitions of categories see Box 15.1 [Data and classification from Magor et al. (2008).]*

mate course of action for the Church was to anathematize or curse them. Another problem then arose: were the locusts sent by Satan or by God? If the former then anathematization was fine, but it would be a sacrilege if the creatures were heaven-sent (in this case, it would be OK to ask them to move from the crops when they were ready into a designated safe area). Therefore, locusts were put on trial in an ecclesiastical court to determine their origin. In one such trial in the Tyrol in 1338 the locusts were given legal council for their defence. Unfortunately, the locusts lost their case and were ordered to vacate the countryside within six days or be anathematized. They are reported to have departed. Locusts continued to be put on trial in Europe up to the middle of the nineteenth century. In Slovakia locusts were prosecuted during 1866; one of the largest individuals was seized, tried, and put death by drowning, with an anathema placed on the whole horde.

Early locust control was little more effective than exorcism. Lacking effective pesticides, mankind could drive hoppers into trenches and burn or bury them, or use drums and sticks to keep the swarms from settling on crops, and stamp on them or beat them when they did. They could also eat the locusts (Box 15.2). All of this could have been effective on a small scale, but was useless for plague control. The first effective chemical against locusts was sodium arsenate, which started to be used in bait from the later nineteenth century. Until the development of synthetic insecticides this was the mainstay of locust control.

As the locust hopper bands are concentrated and slow-moving, it is possible to mount a ground campaign against them, and laying bands of poison bait across their path can be effective. However, this is very expensive, requiring a huge amount of manpower and equipment. For example, during the 1950s, 45 000 tonnes of bait were used in a single season in Algeria and Morocco; and a much smaller operation in South Africa in 1953 against the migratory locust required 83 vehicles and 1000 men to lay only 1000 tonnes of bait. The preparation of this poison bait was also hazardous to the workers: with minimal protective clothing, sodium arsenate was mixed with bran and molasses (making the bran more attractive to the locusts), and allowed to dry. This was then moistened before being broadcast in a thin layer, by hand, in the face of the invading hopper band. A high concentration of arsenic (over 1% by weight) was used so that the hoppers died

BOX 15.2   WHY NOT EAT LOCUSTS?

Although in this chapter we have largely considered locusts as pests that should be controlled, through the ages some populations have thought locust plagues a blessing and a provider of highly desirable foodstuffs.

Locusts were eaten both at times of hunger and as a desired food: for example, a royal banquet scene from the palace of Ashurbanipal (c. 669–627 BC) for the last great Assyrian king shows servants bringing locusts on sticks for guests to eat. The Old Testament's book of *Leviticus* (11:23) gives a special dispensation to allow the eating of locusts while 'all other flying creeping things … shall be an abomination to you'.

Locusts are still sold and eaten today, but this trade has come into conflict with locust control. For example, the consumption of sprayed locusts has caused illness in several countries, control of the migratory locust in Zimbabwe is reported to have deprived people of a major source of food, and consumption

Box 15.2 Fig. 1   *Locusts on skewers being offered to the Assyrian king Sennacherib at a banquet* [Redrawn from a Ninevah bas-relief about eighth century BC.]

of the grasshopper *Oxya velox* or inago, once widely eaten in Japan, was reduced by post-war pesticide use. However, recent reductions in pesticide use have led to an increase in consumption, and similar has been reported from the Philippines. The Thai government even promoted locust eating during plagues as pest control was ineffective, and as a lucrative export trade to Bangkok has developed some farmers have now stopped spraying against locusts in order to sell them as food instead.

However, there is a long-standing western reluctance to eat insects: Kelhoffer (2005) suggests that this appeared in the early centuries AD. For example, early Greek authors such as Herodotus (fifth century BC) and Aristotle mention locust eating in the eastern Mediterranean. In the first century BC, Diodorus gives an account of a race of Ethiopians who choke advancing swarms of locusts in smoke from a brush fire, kill them and process them as food for storage. Diodorus finds it remarkable, but not repellent, that this tribe subsists on locusts alone. However, Strabo, who repeats this story slightly later, emphasized the inadequacy of this food, suggesting that it led to a shorter life-span; and Pliny, repeating this again in the first century AD, emphasized the negative aspects of locust consumption. From that date other Roman authors attest to a dislike of eating locusts.

Throughout the world, attaining a western life-style tends to lead to the elimination of insects from the diet, without, however, often the means to replace the nutrition lost.

Locusts have been prepared for eating in a number of ways:

o Fried, and as locust fritters
o Roasted with pepper and vinegar
o Dried in the sun, ground into flour, made into cakes and baked, or added to soups or milk
o Grilled: slit locust abdomen and stuff with a peanut
o Boiled and then preserved in vinegar and pepper
o Stewed: mix with onion, garlic and chilli powder and boil
o Boiled slowly for 3–4 hours, seasoned with butter, salt and pepper
o Boiled briskly, seasoned with salt and pepper and grated nutmeg, stirred occasionally, then pounded in a mortar with fried bread or a rice puree – place back in saucepan, and heat (do not boil) and thicken to a broth, strain and add a few croutons
o Eaten raw: salt, serve as a dessert with coffee

rapidly: to maintain morale it was important that the local workmen could quickly see the results of their labours. Although many locusts were undoubtedly killed by this method, it is unlikely that it had an effect on the development of plagues; and when the bait dried out, or the locusts happened not to be feeding (because they were about to moult), they would escape the poison.

Sodium arsenate was also tried in the first aerial spraying against locust swarms in the 1930s, but with little success. Aerial attack against locusts had been tried since the 1920s, but to be successful had to wait for better aircraft and chemicals, both developed during the Second World War. An early synthetic chemical used was DNC (di-nitro-ortho-cresol), a general poison, applied mixed with diesel. Trials were carried out in an air-force bomber converted for anti-malarial spraying and fitted with a large tank in its bomb bay. Spray technology was crude – the DNC mixture was discharged from a 6 cm diameter pipe into the slipstream of the aircraft, where it shattered into various-sized droplets which sank to the ground at different rates. Trials were first carried out against locusts on the ground, but were abandoned as it was difficult to find settled locusts without costly ground crews. Thus, in 1945, air-to-air spraying was tried, using the same idea as in bait laying – to produce an aerial curtain of DNC droplets into which the locusts would fly. The advantage of this was that a ground party was not needed – locust swarms can be spotted from the air 100 km away – but it was hard to know where to lay the curtain. Little was known about locust swarm behaviour (no one knew how many locusts were in a swarm, and even locust flight speed was unknown at the time), and it was difficult to predict the direction that even a slowly flying swarm would take.

Later more concentrated solutions of DNC and the organochloride insecticide γ-BHC were developed. These required a smaller droplet size (a large droplet would contain a wastefully excessive dose to kill one locust), and thus open pipes could not be used for discharge. Smaller, slower aircraft were now used, first with costly and unreliable conventional spray booms, but from the late 1950s a simple and robust spinning cage atomizer was used, producing very fine droplets. The best method of spraying a swarm was found by chance during the early 1950s – fly across the wind through the top of the densest part of the swarm and spray. Flying through a locust swarm was potentially dangerous, and aircraft had to be specially strengthened and adapted to cope with impacts from thousands of insects.

Thus, by the late 1950s an effective low-cost (and by using very low volumes of insecticide, low-risk) method of spraying locust swarms had been developed, and although the insecticides have since changed, the same methodology is used today.

Into the 1950s, control of hopper bands still relied on sodium arsenate-baited bran strips. Gamma-BHC was then used; this remained effective for a long time as it was active in dry bait. Organochloride insecticides were particularly suited to locust control, being contact acting, long-lasting and cumulative (suitable for low-dose hopper control), cheap, and relatively safe for users. From the late 1950s another organochloride, dieldrin, started to be used. This was particularly effective in a newly developed sprayer: an atomizer driven from the exhaust of a four-wheel-drive vehicle. Robust and easily fitted, this enabled insecticide to be quickly laid in front of hopper bands, and it dispensed with the logistical problems of moving tonnes of bait. Bands of dieldrin could also be laid from the air.

THREE PLAGUES

Most locusts, such as the African migratory locust *Locusta migratoria migratorioides* (and the extinct Rocky Mountain locust; see Box 15.3), are termed 'outbreak

BOX 15.3   THE LOCUST THAT WENT

Pioneers into the American mid-west during the middle of the nineteenth century had to cope not only with weather conditions, disease and attacks from Native Americans, but also the continent's only locust species: the Rocky Mountain Locust (RML) *Melanoplus spretus*. In an outbreak, swarms of locusts would descend from the northern Rockies in early June and spread south and east, mate and lay eggs. Embryos would mature through the summer and hibernate until next spring. The next generation would then emerge, the ground boiling with the emerging hoppers, and the plague would continue. After 3–4 years a portion of the population would return to the Rockies and the plague would cease. This species flourished during droughts and also needed a constant southerly wind to maintain the coherence of the swarm. This was provided by the Great Plains low-level jet, a 350 km-wide flow of air moving at almost 20 km/h during the day and active from late spring to early autumn.

The RML plague of 1874–7 was particularly destructive, and was estimated to have caused US$ 200 million damage to crops (equivalent to about US$ 116 billion today). In 1875 the size of a swarm that passed over Nebraska was estimated at 330 000 km²: it took five successive days to pass over one spot, and was estimated to have contained upwards of 3.5 *trillion* locusts. Even allowing for overestimation, this would still be the largest ever recorded locust swarm.

The effects of the RML outbreaks on the settlers were not overstated: crops were destroyed and many settlers died of starvation. Federal assistance was called for, and the governors of Minnesota and Dakota called for a day of prayer to deliver the population from the locusts. Further medieval tactics were used to try to repel the locusts, including beating, crushing, flooding and burning. Subsequently, ingenious horse-drawn machines were developed to crush, scoop up, or incinerate the locusts. Although some may have a small local effect, they were useless in controlling the plague.

An entomological commission started work on the problem in 1877, just as the locust plague began to subside. It suggested a diversification of agriculture away from a wheat monoculture (favoured by the locusts) to more diversified cropping, including peas and beans which were less vulnerable to the locust. The occasional RML swarm was recorded in the 1880s and early 1890s, but no further plagues occurred, and the last RML was recorded in 1902.

At first no one could believe that the locust had disappeared, and were expecting it to reappear at any time, but by the 1930s it became accepted that it had disappeared and workers began to look for reasons. Uvarov's phase transition theory of locusts was now accepted and thus it was suggested that the solitary form of the locust might still exist (there are several North American *Melanoplus* species that are solitary grasshoppers), but none were found to match the RML. Other theories, involving the widespread growth of alfalfa (a poor food for nymphs), the loss of bison, climate change and changing patterns of fire were popular for a time, but all were eventually discarded—either their chronology was incorrect or their effects would have been insufficiently widespread to cause such an extinction.

In the mid-1980s, Jeffrey Lockwood, an entomologist at the University of Wyoming, started to look into the locust. He first examined remains in the melting and retreating Rocky Mountain glaciers and found that the locust had been in abundance for several centuries before the settlers appeared, but there was no clue to its extinction. However, by 1990 he had formulated a hypothesis to account for this. The RML was an outbreak-area species, like the migratory locust, *Locusta migratoria*, and had a relatively small recession area in which it bred permanently. This was in the stream-sides, flood-plains, grasslands and sunny slopes of the foothill valleys of the Rocky Mountains. These areas were settled by pioneers from the 1860s and the landscape converted to agricultural production. Crops were grown, such as alfalfa, which needed constant irrigation and periodic flooding. Land was ploughed and harrowed and cattle were introduced. Their trampling and breaking down of the river banks caused rivers to flood when swollen with spring melt water. All these actions would lead to the destruction of RML egg masses in the soil, and the latter in particular would hit the eggs at their most vulnerable, just prior to hatching in the spring.

Therefore, Lockwood suggests that by these actions the settlers unwittingly destroyed the locust in its recession area, its heartland. This theory is by far the most cogent to have been put forward to account for the extinction of this locust, an accidental act which left North America the only continent without a locust species, and the RML as probably the only agricultural insect pest to have been unknowingly driven to extinction by mankind, and all without pesticides.

area species' because they have a defined area in which breeding, concentration and gregarization occurs, and from which swarms develop. For L. *migratoria* this outbreak area is in the Niger flood plain in Mali (Fig. 15.8). Since 1890 there have been two major plagues of this species each lasting about 15 years (see Fig. 15.6), with swarms spreading through west Africa into east and central regions, and infesting over 30 territories. Uvarov began a systematic collection of records of L. *migratoria* in Africa in 1929, and in 1948 a Migratory Locust Control Centre was set up. Once the outbreak area of this locust had been found, with sufficient monitoring and suitable control measures outbreaks could be controlled. No migratory locust plagues have arisen since the 1940s (see Fig. 15.6), although control of localized outbreaks is still necessary.

The desert locust, however, has no one settled outbreak area and can breed throughout a large area through which it moves during the year (see Fig. 15.4). Monitoring locust populations in such an area, much of which is among the most inaccessible and inhospitable habitat in the world, let alone timely control of outbreaks, is very costly and requires great international cooperation.

By the early 1960s, modern methods for control of the desert locust were available for the first time, but looking back, has mankind made progress against the desert locust? The answer is probably yes, but one cannot be sure. Certainly, plagues have become much less frequent and of shorter duration and most importantly have invaded fewer territories (major areas in western and eastern Africa and the Middle East have been swarm-

Fig. 15.8 *The main breeding area of the African migratory locust,* Locusta migratoria migratorioides, *is the Middle Niger in west Africa (red, arrowed). It is the escape of swarms of gregarious locusts from this area that has led to plagues that have spread throughout much of Africa (yellow). Swarms of the migratory locust have also formed in other parts of Africa (orange) but none are as important as the Middle Niger [Redrawn from Steedman (1990).]*

free since 1965), but is this due to control or to weather conditions? To decide the answer to this question, we will now examine the three desert locust plagues that have occurred since the mid-1960s.

The desert locust had been in recession (a period lacking swarms) since 1963, when from late in 1966 very heavy rain began to fall in many parts of Africa and the southern Arabian Peninsular (Fig. 15.9). During the next year there were four periods of heavy rain, some areas receiving over 200 mm. This encouraged locust breeding, and by the autumn of 1967 large numbers of locusts were present in the Arabian Peninsular, and gregarization was starting there and in parts of Africa. In late 1967 there was a major locust redistribution and significant migration. The heavy rainfall persisted during the winter of 1967–8, with breeding and gregarization continuing, and by April 1968 locusts were breeding over an area of 100 000 km² in Saudi Arabia; subsequently swarms began to cross the Red Sea into Africa. By the summer of 1968 swarm breeding was taking place and a plague now existed. Poison baiting was still being used in Sudan against the hoppers, but was ineffective in the 11 000 km² infested, and many swarms left this area in the autumn and redistributed. In the winter of 1968–9 the majority moved west, where almost all were killed in a large control campaign. Somalian–Ethiopian swarms were also partially controlled by spraying, but many moved into the Ethiopian highlands where they died without breeding, and by the spring of 1969 the outbreak had ceased.

This plague had followed the standard pattern of outbreaks and movements (see Fig. 15.9), starting with an unusually wet period allowing several generations to increase in numbers successively, and then swarming on the wind into the inter-tropical convergence zone (see Fig. 15.4). However, the end was abrupt, and the plague was short-lived compared to the previous one. This was thought to be the first plague that had been controlled by human intervention (although the weather had played at least some part), and was studied intensively.

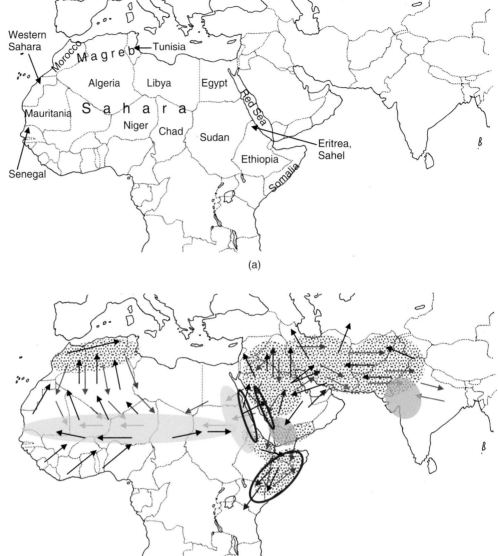

Fig. 15.9 The desert locust. **(a)** *Areas mentioned in the text.* **(b)** *Seasonal breeding areas and major movements common up to the late 1960s. Spring (March–June), red stippled areas and red/orange arrows; summer (August–September), orange area, black arrows; winter (October–January), blue ovals and arrows [Data from Bennett (1976).]*

The conclusions reached shaped control strategies for future outbreaks. Much was made of the success late in the plague of controlling the small, tightly packed swarms by aerial spraying, in contrast with ineffective control on the ground of the much more widespread and diffuse hopper bands and the large scattered populations of solitarious locusts.

After the 1966–9 plague subsided, the desert locust was quiet for several years. Thus, by 1984–5 desert locusts were at their lowest levels for 50 years. However, for three years from the summer of 1985 there was repeated successful breeding, gregarization, and a plague developed after heavy rains: in 1987–8 there was almost continuous breeding in western and north-western Africa (Fig. 15.10). FAO bulletins in October and November 1987 noted this heavy breeding, and swarms invading Mali and Mauritania, but there was no international mobilization for control in the Magreb: donors said they needed more precise field information before committing resources. Thus, control did not start until February 1988, by which time the first large swarms had already travelled from Mauritania and Western Sahara and crossed into Algeria and Morocco on

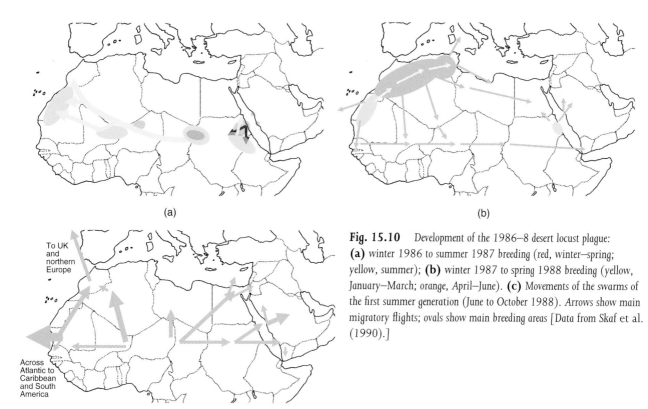

(a)

(b)

To UK
and
northern
Europe

Across
Atlantic to
Caribbean
and South
America

(c)

Fig. 15.10 *Development of the 1986–8 desert locust plague:*
(a) *winter 1986 to summer 1987 breeding (red, winter–spring;*
yellow, summer); **(b)** *winter 1987 to spring 1988 breeding (yellow,*
January–March; orange, April–June). **(c)** *Movements of the swarms of*
the first summer generation (June to October 1988). Arrows show main
migratory flights; ovals show main breeding areas [Data from Skaf et al.
(1990).]

warm and strong winds. Control started late and was also hampered by the inaccessibility of certain breeding areas in Eritrea, Chad and Mauritania owing to wars and guerrilla operations.

During September 1988, workers thought that they were loosing the battle and that a full plague, the worst this century, would occur the following year. However, during the last quarter of 1988 and the first quarter of 1989 a combination of adverse weather and control led to a dramatic decline in the plague: a failure of the winter rains in the Red Sea basin helped stop breeding and large numbers of locusts also drowned in the Atlantic. This latter was due to the majority of swarms being caught up in Atlantic cyclones and transported during October 1988 out over the Atlantic, with repeated invasions of the Cape Verde Islands, and large-scale invasion of the eastern Caribbean and South America (Fig. 15.10c). This was the first time the desert locust had moved across the Atlantic, and unpredicted migrations to Mediterranean countries including Italy, Greece and Turkey also occurred.

Control of this plague was carried out. Up to May 1988, 7.6 million hectares were treated, and another 9.2 million hectares between June 1988 and June 1989. A total of 11–15 million litres of pesticides and 2700 tonnes of dusts were used; and US$ 200–300 million in aid was given. Yet it had little effect on the plague until the end of 1988. It was estimated that locust mortality at the end of 1988 had been due to 30% of the locusts drowning in the Atlantic, 30% killed by low temperatures, 20% by drought and 20% by control measures.

The small effect of control operations came as a shock to locust workers, aid agencies and the FAO. A number of investigations were carried out and donor countries started to question the efficacy of strategies, organization, and the real economic importance of locusts. For example, the US Congress Office of Technology Assessment published an influential report in 1990 which concluded that the results of locust and grasshopper control were disappointing and future control efforts must move away from crisis management by chemical sprays.

These reports revealed that in the long desert locust recession between 1969 and 1986 many of the problems that beset control of the previous plague had not been addressed and locust control capability had actually decreased.

Although desert locust control was predicated upon finding developing gregarizing populations and controlling them before an outbreak occurred, this failed due to lack of preparation and lack of security in the outbreak areas. For example, parts of Eritrea were still mined, and Polisario guerrillas fired rockets that hit two US spray aircraft en route to Morocco from Senegal after completing a spraying mission: one of the aircraft crashed with the loss of the crew of five.

There had been little development in locust control methods since the last outbreak, with only a slow move to ULV spraying; and when control was attempted organizations found that in the meantime their effective organochloride insecticides, which were stockpiled, had been banned or their use severely restricted, and had to be replaced with more expensive and less effective organophosphates.

Research after this plague concentrated on three areas: better forecasting, improved organization, and the development of more effective and environmentally benign control methods. Improving forecasting involved access to better and more timely data and also the ability to process it. Thus, a geographical information system, SWARMS, was developed in the early 1990s, run by FAO in Rome. Research into remote sensing using vegetation mapping from satellite imaging also started, but communication in the field remained a major problem.

To strengthen monitoring and control organization, the large EMPRES (Emergency Prevention System) programme was initiated by FAO in 1996, and it aimed to develop regional cooperation and re-establish regional desert locust control centres. However, efficient control measures were needed. With the banning of the persistent organochloride insecticides, workers looked for alternatives that were also persistent, but with a lesser effect on the environment. New chemicals, such as insect growth regulators, were also tried. However, these were generally less persistent than traditional chemicals and did not have a cumulative effect on the locust, and thus had to be applied in higher concentrations to have a lethal effect.

Belatedly, biological control was investigated. Although *Nosema locustae* (a microsporidian specific to Orthoptera; see Box 5.4), was available for locust and grasshopper control and had widespread use in China, its effects were mainly sublethal and could be inconsistent. Thus, it was thought a poor candidate for desert locust control where reliable kill was important. Therefore, to find an agent that would effectively and quickly kill locusts, entomopathogenic fungi (see Box 5.3) were investigated through the LUBILOSA programme set up in 1989 by CABI (Commonweath Agricultural Bureau) in the UK, IITA (International Institute for Tropical Agriculture) and agencies in Benin and Niger. They collected 180 fungal isolates of hyphomycete fungi from Orthoptera, and the majority of the virulent strains collected from Africa were of *Metarhizium anisopliae* var. *acridum*. One strain of this species was selected from locusts in Niger for further development and trialling, and other strains of this fungus have been used in biocontrol products against locusts elsewhere, for example Australia (Box 15.4).

By 2003, EMPRES was active in nine countries around the Red Sea, known as the central region. This was able to coordinate rapid control of locusts detected that year; 200 000 ha were treated in late 2003 and early 2004 and this

BOX 15.4   LOCUST BIOCONTROL

Although an entomopathogen may be very infective, producing a successful pesticide from it is difficult. One of the problems is to ensure that the conidia (the infectious spore stage of the fungus; see Box 5.3) remain viable for as long as possible, both on the shelf and also when applied in the field. Once a successful formation of *Metarhizium anisopliae* var. *acridum* had been developed, it needed to be manufactured cheaply and in bulk. No commercial company was able or willing to mass-produce the fungus (given the trade name Green Muscle® due to the dark-green colour of the spores in the oil), and thus LUBILOSA had to build a spore production pilot plant in Benin. The fungus was grown in an initial liquid phase in sucrose and brewer's yeast followed by solid phase in rice in plastic bags. This plant was not cost effective for large-scale production, and thus after trying, and failing (due to contamination) to develop low-cost 'artisanal' production methods that could be used by developing countries, LUBILOSA entered into partnership with some private sector companies to produce Green Muscle® using high-tech large-scale solid fermentation methods.

Compared to chemical insecticides, Green Muscle® is slower to have an effect as the fungus has to penetrate and grow within the locust, but does have a long-lasting effect. One problem that could not be overcome is that in some situations Green Muscle® is operating near its thermal maximum, and some Orthoptera (including the desert locust) are able to produce a behavioural fever when infected that pushes their temperature above that suitable for fungal development, which is then either unable to kill the locust or takes longer to do so.

Although the commercialization of this mycopesticide for desert locust control has been slow, an offshoot of that project has had considerable success, in Australia.

The Australian plague locust (APL), *Chortoicetes terminifera*, is an important pest species which, like the migratory locust (which also occurs in Australia), has defined outbreak areas in which breeding occurs. A plague can develop within one year if there is a series of rains in these areas. It is usually a one-season pest with swarms dying out in the summer without egg-laying. The earliest APL swarm in Australia was recorded in 1844, and since 1900 the swarms have been more frequent and are now a regular feature of eastern Australian agriculture. The hopper bands damage mainly pasture, where losses equivalent to a 10% increase in stocking rate have been recorded; while swarms damage young winter cereals. In 1984 the largest outbreak of APL for 50 years caused A\$5 million of losses, yet it was estimated that without control (costing A\$3.4 million) losses would have been A\$105 million.

Control was based on outbreak prevention, using chemical insecticides, similar to desert locust control. However, a switch by farmers during the mid-1990s to the production of organic beef for the Asian market promoted an intensive search for an acceptable alternative to chemical control. The slower action of biological control in a pasture system would be acceptable, whereas it may not be in an arable system where immediate economic damage to the crop can occur. Thus, LUBILOSA assisted in the development of a strain of *Metarhizium anisopliae* var. *acridum*.

An indigenous isolate was chosen which was virulent to a wide range of acridids and amenable to mass production. First trials were carried out on wingless grasshoppers, and then the APL when it became available. By 2000, trials had progressed to 50 ha plots and the mycopesticide was registered as Green Guard®. The agreement of the Australian Plague Locust Commission to purchase A\$200 000 worth of the product per year for three years ensured that it was commercially viable to develop suitable production methods, resulting in higher fungus yields and reduced product costs. A yeast extract/dextrose broth culture of the fungus is used to inoculate bags of par-boiled and sterilized long-grain rice, which are incubated and then air-dried and the conidia sieved out. The conidia are formulated in corn oil (selected to inhibit settling of spores, suitability for organic systems, and enabling the product to remain viable for 18 months when stored at 4°C) and then diluted in the field with a mineral oil for ULV spraying.

The APL is particularly sensitive to this mycopesticide, and low doses can give over 90% control within 14 days of application. In late 2000, Australia was threatened with the possibility of the largest ever APL plague and Green Guard® was applied over 23 000 ha between October 2000 and January 2001. Since then it has been used to treat 15 000 ha in 2003–4 and 35 000 ha in 2004–5.

A major problem is how to ensure supply is available quickly for unpredictable locust outbreaks. Green Guard® was originally available for use only on special government licence, but in 2005 a world-wide agreement was signed with a manufacturer to produce and sell it to farmers directly. It is also being marketed as a control for other Orthoptera that have regular outbreaks, with the aim to have one or more producers that will be able to supply the product at short notice. Furthermore, trials in Australia and China since 1998 have shown that it is effective against the migratory locust, and from 2005 the Chinese government has used it for locust and grasshopper control.

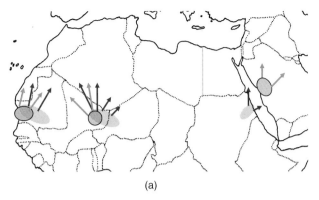

(a)

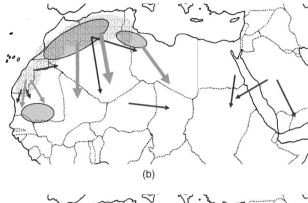

(b)

Fig. 15.11 *Development of the 2003–5 desert locust plague which was concentrated in north-western Africa:* (**a**) *October 2003 (red) and January 2004 (orange);* (**b**) *April 2004 (red) and July 2004 (orange);* (**c**) *October 2004 (red), remaining populations January 2005 (orange). Ovals show main breeding populations; arrows show main migrations [Data from Lecoq (2005).]*

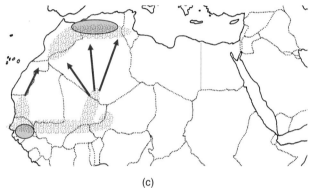

(c)

probably prevented an upsurge here. However, western African countries could not afford to join EMPRES, and in the summer of 2003 exceptional rains fell in the Sahara zone over a large area between Mauritania and Sudan. It was clear in August 2003 that the desert locust was likely to upsurge here, but again poor surveying and prevention due to lack of resources and trained personnel led to the situation worsening, and in October 2003 the upsurge developed into a plague (Fig. 15.11a). Unpredicted widespread and heavy rains from Senegal to the Western Sahara–Morocco border meant that the swarms that invaded north-west Africa now had a good breeding area.

Lack of funding meant that this outbreak could not be controlled in 2003 and the swarms invaded north-west Africa and bred between March and June 2004 and now needed large-scale control (Fig. 15.11b). This was insufficient and did not control swarms re-invading south in the summer of 2004. These locusts bred quickly in western Africa throughout May to July 2004, still controlled ineffectively, and again invaded northern Africa in Autumn 2004 (Fig. 15.11c). The swarms that invaded the Magreb in early autumn persisted through the winter and moved little; almost 4 million litres of insecticide was sprayed on these, and probably destroyed most of them. The plague finally regressed in 2005 largely due to poor breeding weather.

Again, donor assistance was slow. The FAO made its first alert in October 2003 and asked for US$ 9 million in February 2004; by September that year it had received only US$ 2 million. Although US$ 60 million had been pledged by November 2004, control costs then were estimated at up to US$ 100 million. The total cost of control of the 2003–5 outbreak was estimated at US$ 280 million.

Organophosphates were again the main control chemical and, apart from EMPRES, and SWARMS to manage and analyse the data, the control measures developed after the 1986–8 plague were not used: insect growth regulators and biopesticides had not been operationally tested against the desert locust[3], and GPS and remote sensing were ineffective. Satellite imaging, though with a much improved resolution, still cannot detect the thin green vegetation containing the desert locust and often gives false negatives, recording areas as dry when in reality they are green and suitable for the locust.

CONCLUSIONS

Although one locust has disappeared recently (see Box 15.3), and other outbreak area species such as the migratory locust are probably under control, we have some way to go to control the desert locust. It seems that technology and science alone cannot control it. Effective organization is necessary; indeed Lecoq (2005) states: 'the locust is no longer the main problem, humans are the real problem', and the failure to control early the 2003–5 plague has led to further argument about the best control strategy for the desert locust. Central to this is the question of when it is best to control an outbreak. Controlling as soon as the outbreak starts might seem like the best idea (small-scale operations and not time-critical), but this relies on accurate spotting of the small developing populations, and in most cases populations at this stage do not lead to outbreak swarms and die out naturally. On the other hand, if control is left until a plague has formed it has to be swift, with a very high percentage kill. Although early control has been the policy, in practice each of the three plagues above has only been controlled late, in a reactive manner.

Some workers have gone further and question whether we should be controlling the desert locust at all. Their arguments are that it is difficult to quantify the damage that it causes, and there have been allegations that countries have inflated this cost to obtain extra donor funds. It may not even be cost-effective to control the locust if it is damaging; for example, it is reported that Morocco spent US\$ 30 million in 2004 to defend its agricultural production (worth US\$ 7 million) against locusts, although it can be argued that the cost to the whole region of allowing these locusts to spread and breed would be much greater. Although such arguments may be justified on a macroeconomic scale, they must appear crass and insensitive to the farmers seeing their crops destroyed, and the suggestion that these farmers should take out insurance against locust plagues is not feasible.

While the international locust research community argues over the best way to control locusts, if at all, the humble 'grasshopper laughs in the merry scene'.[4] As locusts have become less abundant in the last 60 years, so grasshoppers have become much more important crop pests, and it is them, not the more charismatic locust, that causes the most crop damage world-wide and needs greater control.

FURTHER READING

For a general introduction to locust biology see Chapman (1976), and Haskell (1992) is a useful short introduction. For more detail see Chapman and Joern (1990) for grasshopper biology and Rainey (1989) for locust migration. Despland (2004) is a useful summary of locust phase change, and see Miller *et al.* (2008),

[3]The first large-scale application of Green Muscle was not until 2009, when over 10 000 ha were sprayed against the red locust in Tanzania rather than the desert locust.

[4]'Laughing Song', William Blake, *Songs of Innocence* (1789).

Anstey *et al.* (2009) and Stevenson (2009) for the latest developments. Steedman (1990) is the standard work on locust control, especially for the desert locust, and see Hassall (1990) for pesticides. For the pest status of Orthoptera in general see Jago (1998), and Brown (1990) has keys to the species of Orthoptera in the UK.

See Baron (1972) and Bennett (1976) for the 1966–9 plague, and a set of articles in the *Philosophical Transactions of the Royal Society of London, Series B* (vol. 287, 1979). For the 1986–8 outbreak see Skaf *et al.* (1990), and Krall *et al.* (1997) and Lecoq (2001) for developments arising from it. Henry (1981) covers early work on microsporidia for locust control; Douthwaite *et al.* (2001), Lomer *et al.* (2001) and Moore (2008) describe the LUBILOSA project. For the 2003–5 plague see Enserink (2004), Lecoq (2005) and Matthews (2007). For recent discussions on control tactics see van Huis *et al.* (2007), Magor *et al.* (2008) and Symmons (2009), and for an alternative argument see Krall (1995).

Evans (1906) is an incomparable source for locust prosecutions, and Bodenheimer (1951) is a comprehensive treatise on eating insects with much on locusts, but is hard to find. DeFoliart (1999) is useful, and Kelhoffer (2005) gives the most comprehensive account of locust eating in antiquity. Holt (1885) is an interesting polemic on why we should eat insects, and has been reprinted several times. There are several insect cookbooks available, which have some locust recipes.

For the Australian plague locust see Casimir (1962), Farrow (1977) and Wright (1986), and for Green Guard® see Milner (2002) and Hunter (2004, 2005).

The history of the Rocky Mountain locust is given in Lockwood's (2004) excellent book, which also has much of interest on locusts in general, and Lockwood and DeBrey (1990) give a shorter account of this locust's biology and their explanation for its extinction.

The current pest status of the desert locust can be checked on the FAO website (www.fao.org/ag/locusts) which also contains monthly bulletins on the locust.

REFERENCES

Anstey, M.L., Rogers, S.M., Ott, S.R., Burrows, M. and Simpson, S.J. (2009) Serotonin mediates behavioural gregarization underlying swarm formation in desert locusts. *Science* **323**, 627–630.

Baron, S. (1972) *The Desert Locust*. Methuen, London.

Bennett, L.V. (1976) The development and termination of the 1968 plague of the desert locust, *Schistocerca gregaria* (Forskål) (Orthoptera, Acrididae). *Bulletin of Entomological Research* **66**, 511–522.

Bodenheimer, F.S. (1951) *Insects as Human Food*. Junk, The Hague, Netherlands.

Brown, V.K. (1990) *Grasshoppers*, revised edn. Naturalists' Handbooks no. 2, Richmond, Slough, UK.

Casimir, M. (1962) History of outbreaks of the Australian plague locust, *Chortoicetes terminifera* (Walk.), between 1933 and 1959, and analysis of the influence of rainfall on these outbreaks. *Australian Journal of Agricultural Research* **13**, 674–700.

Chapman, R.F. (1976) *A Biology of Locusts*. Studies in Biology no. 71, Edward Arnold, London.

Chapman, R.F. and Joern, A. (eds) (1990) *Biology of Grasshoppers*. John Wiley & Sons, New York.

DeFoliart, G.R. (1999) Insects as food: why the western attitude is important. *Annual Review of Entomology* **44**, 21–50.

Despland, E. (2004) Locust transformation. *Biologist* **51**, 18–22.

Douthwaite, B., Langewald, J. and Harris, J. (2001) *Development and Commercialization of the Green Muscle Biopesticide*. International Institute of Tropical Agriculture, Ibadan, Nigeria.

Enserink, M. (2004) Can the war on locusts be won? *Science* **306**, 1880–1882.

Evans, E.P. (1906) *The Criminal Prosecution and Capital Punishment of Animals*. Heinemann, London.

Farrow, R.A. (1977) Origin and decline of the 1973 plague locust outbreak in central western New South Wales. *Australian Journal of Zoology* **25**, 455–489.

Farrow, R.A. (1990) Flight and migration in acridoids. In: *Biology of Grasshoppers* (eds R.F. Chapman and A. Joern), pp. 227–314. John Wiley & Sons, New York.

Haskell, P.T. (1992) The locust. *Biologist* **39**, 111–117.

Hassall, K.A. (1990) *The Biochemistry and Uses of Pesticides*, 2nd edn. Macmillan, Basingstoke, UK.

Henry, J.E. (1981) Natural and applied control of insects by protozoa. *Annual Review of Entomology* **26**, 49–73.

Holt, V.M. (1885) *Why Not Eat Insects?* London.

Hunter, D.M. (2004) Advances in the control of locusts (Orthoptera: Acrididae) in eastern Australia: from crop protection to preventative control. *Australian Journal of Entomology* **43**, 293–303.

Hunter, D.M. (2005) Mycopesticides as part of integrated pest management of locusts and grasshoppers. *Journal of Orthoptera Research* **14**, 197–201.

Jago, N.D. (1998) The world-wide magnitude of Orthoptera as pests. *Journal of Orthoptera Research* **7**, 117–124.

Kelhoffer, J.A. (2005) *The Diet of John the Baptist*. Mohr Siebeck, Tübingen, Germany.

Krall, S. (1995) Desert locusts in Africa: a disaster? *Disasters* **19**, 1–7.

Krall, S., Peveling, R. and Ba Diallo, D. (eds) (1997) *New Strategies in Locust Control*. Birkäuser Verlag, Basel, Switzerland.

Lecoq, M. (2001) Recent progress in desert and migratory locust management in Africa. Are preventative actions possible? *Journal of Orthoptera Research* **10**, 277–291.

Lecoq, M. (2005) Desert locust management: from ecology to anthropology. *Journal of Orthoptera Research* **14**, 179–186.

Lockwood, J.A. (2004) *Locust: the devastating rise and mysterious disappearance of the insect that shaped the American frontier*. Basic Books, New York.

Lockwood, J.A. and DeBrey, L.D. (1990) A solution for the sudden and unexplained extinction of the Rocky Mountain grasshopper (Orthoptera: Acrididae). *Environmental Entomology* **19**, 1194–1205.

Lomer, C.J., Bateman, R.P., Johnson, D.L., Langewald, J. and Thomas, M. (2001). Biological control of locusts and grasshoppers. *Annual Review of Entomology* **46**, 667–702.

Magor, J.I., Lecoq, M. and Hunter, D.M. (2008) Preventive control and desert locust plagues. *Crop Protection* **27**, 1527–1533.

Matthews, G. (2007) Are we getting on top of locust control? *Outlooks on Pest Management*, February, 31–32.

Miller, G.A., Islam, M.S., Claridge, T.D.W., Dodgson, T. and Simpson, S.J. (2008) Swarm formation in the desert locust *Schistocerca gregaria*: isolation and NMR analysis of the primary maternal gregarizing agent. *Journal of Experimental Biology* **211**, 370–376.

Milner, R. (2002) Green Guard®. *Pesticide Outlook*, February, 20–24.

Moore, D. (2008) A plague on locusts: the LUBILOSA story. *Outlooks on Pest Management*, February, 14–17.

Rainey, R.C. (1989) *Migration and Meteorology: flight behaviour and the atmospheric environment of locusts and other migrant pests*. Clarendon Press, Oxford.

Skaf, R., Popov, G.B. and Roffey, J. (1990) The desert locust: an international challenge. *Philosophical Transactions of the Royal Society of London B* **328**, 525–538.

Steedman, A. (ed.) (1990) *Locust Handbook*, 3rd edn. Natural Resources Institute, Chatham, UK.

Stevenson, P.A. (2009) The key to Pandora's Box. *Science* **323**, 594–595.

Symmons, P. (2009) A critique of 'preventive control and desert locust plagues'. *Crop Protection* **28**, 905–907.

Van Huis, A., Cressman, K., Magor, J.I. (2007) Preventing desert locust plagues: optimizing management interventions. *Entomologia Experimentalis et Applicata* **122**, 191–214.

Wright, D.E. (1986) Economic assessment of actual and potential damage to crops caused by the 1984 locust plague in south-eastern Australia. *Journal of Environmental Management* **23**, 293–308.

Plague

PLAGUE IS ONE of only two epidemic diseases subject to the International Health Regulations. The other is cholera; a third, smallpox, was excluded in 1981 as it had become extinct. All cases of plague must be investigated by public health authorities and confirmed cases reported to the World Health Organization in Geneva. Yet, only a couple of thousand cases of this disease and a few hundred deaths are recorded annually. In this chapter we will discuss the biology and history of this disease and the real and possibly imaginary reasons why it is so feared and why so much effort has been expended in trying to turn it into a weapon.

YERSINIA PESTIS

The plague we are discussing in this chapter is caused by the bacterium *Yersinia pestis*. This is a non-motile non-spore-forming coccobacillus and member of the Enterobacteriaceae, in which family it is unique in being blood-infecting and flea-transmitted. *Yersinia pestis* is closely related to *Y. pseudotuberculosis*, and probably recently arose as a clone of it (Box 16.1). *Yersinia pestis* can infect many mammal species, especially rodents. Its effects vary from acute to chronic, and in some species it is able to form a sylvatic disease reservoir. This is termed the enzootic cycle (Fig. 16.1), a self-sustaining infection maintained at a relatively high level in the animal population. Today, sylvatic reservoirs forming plague foci in wild rodents occur on all continents except Australasia (Fig. 16.2), through a broad belt in tropical and subtropical latitudes between 55°N and 40°S.

Yersinia pestis is spread between these rodents by fleas. Bacteria are taken up in blood by the flea and settle in the proventriculus, a structure separating the flea's oesophagus from its midgut (Fig. 16.3). The proventriculus has a number of spines, which function as a one-way valve. The proventricular spines are hydrophobic and Y. pestis forms a biofilm (an example of internal biofouling; see Chapter 13) upon them and multiplies. Eventually, after 2–3 weeks the proventriculus becomes blocked by the bacterial biofilm: it is not now thought that the blockage is caused by a blood clot. This blockage prevents blood from entering the flea's midgut, but does not prevent the flea from trying to feed; indeed its feeding efforts may be increased as it slowly starves. However, when the flea stops trying to feed, the oesophagus recoils and blood containing bacteria from the proventricular blockage returns to the mammalian host, thus potentially infecting it.

Whether the mammal becomes infected is a numbers game. The flea's digestive system is tiny, and it takes a fraction of a millilitre in blood per meal.

Biological Diversity: Exploiters and Exploited, First Edition. Paul Hatcher and Nick Battey.
© 2011 John Wiley & Sons, Ltd. Published 2011 by John Wiley & Sons, Ltd.

BOX 16.1	THE EVOLUTION OF *YERSINIA PESTIS*

Analysis of its genetic structure demonstrates that *Yersinia pestis* evolved from *Y. pseudotuberculosis* sometime between 1500 and 20 000 years ago. As *Y. pseudotuberculosis* is a mammalian enteric pathogen transmitted by the faecal–oral route that rarely leads to death, how did *Y. pestis*, with a quite different mode of transmission, evolve from it?

All bacteria contain a circular chromosome, and can contain extrachromosomal DNA molecules in the form of plasmids, which can replicate autonomously within the bacterial cell, and can be transmitted between them. *Yersinia pseudotuberculosis* already contains some genes vital to the mode of transmission of *Y. pestis*: the *hms* gene necessary for biofilm formation, and thus the production of the blockage in the flea, and the pCD1 plasmid that is required to overcome host defences and prevent early fatality of the host.

Subsequently, the precursor to *Y. pestis* has acquired two further plasmids. One, pFra, codes for phospholipase D, and is essential for survival in fleas as it helps the bacterium withstand host defences. The second plasmid, pPst, encodes a plasminogen activator and is needed for high virulence in mammals as it promotes dissemination of the bacteria from the site of the flea bite by colonization of adjacent lymphoid tissues. Both plasmids were probably acquired from co-occurring bacteria in the host's gut.

Acquisition of these plasmids would have allowed the bacterium to colonize fleas and infect via flea bites. However, biofilm formation in *Y. pseudotuberculosis* is poor and erratic, and cannot occur in fleas, and *Y. pestis* has evolved changes in the regulation of the biofilm formation genes to produce a stable biofilm. This remodelling of gene regulation would have allowed *Y. pestis* to commit to its flea vector and a blood-transmitted

life-cycle. Subsequently, the enteric life-cycle is thought to have been lost: 13% of *Y. pseudotuberculosis* genes do not function in *Y. pestis*. Among those inactivated are adhesin and invasin genes that enable *Y. pseudotuberculosis* to adhere to the surface of the gut wall and invade living epithelial cells.

The selective pressures that led to the evolution of *Y. pestis* are unknown, but could have been linked to a change in the environment that led to increased rodent populations. One such event was the development of agriculture and animal domestication that took place during the Neolithic revolution about 10 000 years ago, and which led to the increase in human settlements and populations (see Chapter 7). The significantly increased food supply is likely to have led to the increase in rodent populations and the development of the commensal lifestyle of many rodents.

Yersinia pestis is still evolving. In 1995 the first multiply drug resistant (MDR) *Y. pestis* strain was found in Madagascar. This has a plasmid that confers resistance to eight of the recommended anti-plague drugs. It has a near-identical backbone to MDR plasmids found in other bacteria including many MDR enterobacteria collected from meat samples. Thus, it is widely spread through MDR zoonotic pathogens (i.e. those able to be transmitted from non-human to human hosts) associated with agriculture. This is particularly worrying, as although rare at the moment, it means that a reservoir of mobile resistance determinants is available, and has the potential to enter *Y. pestis* again, potentially leading rapidly to a high level of MDR. Therefore, because there is no useable plague vaccine at the moment there is an urgent need for surveillance of *Y. pestis* to check for the development of resistance.

A minimum of 10^4 bacteria are needed to infect a flea, and this requires the very high concentration of 10^8 bacteria per millilitre in the blood ingested. To be a good plague vector a flea must be able to ingest enough bacteria, must live long enough for them to multiply, must block easily and be able to transfer the bacteria to a host, and must be present in large enough numbers to maintain infection in the local rodent hosts. Most fleas, including the human flea *Pulex irritans* (over 80 species have been recorded infected with *Y. pestis*), are poor vectors. Even in the efficient ones, such as the rat flea *Xenopsylla cheopis*, only about 50% become blocked and only about 50% of those transmit the infection, of around 100 bacteria per bite. Much is still unknown about this stage of the disease; for example it has only recently been discovered that unblocked fleas can also transmit the bacterium, although with what efficiency is not yet known.

The flea vectoring stage is the bottleneck in *Y. pestis* infection, yet the bacterium has evolved a couple of tricks to increase success. First, it is very infectious – it has been calculated that only ten bacteria are necessary to start an infection in a susceptible mammal. Second, *Y. pestis* is able to attain a very high concentration in the bloodstream of susceptible mammals for a time. The reasons for this are complex but probably include the disablement of the immune response (see Box 16.1).

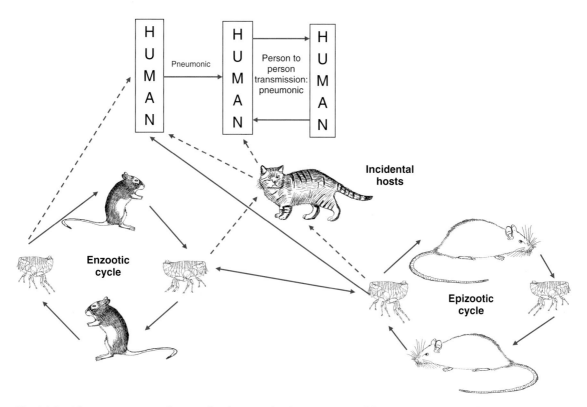

Fig. 16.1 The Yersinia pestis *disease cycle. The principal pathways are given solid arrows, rarer pathways dashed arrows*

Fig. 16.2 *Spread of plague in the nineteenth and twentieth centuries and its legacy. First outbreaks are given in selected ports; red lines indicate recorded routes of infection. Current plague sylvatic foci are given in red. Countries that have recorded plague cases in the twenty-first century are given in orange, those with last cases in the 1990s in yellow, and those with their last cases before 1990 in green [Based on Dennis et al. (1999), with data also from WHO (2010) and various sources.]*

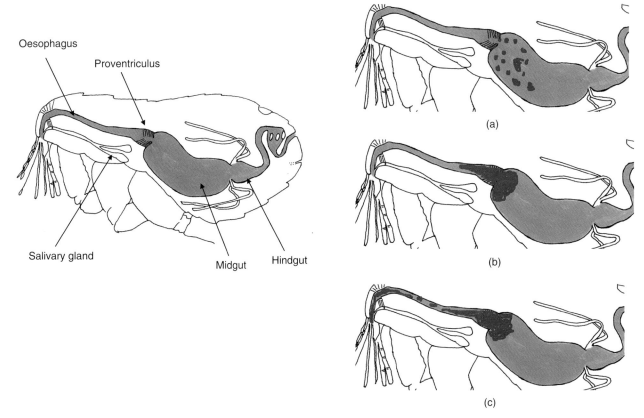

Fig. 16.3 *The flea's alimentary canal and stages in blocking of the proventriculus.* **(a)** Yersinia pestis (red) *starts to clump in the midgut.* **(b)** Y. pestis *has now formed a plug in the proventriculus, extending into the oesophagus. The flea is now said to be blocked.* **(c)** *Bacteria break off from the proventriculus of the blocked flea, while it is trying to feed, and pass out of the oesophagus into the host [Flea diagram based on Marshall (1981).]*

Some mammals, however, are not resistant to Y. *pestis* infection. These highly susceptible species, again largely rodents, form the epizootic (amplification) cycle (see Fig. 16.1), analogous to an epidemic in humans and resulting in high mortality. Fleas living on these animals will then leave their hosts. Unfortunately for us, several species of commensal rodent (i.e. those living in close proximity to humans), especially the black rat (*Rattus rattus*) and brown rat (*R. norvegicus*), are highly susceptible to Y. *pestis* and usually have only a low resistance. They also have one of the most efficient plague vectors, the rat flea X. *cheopis*, in the tropics. Humans can then become infected via infected fleas leaving dead rats, often in or near human habitation. In 1901, Koch accurately summarized plague as a disease of rats in which man participates.

The rat flea cannot jump very far, and thus often bites the ankles or lower legs. At the bite site there is local cutaneous proliferation of bacteria, not usually visible, before they spread via the lymphatic system to the regional lymph nodes. Here inflammation and swelling (buboes) can occur. *Yersinia pestis* probably travels to these lymph nodes by surviving and multiplying in macrophages, and there switches to extracellular replication induced by the change in temperature from about 26°C in the flea to 37°C in mammals. After 2–6 days of incubation and the formation and increase of necrotic foci containing extracellular Y. *pestis*, there is a sudden onset of illness, headache, chills and fever, total loss of energy and acute pain in the affected lymph node. The disease can be survivable at this stage: the concentration of plague bacteria in the human's blood is low and someone suffering from *bubonic plague*, as this form is termed, is not infectious. However, sometimes the disease spreads beyond the lymph nodes. This is termed *septicaemic plague* and results in a self-perpetuating immunological cascade. This leads to multiple organ failure and rapid death in almost all untreated cases. It is possible that someone suffering from septicaemic plague could be infectious via a flea vector, if the person lived for long enough to build up a sufficiently large bacterial concentration in the blood, and happened to be bitten by an efficient vector: both are unlikely.

However, once in the blood Y. *pestis* can spread to the lungs. The bacteria multiply in the pulmonary tissue and spill over the alveolar spaces, forming secondary *pneumonic plague*. Without treatment this is almost always fatal, often within hours of symptoms appearing. The disease progresses rapidly, engulfing the lungs, leading to respiratory failure and circulatory collapse. Bacteria can be expelled during coughing: however, most patients die before they can produce enough sputum, or are too feeble to produce a cough. Primary pneumonic plague results from inhalation of Y. *pestis* and this is the only means of direct human–human transmission. Severe outbreaks of pneumonic plague have occurred in humans, as we will discuss below, and untreated primary pneumonic plague is invariably fatal within 24 hours.

Since the 1940s, treatment of plague in humans has relied on the use of streptomycin, which is still the drug of choice listed by the WHO especially against pneumonic plague. Other drugs are also effective against some types of plague, and chloramphenicol was used, especially when plague had reached areas of the body that were not treatable by streptomycin. It is rarely used now because of its side-effects, including bone marrow toxicity.

An anti-plague serum and vaccine were rapidly developed during the nineteenth-century plague outbreak in China and India, and have been refined since. A whole-cell vaccine contains heat-killed bacteria and is protective against bubonic (but not pneumonic) plague, but needs multiple doses with six-monthly boosters, takes a month to become effective and has unpleasant, often unacceptable, side-effects. This vaccine has been used by military forces in areas of high plague risk, for example, Vietnam in the 1960s and 70s and more recently in the Gulf conflicts in the 1990s, but was withdrawn in 1999.

Since the 11 September 2001 terrorist attacks in the USA large grants have been available for anti-bioterrorism work, including plague vaccines. Before this there had been little incentive to develop effective vaccines or even to keep existing ones in production for a disease that rarely strikes Western populations. Currently, a recombinant vaccine is being developed that expresses two virulence factors in the plague bacterium, to generate protective host antigens. Preliminary non-clinical trials have shown that it can protect also against pneumonic plague. However, it cannot undergo standard clinical trials for efficacy because natural exposure to Y. *pestis* is rare and intentional exposure may be untreatable. In the future it is hoped that large-scale immunization of populations against plague with a dry-powder inhaler may be possible.

PANDEMICS

> Dead rats in the east,
> Dead rats in the west!
> As if they were tigers,
>   Indeed are the people scared.
> A few days following the death of the rats,
>   Men pass away like falling walls!
>
> From *Death of Rats*, Shi Daonan[1]

Fig. 16.4 *The start of the Yersinia plague pandemic in China.* **Main map**: *Spread in the nineteenth century: trade routes (1840) in green, spread of plague with dates of first record and supposed route of spread in red.* **Inset**: *Beginning of the pandemic in the eighteenth century: caravan routes in brown, first records and spread of plague in red. Red shaded areas indicate sylvatic foci of plague [Based on Benedict (1996).]*

In May 1894 a new disease appeared in Hong Kong, one that superficially resembled the medieval Black Death, and one about which scientists knew very little. This disease was to develop into a pandemic that lasted at least 50 years and claimed over 15 million lives, including 12 million in India, 30 000 in central and southern America, 7000 in Europe and 500 in the US.

The disease was plague, but as plague is a general term for an overwhelming outbreak of a pestilence (e.g. locusts; Chapter 15) it is better termed rat plague or

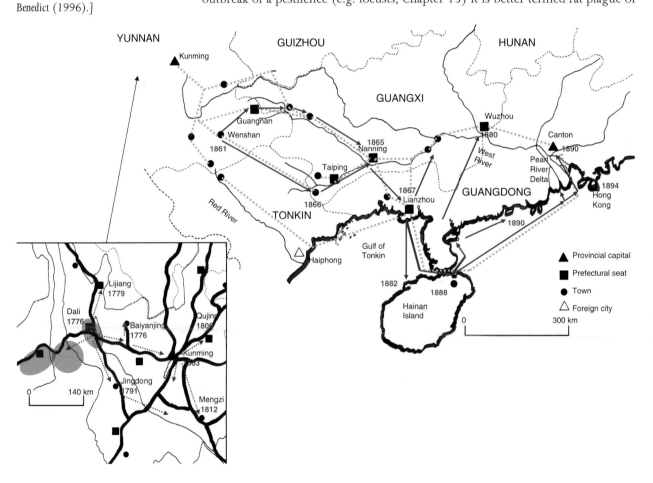

[1]Shi Daonan (1765–92) from Zhaozhou County, China. This poem describes the first outbreaks of bubonic plague in China. Cited from Benedict (1996), p. 23.

Yersinia plague. Yet this disease was not new, it had existed in the Chinese mainland for many years. Indeed, Benedict (1996) suggests that this disease first erupted in a series of epidemics in the Yunnan district between 1772 and 1825 in a natural plague focus area (Fig. 16.4). The disease was initially confined to the west of the area but gradually moved east and south-east, away from the sylvatic foci, aided by human movement. Benedict suggests that this outbreak occurred due to increasing human population density in Yunnan from 1750 following immigration into the area to work in the rapidly expanding copper mines. Bustling mining camps grew into small towns and the area was transformed from predominantly rural to urban. Plague reports from this area decrease from 1825; this may have been due to a slowing down of trade as the copper industry contracted. Then from the 1850s incidents increased, this time due to troop and refugee movements during the Muslim Rebellion in the area between 1856 and 1873. In the 1870s and 1880s plague moved to the coast, to Hainan Island and also along the south Guangdong coast, and by the 1890s had reached the Pearl River Delta. Movement of plague here was helped by a significant change in the pattern of trade, in particular certain opium trade routes along the West River, as well as overland movement of opium grown in Yunnan to the coast. From here plague reached Canton in January 1894, the springboard for its infection of Hong Kong.

The plague outbreak in Hong Kong immediately sparked global concerns and several scientists arrived to investigate, to seek the cause and also fame. Only four weeks after news of the outbreak reached Japan a well-equipped government team of six researchers, led by Prof. Kitasato who had worked in Koch's Berlin Institute, arrived in Hong Kong. Kitasato spotted the bacterium causing the disease within a couple of days and a full project report was published in *The Lancet* in August 1894. Another researcher, Alexandre Yersin, arrived in Hong Kong with two assistants, virtually unannounced three days after the Japanese team. Yersin had worked in the Pasteur Institute in Paris up to 1890, but was refused facilities in Hong Kong and had to set up his laboratory in a straw hut he constructed himself in hospital grounds and had to bribe soldiers to bring him plague cadavers. Yet he soon found the plague bacillus and succeeded in reinfecting guinea pigs with it. Although the Japanese team probably observed the bacilli six days before Yersin and published three weeks earlier, only Yersin correctly described the bacteria. He also was the only one to work with uncontaminated cultures and was first to mention the importance of rats. But for three decades British and US textbooks continued to give prominence to Kitasato, although this was not the same in France!

In Hong Kong the first plague epidemic abated later in 1894, but arose again in 1896 and then almost annually until 1929. From Hong Kong, plague soon travelled to other ports along well-established shipping routes, reaching for example Bombay in September 1896, Japan in 1898, South America, Hawaii and Europe in 1899, and the US, South Africa and Australia in 1900 (see Fig. 16.2). By 1903 it had spread to 77 ports, including some in the UK, on five continents (Box 16.2).

The reaction to, and attempts to control the spread of, this pandemic can be seen as a battle of scientific method and rational thought against ingrained superstition and irrationality. But this misses the point. Rather, the treatment of the pandemic and the still-nascent sciences of human pathology and immunology developed together and revealed, in hindsight, not only how fast science can develop but also how it can be hindered by understandable concerns and confusions.

| BOX 16.2 | PLAGUE IN THE UNITED KINGDOM |

Between 1900 and 1918, *Yersinia* plague was reported from several of the major ports in the UK. In most cases plague was confined to ships arriving from infected ports, and did not spread ashore or form sylvatic foci, but in a couple of cases this did occur.

The first authenticated case of plague in the UK occurred during 1900, in one of the poorest and most run-down areas of Glasgow. A child and her grandmother were taken ill on 3 August and died within a week. A wake was held on 12 August and the grandfather fell ill afterwards. Four members of a family that visited at this time also became ill and one died. Initially enteric fever was suspected; it was not until 25 August that plague was thought the likely cause, and a search of the infected houses found three more cases. Over 100 people who had been in contact with the infected family were taken into quarantine and given anti-plague serum. New cases were being recorded up to the end of September, including two further deaths.

The following year there was another outbreak with one death in the same part of the city. This was thought to be due to large numbers of rats driven out of tenements demolished to make way for a railway station. *Yersinia pestis* was found in this area in 1901 and in another outbreak in 1907. Thus, unless rats were repeatedly becoming reinfected from ship-borne rats, it is probable that plague existed within the rat population during this time.

The only verified case of plague striking a rural community in the UK occurred in Suffolk in the early twentieth century, and is important in demonstrating how plague may spread in northern Europe. On 13 September 1910 a child was taken ill in a cottage on the outskirts of the village of Freston, 6 km from Ipswich (Fig. 1). She died on 16 September, and her mother became ill on the 21st and died two days later. Her stepfather and a neighbour who had nursed the mother became ill on the 26th and both died on the 29th.

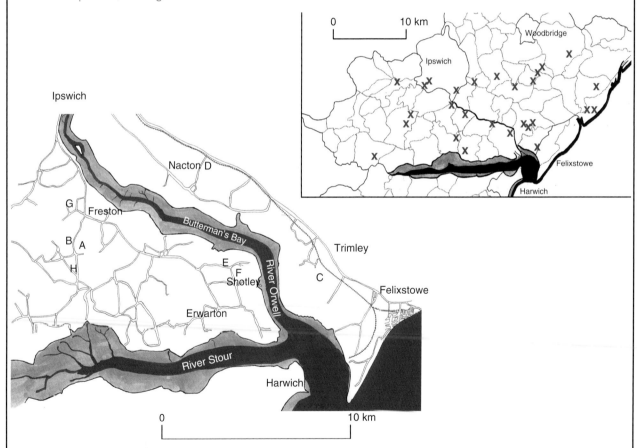

Box 16.2 Fig. 1 *Plague in England.* **Main map**: *cases of plague in Suffolk, 1906–1918;* **inset:** *location of rats infected with plague found during the July–October 1911 survey. A, primary cases of human illness at Freston; B, secondary case at Freston; C, primary cases at Trimley; D, secondary case at Nacton; E, primary cases at Shotley; F, secondary cases at Shotley; G, wood in which first rat dead of plague found; H, wood in which first hare dead of plague found. Grain was often transhipped in Butterman's Bay* [Maps based on HMSO (1911) and Eastwood and Griffith (1914).]

Plague (probably pneumonic) was not diagnosed until 28 September, too late to treat the patients with serum, and subsequent investigations revealed that for several years past large numbers of rats had been dying in the neighbourhood. Furthermore, there had probably been an outbreak of pneumonic plague in the village of Shotley, 9 km from Freston, in December 1906 that killed six, and of bubonic plague at Trimley and Nacton in December 1909 to January 1910 (infecting seven and killing four or five) in similar circumstances – isolated dwellings close to farm outbuildings and barns.

Just as the pneumonic plague outbreak in Manchuria that year caused much more public interest than the far greater death toll from bubonic plague in India, so these few cases in Suffolk caused far more interest than either. The medical papers and serious newspapers were quick to try to dispel panic, stating, rightly, that this plague was different from the Black Death and that the return of an epidemic of that ferocity was not possible. This seems to have worked, and while there was great interest shown, no panic was reported. The *British Medical Journal* was quick to castigate the victims of the plague: 'the disease had picked out two very dirty households … of all the people who had come into contact with the patients, only those who had but little respect for hygiene were infected',[a] a common class-riddled assertion of the time.

The outbreak was treated by quarantining contacts and cleaning affected houses, and a programme of rat destruction over a wide area was started. Although it was stated that this was 'willingly assisted by the public',[a] owing to the lack of cooperation

[a]Sleigh (1910), p. 1489.

of some landowners a Rat Destruction Order had to be issued in November 1910 enabling district councils to undertake rat control on private land. It was thought that the plague had arrived in rats from grain ships that had either transhipped some of their load in the river Orwell, nearby the affected villages, or from the port of Ipswich itself.

Three surveys of rats in the area were carried out. In the first during November and December 1910, 17 areas were investigated, not where rat deaths were recorded but rather where access was granted. Almost 600 rats were examined and 17 infected ones in six locations were found, along with two infected rabbits. A prohibition on catching and transporting rabbits was subsequently put in place in the Ipswich area. In January and February 1911 a wide area of West Suffolk and North Essex was examined and, although 6000 rats were examined, no *Y. pestis* infection was found. This area was re-examined in the summer of 1911 when local police constables were asked to coordinate the survey, receiving (and paying 2d each for), disinfecting, and packaging up rats for transport to Ipswich. In four months over 15 000 rats were examined, and 28 firm cases of plague infection were recorded. No black rats or *X. cheopis* fleas were recorded, but the commonest rat flea *Nosopsyllus fasciatus*, was found to bite man, and some from infected rats contained the plague bacilli. However, the density of fleas found on Suffolk rats was low compared to India.

In 1918 two people died from pneumonic plague in the small village of Erwarton, close to Shotley. This was the last outbreak of plague in the UK. Thus, it is likely that a localized enzootic plague infection existed in rodents in this area between about 1906 and 1918.

The immediate reaction of authorities in many plague-stricken ports was to isolate cases of the plague, often with very intrusive searching and forced removal of suspected cases to plague hospitals. Because in most cases bubonic plague was involved this isolation was unnecessary, but conversely in outbreaks of pneumonic plague, sometimes uninfected people were prevented from leaving infected buildings – effectively a death sentence. Treatment of the cause of the outbreaks was initially limited to disinfecting houses: in Bombay in the 1890s millions of gallons of carbolic acid were sprayed, and in Honolulu large areas of the city with concentrations of the plague were burnt by the fire services operating in *Fahrenheit 451* mode. All these activities probably only served to increase the rate of spread of the disease, by flushing infected rats into other areas of the city.

Gradually, the biology of the disease became better understood. Vaccination, although not without risks itself, was the first success, being first used extensively during the 1899 outbreak in Oporto, Portugal. Slower was the determination of the mode of spread of the epidemic, with conclusions from scientific investigations coming up against ingrained ideas that human-to-human spread was of primary importance. Through sterling work by successive teams of investigators from the Indian Plague Commission set up in 1905, and pioneering work in Australia and the US, it gradually became accepted, first, that rats and other rodents were the main carriers, and later that fleas were the main transmitters of plague between

MONGOLIA

INNER
MONGOLIA

SIBERIA

MANCHURIA

Trans-Siberia Railway

Manchouli
13.9.10

21.1.21

22.10.20

27.10.10
Hailar

27.10.10
22.1.21
Harbin

31.12.10
Changchun

Kirin

31.1.21

9.4.21

Vladivostok

5.1.11

29.3.21
Mukden

Peking

Tientsin

4.1.11
21.1.11
Dairen
Chefoo

3.5.21

Seoul

SEA OF
JAPAN

Tsingtao

YELLOW
SEA

KOREA

0   500 km

Fig. 16.5 Yersinia plague in Manchuria. First occurrence and spread during the 1910–11 (red) and 1920–21 (purple) pneumonic plague outbreaks are shown. Contemporary town names are used. **Inset**: The tarbagan, Marmota sibirica [Sketch map based on Chernin (1989); dates from Lien-Teh (1922).]

[2] The numbers of tarbagan have been reduced so much by centuries of hunting and plague that in 2008 it was classified as endangered by the IUCN.

rodents and humans. The latter was first reported by Paul-Louis Simond in 1898 but was not fully accepted, even after being experimentally demonstrated, for over a decade. Once the life-cycle of the disease had been determined then appropriate measures could be taken, controlling rodent populations near habitations, and later, with effective insecticides, control of the fleas themselves. The systematic spraying of the inside of houses against malarial mosquitoes from the late 1940s almost certainly also helped (Chapter 12).

In most cases the plague reported was bubonic, but it was recognized early that another, pneumonic, could occur, and it was later realized that this was far more infectious and could, unlike bubonic plague, be transmitted from human to human. In more or less exceptional circumstances, serious outbreaks of pneumonic plague could occur, the most famous being those in Manchuria (Fig. 16.5). These outbreaks occurred in areas with natural sylvatic foci for plague, the main reservoir probably being the tarbagan, Marmota sibirica, a 4–5 kg burrowing rodent.[2] This had long been hunted for its fur, but indigenous hunters had learnt to avoid sick animals. However, in the early twentieth century it was discovered in the US and Europe that, with appropriate dying, tarbagan fur could be made to resemble much more expensive sable and seal skin, and a great demand for it arose. By 1909 the price paid for a tarbagan skin had risen six-fold, and this spurred many people from central China to migrate to tarbagan areas to hunt them. The centre of this trade was the railway town of Manchouli, from which it is estimated that the skins of two million tarbagans, caught by 10 000 hunters, were exported in 1910.

In this inhospitable environment, with winter daytime temperatures averaging −20°C, the tarbagan hunters lived in overcrowded, badly ventilated semi-underground inns. The tarbagans they caught were skinned here, so it is not surprising that, with the main flea species on this rodent quite capable of biting humans, plague should develop (although we are not certain of the series of events that led to the development of pneumonic plague). Six hundred people soon died from plague in Manchouli. Those who could fled the area, often via overcrowded trains, and within weeks pneumonic plague had broken out at Harbin, where 10% of the population was killed, and continued further along the railway system (Fig. 16.5). In all, this outbreak is thought to have killed 60 000 people.

There was a smaller outbreak in the same area in 1920–1, again spread though workers, this time miners, huddled in semi-underground huts. The

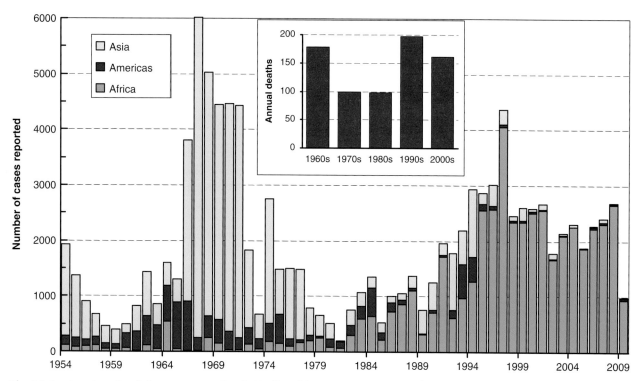

Fig. 16.6 *Annual number of plague cases reported to the World Health Organization since 1954 for Asia, the Americas and Africa.* **Inset:** *Mean annual number of reported deaths from plague [Data from Dennis et al. (1999) and WHO (2010).]*

infectiousness of this disease is underlined by the fact that, although strict quarantine measures were put in place, a couple of infected soldiers who escaped from the centre of the outbreak at Hailar were able to initiate the epidemic, with a total of 9300 lives lost. For example, in the mining community of Dalainor an infected person arrived on 2 January 1921 and settled in a bunk-house with 17 other occupants. Within a fortnight all were dead from pneumonic plague; and in a neighbouring hut 61 out of 64 workers died in three days from pneumonic plague. In total 1017 deaths were recorded here from a population of 6000. Small human outbreaks of plague occur to this day in tarbagan hunting areas in Mongolia and Manchuria.

The lasting legacy of this pandemic is sylvatic foci of plague located throughout the world (see Fig. 16.2), from which reinfection of humans occurs at intervals (Fig. 16.6). Plague is still caught from fleas transferring from dead or hunted rodents, and in some countries (e.g. western USA) as the population expands into plague foci so the numbers of cases rise and some more unusual modes of transmission are found. For example, between 1977 and 1998 domestic cats (unusually for a carnivore very susceptible to *Y. pestis*) were the source of 23 cases of plague in the US, mainly caused by bites or scratches, but in five cases pneumonic plague was caused by inhalation of cat cough. Transmission of plague to humans from pets is a rare and unusual – but not impossible – scenario which made it suitable for the attention of Dr Gregory House in an episode of the television drama series *House*. In this case the supposed route was via infected fleas from a prairie dog to a pet dog and then to its owner.

However, in some parts of the world plague is not a medical curiosity, but is still a significant problem. In India in 1994[3] and Algeria in 2003[4] plague reappeared in humans after many years absence, and in Madagascar and the Democratic

[3]In August 1994 an unusual number of rats were recorded dying in the village of Mamla, 150 km south-east of Surat, in Western India, but this was ignored, and three weeks later people began to die of bubonic plague. This spread to Surat in September 1994, where after starting bubonic it became pneumonic. There was widespread panic and upwards of 300 000 people fled the city, the largest movement of people in India since independence, spreading plague to four neighbouring states. Hospitals were overwhelmed with patients reporting plague-like symptoms, and it took two months to bring the outbreak under control, with almost 900 cases detected and the loss of 56 lives, mainly in Surat. This plague outbreak, the first in India since 1966, cost the Indian economy US$ 1.7 billion as panic set in and countries imposed stringent travel and trade restrictions. The origin of the outbreak is still unclear. One theory is that an earthquake in the area in October 1993 caused people to flee Mamla, leaving large quantities of grain – food for domestic rats – causing a population explosion and rodent epizootic. The outbreak triggered improvements in the Indian plague control system, and a subsequent outbreak in Shimlu (Himachal Pradesh) in 2002 was contained quickly with only 16 cases and four deaths recorded.

[4]Plague struck Algeria in 2003 after being absent from the country since 1950. In June and July, 18 cases (one fatal) of (continues over)

Republic of the Congo (DRC) plague is an ever-present and increasing threat. For example, in 2005, 130 cases with 57 deaths from pneumonic plague were reported from a diamond mine in the DRC, and the following year over 1000 cases including 50 deaths were reported from another area in the country.

WHAT WAS THE BLACK DEATH?

As soon as Yersin found the causal agent for the Hong Kong plague in 1894 he proclaimed that he had found the cause of the Black Death. This view soon became accepted, although it was only based on the broad similarity in symptoms, ones not exclusive to either disease, and by 1900 the *British Medical Journal* was able to state:

> It is, we believe, now generally accepted that the pestilence which over-ran the Roman Empire in the sixth century, the Black Death, which caused so awful a mortality in the fourteenth century, and the plague of the seventeenth century, were identical with the plague (*pestis*) which has in recent years caused so many deaths in India, and is now threatening Western Europe and both North and South America.[5]

The Black Death caused 50 million deaths between 1347 and 1350, half of them in Europe, and was the start of a series of plague outbreaks in Europe that lasted into the eighteenth century, and Justinian's Plague between 542–6 in Asia, Africa and Europe claimed at least 100 million lives.

Cohn (2002) makes the case that this assumption that previous plagues and that erupting in Asia were identical hindered the study of present-day plague. It is as if a curious cross-infection through time had taken place: this is a literary device used to great effect in the novels of Barry Unsworth and Peter Ackroyd, among others, but with serious consequences if it occurs in science and scholarship. For example, in contemporary records Black Death was not associated with rats, but was very contagious, with the assumption of high person-to-person infectivity. This could have delayed the acceptance of scientific evidence that rodents were the main carriers of the modern plague, and that the rodent-to-human vector was the flea. The human health implications of this delay were severe; with the assumption of high infectivity between humans, people were isolated, and the early emphasis on cleaning or destroying buildings in infected areas, rather than rat or flea control, could have given *Yersinia pestis* the time needed to spread through the commensal rodent population and to form new sylvatic foci, for example in North America. The greatest legacy of this assumption has been mankind's desire to weaponize *Y. pestis*, discussed below.

Eventually, the scientific study of *Yersinia* plague did win out and the modern disease was characterized, although even into the 1930s there was some concern that it did not match what was known about the Black Death. However, the assumed identicality then meant that previous plagues had to take the form of the modern plague, in the classic form worked out in India. Thus, the

Fig. 16.7 *It has been an unwritten rule that books on the Black Death should include an engraving of a rat. Here is one that has been used. Thomas Bewick produced this wood engraving in the late eighteenth century, over 100 years after the last plague occurrence in the UK*

8   The Rat

bubonic plague were recorded from two poor rural settlements near Oran. Subsequently a persistent zoonotic focus of *Y. pestis* in *X. cheopis* fleas was discovered nearby, possibly representing a new plague focus. The last outbreak, in Oran, was in 1946, at the time Albert Camus was writing *La Peste* (The Plague) a fictional account of a plague outbreak in that city.

[5]Anon (1900), p. 1247.

Black Death became a disease in which the black rat and rat flea *Xenopsylla cheopis* were pivotal. By the second half of the twentieth century this was the accepted view on the cause of the historical plague outbreaks, accompanied in books with an obligatory engraving of a rat (Fig 16.7).

However, in the last 25 years there has been increasing disquiet with this accepted view, particularly articulated by Twigg (1984, 2003), Cohn (2002) and Scott and Duncan (2004). Between them these authors have challenged most of the assumptions underlying the description of historic plague, especially the Black Death in Europe, as *Yersinia* plague. We will examine some of their arguments below; a more radical proposal is summarized in Box 16.3.

One of the striking features of modern bubonic plague outbreaks is that they are preceded by mass deaths of commensal rodents. In Asia, where *Yersinia* plague has been endemic for centuries, people learnt to leave their towns and villages when these deaths occurred, to avoid infection; and in the nineteenth century the contemporary name for the plague in China was *shuyi*, 'rat epidemic'. Yet there are almost no records of this occurring prior to outbreaks of historic plague in Europe. Perhaps people were not interested in rodents and did not record them, but then why should other cultures be more observant?

For medieval plague outbreaks to be so pervasive throughout Europe, it has been necessary to suggest that rats were very common throughout Europe during this period. We can ignore the brown rat, as this did not spread throughout Europe, from central Asia, until the eighteenth century, after historic plague had ceased. Thus the black rat is implicated. Today, although common in southern Europe, it is rare in northern Europe, and in the UK it is thought to be facing extinction with less than 1500 individuals left. The history in the UK of the black rat, native to India and south-east Asia, still uncertain, but it is now thought that it was introduced by the Romans and became established in at least some ports and major cities. It then died out in the centuries after the Romans left until being re-introduced by the ninth century, and then occurs in many medieval archaeological surveys. However, it is not at all clear that the black rat was ever common outside of ports and large towns, where rat populations were continually replenished by immigrants from shipping, and this sedentary and non-migratory species was unlikely to have been common enough to account for the spread of the Black Death through the British countryside.

This problem of how plague infected a high proportion of a sparsely distributed population in a cold climate is exemplified by Iceland, which was visited by two plague epidemics in the fifteenth century. These had the same characteristics as those occurring elsewhere, yet there were no rats present on Iceland at the time.

It is also unlikely that the rat flea X. *cheopis* was present in large numbers in northern Europe: only a few have been recorded in England, and these were associated with rats in large ports. The northern European climate would have been too inclement for this semi-tropical species to thrive and breed. Of course Y. *pestis* can be transmitted by other flea and rodent species, but none of the flea species present in Europe is nearly as efficient a vector as X. *cheopis*, and the common European rat flea *Nosopsyllus fasciatus* is rarely encountered on humans. It is perhaps most telling that Europe never developed a plague focus after 400 years of supposed infection; and – unlike parts of Africa, North America and South America – Europe never developed a persistent sylvatic focus of Y. *pestis* as a result of the recent spread of the disease. This suggests the lack of suitable conditions, fleas or rodents.

Thus, while it is quite possible that *Yersinia* plague was present in the warmer Mediterranean ports, and could have produced local outbreaks throughout history,

BOX 16.3 IT CAME FROM OUTER SPACE?

Contemporary writings on the Black Death often mentioned that it came from the skies. For example Boccacio, writing *The Decameron* during the Black Death between 1348 and 1353, stated:

> In the year 1348..., that most beautiful of Italian cities, noble Florence, was attacked by deadly plague. It started in the East either through the influence of the heavenly bodies...[a]

However, this is not meant to be taken literally as he finishes the sentence:

> ... or because God's just anger with our wicked deeds sent it as a punishment to mortal men;

Thus, the plague of Black Death is a visitation similar to that of locusts (Chapter 15).

More recently, however, this has been taken more literally. In the later nineteenth century an ancient idea that the 'seeds' of life exist all over the universe, and that life on Earth originated from space, was revived by several notable scientists including Kelvin, von Helmholtz and Arrhenius. Although popular for a while this panspermia theory was largely discredited by the 1930s when the intense radiation present in outer space was discovered and thought to preclude the survival of life there.

The theory did survive, however, as an underground thought, emerging occasionally through alternative religions, and with Joni Mitchell's assertion of the Woodstock Generation 'We are stardust' in the late 1960s. It reappeared more fully developed in Sir Fred Hoyle's and Chandra Wickramasinghe's articles and books from the mid-1970s. They articulated the views that not only are interstellar dust grains bacteria, but pandemics such as influenza or HIV appeared from comets. As a hypothesis this is neither easily provable nor falsifiable, and thus has little explanatory value. Certainly there is little evidence in support of it, and it has been discounted by most scientists.

In 2006, however, Baillie suggested a link between panspermia and the Black Death. He investigated several preserved environmental indicators of the period; for example, tree rings, radiocarbon evidence and chemicals in ice cores. This suggested to him that the atmosphere had become corrupted with a poisonous gas, possibly ammonia, by early 1348 and that

this led to the reduced growth of trees in many parts of the world and a massive ocean turnover event. The cause of this, he suggests, based on contemporary written records and comparison with modern occurrences, was a large comet impact.

Thus, he suggests that the Black Death was caused largely by poisonous vapours released by the comet, with 'just the possibility of some biotic material from space'.[b] His re-analysis of the spread of the Black Death through Europe (Fig. 1) provides support for this.

At present it is hard to know what to make of this ingenious theory: apart from correlations there is little evidence for it and it awaits serious scientific review. On the one hand it raises more questions than it answers. For example, evidence for this comet impact is needed. Why did its effects bypass certain areas? And if the initial Black Death outbreak was due to 'bad air', which after a comet impact would be relatively short-lived, what caused the repeated outbreaks of plague for the next couple of centuries? On the other, we are becoming increasingly aware of the possible effects of climate on plague, and any stimulation of studies on climatic change before and during the major pandemics is likely to take us closer to the reason for their outbreak.

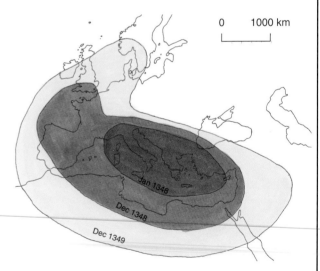

Box 16.3 Fig. 1 *Baillie's suggested spread of the Black Death, assuming a true start in January 1348 (compare with Fig. 16.8). Baillie suggests that this is consistent with a pathogen descending from the atmosphere* [Based on Baillie (2006).]

[a]Boccaccio (1954), p. 23.
[b]Baillie (2006), p. 198.

without a sylvatic focus, widespread effective amplifying host or vector, it is unlikely that it could have spread throughout Europe in the manner of the Black Death. Yet, even if these two agents had been present, Black Death produced a far higher mortality rate than *Yersinia* plague in India, the country where it persisted for longest during the nineteenth and twentieth centuries, and in which the majority of the deaths during this pandemic occurred. Even though the mortality from the Indian pandemic has often been downplayed, it never approached the 20–40% estimated from the Black Death: 3% maximum, and usually much less, was recorded. Furthermore, Black Death, at a time of slower transport, spread distances in days that took the *Yersinia* plague years to cover in China in the nineteenth century (compare Fig. 16.8 and Fig. 16.4).

Pneumonic plague is much more infectious than bubonic plague, and it can spread very quickly if unchecked. Although in the twentieth century Manchurian outbreaks mortality was usually less than 3%, in some towns and settlements 30% mortality was recorded. These factors have led some to suggest that the Black Death was pneumonic plague. However, pneumonic plague outbreaks start from bubonic plague cases, and still need rodents, fleas, and sylvatic foci to occur. Most outbreaks have been short-lived, and it would be difficult to sustain a long-term pandemic of pneumonic plague without these other players in the plague cycle.

Although often criticized as inadequate for epidemiological studies, medieval records of plague outbreaks suggest a disease rather different from that of modern *Y. pestis*. For example, deaths from Black Death peaked in the summer, while those of modern plague usually peak in the spring and autumn. Over time there is the suggestion that adults became resistant to the medieval plague, and it became a disease of children and newcomers to a plague area. Scott and Duncan (2004) have examined the detailed parish records from Penrith, on the edge of the English Lake District, during its late sixteenth-century plague outbreak and have calculated its epidemiology (Fig. 16.9). This emerges as quite unlike *Yersinia* plague, and the key to the success of the medieval plague lay in its very long incubation period – it could be spread widely, and many people could be infected before symptoms appeared. According to their calculations, the medieval quarantine period of 40 days for visitors from plague-infected areas appears highly appropriate.

On the other hand, it can be reasonably suggested that medieval human populations were not the same as modern ones, and did not react in the same way to *Y. pestis*. *Yersinia pestis* is also a recently evolved pathogen (see Box 16.1), so perhaps it has changed and was much more virulent in the past?[6] We are unlikely to be able to answer these questions definitively, but one can seek concrete evidence of *Yersinia* plague in medieval people, as well as evidence that it can behave

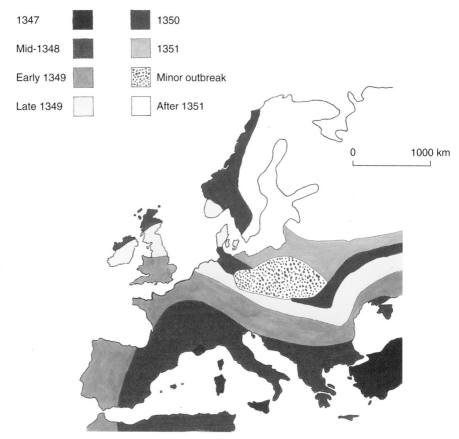

1347	1350
Mid-1348	1351
Early 1349	Minor outbreak
Late 1349	After 1351

0   1000 km

Fig. 16.8 *Spread of the Black Death throughout Europe [Based on WHO (2007).]*

[6]Yet the mode of transmission via tiny numbers from a flea would lead to the maintenance of high virulence in mammals. Furthermore, as humans are a dead-end host for the bacterium there would be no selective pressure for a reduction in virulence in humans, as there is in human-to-human transmitted diseases. No reduction in *Y. pestis* virulence has been demonstrated.

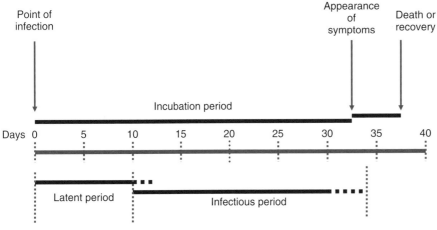

Fig. 16.9 *Time scale and sequence of events for the plague that struck Penrith in the late sixteenth century, as calculated by Scott and Duncan. The latent period is the time during which the germs multiply before the person becomes infectious. Note the long incubation period and infectious period before symptoms appear, quite unlike Yersinia plague [Based on Scott and Duncan (2004).]*

in the manner necessary to have caused the historic pandemics. Such evidence has recently been sought. In 1998 a French laboratory claimed to have detected the bacterium from dental pulp DNA from three skeletons of suspected plague victims buried in Marseilles in 1722 and two corpses buried in southern France during 1590. Subsequently, this laboratory reported finding the bacterium in 20 out of 23 teeth examined from three fourteenth-century French skeletons, and they stated 'we believe that we can end the controversy: Medieval Black Death was plague'.[7] Of course their results cannot demonstrate this, only that Y. *pestis* was present in southern France at those times – which is not unexpected. Furthermore, these results have been heavily criticized, by Gilbert *et al.* (2004) in particular. Using similar techniques they examined 108 teeth from 61 individuals from five probable plague sites in three European countries and could find no evidence of Y. *pestis*. They suggest that the high success rate in finding Y. *pestis* in such old specimens is questionable, and probably due to contamination from modern Y. *pestis*, with which the French laboratory had worked.

Detecting archaeological DNA is in its infancy, and has numerous problems, contamination being but one. A recent technique to detect signs of ancient Y. *pestis* avoids some of these problems by instead using a diagnostic protein test for plague usually used in living patients. Using this technique, Bianucci *et al.* (2009) detected signs of Y. *pestis* in the skeletons of four Benedictine nuns and two priests exhumed from two sixteenth- to eighteenth-century burial sites in central France. This laboratory has used this test to detect plague in other French burial sites and we wait with interest confirmation by other laboratories of the veracity of these tests.

Now that it is becoming accepted that there is a problem in reconciling Black Death with modern plague, researchers have begun to reinvestigate the transmission of modern plague. For example, Drancourt *et al.* (2006) have proposed that human ectoparasites enable human-to-human transmission of bubonic plague, but the vectoring capabilities of these parasites remain unproven. Furthermore, it has been suggested that the soil may act as a reservoir for Y. *pestis*, from which mammals may become infected. However, although recent work demonstrates that Y. *pestis* can survive in soil in an infectious state for at least a month in natural conditions, there is as yet no evidence that these soil-borne bacteria can naturally reinfect organisms.

If not *Yersinia* plague, then what caused the historic outbreaks of plague? Here we can be less sure, but Scott and Duncan (2004) suggest that it was a haemorrhagic viral infection, similar to modern Ebola or Marburg RNA Filoviruses. Ultimately, we are unlikely ever to know. However, we have to move away from the concept that the Black Death or the other historic plague pandemics were caused by any one disease, to a view that recognizes that a whole suite of infections could have contributed to these disasters.

[7]Raoult *et al.* (2000), p. 12800.

YERSINIA AS A BIOWEAPON

Almost as soon as *Yersinia* plague had been suggested as the cause of the Black Death, ideas to use it as a military weapon surfaced. For example, in September 1916 a German officer suggested dropping vials of plague bacilli from Zeppelin airships onto British ports. His proposal was rejected out of hand. The Geneva Protocol of 1925 included a ban on the development of offensive biological weapons, but left the door open for countries to develop a defensive biological warfare capability. Several countries, including Britain, the USSR, the USA and Canada carried out limited work on plague bioweapons, often fuelled by mutual paranoia.

Most of this work only involved laboratory attempts to produce a stable aerosol plague formulation, but the US developed a virulent plague strain and were about to build a small-scale pilot plant as the Second World War ended. The USSR carried out field tests of plague on isolated islands, and possibly also carried out work on weaponizing the flea vector. Much of the Soviet capability was reduced by the late 1930s Stalinist purge of scientists, but there are unconfirmed German reports of Soviet tests of plague on prisoners – in one case asserting that chained prisoners were exposed to plague-infected rats, and also that one prisoner escaped and a plague epidemic erupted in Mongolia shortly afterwards, causing 3000–5000 deaths.

Before the Second World War there was great concern in the UK that Germany was developing bioweapons; for example the respected scientist J.B.S. Haldane pessimistically stated in a book on air-raid precautions in 1938:

> Possibly pneumonic plague or some other air-borne disease might be started by a dust-bomb. … A million fleas weigh very little and could easily be dropped. In theory they could be infected with plague. … Some may very well be tried, if only to create a panic.[8]

In fact, Germany carried out little research into biological weapons before or during the war, but did produce a large quantity of plague vaccine against a possible Soviet plague attack. German scientists concluded, reasonably, that even aerosol Y. *pestis* application was unlikely to produce massive epidemics in countries with good hygiene, but they are reported to have experimented on the use of the rat as a plague vector. This involved determining how far rats could swim, as part of an idea to release plague-infested rats near the shore from U-boats.

However, one country, Japan, did make a concerted effort to weaponize plague. Along with the US, Japan did not ratify the 1925 Geneva Protocol, and Japanese scientists began work on plague weapons in the early 1930s in Manchuria (which they had just invaded). Subsequently, several biowarfare research facilities were built, including possibly the largest ever biowarfare facility near Harbin (see Fig. 16.5), which included more than 150 buildings housing over 3000 scientists, a private railway station, airport and fleet of aircraft. In 1941 this became the notorious Unit 731, and the Japanese bioweapon network grew throughout Asia with at its peak up to 20 000 scientists, doctors and nurses working in two dozen satellite facilities. Research of the most barbaric kind was undertaken here, testing plague and its vectors on humans, often prisoners of war, who were sometimes chained to stakes, exposed to plague bioweapons (e.g. aerial bombs) and then dissected alive to determine the progress of the disease. By 1939 these facilities had the capability of producing 300 kg of Y. *pestis*, using thousands of captive rats.

[8]Haldane (1938), p. 38.

The Japanese army then proceeded to use plague for the first (and hopefully the last) time as a bioweapon, during the Japanese advance through China in the late 1930s. First, bacterial bioweapon artillery shells were fired at Soviet troops in 1939, and plague erupted on both sides soon afterwards. Plague-infected rats were also let loose in densely populated communities, and again plague soon appeared. In Ningbo Region (south of Shanghai) in 1940, plague-infected fleas were sprayed from the air, and this was repeated in 1941 against Changteh City, when over 100 million fleas (including the human flea *Pulex irritans*) were released. In both cases deaths from plague were reported afterwards. In the worst incident, the Japanese carried out a multiple pathogen attack in the summer of 1942 (including anthrax, paratyphoid, cholera, dysentery, typhoid and plague) on the cities of Yushan, Kinhwa and Futsing. However, this backfired, and there were up to 10 000 Japanese casualties. Despite this, Japan never succeeded in developing a totally effective plague delivery system, and the number of deaths actually resulting from plague weapons remains open to doubt – they were being used in an area that was already a plague focus, and the outbreaks could have naturally erupted. But the death toll from Japanese plague and bioweapon experimentation, which runs into the thousands, cannot be disputed.

In September 1945, Soviet troops overran Unit 731 in Manchuria and their documentation was sent back to Moscow. This enabled the USSR to resume plague research, with the aim of matching and surpassing the Japanese output. By the 1980s the USSR possibly had a capability for manufacturing several hundred tonnes of Y. *pestis* per year, and was working on a transgenic plague bacillus. Luckily, the Soviet Union collapsed before this could be weaponized, and Russia reported in 1992 that it had destroyed its bioweapon capability.

Although there were at least five post-war US investigations into Japanese biowarfare, no charges were ever brought against the principal figures involved. It has been alleged that this was because immunity from prosecution was granted to those who provided biowarfare information to the American military. Some US work on plague was undertaken in the 1940s and 1950s, and although by the early 1950s it was their bioweapon of choice for militarization, progress was so slow that they turned to anthrax instead. In 1969, the USA unilaterally renounced offensive biological warfare, and dismantled its programme, and in 1975 it finally ratified the Geneva Protocol.

The UK also had an offensive bioweapon programme, and plague was one of the agents investigated. During the Second World War the Biology Department at Porton Down, Wiltshire, worked on aerosolizing Y. *pestis* and demonstrated that this could kill small mammals in the laboratory, and two trials with an 'avirulent plague' bacillus were carried out on Porton Ranges in 1944, and trials continued into the 1950s with the aim of developing an anti-personnel bioweapon with comparable effect and importance to the atomic bomb (Box 16.4).

The repeated attempts at weaponizing plague are the worst legacy of the assumption that Y. *pestis* caused the Black Death. Unlike the anthrax bacillus, Y. *pestis* does not form spores, and survives for only a short time outside its host even as an aerosol. Thus, it is not surprising that no one has developed an effective bioweapon using aerosolized plague bacteria. However, the threat of terrorist-caused outbreak of plague is still taken seriously (the US Centers for Disease Control and Prevention, Atlanta, consider Y. *pestis* as a type A bioterrorism threat) and risk planning for this eventuality has taken place in the US.

BOX 16.4	OPERATION CAULDRON AND THE *CARELLA*

Between May and September 1952, *Yersinia pestis* bioweapons were tested by the British military off the north-east coast of the Isle of Lewis, Outer Hebrides, in Operation Cauldron. Here, plague munitions (aerosol sprays and bombs) were tested against guinea pigs and monkeys on a large floating pontoon. Although considered a success at the time, a year later these trials were recorded as a failure, and plague bioweapons were not tested again outdoors in the UK.

On the last day of the trials, the Hull steam trawler *Carella* with 18 crew strayed into the exclusion area around the tests, despite repeated warnings. However, anxious to complete trials in favourable weather, the last plague release was carried out regardless, and the trawler passed within 3 km of the test pontoon and into the path of the cloud of plague bacilli. The trawler arrived at Fleetwood a day later and her crew were allowed to disembark. Because of the secrecy surrounding the trial it was decided not to intervene and treat the crew with antibiotics and not to inform anyone of the possibility of infection, as any treatment, cleaning, and destruction of rats was thought likely to draw attention to the nature of the trials. But after the crew disembarked, the Chancellor did write to the Admiralty emphasizing that he was relying on them to deal with any plague outbreak that might result.

The *Carella* left Fleetwood for Icelandic waters a couple of days after and was shadowed by a Royal Navy destroyer, waiting for a distress call signalling crew illness. The destroyer carried a doctor who had been briefed on the situation and was supplied with antibiotics, but orders were given that the trawler was to be prevented from landing in Iceland, or anywhere except back in Fleetwood. Luckily, none of the crew became ill, and as the trawler arrived back in Fleetwood, the shadowing was called off as the supposed incubation period of the disease had passed. Most of the documents relating to this incident were destroyed soon afterwards and the trawlermen involved only became aware of it over 50 years later when approached by BBC documentary makers.

As an example of the potential effects of extreme secrecy this incident is salutary. In Joseph Losey's disturbing British anti-war film *The Damned* (1961) some adults stumble across radioactive children being reared in a secret underground military installation. Any creature that comes in contact with the children dies from radiation sickness, which the children call 'the black death'. The film ends with a powerful and depressing long aerial shot of the motorboat carrying the hero and heroine, now dying from radiation sickness, making its way into the English Channel. The boat is shadowed by a helicopter which has instructions to sink it after the crew have died. The image of a trawler with potential victims of another black death sailing off into the North Atlantic shadowed by a destroyer is equally vivid, more so for being fact not fiction.

CONCLUSION

We are in a curious situation with plague at present, with much of the global concern over it, and the effect of potential terrorist use, based on the supposition that it is essentially the same disease as that which caused the Black Death. This is looking increasingly uncertain, but is unlikely ever to be proven to the satisfaction of all. At least this possibility is being considered in a number of recently published books on historic plague, for example Gummer (2009).

However, one should not underestimate plague either. Incidents of plague have been increasing in recent years (see Fig. 16.6), and sylvatic foci have been increasing in area. Cases in China have increased 25-fold in the 20 years to 2000, and a large increase in the numbers of rats on ships has been reported in the last ten years, along with a re-emergence of the black rat. Changing climate may affect plague: for example, in Kazakhstan recent warmer springs and wetter summers have led to an increase in prevalence of plague in its main hosts, the great gerbil *Rhombomys opimus*. It has been projected that a 1°C increase in spring temperature could result in 50% increase in Y. *pestis* in its enzootic hosts. All this has led the WHO to reclassify plague in 1996 as a re-emerging disease rather than a dormant one.

A pestilence isn't a thing made to man's measure; therefore we tell ourselves that pestilence is a mere bogey of the mind, a bad dream that will pass away. But it doesn't always pass away and, from one bad dream to another, it is men who pass away.[9]

FURTHER READING

Between 1900 and 1930 the *Journal of Hygiene* published much of the groundbreaking work on plague, from early reports from India, the USA and Australia to the Indian Plague Commission, and commissions of enquiry in Manchuria and the UK. A comprehensive account of plague biology is given by Dennis *et al.* (1999), shorter summaries by Gage and Kosoy (2005), Prentice and Rahalison (2007) and Stenseth *et al.* (2008). Current work on plague vaccines is reviewed by Williamson (2009) and Eisen and Gage (2009), and Wimsatt and Biggins (2009) highlight gaps in our knowledge of the plague cycle. For the possible effects of climate on plague see Stenseth *et al.* (2006), and for the current plague situation see WHO (2008, 2010) and www.who.int/.

The spread of the nineteenth-century plague pandemic is described by Benedict (1996) and Echenberg (2007). See Chernin (1989) and Gamsa (2006) for pneumonic plague in Manchuria, and in Algeria Bertherat *et al.* (2007).

Justinian's plague is well covered by Rosen (2007). There are many studies on the Black Death, for classical studies (which need to be read with the caveats discussed above) see Ziegler (1969) and Benedictow (2004), and for an up-to-date account of Black Death in the UK see Gummer (2009). For plague in Iceland see Karlsson (1996), and for the black rat in history see Armitage (1994), Yalden (1999) and McCormick (2003). The three main alternative accounts of the Black Death are Twigg (1984), Cohn (2002) and Scott and Duncan (2004), summarized in Duncan and Scott (2005). Recent work by Theilmann and Cate (2007) and Welford and Bossak (2009) are also important. For the ancient Y. *pestis* DNA controversy see Raoult *et al.* (2000), Drancourt and Raoult (2002), Wood and DeWitte-Aviña (2003), and Gilbert *et al.* (2004), and for the use of immunodetection of ancient plague see Bianucci *et al.* (2009). For the possibility of Y. *pestis* being soil-borne see Drancourt *et al.* (2006), Eisen *et al.* (2008) and Ayyadurai *et al.* (2008).

For the weaponization of plague, Lockwood (2009) is a good general account; for more detail see Geissler and van Courtland Moon (1999) and Wheelis *et al.* (2006). For the UK specifically see Hammond and Carter (2002), although for details of operation Cauldron and the *Carella*, one needs to read Balmer (2004). For a recent example of risk planning for coping with a plague bioattack, see Casman and Fischhoff (2008).

For the evolution of *Yersinia pestis* see Achtman *et al.* (1999) and Zhou and Yang (2009). For bacterial evolution in general see Pallen and Wren (2007). For plague in Suffolk, see HMSO (1911), Eastwood and Griffith (1914) and Twigg (1984). For an up-to-date introduction to panspermia see Warmflash and Weiss (2005).

REFERENCES

Achtman, M., Zurth, K., Morelli, G., Torrea, G., Guiyoule, A. and Carniel, E. (1999) *Yersinia pestis*, the cause of plague, is a recently emerged clone of *Yersinia pseudotuberculosis. PNAS* **96**, 14043–14048.

Anon (1900) Pandemic plague. *British Medical Journal* **2**, 1247–1255.

Armitage, P.L. (1994) Unwelcome companions: ancient rats reviewed. *Antiquity* **68**, 231–240.

Ayyadurai, S., Houhamdi, L., Lepidi, H., Nappez, C., Raoult, D. and Drancourt, M. (2008) Long-term persistence of virulent *Yersinia pestis* in soil. *Microbiology* **154**, 2865–2871.

Baillie, M. (2006) *New Light on the Black Death: the cosmic connection.* Tempus, Stroud, UK.

Balmer, B. (2004) How does an accident become an experiment? Secret science and the exposure of the public to biological warfare agents. *Science as Culture* **13**, 197–228.

Benedict, C. (1996) *Bubonic Plague in Nineteenth-Century China.* Stanford University Press, Stanford, CA, USA.

Benedictow, O.J. (2004) *The Black Death 1346–1353: the complete history.* Boydell Press, Woodbridge, Suffolk, UK.

Bertherat, E., Bekhoucha, S., Chougrani, S. *et al.* (2007) Plague reappearance in Algeria after 50 years, 2003. *Emerging Infectious Diseases* **13**, 1459–1462.

Bianucci, R., Rahalison, L., Peluso, A. *et al.* (2009) Plague immunodetection in remains of religious exhumed from burial sites in central France. *Journal of Archaeological Science* **36**, 616–621.

Boccaccio, G. (1954) *The Decameron: the first five days* [transl. R. Aldington.]. Folio Society, London.

Camus, A. (1947) *The Plague* [transl. S. Gilbert]. Folio Society, London (1987).

Casman, E.A. and Fischhoff, B. (2008) Risk communication planning for the aftermath of a plague bioattack. *Risk Analysis* **28**, 1327–1342.

Chernin, E. (1989) Richard Pearson Strong and the Manchurian epidemic of pneumonic plague, 1910–1911. *Journal of the History of Medicine and Allied Sciences* **44**, 296–319.

Cohn, S.K. (2002) *The Black Death Transformed: disease and culture in early Renaissance Europe.* Arnold, London.

Dennis, D.T., Gage, K.L., Gratz, N., Poland, J.D. and Tikhomirov, E. (1999) *Plague Manual: epidemiology, distribution, surveillance and control.* World Health Organization, Geneva, Switzerland.

Drancourt, M., Houhamdi, L. and Raoult, D. (2006) *Yersinia pestis* as a telluric, human ectoparasite-borne organism. *Lancet Infectious Diseases* **6**, 234–241.

Drancourt, M. and Raoult, D. (2002) Molecular insights into the history of plague. *Microbes and Infection* **4**, 105–109.

Duncan, C.J. and Scott, S. (2005) What caused the Black Death? *Postgraduate Medical Journal* **81**, 315–320.

Eastwood, A. and Griffith, F. (1914) Report to the Local Government Board on an enquiry into rat plague in East Anglia during the period July–October 1911. *Journal of Hygiene* **14**, 285–315.

Echenberg, M. (2007) *Plague Ports: the global urban impact of bubonic plague, 1894–1901.* New York University Press, New York.

Eisen, R.J. and Gage, K.L. (2009) Adaptive strategies of *Yersinia pestis* to persist during inter-epizootic and epizootic periods. *Vetinerary Research* **40**, 01.

Eisen, R.J., Petersen, J.M., Higgins, C.L. *et al.* (2008) Persistence of *Yersinia pestis* in soil under natural conditions. *Emerging Infectious Diseases* **14**, 941–943.

Gage, K.L. and Kosoy, M.Y. (2005) Natural history of plague: perspectives from more than a century of research. *Annual Review of Entomology* **50**, 505–528.

Gamsa, M. (2006) The epidemic of pneumonic plague in Manchuria, 1910–1911. *Past and Present* **190**, 147–183.

Geissler, E. and van Courtland Moon, J.E. (eds) (1999) *Biological and Toxin Weapons: research, development and use from the Middle Ages to 1945*. SIPRI Chemical and Biological Warfare Studies no. 18. Oxford University Press, Oxford.

Gilbert, M.T.P., Cuccui, J., White, W. et al. (2004) Absence of *Yersinia pestis*-specific DNA in human teeth from five European excavations of putative plague victims. *Microbiology* **150**, 341–354.

Gummer, B. (2009) *The Scourging Angel: the Black Death in the British Isles*. Bodley Head, London.

Haldane, J.B.S. (1938) *A.R.P.* Victor Gollancz, London.

Hammond, P. and Carter, G. (2002) *From Biological Warfare to Healthcare: Porton Down 1940–2000*. Palgrave, Basingstoke, UK.

HMSO (1911) Reports and papers on suspected cases of human plague in East Suffolk and on an epizootic of plague in rodents. *Reports to the Local Government Board on Public Heath and Medical Subjects* New Series **52**, 1–87.

Karlsson, G. (1996) Plague without rats: the case of fifteenth-century Iceland. *Journal of Medieval History* **22**, 263–284.

Lien-Teh, W. (1922) Plague in the Orient with special reference to the Manchurian outbreaks. *Journal of Hygiene* **21**, 62–76.

Lockwood, J.A. (2009) *Six-legged Soldiers: using insects as weapons of war*. Oxford University Press, New York.

Marshall, A.G. (1981) *The Ecology of Ectoparasitic Insects*. Academic Press, London.

McCormick, M. (2003) Rats, communications, and plague: towards an ecological history. *Journal of Interdisciplinary History* **34**, 1–25.

Pallen, M.J. and Wren, B.W. (2007) Bacterial pathogenomics. *Nature* **449**, 835–842.

Prentice, M.B. and Rahalison, L. (2007) Plague. *The Lancet* **369**, 1196–1207.

Raoult, D., Aboudharam, G., Crubézy, E., Larrouy, G., Ludes, B. and Drancourt, M. (2000) Molecular identification by 'suicide PCR' of *Yersinia pestis* as the agent of Medieval Black Death. *PNAS* **97**, 12800–12803.

Rosen, W. (2007) *Justinian's Flea: plague, empire, and the birth of Europe*. Viking, New York.

Scott, S. and Duncan, C. (2004) *Return of the Black Death: the world's greatest serial killer*. John Wiley & Sons, Chichester, UK.

Sleigh, H.P. (1910) Four cases of pneumonic plague. *British Medical Journal* **2**, 1489–1490.

Stenseth, N.C., Atshabar, B.B., Begon, M. et al. (2008) Plague: past, present, and future. *PLoS Medicine* **5**(1), e3.

Stenseth, N.C., Samia, N.I., Viljugrein, H. et al. (2006) Plague dynamics are driven by climate variation. *PNAS* **103**, 13110–13115.

Theilmann, J. and Cate, F. (2007) A plague of plagues: the problem of plague diagnosis in medieval England. *Journal of Interdisciplinary History* **37**, 371–393.

Twigg, G. (1984) *The Black Death: a biological reappraisal*. Batsford, London.

Twigg, G. (2003) The Black Death: a problem of population-wide infection. *Local Population Studies* **71**, 40–52.

Warmflash, D. and Weiss, B. (2005) Did life come from another world? *Scientific American*, November, 40–47.

Welford, M.R. and Bossak, B.H. (2009) Validation of inverse seasonal peak mortality in medieval plagues, including the Black Death, in comparison to modern *Yersinia pestis*-variant diseases. *PLoS ONE* **4**(12), e8401.

Wheelis, M., Rózsa, L. and Dando, M. (eds) (2006) *Deadly Cultures: biological weapons since 1945.* Harvard University Press, Cambridge, MA, USA.

WHO (2007) *The World Health Report: a safer future: global public health security in the 21st century.* World Health Organization, Geneva, Switzerland.

WHO (2008) *Interregional Meeting on Prevention and Control of Plague, Antananarivo, Madagascar 1–11 April 2006.* WHO/HSE/EPR/2008.3. World Health Organization, Geneva, Switzerland.

WHO (2010) Human plague: review of regional morbidity and mortality, 2004–2009. *Weekly Epidemiological Record* **85**, 40–45.

Williamson, E.D. (2009) Plague. *Vaccine* **27**, D56–D60.

Wimsatt, J. and Biggins, D.E. (2009) A review of plague persistence with special emphasis on fleas. *Journal of Vector Borne Disease* **46**, 85–99.

Wood, J. and DeWitte-Aviña, S. (2003) Was the Black Death yersinial plague? *Lancet Infectious Diseases* **3**, 327–328.

Yalden, D. (1999) *The History of British Mammals.* T. & A.D. Poyser, London.

Zhou, D. and Yang, R. (2009) Molecular Darwinian evolution of virulence in *Yersinia pestis.* *Infection and Immunity* **77**, 2242–2250.

Ziegler, P. (1969) *The Black Death.* Collins, London.

A GRAPH SHOWING THE FALL & RISE OF THE RED KITE POPULATION IN BRITAIN DUE TO CHANGING HUMAN ATTITUDES

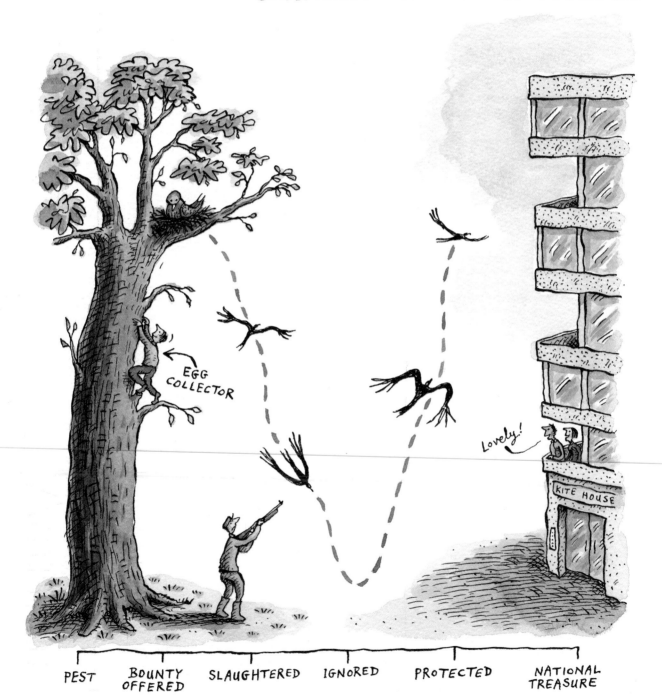

PEST   BOUNTY OFFERED   SLAUGHTERED   IGNORED   PROTECTED   NATIONAL TREASURE

The Red Kite

A S I AM WRITING this, on the outskirts of Reading, a red kite is circling outside the first-floor window. It holds its wide fringed wings stiff and immobile all the while it passes across my field of view, only changing direction by twisting its forked tail in the fresh breeze – more like a glider than a kite. Several kites appear most days here, from mid morning to mid afternoon; yet five years ago this would have been a rare occurrence, and 15 years ago unheard of. However, kites would have been a common sight over the parkland and heath present here 200 years ago. Why did the kite disappear, and how has it returned? In answering these questions we will need to consider topics including changes in farming practices and the use of rodenticides, and the rise in red grouse management.

THE HUNTER

The red kite, *Milvus milvus*, is a bird of prey. Birds of prey proper (raptors) include four orders. Three are diurnal (day-hunting):

- Accipitriformes – broad-winged hawks, kites, vultures and eagles
- Falconiformes – narrow-winged falcons
- Cathartiformes – 'new world' vultures;

and one order is nocturnal – the owls (Strigiformes).

These orders have evolved independently and convergently, and share certain characteristics. These include a powerful hooked beak for tearing flesh, and grasping feet with toes ending in sharp talons to kill prey and hold it down while feeding.[1] The diurnal birds of prey rely on their good eyesight rather than hearing or smell, and have far better distance perception and resolution than humans (Fig. 17.1).

The red kite is a member of the hawk family, Accipitridae, which has over 220 species, and is confined to central and southern Europe with small populations in North Africa and the Middle East (Fig. 17.2). There is only one other species in the genus, the black kite, *Milvus migrans*. This is a much darker and slightly smaller bird, with shorter wings and less deeply forked tail, and is a rare summer visitor to the UK. However, the black kite is very common over much of Europe and has a world-wide distribution, overall probably being one of the most abundant birds of prey in the world.

There are several aspects of the red kite's biology and ecology that make it both very adaptable to a habitat created by humans, but also very vulnerable.

Fig. 17.1 *The red kite, Milvus milvus*

[1]Golden eagles, *Aquila chrysaetos*, were trained in Russia and Asia to hunt wolves, and can allegedly break a wolf's back, grasping it with its talons and exerting a force of over one tonne.

Biological Diversity: Exploiters and Exploited, First Edition.   Paul Hatcher and Nick Battey.
© 2011 John Wiley & Sons, Ltd. Published 2011 by John Wiley & Sons, Ltd.

Fig. 17.2 *The status of the red kite in Europe. Red, resident breeding population; dark brown, wintering population; light brown, summer visitors. Numbers indicate estimated mean number of breeding pairs from 2004 survey (data mostly from mid-1990s onwards). (+), (−) and (=) indicate population increasing, decreasing or remaining stable since 1990. (?) with numbers indicates uncertainty over numbers; (?) without figures indicates uncertainty over presence of a breeding population [Based on Carter (2007).]*

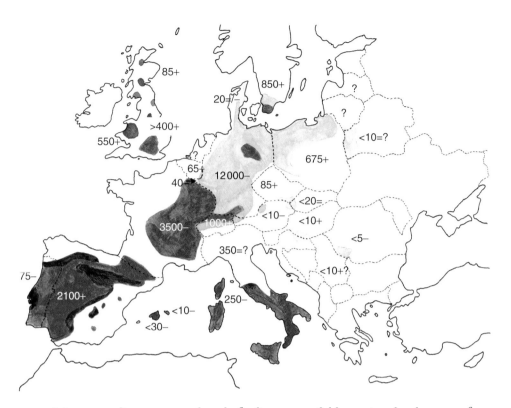

It is a generalist scavenger, largely feeding on available carrion; but because of its rather weak bill and talons (Fig. 17.3) it must often wait for foxes and ravens to open up larger carrion, such as sheep and cattle, before it can feed. Smaller animals (voles and young rabbits, nestlings and fledgling birds, and invertebrates) may be caught alive, on the wing or from the ground. The red kite's most noticeable adaptations for this mode of feeding are its wings and flight behaviour. It takes advantage of upcurrents of air formed over changes in topography, or thermals produced by rising air from heated ground. When prey is spotted, the kite will descend in ever tightening circles until only a few metres above the ground, and would rather not land if possible; snatching prey and eating it in flight. Red

Fig. 17.3 *The red kite:* **(a)** *dropping a piece of string, probably collected to adorn its nest;* **(b)** *eating on the wing. Note the relatively weak talons and beak for a bird of this size*

(a)

(b)

kites can be seen 'eyeing up' rabbits in this way, and will approach very close to a stationary rabbit, but will back off a large rabbit if it shows signs of movement. Unlike some smaller hawks, the red kite only rarely hovers by flapping its wings, but by facing into a breeze it can remain motionless, only a metre or so above the ground (Fig. 17.4). Coming across such a large bird as the red kite in this situation is unnerving and also rather beautiful: the bird is not bothered by your intrusion and will continue to hover only metres away. This lack of fear of humans has been, and continues to be, a problem for the species' survival.

For such hunting behaviour the red kite needs to be able to fly slowly without stalling. This requires a low wing loading (weight per wing area) and a low aspect ratio (ratio of wing length to breadth). Therefore, although at 1–1.5 kg it is not a heavy bird, it has an impressive wingspan of up to 1.7 metres[2]. The tail adds to the wing area and can change shape, aiding manoeuvrability, as do the moveable feather tips at the apices of the wings, which produce slots, reducing stalling speed (see Fig. 17.1).

The red kite is a long-lived bird and a slow breeder. In Britain most first breed when two or three years old, and may live for 20 or more years in the wild. Kites often pair for life and will return to the same nest area, usually the area in which they were reared (termed natal philopatry, Fig. 17.5), and often to the same nest year after year. The pairs do not defend large territories, but do defend small areas around the nest. These nests are built 4–30 m above ground in

Fig. 17.4 *A red kite about 150 cm above the ground: by flying slowly into the wind it appears almost to hover*

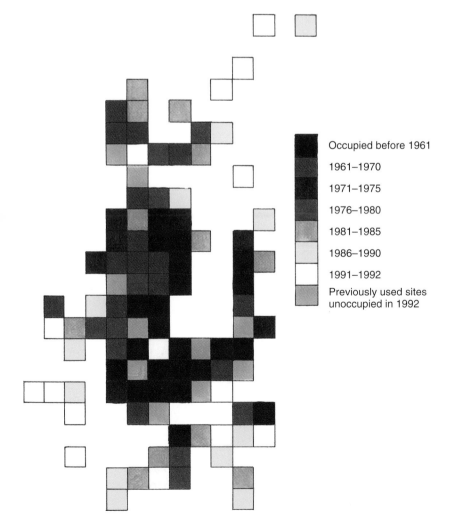

Occupied before 1961

1961–1970

1971–1975

1976–1980

1981–1985

1986–1990

1991–1992

Previously used sites unoccupied in 1992

Fig. 17.5 *Red kite natal philopatry in action. The spread of red kite breeding territories in Wales, 1946–1992. As the population increased (see Fig. 17.9) birds moved only into neighbouring areas. Boxes represent 5 km squares, and years refer to the first occupancy of these squares [Redrawn from Newton et al. (1994).]*

[2]Some other raptors with a similar wingspan, such as the Egyptian vulture *Neophron percnopterus* and the short-toed eagle *Circaetus gallicus*, weigh up to twice as much.

(a)

(b)

Fig. 17.6 *Red kite country.* **(a)** *Chiltern beechwood.* **(b)** *Nest 10–15 m up a beech tree in the Chilterns*

woodland (Fig. 17.6) in March to early April, and a clutch usually of two or three eggs is laid. These hatch after about 30 days of incubation and the chicks are brooded by the female for the first couple of weeks, while the male hunts and provides most of the food. After 3–4 weeks the chicks are no longer fed directly and food is just left for them in the nest; they start to flap their wings at 5–6 weeks and may take their first flight a fortnight later. They are dependent on their parents for food for several weeks more, reaching full independence 3–4 weeks after their first flight, when about three months old. There appears to be little feeding behaviour learnt from their parents: it seems that scavenging behaviour does not require it.

The juveniles often disperse for hundreds of kilometres in the autumn or the following spring, and it is at this time that they are most vulnerable, before they usually return to the area of their birth when ready to breed. Large roost gatherings (one of over 200 birds has been recorded in the Chilterns) form especially in winter from late afternoon onwards: the younger birds here often engage in mock combat, chases and other behaviours with each other.

RISE AND FALL

One of the problems inherent in determining the history of the red kite is that it is often not clear which species is being referred to. In much of southern Europe both red and black kites would have been common, and yet they are rarely distinguished in early records. Below we refer to 'kites' when we are not sure of the species involved. Some have interpreted the records of Aristotle on kites as referring to the black kite, but this places too great a reliance on the accuracy of his recordings, which may be unjustified. This conclusion is then used to infer that the kites of medieval London, which were reported to have similar characteristics, were also black kites. This can turn into a circular argument. It is likely that the black kite has always only been at best an uncommon migrant to the UK, and all archaeological remains of kites from medieval and earlier towns in the UK have been of a size commensurate with the red, not black, kite.

However, we still cannot be sure that the historical term 'kite' means the red kite. For example, Lovegrove (2007) notes that the term 'kite' (from Anglo-Saxon 'cyta' from the bird's call) in seventeenth- to nineteenth-century vermin lists also probably included buzzards, *Buteo buteo*, which are rarely mentioned separately, and 'kite' could possibly include any large bird of prey. The local and still older name 'glede' or 'glead' from the gliding flight of the bird was used exclusively for the red kite in Cumberland, but elsewhere in northern England also included harriers, and extended to buzzards in Scotland. Similarly 'puttock' or 'puddock' could refer to the kite or buzzard in the Midlands and eastern counties. We should also be careful in ascribing place names to the kite. While names based on glede (e.g. Gledhill) or kite (e.g. Kitley Hill in south Devon = kite wood hill) may refer to kites, puddock is also Middle English for frog or toad.

Unlike many organisms, including the wolf (Chapter 19), which seem to have been in continual retreat from the advance of humans across the landscape, the kite would have initially benefited from changes brought about by people. The original wildwood covering much of Britain after the last glaciation (Chapter 10) would not have favoured a species that required both woodland for breeding and open ground for hunting. However, woodland clearance from Neolithic times and the introduction of sheep and cattle from the fourth millennium BC would have favoured the kite, and few remains of kites from before this time have been found. A patchwork landscape of small fields with mice and voles, carrion from livestock, and farmsteads with abundant vermin and poultry would have been ideal (Fig. 17.7). The spread of introduced species such as the rabbit (Chapter 11) would have been an additional benefit, as would the growth of towns with their associated waste and rubbish – kites were noted as scavengers in ancient Greek cities. Therefore, by the Middle Ages it has been estimated that the red kite was the most abundant bird of prey in Britain, and visitors to London in the fifteenth and sixteenth centuries have remarked on the numbers of kites present.

The kite was viewed with mixed feelings, not entirely negatively as with the wolf, but neither very positively. In towns, kites along with ravens had a useful function for a while in removing animal waste from the streets. This is said to have been particularly the case in London where they were protected by law (probably the first legislation to protect a wild bird), although it is unclear how much of an effect they would have had in the narrow overhanging streets of that time. In London it is likely that kites were favoured only between the end of the thirteenth century, when the wild pigs that functioned as scavengers on the streets were removed, and the middle of the sixteenth century, when laws were enacted to attempt to clean up the streets. The London kites gradually declined, with the last breeding pair recorded in London at Gray's Inn in 1777.

In other respects, and certainly in the countryside, kites were considered at best a nuisance and often as a pest. Their bold thieving habits were disliked, with ancient Greek records of them robbing men in the street, carrying off chickens and removing meat intended for sacrifices. In medieval bestiaries they are also noted for taking domestic birds. In 1500, a Venetian nobleman who accompanied his ambassador on a visit to London noted:

> [T]he kites, which are so tame, that they often take out of the hands of little children, the bread smeared with butter, in the Flemish fashion, given to them by their mothers. And although this is general throughout the island, it is more observed in the kingdom of England, than elsewhere.[3]

They would also take clothes left out to dry to add to their nests, as noted by Shakespeare in London, his 'city of crows and kites'.

Concerted action against kites can be dated to 1534 and the passing of an Act by Henry VIII for the 'destruction of noisome foule and Vermine'.[4] This listed a

(a)

(b)

Fig. 17.7 *Red kite country, Chilterns.* **(a)** *Short chalk grassland maintained by rabbits and sheep, a good source of carrion.* **(b)** *Patchwork of an enclosed landscape*

[3]Sneyd (1847), p. 11.
[4]Bircham (2007), p. 40.

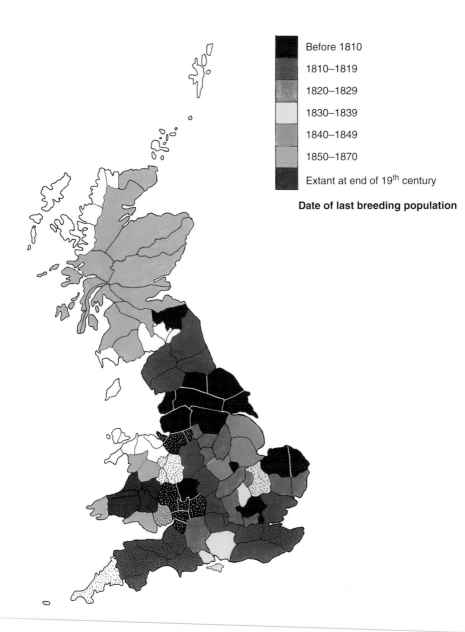

Before 1810
1810–1819
1820–1829
1830–1839
1840–1849
1850–1870
Extant at end of 19th century

Date of last breeding population

Fig. 17.8 *Disappearance of the red kite from Great Britain. Last recorded breeding in the counties, from Shrubb (2003) for England, and Holloway (1996) for Scotland and Wales. These dates are estimates, especially for Scotland; some very limited breeding may have taken place shortly after. Data not available for uncoloured counties. Lovegrove (2007) recorded bounty payments for red kites from the stippled counties*

range of species that were to be controlled, including many small mammals and birds, and most birds of prey, including the red kite. The act was renewed several times during the sixteenth century, and not repealed until 1863. By the middle of the sixteenth century this Act required all parishes to raise a levy from landowners, tenants etc. to fund a bounty for each pest submitted (e.g. one penny for each kite). For a time there was a conflict between pest control and the preservation of some scavengers in towns, and the Act also states that kites and ravens could not be killed in 'anie citie or towne corporate or within two miles of the same.'[4]

Lovegrove (2007) has analysed many of the parish records resulting from this Act and presents much evidence on the scale of kite killing in England. The pattern of persecution was localized (Fig. 17.8) and concentrated in the western and south-eastern counties. Where kite control did occur it could be extreme, and sometimes the authorities issued 'special offers'. For example, for two months in 1756 the parish of Otterton in Devon offered double the usual rate of two pence for a kite. Devon and Cornish parishes were the most prolific in kite persecution. For example, in 54 years between 1755 and 1809 the Cornish parish of Lezant paid for 1245 kites; and over 22 years between 1727 and 1854 Altarnun in Cornwall paid for about 300 kites.

By the early nineteenth century parishes stopped paying for vermin control, but this role was taken over with renewed vigour by gamekeepers for the sporting estates that were by then developing over much of lowland Britain. This led to the rapid extinction of the kite during the early nineteenth century throughout England (see Fig. 17.8). Whereas Gilbert White can mention seeing several kites over the South Downs in the 1770s and John Clare was familiar with the kite in Northamptonshire during the early nineteenth century (Box 17.1), by the time the great artist-naturalist, Henry Hudson, was writing in southern England from the last decades of the nineteenth century there were no kites left there.

BOX 17.1   JOHN CLARE AND THE KITE

John Clare was described by James Fisher as 'the finest poet of Britain's minor naturalists and the finest naturalist of all Britain's major poets'.[a] Indeed, only Ted Hughes (see Box 9.2) has subsequently come close as a naturalist-poet, and yet while Hughes *describes*, Clare *is*.

John Clare was born on 13 July 1793, the son of a thresher in the Northamptonshire village of Helpstone. He left School at 12 and took a variety of labouring jobs in his youth, including thresher and ploughboy, and was writing poetry from an early age. His first volume of poetry, *Poems Descriptive of Rural Life*, was published in 1820 and, feted by London literary society as a 'peasant genius', was successful. Further volumes of increasingly accomplished work followed, but were decreasingly commercially successful. This probably contributed to Clare's increasing mental instability, and he was placed in a private asylum at Epping Forest in 1837. He escaped in 1841 and walked home to try to find his child-hood sweetheart. The diary of his walk of 150 km in four days, starving and hallucinating, makes for some of the most startling and moving autobiographical writing in English. Six months later he was placed in Northampton general lunatic asylum, where he died, largely forgotten, in 1864.

Unlike the romantic poets such as Wordsworth with which he is generally included, Clare does not use nature as a route for an explanation of his emotions, or lessons learnt, and classical allusion; rather he is interested in real wildlife and real village labourers. His picture, and poetry, is unique in this honesty to the realities of the world.

Clare uses both 'kite' and 'puddock' in his poems, which could refer to red kite or buzzard. In his manuscript list of birds Clare is careful to distinguish between kite and buzzard, noting the difference in flight behaviour between the two. He also notes that the buzzard builds a smaller nest and lays shorter eggs than the kite. In most cases it is probable that 'kite' and 'puddock' refer to the red kite (Fig. 1).

The kite was familiar to Clare and occurs in many of his poems, sometimes to give colour:

> The old hen leads her flickering chicks abroad,
> Oft scuttling 'neath her wings to see the kite
> Hang wavering o'er them in the spring's blue light.[b]

and

> And other losses too the dames recite,
> Of chick, and duck, and gosling gone astray,
> All falling prizes to the swopping kite:[c]

Several poems on autumn and the approach of winter contain the kite, for example in 'Autumn'. This starts with 'The summer-flower has run to seed' and autumn gradually turns into winter. In the penultimate verse:

> Slow o'er the wood the puddock sails;
> And mournful, as the storms arise,
> His feeble note of sorrow wails
> To the unpitying, frowning skies.[d]

A harbinger of winter, 'and desolation shuts the scene', ends the poem. Likewise, in 'Winter':

> And feebly whines the puddock's wail,
> Slow circling naked woods around,[e]

Central to Clare's poetry is the sense of loss and the retreat of Eden; the kite appears as a symbol of this. For example, after moving to a neighbouring village in 1832 Clare could not recognize the birdsong there and he looked back to his previous village:

> The sailing puddock's shrill 'peelew'
> On Royce Wood seemed a sweeter tune.[f]

6   The Kite

Box 17.1 Fig. 1 *The red kite, from Thomas Bewick's A History of British Birds, volume 1, containing the history and description of land birds (London: Longman & Co., 1826). Clare could not afford a copy of this book, but he did write to some of his friends who he thought might have a copy, asking if he could borrow it. It is not certain whether he ever saw it*

[a]Fisher (nd), p. 26.
[b]'Home Pictures in May', written 1821–4; Symons (1908), p. 153.
[c]'The Village Minstrel', published 1821; Tibble (1935a), p. 140.
[d]'Autumn', publ. 1821; Tibble (1935a), p. 174.
[e]'Winter', written 1821–4; Tibble (1935a), p. 360.
[f]'The Flitting', written 1832–5; Symons (1908), p. 118.

Later, describing the local fens he decries the lack of familiar birds:

Ah, could I see a spinney nigh,
A puddock riding in the sky
Above the oaks with easy sail
On stilly wings and forkèd tail,[g]

This poem also laments enclosure of the wasteland, 'Gain mars the landscape every day', but falls short of making an explicit link between lack of kites and enclosure. By the 1830s, when this poem was written, the kite was becoming uncommon in Northamptonshire (see Fig. 17.8).

Clare wrote continuously (over 10 000 manuscript pages survive) and probably wrote more about birds than any other subject, and is likely to have composed more bird poems than any other British writer. Fewer than a quarter of his poems were published in his lifetime, including none of those he composed in Northampton asylum, where Clare still remembered the kite as a symbol of autumn and places 'Where the kite peelews and the ravens croak'.[h]

[g] 'The Fens', written 1832–5; Tibble (1935b), p. 280.
[h] 'Autumn Change'; Tibble (1935b), p. 413.

Were vermin control laws and gamekeeping entirely responsible for the extinction of the red kite in England? Shrubb (2003) has suggested that the process of parliamentary enclosure, most active between 1760 and 1850, was also important. He suggests that this process, by which about a fifth of English farmland was 'enclosed', and changed from a communal open field system to small enclosed fields owned by one landlord, led to a decrease in livestock and thus less carrion for kites. It is probable that enclosure had an effect on kites, especially through enclosing and improving 'wastes' – uncultivated heaths, moors and grassland that would have been important hunting grounds – but enclosure only accelerated changes already taking place. The effect was also not uniform. In some areas, such as the west and central Midlands, enclosure extended the area under grass for livestock, while in the chalklands of Wiltshire, Hampshire and Dorset it tended to increase arable cultivation at the expense of livestock and sheep pasture.

The destruction of the red kite in Scotland started in earnest with the conversion of many upland areas to sheep walks in the early nineteenth century (kites were mistakenly thought to kill lambs), and increased subsequently with the rise in management of the red grouse (*Lagopus lagopus*) for shooting. This was encouraged by changes in the law during the early nineteenth century which permitted non-locals and non-estate holders to hunt game with the owner's permission; improved communication, with railways reaching Edinburgh in 1843 and Inverness in 1863; and the development of reliable, accurate, breech-loading shotguns during the 1850s which afforded a good chance of killing birds in the air, and which could be reloaded quickly. Finally, after Queen Victoria bought the Balmoral Estate in 1852, it became fashionable to own and be seen on an upland shooting estate.

As sheep-walk was increasingly converted to grouse moor during the nineteenth century, wildlife such as the red kite that was thought to pose a threat to the grouse was exterminated. For example, in Glen Garry, Inverness-shire, 275 kites were trapped from 1837 to 1840, and Pearsall (1971) estimated that for every one eagle or two foxes killed 9–10 kites were killed on Scottish estates during the nineteenth century. The effects of this, aided by egg collectors, were rapid: the kite disappeared so quickly from Scotland during the second half of the nineteenth century that accurate records do not exist (see Fig. 17.8). By 1900 the red kite was extinct in Scotland also.

The eradication of raptors such as the red kite was soon blamed for the two great field vole (*Microtus agrestis*) outbreaks during 1875–6 and 1890–2 in the southern uplands of Scotland. Although voles are prone to outbreak, nothing like these outbreaks had been seen for 150 years. The latter plague affected over 20 000 hectares in Dumfriesshire alone, and caused considerable hardship to sheep farmers in the area. Although later investigators suggested that weather conditions and changes in grass quality were the main factors leading to the vole outbreaks, the lack of predators including the red kite was likely to have contributed to the severity of the plague.

CONSERVATION AND RECOVERY

By 1900 the red kite in Britain only remained in central Wales, and within a few years was reduced to fewer than 12 pairs there. During the nineteenth century there had been increasing awareness of the need to protect wildlife in the Britain, with the Society for the Protection of Birds founded in 1889 and the passing of the Wild Birds Protection Act 1880. Furthermore, an Act in 1894 gave local authorities the power to apply for the protection of areas and certain species. But this was too little and too late to save the red kite. Subsequently, the Protection of Birds Act 1954 gave UK-wide protection for all birds of prey except the sparrowhawk, and the EEC Council Directive 1979 on the conservation of wild birds (known as the Birds Directive) gave the strongest protection to any group of species in the world. However, Stroud (2003) still noted 'widespread failure in the effectiveness of direct conservation action in support of these treaties'.[5]

The story of red kite conservation in Wales since the early 1900s is well told by Lovegrove (1990) and is one of the longest-running bird conservation efforts anywhere. It centred on protecting known nests from egg-collectors with bounty payments to landowners for the breeding success of a nest on their land. The RSPB became involved in 1903, and in 1911 a paid kite watcher was employed; unfortunately this person was also an egg collector and was thought to have taken several clutches of red kite eggs. Nestlings were also taken and offered for sale. There were other problems, and Lovegrove (1990) notes that red kite conservation in Wales at this time was plagued by 'bitter division, acrimonious clashes of personality, jealousies, intrigue and deception'.[6] It was not until after 1945, and especially from the 1960s, that record-keeping, coordination, and effectiveness of conservation efforts improved.

The Welsh red kites occupied an area of only 70 by 45 km (see Fig. 17.5), comprising sessile oak (*Quercus petraea*) woodland on the valley slopes in which they bred, marginal hill land over the lower hills and mountain sheep-walk (over which they foraged during the summer), and lowland valley farmland used for hunting in the winter. Despite conservation efforts, the numbers of breeding pairs remained low until the 1960s (Fig. 17.9a), with Walters Davies and Davis (1973) noting that 'considering the effort which has gone into protection during these years, the increase has been disappointingly slow'.[7] The breeding rate of the kites in Wales was lower than almost anywhere else in Europe, probably due to poor food supply and frequent cool and damp weather during the breeding season. Thus, setbacks to breeding due to egg thefts, illegal poisoning, or the effects of myxomatosis on rabbit populations in the mid-1950s (Chapter 11) had great effects on the red kite population.

[5]Stroud (2003), p. 71.
[6]Lovegrove (1990), p. 55.
[7]Walters Davies and Davis (1973), p. 204.

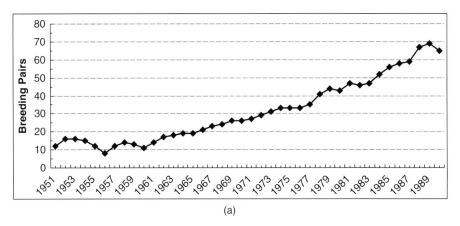

(a)

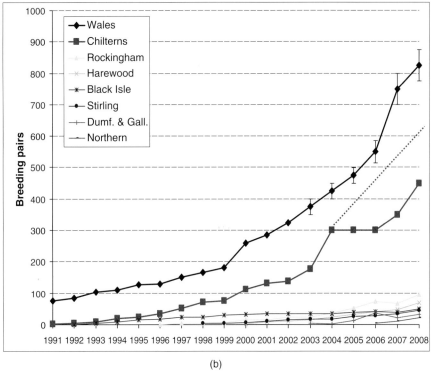

(b)

Fig. 17.9 *Numbers of breeding pairs of red kites in Britain:* **(a)** *Welsh population up to 1990 (Lovegrove, 1990);* **(b)** *populations from 1990 (Welsh Kite Trust). From 2004, Chiltern populations have been been estimated from incomplete counts, the likely population is given by a dotted line. From 2002, the Welsh populations have been estimated, mean and range of estimate given*

Although the Welsh red kite population started to steadily increase from the 1970s (Fig. 17.9), mainly due to greater survival of adult birds, the population remained in the same location, only gradually increasing its area. Much of the UK was thought to be suitable for the kite, but Welsh kites were not moving out of their area of birth to breed (see Fig. 17.5). Therefore, a group led by the Nature Conservancy Council and the RSPB decided to reintroduce the species into England and Scotland (Box 17.2).

The reintroduction programme started in 1989 when 4- to 6-week-old chicks from southern Sweden were flown to specially prepared aviaries in the Chilterns of central southern England and the Black Isle, northern Scotland. Birds of this age were chosen based on existing falconry techniques, being old enough to thermoregulate without parental brooding and to feed themselves, and were taken from nests containing at least three offspring. They were released into the wild in early July to August when 10–12 weeks old, and food stations were maintained in the release areas for the next 3–4 weeks. A total of 93 birds from Sweden were introduced into Scotland up to 1993. The same number was also released in the Chilterns – seven from Wales, four from Sweden and the rest from Navarra and Aragon in Spain (Fig. 17.10). Spanish birds were thought to be more suited to southern English conditions that Scandinavian ones.

Although the programme was initiated as a trial of reintroduction methods, red kites in both areas began to breed in 1992 and the populations increased sufficiently rapidly that no further translocations were necessary. From 1995 reintroductions using the same procedure have taken place in other English and Scottish areas (see Fig. 17.10). In 1997, the Spanish authorities decided that it would be difficult for them to supply further red kite chicks for reintroduction as their population in Spain was declining. Thus, after deciding that the effects on the established British populations would be small, ten young from the Chilterns population, along with ten from Spain were released

| **BOX 17.2** | **SPECIES REINTRODUCTION** |

Species reintroduction – by which we mean the release of an organism into the wild to restore it to an area in which it formerly occurred naturally, but now is absent – is the most drastic of all conservation measures, and the one fraught with the greatest controversy.

The reintroduction of the red kite into England and Scotland is a model example of this technique, and considerable research was needed before reintroduction was agreed. First, relevant national legislation had to be adhered to. The relevant British law was the Wildlife and Countryside Act 1981, which is silent about such reintroductions, stating only that it is illegal to release or allow to escape an animal which 'is of a kind not ordinarily resident in and is not a regular visitor to Great Britain in a wild state' (Section 14), or is one of the three birds (Capercaillie, *Tetrao urogallus*; white-tailed eagle, *Haliaeetus albicilla*; or barn owl, *Tyto alba*) listed in Part 1 of Schedule 9 of the Act.

The UK has adopted the internationally recognized guidelines for the reintroduction of species drawn up by the International Union for the Conservation of Nature (IUCN), and the criteria used at the planning stage for red kite reintroduction state (Green, 1979):

1. There should be good historical evidence of former natural occurrence.
2. There should be a clear understanding of why the species was lost. In general only those lost through human agency and unlikely to recolonize naturally should be candidates for re-establishment.
3. The factors causing extinction should have been rectified.
4. There should be suitable habitats of sufficient extent to support the re-established population and allow it to expand.
5. The donor population for which individuals are translocated should be as close as possible genetically to that of the original native population.
6. The loss of individuals taken for re-establishment should not prejudice the survival of the population from which they were taken.

It is clear that points 1, 2 and 4 were met for the red kite. Point 3 was thought to have been rectified by legislation and a change in public opinion (but see later in this chapter). Points 5 and 6 are harder to determine: there was an insufficient population in Wales to translocate birds from, and some studies suggested that there was unlikely to be major genetic differences between British and continental red kite populations. Birds were only translocated from the largest populations.

Once it was agreed that the red kite should, and could be reintroduced into England and Scotland, suitable release sites had to be found. The project team formulated the following guidelines. Each release site should be (Evans *et al.*, 1997):

1. In a sufficiently large area of suitable habitat and not an isolated patch of good habitat in an otherwise unsuitable area.
2. In a region with sustainable food supply in both winter and summer.
3. In an area where the breeding success of common buzzards (if present) indicated an abundant food supply.
4. In an area free of illegal poison baits.
5. In an area of low rainfall, below an altitude of 350 m above sea level.
6. In an area where the local community was favourable to the project.

Point 5 was included because of the perceived effects of high rainfall on the breeding success of Welsh red kites.

Even after the successful reintroduction of the red kite and the white-tailed eagle (*Haliaeetus albicilla*, which became extinct in Britain in 1918 and has been successfully re-established in western Scotland after large-scale releases of Norwegian birds from 1975), species reintroductions are still controversial. On the one hand there are biological concerns over the effects that such reintroduced species may have on established ecosystems, and the thought that the large amounts of money spent on just a couple of species could be better spent protecting whole habitats. There can also be philosophical objections to continued human interference in trying to recreate the past, and the thought that efforts to save species endangered in an area may be weakened if it is thought that they could be easily reintroduced later.

However, reintroductions of highly visible and popular species such as the red kite serve as flagships for conservation, and can highlight to the public important issues such as continuing raptor persecution. Species reintroduction is not the sole conservation strategy, and overall is not the most important, but carried out carefully can have benefits far in excess of the addition of one species to the country's list.

in the Midlands reintroduction area. Since then Chilterns chicks have been translocated also to Yorkshire, Gateshead, Dumfries and Galloway and Aberdeenshire. The Welsh population continued to recover, and in 2004 Welsh red kites started to recolonize western England, rearing a chick across the border in Herefordshire, and in Shropshire the following year. In 2007, red kites from Wales started to be released in Co. Wicklow, Ireland, and in 2008 in Co.

Balnagown Estate

Black Isle

Aberdeen*

Stirlingshire

Dumfries & Galloway

Gateshead

Harewood House

Rockingham Forest

Chilterns

Co. Down

Co. Wicklow

Wales

Fig. 17.10 *Red kite reintroduction in the UK. Red shaded areas represent approximate areas of reintroduction (except for the native population in Wales) and give the dates of reintroduction. Solid arrows indicate birds introduced from abroad (numbers translocated and origin), dashed arrows indicate introductions from UK populations, with numbers translocated (note the numbers translocated from the Chilterns population). (*) 57 chicks were translocated from Chilterns, Black Isle, and Stirlingshire [Data from Carter (2007), updated to 2009 releases.]*

Down, Northern Island. In 2010, the first red kite chicks hatched in Ireland for over a century.

CURRENT CONCERNS

Many of the reintroduced red kite populations in Britain are now thriving and increasing steadily (Fig. 17.11), and birds from the Chilterns population have started to breed elsewhere in southern England. Although the release sites for red kites were initially kept secret, these large and not at all shy birds have recently become visitor attractions in England, Scotland and Wales, with several red kite-based tourist initiatives launched. The reintroduction of the red kite into England and Scotland has been hailed as one of the great conservation success stories of recent years, and yet there are still threats to the long-term future of the species in Britain, particularly in Scotland, and for several years concerns have been raised that these populations are not expanding as rapidly as they should.

The situation in continental Europe for the red kite is mixed. Many red kites on the continent are migratory, overwintering in southern France and Spain and returning north to Germany and Sweden to breed (see Fig. 17.2). This mobility gives increased chances for recolonization, unlike the mainly sedentary populations in Britain, and this has occurred from the 1970s in Denmark and Belgium; populations in other countries have increased recently also. However, the three main countries for the red kite – France, Germany and Spain – have all experienced recent reductions in populations. In Germany, a 50% decline since 1990 has been blamed on agricultural intensification in the former GDR after reunification and the reduction of rubbish dumps. In France, poisoning is important, with massive secondary poisoning occurring from anti-coagulant rodenticides used against voles.

In Spain, there was a 43% decline in the red kite breeding population between 1994 and 2002/3. This has been linked to the reduction in the rabbit population brought about by myxomatosis and more recently rabbit haemorrhagic disease (RHD; see Chapter 11). Rabbits are one of the main game species in Spain, and at the same time as myxomatosis reduced rabbit numbers a government-sponsored vermin eradication programme started, which killed 10 000 red and black kites between 1954 and 1961. As rabbit numbers recovered, legal killing of predators ended in 1973 and kite conservation started. However, RHD killed between a third and two-thirds of Spanish rabbits between 1988 and 1993, and subsequently there were increased demands for predator control. A survey thereafter found that red kites had disappeared or become scarce from areas with high rabbit populations, and the most secure kite populations remained in areas with low rabbit numbers. This distribution was linked to the poisoning of red kites in areas where rabbits were important game: 408 poisoned red kites were found between 1990 and 2000 in Spain – there have been only about 15 successful prosecutions for the illegal poisoning of any raptors in Spain during this period. A recent survey estimated that several thousand Spanish kites may be shot each winter on their winter roosts.

Fig. 17.11 *Red kite flying over Reading, Berkshire, UK, in spring 2010. Some red kites are tagged to aid recording: the tag on the left wing records the site of birth and that on the right the year of birth. This red kite was born in 2009 in the Rockingham Forest, East Midlands site*

The careful monitoring of the British populations has revealed much about the main mortality factors affecting the kite, and this has supplied useful information to aid the conservation of other birds of prey. Since the loss and reintroduction of the red kite the landscape has gained powerlines, roads with fast vehicles and railway lines, all of which claim some red kites. However, humans are still the greatest threat, with accidental secondary poisoning from rodenticides and deliberate illegal poisoning the main mortality factors for the red kite in Britain. Lovegrove (2007) lists a total of 113 red kites killed in Britain by misuse or illegal use of pesticides between 1991 and 2004; inevitably just the tip of the iceberg, as most poisoned birds are not found.

Poisons have long been used to kill red kites, one of the earliest being strychnine (Box 17.3). This was the usual poison found in dead Welsh red kites until the 1980s, and is still being used in parts of Europe. Strychnine was used to kill rodents, until safer (to the operator) poisons were developed. One of these, warfarin, became the staple of the rodent control industry for many years.

BOX 17.3 STRYCHNINE

During the last 150 years one of the greatest threats to the red kite has been from poisoning, and during much of this time the poison of choice has been strychnine. This has been used as a poison of many other vertebrates, including other birds of prey and mammals such as wolves (see Chapter 19), rabbits (see Chapter 11), rats, mice and moles.

Strychnine is an alkaloid, a group of nitrogen-containing compounds which includes quinine, nicotine, cocaine, morphine and caffeine. All are found in plants, where many function as deterrents to vertebrate and invertebrate herbivores. Whereas some alkaloids were known to the ancient Greeks, strychnine was discovered rather later. The poisonous effect of the seeds of some Asian trees, now of the genus *Strychnos* (Loganaceae), seems to have been known in Europe from the sixteenth century. Most supplies of strychnine came from the seeds of the evergreen tree *Strychnos nux-vomica*, found especially in parts of India, and later cultivated for strychnine elsewhere in Asia. This tree forms fruits the size of a large apple, orange when ripe, which contain about five disc-shaped seeds about 2.5 cm diameter and 0.5 cm thick, embedded in a white pulp. The seeds are dried and ground before use.

The active ingredient of these seeds was first isolated in 1818, but it was not until 1954 that the structure of the chemical strychnine was determined. For a small molecule it has a complicated structure, and like quinine (see Box 12.3), complete artificial synthesis of it is not cost-effective. Thus, most strychnine sold was in fact the powdered seeds (also called 'nux vomica' or 'faba sancti ignatii', depending on the source), containing up to 1.5% strychnine, along with other less effective alkaloids.

Some years after its introduction as a vermin poison, it was thought that low doses of strychnine had a tonic action in humans, increasing respiration and digestion, and it began to be prescribed as such. From at least the nineteenth century, over-the-counter tonics containing strychnine, quinine and iron salts could be purchased, including the famous 'Easton's' tablets and tonic. These were available until the 1970s. Strychnine,

like quinine, is intensely bitter, and up to the later nineteenth century was added to beers and spirits to increase bitterness, and to mask off flavours. More recently it has been found as an adulterant in street drugs such as heroin.

Strychnine is a neurotoxin and acts to block the binding of the amino acid glycine to receptor sites on some nerves, especially in the spinal column. Glycine's action here is to inhibit nerve impulse transmission. Therefore, strychnine does increase nervous excitation; but any stimulatory effect only occurs at toxic doses, not the much lower concentrations present in tonics. However, it could be, and often was, fatal: in tonic syrups strychnine was only in suspension, and thus if not shaken well, the last few drops of the bottle could contain fatal doses, and if mixed with other liquids the strychnine could precipitate out, again leading to fatal doses. Sugar-coated strychnine tonic tablets were particularly inviting to children, leading to further fatal poisonings.

Doses of over 100 mg strychnine in adults and 20 mg for children are generally fatal, and this is one of the most unpleasant ways of dying. The victim is conscious throughout, indeed with heightened consciousness, and experiences muscle stiffness within minutes, with tremors following. Within an hour or so the victim will suffer a series of violent convulsions, each lasting at least one minute at intervals of 10–15 minutes. In these intervals the victim has complete muscular relaxation, and the slightest stimulus, such as a draft of air, can start another convulsion. People rarely survive more than five such convulsions, often dying of asphyxia. There is no antidote; the only treatment is heavy sedation to avoid fits, until the strychnine has passed through the body.

It is unlikely that the effects of strychnine on other vertebrates are any different from its effects on humans, and also unlikely that any of those using strychnine as a poison observed the death of these animals. Would they have continued to use it had they done so? A complete ban on the use of strychnine in the European Union did not begin until 2006.

Originally developed as a human therapeutic drug (and still used as such), warfarin blocks the recycling of vitamin K (required for the synthesis of blood clotting factors) in liver cells. Poisoned animals die from haemorrhaging after 10–14 days. This slow effect means that rodents do not learn to associate the bait with its effects and thus do not show learned bait avoidance. Warfarin and its derivates, the so-called first-generation anti-coagulants, were extensively developed between 1950 and 1970. They have a fairly high LD_{50} (the milligram dose per kilogram body weight needed to kill 50% of an experimental population) and need repeated application to kill. Thus, carrion killed by this poison is probably little threat to birds of prey.

Increasing resistance to warfarin in rats spurred the development of replacement second-generation anti-coagulants which became available in the 1970s. These chemicals are up to 100 times as toxic as warfarin, and secondary poisoning of scavenging birds of prey becomes more likely. Red kites are particularly vulnerable to secondary poisoning because they take small carrion, and hunt around farm buildings where such poison is likely to be legally used. There is no ready solution to this, but the rapid removal of rodents killed by these compounds would help.

In some areas of Britain and Ireland red kites are still actively persecuted; some are shot (one red kite was shot in Ireland in 2007 within six weeks of release), but most are poisoned deliberately and illegally. Although this takes place throughout Britain it is particularly of concern in Scotland, where it seems to be increasing – in 2007 there were 11 deliberate cases of poisoning in Scotland, and the Scottish Executive is considering new tougher penalties for wildlife crime of this nature. Smart *et al.* (2010) conclude that the northern Scottish red kites have a breeding rate comparable to the faster-growing English populations, and that if the mortality from illegal poisoning is excluded, there would now be as many red kites in Northern Scotland as in the Chilterns.

Much of the Scottish poisoning of raptors has been linked to the management of grouse moors. Carcasses are laced with rodenticides or pesticides, although the laying of poison baits against birds has been illegal in Britain since the early twentieth century.

Although there has been a decrease in poisoning incidents in sheep-rearing areas as shepherds realised that birds of prey do not affect sheep numbers, there is still a view among gamekeepers that raptors affect game birds. However, in 1867 a paper was presented to the British Association by Alfred Newton stating that there was no relationship between the population levels of birds of prey and game birds, with these removing the least fit game, and for the next 100 years this was the accepted view. Current research suggests that some birds of prey such as hen harriers (*Circus cyaneus*) and peregrines (*Falco peregrinus*) may have an effect on red grouse numbers (although this may be minimized in well-managed heather moorland with plenty of cover), but there is no evidence that red kites have an effect. However, there is a view that economically viable grouse moors are incompatible with uncontrolled raptor populations.

The attitude of a section of the hunting industry towards birds of prey has always been deeply ambiguous. For example, Henry Williamson could both mourn the loss of the red kite while praising the abilities of the gamekeepers, the 'compassionate guardian of the forest',[8] who were involved in this, and this is seen in other early twentieth-century nature writers. Even after the Second World War, Brian Vesey-Fitzgerald (1946), a vice-president of the Gamekeeper's Association, in the second of the New Naturalist series *British Game*, listed the red kite as an enemy of game, but in the same section notes 'should be protected throughout the year in the British Isles, but unfortunately is not'.[9] Sixty-one years and ninety-nine New Naturalist volumes later, the late Derek Ratcliffe (2007) was unambiguous and forthright:

> The ritual slaughter of large numbers of sentient creatures, for little more purpose than to demonstrate the skill, status and wealth of a select

[8]*Wood Rogue*. In Williamson (1923).
[9]Vesey-Fitzgerald (1946), p. 200.

group of people, is in itself increasingly abhorrent to the public at large today. The argument that it is a traditional sport cuts no ice at all: many practices once considered traditional were abandoned as society became more civilised.[10]

Although the red kite reintroduction project has had its critics, one of its greatest successes must be in alerting us to aspects of our behaviour in which we need to show that greater civilization.

FURTHER READING

Carter (2007) and Lovegrove (1990) are indispensable books on the red kite, and give a good introduction to the literature. Both the Welsh Kite Trust (welshkitetrust.org) and the Chilterns Conservation Board (www.chilternsaonb.org) have useful websites, and also publish well-illustrated booklets on the kite. Carter *et al.* (1995), Carter and Burn (2000) and Snell *et al.* (2002) give accessible accounts of aspects of the reintroduction programme. Thompson *et al.* (2003) present a variety of papers on raptor conservation in the UK, including some on the red kite.

Lovegrove (2007) is very useful for details of the persecution of the red kite. For continuing conflicts between red kites and gamekeeping see Whitfield *et al.* (2003), Valkama *et al.* (2005) and Watson and Moss (2008), and for red kites and rabbits in Spain see Villafuerte *et al.* (1998) and Chapter 11. For rodent control, Buckle and Smith (1994) is an introduction, and see Elton (1942) for details of rodent plagues.

Bate (2003) is the standard biography of John Clare. His natural history prose can be found in Grainger (1983) and also in Robinson and Fitter (1982); the latter also contains a couple of manuscript poems on kites. The reintroduction criteria used for the red kite are discussed in the red kite references above; see also Hodder and Bullock (1997) and Carter and Newbery (2004) for discussions on reintroduction policy in general and Simberloff (1998) for a discussion on whether we should continue to conserve single species. References to strychnine are unfortunately scattered through the pharmacological, herbal and biochemical literature, but see Bisset (1974) and Philippe *et al.* (2004) as starting points.

REFERENCES

Bate, J. (2003) *John Clare: a biography*. Picador, London.
Bircham, P. (2007) *A History of Ornithology*. Collins, London.
Bisset, N.G. (1974) The Asian species of *Strychnos*. III: The ethnobotany. *Lloydia* **37**, 62–107.
Buckle, A.P. and Smith, R.H. (eds) (1994) *Rodent Pests and Their Control*. CAB International, Wallingford, UK.
Carter, I. (2007) *The Red Kite*. Arlequin Press, Shrewsbury, UK.
Carter, I. and Burn, A. (2000) Problems with rodenticides: the threat to red kites and other wildlife. *British Wildlife* **11**, 192–197.

[10]Ratcliffe (2007), p. 326.

Carter, I., Evans, I. and Crockford, N. (1995) The red kite re-introduction project in Britain: progress so far and future plans. *British Wildlife* **7**, 18–25.

Carter, I. and Newbery, P. (2004) Reintroduction as a tool for population recovery of farmland birds. *Ibis* **146** (Supplement 2), 221–229.

Elton, C. (1942) *Voles, Mice and Lemmings: problems in population dynamics.* Clarendon Press, Oxford.

Evans, I.M., Dennis, R.H., Orr-Ewing, D.C. *et al.* (1997) The re-establishment of red kite breeding populations in Scotland and England. *British Birds* **90**, 123–138.

Fisher, J. (no date, *c.* 1956) The birds of John Clare. In: *The First Fifty Years: a history of the Kettering and District Naturalist's Society and Field Club*, pp. 26–69. Kettering, UK.

Grainger, M. (ed.) (1983) *The Natural History Prose Writings of John Clare.* Clarendon Press, Oxford.

Green, B.H. (1979) Wildlife introductions to Great Britain. Report by the Working Group on Introductions of the UK Committee for International Nature Conservation. NCC, London.

Hodder, K.H. and Bullock, J.M. (1997) Translocations of native species in the UK: implications for biodiversity. *Journal of Applied Ecology* **34**, 547–565.

Holloway, S. (1996) *The Historical Atlas of Breeding Birds in Britain and Ireland: 1875–1900.* Poyser, London.

Lovegrove, R. (1990) *The Kite's Tale: the story of the red kite in Wales.* RSPB, Sandy, Beds, UK.

Lovegrove, R. (2007) *Silent Fields: the long decline of a nation's wildlife.* Oxford University Press, Oxford.

Newton, I., Davis, P.E. and Moss, D. (1994) Philopatry and population growth of red kites, *Milvus milvus*, in Wales. *Proceedings of the Royal Society B* **257**, 317–323.

Pearsall, W.H. (1971) *Mountains and Moorlands*, revised edn. Collins, London.

Philippe, G., Angenot, L., Tits, M. and Frédérich, M. (2004) About the toxicity of some *Strychnos* species and their alkaloids. *Toxicon* **44**, 405–416.

Ratcliffe, D. (2007) *Galloway and the Borders.* Collins, London.

Robinson, E. and Fitter, R. (eds) (1982) *John Clare's Birds.* Oxford University Press, Oxford.

Shrubb, M. (2003) *Birds, Scythes and Combines: a history of birds and agricultural change.* Cambridge University Press, Cambridge.

Simberloff, D. (1998) Flagships, umbrellas, and keystones: is single-species management passé in the landscape era? *Biological Conservation* **83**, 247–257.

Smart, J., Amar, A., Sim, I.M.W. *et al.* (2010) Illegal killing slows population recovery of a re-introduced raptor of high conservation concern: the red kite *Milvus milvus*. *Biological Conservation* **143**, 1278–1286.

Snell, N., Dixon, W., Freeman, A., McQuaid, M. and Stevens, P. (2002) Nesting behaviour of the red kite in the Chilterns. *British Wildlife* **13**, 177–183.

Sneyd, C.A. (transl.) (1847) *A Relation, or Rather a True Account, of the Island of England; with sundry particulars of the customs of these people, and of the Royal revenues under King Henry the seventh, about the year 1500.* Camden Society, London.

Stroud, D.A. (2003) The status and legislative protection of birds of prey and their habitats in Europe. In: *Birds of Prey in a Changing Environment* (eds D.B.A. Thompson, S.M. Redpath, A.H. Fielding, M. Marquiss and C.A. Galbraith), pp. 51–84. Scottish Natural Heritage, Edinburgh.

Symons, A. (ed.) (1908) *Poems by John Clare.* Henry Frowde, London.

Thompson, D.B.A., Redpath, S.M., Fielding, A.H., Marquiss, M. and Galbraith, C.A. (2003) *Birds of Prey in a Changing Environment*. Scottish Natural Heritage, Edinburgh.

Tibble, J.W. (ed.) (1935a) *The Poems of John Clare. Volume 1*. Dent, London.

Tibble, J.W. (ed.) (1935b) *The Poems of John Clare. Volume 2*. Dent, London.

Valkama, J., Korpimäki, E., Arroyo, B. *et al.* (2005) Birds of prey as limiting factors of gamebird populations in Europe: a review. *Biological Reviews* **80**, 171–203.

Vesey-Fitzgerald, B. (1946) *British Game*. Collins, London.

Villafuerte, R., Viñuela, J. and Blanco, J.C. (1998) Extensive predator persecution caused by population crash in a game species: the case of red kites and rabbits in Spain. *Biological Conservation* **84**, 181–188.

Walters Davies, P. and Davis, P.E. (1973) The ecology and conservation of the red kite in Wales. *British Birds* **66**, 183–224, 241–270.

Watson, A. and Moss, R. (2008) *Grouse: the natural history of British and Irish Species*. Collins, London.

Whitfield, D.P., McLeod, D.R.A., Watson, J., Fielding, A.H. and Haworth, P.F. (2003) The association of grouse moor in Scotland with the illegal use of poisons to control predators. *Biological Control* **114**, 157–163.

Williamson, H. (1923) *The Peregrine's Saga and Other Tales*. Collins, London.

MISTLETOE & ITS PREDATORS

AUSTRALIAN MISTLETOE HAS EVOLVED CAMOUFLAGE TO EVADE PREDATORS.

POSSUMS

EUROPEAN MISTLETOE IN AN OAK IS EASILY SPOTTED BY PREDATORS.

OFFICE WORKERS WHO LIKE KISSING

DRUIDS

SUPERSTITIOUS LADIES WHO WANT TO MARRY

Parasitic Plants: Mistletoes

I T'S A CHRISTMAS CUSTOM to bring mistletoe into the house, part of a very ancient tradition of displaying evergreen plants during midwinter festivals. The European mistletoe (*Viscum album*) has also accreted other customs, such as kissing under it, which is said to be Scandinavian in origin; indeed it seems to have unusually wide symbolic importance. This is because its appearance is so extraordinary: green in midwinter, with white berries and hanging as a clump in the bare branches of a tree. We shall see in this chapter that mistletoes – defined as shrubby plants of the order Santalales which parasitize via the stems of other plants – have a wide range of appearances, hosts and ecological associations (with birds, mammals and insects), and this perhaps unexpected importance parallels their significance to human culture.

THE GOLDEN BOUGH

James George Frazer's *The Golden Bough* is considered to be a founding work in the discipline of social anthropology. An amassment of religious customs, rites and superstitions from around the world, its sprawling structure (12 volumes by the third edition in 1911–15) is held together by the mistletoe. Frazer begins by identifying as mistletoe the branch picked from within a sacred grove by the usurper (and slayer) of the priest-king of a cult of the Roman goddess Diana at Nemi, near Rome. He connects this to the following passage from Virgil's *Aeneid*, in which Aeneas discovers, with the aid of two doves, the golden bough which will later protect him in his journey through the underworld:

> Then, when they [the doves] came to the mouth of foul-breathing Avernus,
> Swiftly they soared, went gliding through the soft air and settled,
> The pair of them, on a tree, the wished-for place, a tree
> Amid whose branches there gleamed a bright haze, a different colour –
> Gold. Just as in depth of winter the mistletoe blooms
> In the woods with its strange leafage, a parasite on the tree,
> Hanging its yellow-green berries about the smooth round boles:
> So looked the bough of gold leaves upon that ilex dark,
> And in the gentle breeze the gold-foil foliage rustled.[1]

At the end of *The Golden Bough*, Frazer describes the Norse myth of Balder, who is tricked by Loki (see Chapter 19) and killed by a shaft of mistletoe. It is central to Frazer's argument about the deeper significance of myth that there is a

[1]Virgil, *Aeneid* 6, 201–9, translated by C. Day-Lewis; see Beard (1992).

Biological Diversity: Exploiters and Exploited, First Edition.   Paul Hatcher and Nick Battey.
© 2011 John Wiley & Sons, Ltd. Published 2011 by John Wiley & Sons, Ltd.

link between the classical ritual of the dying priest-king, and the story as it appears in other cultures (such as the Balder myth). Mistletoe provides this link; and although the equivalence of the myths in which it features is now considered very uncertain (Frazer wove stories rather than seeking objectivity), the appropriation of mistletoe was only possible because it kept popping up in mythology from widely differing cultures. The sacredness of mistletoe on oak for the Druids of Gaul is a further example of the impact of the plant on the human imagination, as the Roman encyclopaedist Pliny (AD 23–79) reports:

> The Druids … hold nothing more sacred than mistletoe and a tree on which it is growing, provided it is a hard-oak. Groves of hard-oaks are chosen even for their own sake, and the magicians perform no rites without using the foliage of those trees … but further, anything growing on oak-trees they think to have been sent down from heaven, and to be a sign that the particular tree has been chosen by God himself. Mistletoe is, however, rather seldom found on a hard-oak, and when it is discovered it is gathered with great ceremony … Hailing the moon … they prepare a ritual sacrifice and banquet beneath a tree and bring up two white bulls … A priest arrayed in white vestments climbs the tree and with a golden sickle cuts down the mistletoe, which is caught in a white cloak. Then finally they kill the victims, praying to God to render his gift propitious to those on whom he has bestowed it. They believe that mistletoe given in drink will impart fertility to any animal that is barren, and that it is an antidote for all poisons.[2]

In the next section, we survey the spectacular diversity of mistletoes, and describe the evolutionary history of the group, a history that is intimately related to the spread of continents around the world. Then we investigate the haustorium. This is the structure that connects mistletoe to host, and is therefore a crucial adaptation, allowing the mistletoe to parasitize other plants. Finally, we shall see that, although they are parasites, mistletoes have a more complex interaction with other organisms than the term parasite suggests; in fact, they occupy a crucial, often mutualistic role in communities from Australia to North America.

THE ORIGINS OF MISTLETOES

Plants have allowed other life on Earth through their autotrophism: they make their carbohydrates by photosynthesis, harvesting CO_2 and sunlight to create the carbon compounds on which we all depend. Mineral nutrients and water are typically obtained by the root system. About 1% of flowering plants have, however, become heterotrophic, gaining nutrients, water, and at least some of their carbon compounds from other plants. These parasitic plants are significant because of the elaborate methods they adopt to secure their lifestyle; because they can be pests – *Striga asiatica* (witchweed), for instance, is a serious problem for maize cultivation; and because sometimes, as in mistletoes, they have co-evolved mutualisms with birds, insects or mammals which contribute to ecosystem stability and the maintenance of biodiversity.

Parasitism has evolved independently in 12 angiosperm clades and there are about 4500 parasitic angiosperm species. In most cases (about 60%), the

[2]Pliny, *Natural History* Book XVI. xcv. In Volume IV of the English translation by H. Rackham (William Heineman, London, 1945).

(a)   (b)   (c)

haustorial connection is made to the host root; the remainder connect via the stem. About 80% of parasitic plants retain some chlorophyll, and are therefore known as hemiparasites; the remainder are holoparasites. Rarely, parasitism may be facultative, so that survival is possible in the absence of the host; otherwise it is obligate, the host being essential for the parasite to complete its life-cycle. Some examples of these different kinds of parasitic angiosperms are shown in Fig. 18.1.

The mistletoes are a subset of parasitic plants, including around 1500 species, and they are all obligate stem hemiparasites. They are found within the order Santalales, in which the mistletoe habit has arisen five times independently. There are, however, three genera – *Nuytsia*, *Atkinsonia* and *Gaiadendron* – that are root parasites and yet which are usually considered mistletoes. They have retained the relatively primitive character of root parasitism while the other mistletoes have evolved to stem parasitism.

Some examples of mistletoes from around the world are shown in Fig. 18.2. These are from the families Loranthaceae and Viscaceae, which contain 98% of mistletoe species, and whose world distributions are summarized in Fig. 18.3. The mistletoe habit has evolved independently in these two families, so that many of their similarities, such as 'seeds' (see Box 18.1 for discussion) surrounded by sticky material within showy fruits attractive to birds, are taken to reflect convergent evolution associated with life-style. The Loranthaceae is the larger (73 genera, over 900 species), evolutionarily older family, and developed in the Cretaceous on the super-continent of Gondwanaland, which was made up of the continents we now know as South America, Africa/Madagascar, India, Australia/New Zealand and Antarctica, and located at the South Pole, where at the time the climate was temperate. The dispersed distribution of relictual genera (primitive types endemic to restricted localities in regions of Gondwanaland – Fig 18.3a) indicates that by the early Tertiary (about 50 mya), when South America, Antarctica, Australia and New Zealand had separated from Africa, Madagascar and India, the Loranthaceae family had already diversified to create many of the lineages now known on these continents. These include *Psittacanthus, Phthirusa, Tristerix* and *Ameyema*.

(d)

Fig. 18.1 *Some parasitic plants.* **(a)** *Mistletoe (Viscum album): an obligate stem hemiparasite;* **(b)** *Yellow rattle (Rhinanthus): a facultative root hemiparasite;* **(c)** *Broomrape (Orobanche): a root holoparasite;* **(d)** *Dodder (Cuscuta), or strangleweed: a stem holoparasite [(d) Courtesy of Dr Bob Froud-Williams.]*

Fig. 18.2 *Three mistletoes from around the world.* **(a)** Ameyema cambagei *on* Casuarina (*Australia*); **(b)** Viscum minimum *on* Euphorbia polygona (*South Africa*); (c) Arceuthobium douglasii *on* Pseudotsuga menziesii (*North America*) [Photographs courtesy of Daniel L. Nickrent.]

(a)   (b)   (c)

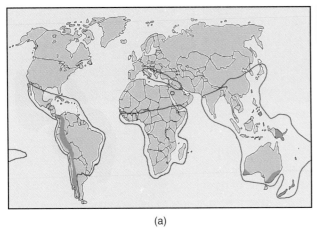
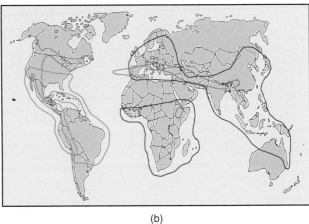

(a)   (b)

Fig. 18.3 *World distributions of* **(a)** *Loranthaceae (red lines; green shading indicating location of relictual genera);* **(b)** *Viscaceae genera Viscum (blue lines), Arceuthobium (orange lines) and Phoradendron/ Dendrophthora (green lines, Dendrophthora falling entirely within the distribution of Phoradendron)* [*Redrawn by Henry Battey from Barlow (1983).*]

The Viscaceae is of Laurasian origin (Laurasia was the supercontinent made up of North America and Eurasia), originating in the late Cretaceous, and probably developed initially in south-east Asia. It is much smaller than the Loranthaceae, with 540 species. Four of the seven genera are restricted to the Old World, including *Viscum*. The New World genera are *Arceuthobium* (also found in the Old World; see Fig. 18.3b), *Phoradendron* and *Dendrophthora*; their progenitors are believed to have found their way to the Americas via the Bering land bridge which connected Asia and America until the Tertiary period.

Misodendrum, the single genus in the South American family Misodendraceae, is considered to represent the first evolutionary experiment with the mistletoe habit, with a divergence date from other extant mistletoes calculated by Vidal-Russell and Nickrent (2008) as 80 mya. These authors have also deduced that the stem parasitic, mistletoe habit evolved 72 million years ago in the Viscaceae, and as recently as 28 million years ago in the Loranthaceae. The earliest ancestors of both families appear to have been root parasites, with more immediate predecessors being able to parasitize via root or stem.

THE HAUSTORIUM: THE ESSENCE OF PLANT PARASITISM

A haustorium is a specialized structure which allows a parasite to absorb nutrients from its host. In the case of parasitic fungi it is a specialized hypha; rust fungi, for example, can use the structure to penetrate the cells of the plant host. In parasitic plants, the haustorium is the bridge between the parasite and the host and allows the flow of water and solutes from host to parasite. It is therefore fundamental to parasitism. In the mistletoes there is considerable variation in the way the haustorium develops. In the first type, a single haustorial connection remains localized at the point of original infection. In some cases, a complex parasite–host interface develops, as the cambium of the host multiplies in response to the development of the xylem strands of the parasite. When the parasite dies, the remaining fluted, ornate structure can be many centimetres in diameter, and is known as a 'woodrose'. *Psittacanthus* and *Tapinanthus* (Loranthaceae), from Central America and Africa respectively, cause woodrose production on their hosts, and these are sold as ornaments.

BOX 18.1   EMBRYOLOGY AND FLORAL CHARACTER IN MISTLETOES

Most angiosperms have flowers made up of green sepals, showy petals, stamens (the male organs producing pollen in anthers); and carpels, the female part of the flower (gynoecium) containing ovules produced from a placenta (Fig. 1). Mistletoe flowers may be large and brightly coloured, as in many of the Loranthaceae, to attract bird pollinators; or they may be very inconspicuous, as in many Viscaceae. But it is in their embryology that the family is united in strangeness. The gynoecium is very reduced and contains a central organ known as the mamelon. Within this tissue arise embryo sacs (the female gametophytes), each containing a small number of nuclei, one of which is the egg nucleus. Whereas in most angiosperms the embryo sac

stays where it is, and waits for the delivery of the male nuclei via the pollen tube, in mistletoes the embryo sacs are much more proactive. They grow into the style, seeking out the male nuclei; in the case of *Helixanthera ligustrina* (Loranthaceae), the embryo sacs actually reach the stigmatic surface. The proembryos resulting from successful union of male and female then grow, in competition with each other, back to the ovary. When the embryo(s) have developed to maturity, they are typically green and located within endosperm; in *V. album* this is surrounded by the fleshy berry composed of sticky viscin (Fig. 2).

The key point here is that there are no ovules in mistletoes. The ovule of normal angiosperms is composed of maternal tissue which surrounds the embryo sac and is delimited by integuments. After fertilization the ovule becomes the seed, the integuments transforming into the hard husk of the seed. This means that mistletoes do not have true seeds. However, because the embryo(s) within the endosperm look like seeds, the term is widely used for convenience.

The unusual embryology of mistletoes is generally considered to reflect their specialized, evolutionarily advanced condition. Consistent with this is the presence of dioecy (male and female flowers borne on different individuals) in the group. In Viscaceae it is common, and appears to be derived from monoecy (separate male and female flowers borne on the same individual). For example, *V. album* is dioecious (typically with a skewing of the ratio of males to females towards females); monoecious species of *Viscum* occur in Africa and India. In the Loranthaceae, dioecy is less frequent and seems to have arisen from the bisexual (hermaphrodite) condition.

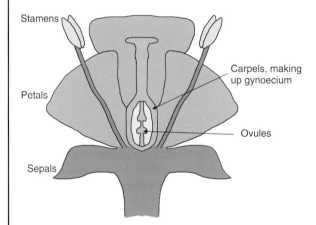

Box 18.1 Fig. 1   *A typical angiosperm flower* [Courtesy of Henry Battey.]

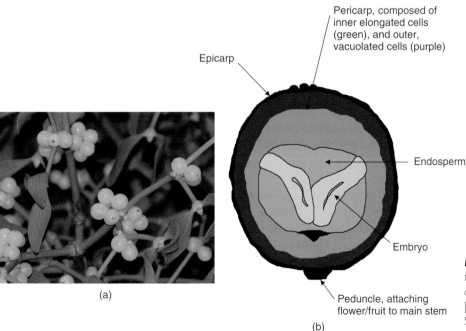

Box 18.1 Fig. 2   **(a)** *Mistletoe berry.* **(b)** *Longitudinal section of a mature berry of* V. album [Redrawn by Henry Battey from Sallé (1983).]

Fig. 18.4 *Tristerix aphyllus:* flowers and fruits on Echinopsis chilensis [*Photograph Courtesy of Guillermo Amico.*]

In the second type of development, the primary haustorium gives rise to strands that grow beneath the bark and allow the parasite to spread within the host plant. In an extreme example, the external part of the seedling of the mistletoe *Tristerix aphyllus* (Loranthaceae) dies away after the haustorium has penetrated its host, a species of columnar cacti from north-central Chile. The parasite then grows only as a system of filaments within the host tissue, until flowering, when its inflorescences emerge over the surface of the cactus (see Fig. 18.4). In most species, though, the original plant remains alive on the surface of the host but propagates itself inside, secondary shoots arising at intervals through the surface of the host. *Viscum album* (the familiar mistletoe of Europe, Fig. 18.1a), and *Viscum minimum* (a minute parasite of *Euphorbia* from South Africa, Fig. 18.2b) both show this form of haustorial development. Economically important examples are the dwarf mistletoes, *Arceuthobium*, which are serious pests of conifers in North America (Fig. 18.2c). Some species of *Arceuthobium* are able to grow in phase with the host in such a way that all new buds are infected from within by the parasite.

In the third type of haustorial development, the host plant is explored via epicortical roots (running along the surface of the host) which give rise to multiple secondary haustoria; species of *Struthanthus* (Loranthaceae) from Central America show this behaviour. The fourth, probably most dramatic type of haustorial development is that found in the root parasite *Nuytsia floribunda*, the Western Australian Christmas tree. This is a tree which parasitizes the root systems of neighbouring plants by means of a clasping haustorium which, having encircled the host root, then develops a tough sclerenchymatous 'prong' of tissue. Differential growth within the haustorium generates downward movement of this prong, which eventually guillotines the root, allowing the parasite tissues to absorb water and solutes from the severed host vascular tissues. So indiscriminate is this parasite that it has been known to sever underground cables in Australia, causing short-circuits.

The key, common event in all these types of haustorial development is the initial, forcible entry of the parasite into the host. This involves parasite attachment, breaking through the bark or other resistant host surface, and the establishment of a physiological connection to allow the parasite to tap host water and solutes. These stages can be seen for *Viscum album* in Fig. 18.5. First, from the seed emerges the hypocotyl (the part of the embryonic stem below the cotyledons); there is no radicle (primary root) and the hypocotyl ends in a flat disc. When it makes contact with the surface of the host, the outer margins of the disc are stimulated to more rapid growth than the inner regions, so that the face of the disc is flattened and brought snugly into contact with the host surface. The domed structure resulting from this process of attachment is known as the holdfast. Contact with the host stimulates the holdfast to secrete large amounts of sticky, lipidic fluid which eventually hardens; at the same time, the epidermal cells on the face of the holdfast form papillae which attach to the host surface. Then, most remarkably, the tissues within the dome above the attached face expand upwards, so that the outer layer of host tissue is torn away by the attached papillae. Further papillae attach themselves to successive layers of host tissue, and by tearing and expansion, disrupt the host surface. While this is going on, a meristematic region becomes established in the centre of the holdfast and forces its way through the disrupted host tissues. This is the haustorium, and it will continue to grow into the host tissues until it contacts the xylem water-conducting system. As we have seen above, in the case of mistletoes showing the *Viscum*

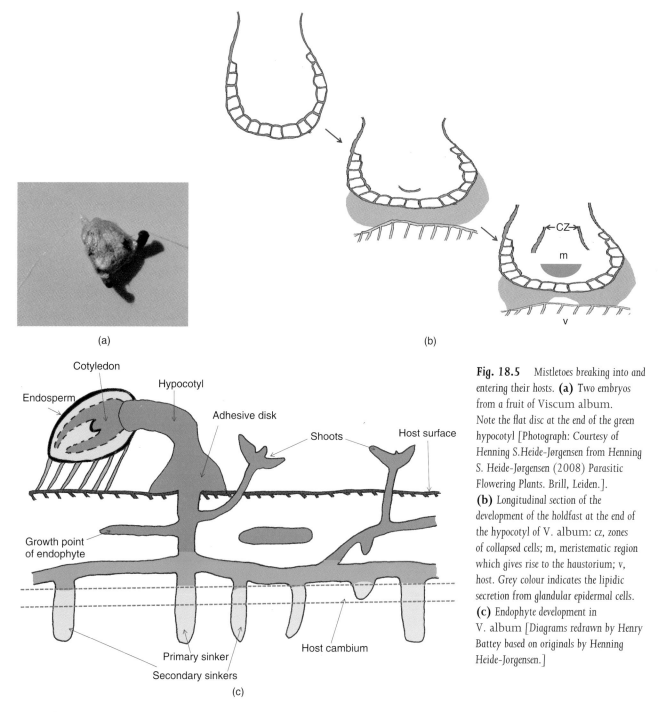

Fig. 18.5 *Mistletoes breaking into and entering their hosts.* **(a)** *Two embryos from a fruit of Viscum album. Note the flat disc at the end of the green hypocotyl [Photograph: Courtesy of Henning S.Heide-Jørgensen from Henning S. Heide-Jørgensen (2008) Parasitic Flowering Plants. Brill, Leiden.].* **(b)** *Longitudinal section of the development of the holdfast at the end of the hypocotyl of V. album: cz, zones of collapsed cells; m, meristematic region which gives rise to the haustorium; v, host. Grey colour indicates the lipidic secretion from glandular epidermal cells.* **(c)** *Endophyte development in V. album [Diagrams redrawn by Henry Battey based on originals by Henning Heide-Jorgensen.]*

type of haustorial development, growth of the haustorium does not cease at this point; strands of parasitic tissue derived from it (the 'endophyte') continue to ramify within the host.

Two additional points should be made. First, the further development of the haustorium within the host is quite variable, depending on the host species, tissue age and other factors. Second, the process of attachment and haustorium formation described here for *Viscum* (Viscaceae) is remarkably similar to that of *Phthirusa* (Loranthaceae), described by Dobbins and Kuijt (1974). The independent origin of the mistletoe habit in these two families implies that they have independently developed, from a root parasitic habit, these very similar modes of breaking into and entering the host stem.

THE MISTLETOE LIFE-CYCLE

Although there is great diversity among the mistletoes, their aerial parasitic life-style has led to certain characteristic features. These can be highlighted by reference to the life-cycle of *V. album* (Fig. 18.6), and are:

Fig. 18.6 *Development of a mistletoe plant, Viscum album ssp. abietis [Redrawn by Henry Battey from Zuber (2004).]*

- initial establishment from seed;
- the mechanisms by which, once established the mistletoe obtains the materials it needs to survive and grow (carbon, water and nutrients), and by which it attains its characteristic growth form;

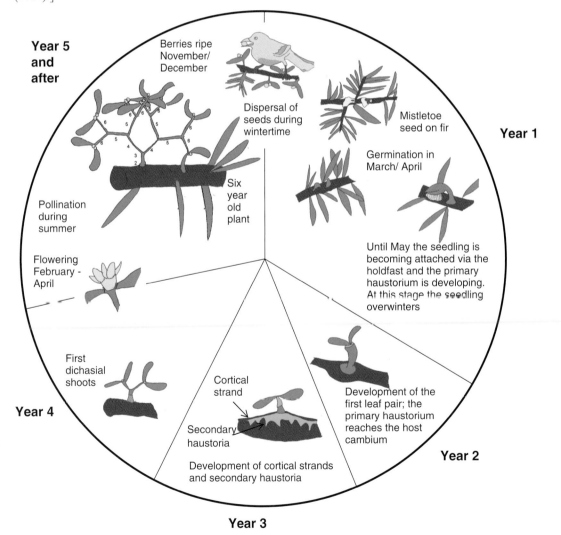

- the character and pollination of its flowers;
- the way in which the mistletoe secures the attention of seed dispersal agents.

The latter topic leads naturally to a consideration of the interaction of mistletoes with their community, which we consider in the next section of this chapter.

There is wide variation between mistletoes in the time taken for germination and establishment of the new plant. This is illustrated by the following comment from Job Kuijt, in 1969:

> I have personally followed the development of *V. album* and *Phthirusa pyrifolia* in Vancouver. The former is of European origin (Baarn, Netherlands); the latter, Costa Rican (San Pedro). The *Viscum* seeds were planted on *Sorbus*, also the host of the mother plants. Germination was visible four months afterward. Two years later, although successfully established, the mistletoes had not grown significantly in size and had not yet formed their first foliage leaves. Contrast this with *Phthirusa*: within two weeks after emplacement the seedling stands erect on its terminal disk, its two large cotyledons spread apart and separate from the collapsed endosperm. Even under the cloudy skies of Vancouver, in a greenhouse, some plants had flowered and borne ripe fruit before two years had elapsed.[3]

There is similar variation in the range of hosts which mistletoes will parasitize. *Viscum album*, for example, is divided into subspecies, and while ssp. *album* is found on a very wide range of dicotyledonous trees (over 100 genera), ssp. *creticum* is found only on *Pinus halepensis* ssp. *brutia* in Crete. In general, however, mistletoes have a wide potential host range which may be restricted in nature by the ecology of dispersal, in particular the feeding habits of birds.

The seed dispersal phase is the non-parasitic part of the mistletoe life-cycle. At this time the parasite must provide its own energy supply, and in the case of *V. album* the seeds and hypocotyl are green and presumed to be photosynthetic. Seed germination requires light and is negatively phototrophic, encouraging the hypocotyl to make contact with the host bark. Ease of penetration varies with host species – for example, the character of the bark of different oak species (principally the thickness of different tissue types) affects the success of *Viscum album*. But perhaps the most striking thing about mistletoe parasitism in general is the lack of a host wound reaction to the initial penetration. This was noted by Thoday (1956) during his classic explorations of the mistletoe haustorium; it still seems noteworthy today, particularly given the dramatic nature of plant responses to pathogens such as fungi and insects.

Moisture is an essential requirement for germination and remains critical until the primary haustorium has made contact with the water-conducting system of the host. This is typically achieved within a couple of months in *V. album* (see Fig.18.6). The connection can in some mistletoe–host combinations involve continuity between host and parasite xylem elements, but more typically it occurs via parenchymatous tissue located between the xylems of host and parasite. It allows a supply to the parasite of water, nutrient elements and carbon compounds from the host xylem system. One of the most striking features of mistletoe physiology is their high leaf transpiration rates relative to those of their hosts. This was

[3]Kuijt (1969), p. 49.

shown for 18 different host–mistletoe pairs investigated from a range of environments during a field tour of Australia reported by Ullman *et al.* (1985). Samples were taken from New South Wales in the south, to the coast of Queensland in the north. Host plants included species of *Acacia* and *Eucalyptus*, and among the mistletoes were several species of *Amyema* and *Lysiana*.

This high transpiration rate is a mechanism for acquiring not only water (in competition with host leaves) but also dissolved nitrogen- and carbon-containing compounds. Studies of the proportion of carbon derived by the parasite from its host by this means suggest that up to 43% is host-derived in *V. album*, and up to 62% in *Phorandendron*. In the dwarf mistletoes (*Arceuthobium* species) it is around 70%, so that these plants fix only 30% of their own carbon. This finding is consistent with molecular analysis of the *Arceuthobium* plastid genome, and morphological studies, all of which suggest that these mistletoes are on an evolutionary trajectory towards holoparasitism.

To many people in Europe and North America the most memorable feature of mistletoe might be its ball-like growth habit. This is true of both the mistletoes valued as Christmas decorations, *V. album* in Europe and *Phoradendron* species in North America. It arises in *V. album* due to a regular dichotomous branching pattern, which begins after the third year from establishment (Fig. 18.7). The consistency of this pattern, with one branch pair growing every year, means that an undamaged mistletoe plant can be fairly accurately aged by visual inspection (see Fig. 18.6). It is this more-or-less spherical shape, along with the persistence of the evergreen foliage when the leaves of deciduous hosts have fallen in winter, which makes the mistletoe so obvious from a distance (e.g. Fig. 18.1a). It is therefore particularly intriguing that a common feature of many of the mistletoes of Australia is their ability to disguise themselves, by mimicking the leaf shape and arrangement of their hosts (see, for example, Fig. 18.2a).

This is not an occasional or exotic phenomenon: a survey by Barlow and Wiens in 1977 showed that the majority of Australian loranthaceous mistletoes showed some degree of host mimicry. They concluded that vegetative mimicry is a response to natural selection, and suggested that the source of selective pressure was herbivorous vertebrates – specifically, the possum. Being a night feeder, the common brush-tailed possum (*Trichosurus*) is likely to cue on leaf shape, and is known to have a particular fondness for mistletoe foliage. Barlow and Wiens also suggested that the serious decline in native mistletoes in New Zealand, where mistletoes are considered primitive relics, and do not mimic host foliage, was associated with the introduction of the possum to

Fig. 18.7 Viscum album, *showing growth form and dichotomous branching*

that country. They explained the almost uniquely Australian character of mistletoe host mimicry as being a consequence of the dominance in its temperate and semi-arid zone forests of *Acacia* and *Eucalyptus*, two genera with notably non-succulent leaves, likely to be unattractive to herbivores. Mistletoes, with their high transpiration rates and fleshy leaves, would be a natural target for hungry possums. It should be noted that there are alternative explanations for the host mimicry of Australian mistletoes; one is that its function is to make fruit-eating bird vectors of the mistletoes search harder for them, thus increasing the dispersal range of the mistletoes.

While loranthaceous mistletoes can disguise their foliage, and viscacean ones do not, when it comes to the reproductive process the situation is reversed. The flowers of Viscaceae are often dull and inconspicuous, while in Loranthaceae they are typically big, showy and nectar-rich (see Box 18.1). This reflects the main pollinators involved: in the tropical and subtropical loranthaceous mistletoes these are birds. Hummingbirds, for instance, are important pollinators of mistletoes in Central and South America; while in the tropics of the Old World some members of the family Dicaeidae are known as 'mistletoe birds' because of their association with mistletoe reproduction. In the Indonesian mistletoe *Macrosolen cochinchinensis*, the insertion of the bill of *Dicaeum* species causes the flower to open violently, sprinkling the bird with pollen. With Viscaceae from more temperate regions, the pollinators are insects; for *V. album*, this means principally flies.

The fruits of mistletoes are generally adapted for bird dispersal. Bright fruit colours attract birds, and the fruit flesh is nutritious, as the sticky viscin coating around the seeds may also be. But the viscin persists as the seeds pass through the gut of the bird (or are regurgitated), and helps the seed to stick to the potential host. This dispersal method has led to a cultural connection between bird excrement and mistletoes; for example, the German word for manure is *Mist*, and for mistletoe, *Mistel*. It is not true, however, that the seeds have to pass through the bird to germinate, even though some bird species are highly adapted to a diet of mistletoe fruits. The grey-sided flowerpecker (*Dicaeum celebicum*), a mistletoe bird of the Old World tropics, shows such adaptation: it has a mechanism which prevents the mistletoe seeds from entering the gizzard, so that they pass through the bird undamaged. The advantage to the bird is presumably that more mistletoe plants become established, providing renewed food for the bird. On the other hand, the European mistle thrush, *Turdus viscivorus*, which gains its name because of an association with *V. album*, does not feed exclusively on its berries. In this case, the relationship is perhaps general, rather than being highly developed.

Although birds are the dominant dispersers of mistletoes, there are some striking non-avian examples of mistletoe dispersal. *Dromiciops gliroides* (monito del monte or 'little mountain monkey') is the sole living member of an ancient marsupial order Microbiotheria, and is the exclusive disperser of the fruits of the loranthaceous mistletoe *Tristerix corymbosus* in the temperate forests of southern Argentina and Chile. This association between marsupial and primitive mistletoe is probably around 70 million years old, pre-dating the break up of Gondwanaland. In contrast, the group of dwarf mistletoes, *Arceuthobium* (Fig. 18.2c), have become independent of any animal vector: they have developed a powerful explosive dispersal mechanism, which fires the seed (about 3 mm long) up to 15 metres from the fruit. A tail of sticky viscin tissue anchors the seed to any substrate which intercepts it. The dwarf mistletoes parasitize tall conifers, and this dispersal mechanism gives the infection the character of slow but remorseless progress outwards from its initial focus; this is in contrast to the patterns shown by many other mistletoes, which are dictated by the feeding habits of birds.

MISTLETOES: PARASITES AND MUTUALISTS

The word 'parasite' draws attention to the method by which mistletoes gain their sustenance. Because of its generally negative connotations, the term suggests that these organisms are primarily a problem – perhaps interesting, unusual and noteworthy, but a problem nevertheless. Recently, however, attention has shifted to the important roles of mistletoes in the communities in which they live. This approach emphasizes the interactions of mistletoes with other species, and their potentially beneficial impact on community diversity. This has prompted a view which, while acknowledging the destructive aspect of the parasitic habit of mistletoes, balances against this their mutualistic interactions. Here we provide three brief examples of diverse mistletoe mutualisms, before offering a concluding view of this fascinating group of plants.

In northern Arizona the mistletoe *Phorandendron juniperinum* grows on juniper (*Juniperus monosperma*), and overwintering birds feed on both mistletoe and juniper fruits. The Townsend's solitaire (*Myadestes townsendi*) is one such frugivore, and is known to be an important disperser of the juniper. Van Ommeren and Whitham (2002) studied the three-way interaction between host, parasite and vector, measuring the abundance of each over a three-year period. They found that the number of avian frugivores was most strongly related to mistletoe fruit production, and that mistletoe abundance was the best predictor of the presence of the birds. The junipers showed very variable year-on-year berry production, due to a masting pattern (see also Chapter 10), whereas the mistletoes were a much more consistent food source for the birds. Thus mistletoes attract the birds that disperse the juniper berries; benefit to the junipers was shown by the presence of twice as many juniper seedlings in stands with high levels of mistletoe infection. This indicates that the mistletoe interaction with juniper is mutualistic, at least within limits; at very high mistletoe densities its effect becomes negative, presumably because the physiological burden of the parasite becomes unsustainable.

We described earlier the extraordinary endophytic lifestyle of the mistletoe, *Tristerix aphyllus*: it exists as a filamentous growth within its cactus host, only appearing to flower and fruit (see Fig. 18.4). In its native habitat in Chile, the fruits of this mistletoe are eaten and the seeds dispersed exclusively by the Chilean mockingbird (*Mimus thenca*), the number of which is positively correlated with the abundance of the mistletoe. In their study of this bird–mistletoe mutualism, Martinez del Rio et al. (1996) found that parasitized host individuals receive more mistletoe seeds than non-parasitized ones, a reinforcing effect that, in the case of the cactus *Echinopsis chilensis*, reduces reproductive effort. Subsequent work showed that the cactus has, however, an evolutionary response to this: selection favours individuals with longer spines, which deter perching by the vector.

The third example of a mistletoe mutualism relates to a more general community effect, resulting from enhanced nutrient cycling. In *Eucalyptus* forests of

Australia, the mistletoe *Ameyema miquelii* was shown by March and Watson (2007) to increase litterfall by up to 189%, due to higher leaf turnover in the mistletoe than in the host tree. A larger amount of ground litter resulted, and higher mistletoe infection rates were associated with greater plant biomass in the understorey, suggesting a benefit of the additional leaf nutrients on plant productivity in this ecosystem. This result mirrors some effects of parasitic plants more generally; in nutrient-limited ecosystems such as those of arctic and alpine regions they enhance nutrient availability through more rapid decomposition, suppressing dominance, and encouraging species diversity. Such effects led Malcolm Press to suggest that parasitic plants could be characterized as botanical Robin Hoods, as well as Draculas, redistributing scarce resources to a wider range of species than would otherwise receive them.

David Watson has not only developed this theme in relation to leaf litter in Australian forests; he has emphasized the key role of mistletoes there and elsewhere in interactions with birds, mammals and insects, and concluded that mistletoes should be regarded as a keystone ecosystem resource. This means that the impact of mistletoes on their communities is large relative to their abundance, and implies a key role in maintaining community diversity. They are a widely used food source: Watson found that 66 families of birds and 30 families of mammals have been recorded as consuming mistletoes worldwide. Fruit, nectar and foliage are all used, and are particularly valuable where other food sources are more seasonal in their availability. Mistletoes are also favoured nesting sites for many animals: seven families of mammals and 43 families of birds were found to use mistletoes for nesting or roosting, with the estimate for bird families increased following subsequent work.

A GOLDEN BOUGH?

The conclusion that mistletoes occupy a special place in ecosystems has led to a new analogy, designed to express their ecological place. This is to dryads, tree nymphs from Greek mythology who, while attached to and dependent on a specific tree for their life, raised to sacred the status of the surrounding woodland (Watson, 2009). Thus the name golden bough may be apt from an ecological as well as a mythological perspective. Nevertheless, it is important to keep a clear head about mistletoes: the destructive effects of *Arceuthobium* in North American pine forests provide a salutary reminder of both the positive and negative dimensions. These mistletoes have been shown to enhance species richness of Colorado ponderosa pine forests, a kind of dryad effect; but the following statement, from a USDA forestry leaflet on dwarf mistletoe, shows the devastating impact of the parasite on timber production:

> Acceptable yields cannot be expected from stands infested when they are young. For example, 100 year-old stands that have been infested for 70 years averaged only 300 cubic feet (8 m³) of wood per acre, while healthy stands of the same age on similar sites averaged 2350 cubic feet (65 m³) of wood per acre.[4]

[4]Hawksworth and Dooling, USDA (1984): *Lodgepole Pine Dwarf Mistletoe*, USDA Forest and Insect Disease Leaflet 18, at: www.fs.fed.us/r6/nr/fid/fidls/fidl18.htm.

FURTHER READING

The topic of parasitic plants was first circumscribed in a single text by Kuijt (1969); more recent sources are the books by Press and Graves (1995) and Heide-Jorgensen (2008). An excellent web resource is 'The Parasitic Plant Connection', by Dan Nickrent (www.parasiticplants.siu.edu). Mistletoes are included in all these sources; a volume devoted specifically to them is Calder and Bernhardt (1983). Key articles covering specific aspects of mistletoe biology, in addition to those mentioned in the main text, are: biogeography and evolution, Barlow (1983); systematics, Nickrent *et al.* (2010); the haustorium, Kuijt (1977), Beyer *et al.* (1989), Kuijt and Lye (2005) and Calvin and Wilson (2006); water relations and nutrient acquisition, Marshall and Ehrlinger (1990) and Press (1995); ecology, Bennetts *et al.* (1996), Watson (2001, 2009), Aukema (2003), Press (1998), Press and Phoenix (2005) and Mathiasen *et al.* (2008).

The monito del monte–*Tristerix* mutualism is described by Amico and Aizen (2000). The leaf mimicry topic has been much discussed and elaborated: see Ehleringer *et al.* (1986), Press (1995) and Schaefer and Ruxton (2009). *Viscum* is covered approachably at Henning Heide-Jorgensen's web page: www.viscum. dk/eng-sider/about_viscum.htm. The biology of *V. album* is reviewed by Zuber (2004), where information relating to the potential medicinal properties of the plant can also be found. A condensed version of *The Golden Bough* by James George Frazer is the abridged edition (Frazer, 1978); useful commentaries on the work are Smith (1978) and Beard (1992).

REFERENCES

Amico, G. and Aizen, M.A. (2000) Mistletoe seed dispersal by a marsupial. *Nature* **408**, 929–930.

Aukema, J.E. (2003) Vectors, viscin, and Viscaceae: mistletoes as parasites, mutualists, and resources. *Frontiers in Ecology and the Environment* **1**, 212–219.

Barlow, B.A. (1983) Biogeography of Loranthaceae and Viscaceae. In: *The Biology of Mistletoes* (eds M. Calder and P. Bernhardt), pp. 19–46. Academic Press, Sydney, Australia.

Barlow, B.A. and Wiens, D. (1977) Host–parasite resemblance in Australian mistletoes: the case for cryptic mimicry. *Evolution* **31**, 69–84.

Beard, M. (1992) Frazer, Leach and Virgil: the popularity (and unpopularity) of the Golden Bough. *Comparative Studies in Society and History* **34**, 203–224.

Bennetts, R.E., White, G.C., Hawksworth, F.G. and Severs, S.E. (1996) The influence of dwarf mistletoe on bird communities in Colorado ponderosa pine forests. *Ecological Applications* **6**, 899–909.

Beyer, C., Forstreuter, W. and Weber, H.C. (1989) Anatomical studies of haustorium ontogeny and the remarkable mode of penetration of the haustorium in *Nuytsia floribunda* (Labill.) R. Br. *Botanica Acta* **102**, 229–235.

Calder, M. and Bernhardt, P. (1983) *The Biology of Mistletoes*. Academic Press, Sydney, Australia.

Calvin, C.L and Wilson, C.A. (2006) Comparative morphology of epicortical roots in Old and New World Loranthaceae with reference to root types, origin, patterns of longitudinal extension and potential for clonal growth. *Flora* **201**, 51–64.

Dobbins, D.R. and Kuijt, J. (1974) Anatomy and fine structure of the mistletoe haustorium (*Phthirusa pyrifolia*). I: Development of the young haustorium. *American Journal of Botany* **61**, 535–543.

Ehleringer, J.R., Ullman, I., Lange, O.L. *et al.* (1986) Mistletoes: a hypothesis concerning morphological and chemical avoidance of herbivory. *Oecologia* **70**, 234–237.

Frazer, J.G. (1978) *The Illustrated Golden Bough*. Macmillan, London.

Heide-Jorgensen, H.S. (2008) *Parasitic Flowering Plants*. E.J. Brill, Leiden, The Netherlands.

Kuijt, J. (1969) *The Biology of Parasitic Flowering Plants*. University of California Press, Berkeley CA, USA.

Kuijt, J. (1977) Haustoria of phanerogamic parasites. *Annual Review of Phytopathology* **15**, 91–118.

Kuijt, J. and Lye, D. (2005) Gross xylem structure of the interface of *Psittacanthus ramiflorus* (Loranthaceae) with its hosts and with a hyperparasite. *Botanical Journal of the Linnean Society* **147**, 197–201.

March, W.A. and Watson, D.M. (2007) Parasites boost productivity: effects of mistletoe on litterfall dynamics in a temperate Australian forest. *Oecologia* **154**, 339–347.

Marshall, J.D. and Ehrlinger, J.R. (1990) Are xylem-tapping mistletoes partially heterotrophic? *Oecologia* **84**, 244–248.

Martinez del Rio, C., Silva, A., Medel, R. and Hourdequin, M. (1996) Seed dispersers as disease vectors: bird transmission of mistletoe seeds to plant hosts. *Ecology* **77**, 912–921.

Mathiasen, R.L., Nickrent, D.L., Shaw, D.C. and Watson, D.M. (2008) Mistletoes: pathology, systematics, ecology, and management. *Plant Disease* **92**, 988–1006.

Nickrent, D.L., Malécot, V., Vidal-Russell, R. and Der, J.P. (2010) A revised classification of the Santalales. *Taxon* **59**, 538–558.

van Ommeren, R.J. and Whitham, T.G. (2002) Changes in interactions between juniper and mistletoe mediated by shared avian frugivores: parasitism to potential mutualism. *Oecologia* **130**, 281–288.

Press, M.C. (1995) Carbon and nitrogen relations. In: *Parasitic Plants* (eds M.C. Press and J.D. Graves), pp. 103–124. Chapman & Hall, London.

Press, M.C. (1998) Dracula or Robin Hood? A functional role for root hemiparasites in nutrient poor ecosystems. *Oikos* **82**, 609–611.

Press, M.C. and Graves, J.D. (1995) *Parasitic Plants*. Chapman & Hall, London.

Press, M.C. and Phoenix, G.K. (2005) Impacts of parasitic plants on natural communities. *New Phytologist* **166**, 737–751.

Sallé, G. (1983) Germination and establishment of *Viscum album* L. In: *The Biology of Mistletoes* (eds M. Calder and P. Bernhardt), pp. 145–159. Academic Press, Sydney, Australia.

Schaefer, H.M. and Ruxton, G.D. (2009) Deception in plants: mimicry or perceptual exploitation? *Trends in Ecology and Evolution* **24**, 676–685.

Smith, J.Z. (1978) *Map is not Territory: studies in the history of religion*. E.J. Brill, Leiden, The Netherlands.

Thoday, D. (1956) Modes of union and interaction between parasite and host in the Loranthaceae. I: Viscoideae, not including Phoradendreae. *Proceedings of the Royal Society of London B* **145**, 531–548.

Ullman, I., Lange, O.L., Ziegler, H., Ehleringer, J., Schulze, E.-D. and Cowan, I.R. (1985) Diurnal courses of leaf conductance and transpiration of mistletoes and their hosts in Central Australia. *Oecologia* **67**, 577–587.

Vidal-Russell, R. and Nickrent, D.L. (2008) The first mistletoes: origins of aerial parasitism in Santalales. *Molecular Phylogenetics and Evolution* **47**, 523–537.

Watson, D.M. (2001) Mistletoe: a keystone resource in forests and woodlands worldwide. *Annual Review of Ecology and Systematics* **32**, 219–224.

Watson, D.M. (2009) Parasitic plants as facilitators: more Dryad than Dracula? *Journal of Ecology* **97**, 1151–1159.

Zuber, D. (2004) Biological flora of Central Europe: *Viscum album* L. *Flora* **199**, 181–203.

The Wolf

IT IS FITTING that in this last chapter we examine the wolf, *Canis lupus* (Box 19.1), as humans and wolves share many characteristics. These include a social organization which, rare amongst mammals, enables animals larger than oneself to be preyed upon, an evolution during the same period (starting about 6 million years ago (mya), at the end of the Miocene) from forest ancestors into the savannahs during a period of great climate change, and a spread to become among the most widely distributed of mammals. Wolves and humans have lived in the same habitats for at least 500 000 years and their bones have been found together from 400 000 BP. This has led to domestication on one hand (Box 19.2), and fear, superstition and

---

BOX 19.1   WOLF TAXONOMY: HOW MANY SPECIES?

A species as widely distributed as the wolf is likely to exist in a number of forms relating to local adaptations and genetic isolation (as we have seen with the honey bee in Chapter 4). The taxonomy of the wolf is controversial: some have suggested that there are over 30 different extant wolf subspecies, while others suggest that as wolves are very mobile and interbreeding is very likely, there are few if any valid subspecies. Which, if any, of these subspecies should be considered as true species also is hotly debated.

Nowak (2003) takes a middle ground and recognizes the following taxonomy. Along with six subspecies in Europe and Asia the following occur in North America (Fig. 1):

o *Canis lupus*: grey wolf.
o Spp. *lycaon*: SE Canada. Eastern timber wolf. Smaller, with longer legs, narrow snout, eyes further apart, large ears. The famous Algonquin Park wolves.
o Spp. *baileyi*: SW USA/Mexico. Lobo or Mexican wolf. The most genetically distinct subspecies. Smallest American wolf, shorter coat, adapted to hotter climates, more pointed ears. Looks more like a coyote than other grey wolves, preys on smaller animals, e.g. deer, antelope, rabbit.
o Spp. *nubilus*: Alaska south. Great Plains wolf. Typical grey wolf.
o Spp. *arctos*: N and E Greenland to Canada. Arctic wolf. White coat, muzzle and legs short, rounded ears. Feeds on musk ox in very large territories. Some may hunt seals.

o Spp. *occidentalis*: W Canada south. Northern wolf. Large, over 60 kg, nomadic, travelling with caribou herds across continent.
o *Canis rufus*: Red wolf. SE USA. Smaller, short coat and pointed ears, long narrow muzzle.

Canis rufus is often considered a subspecies of *lupus* but is becoming increasingly accepted as a species in its own right. If correct, it would become one of the rarest and most endangered mammal species in the world. Some have argued that the red wolf is a only a *lupus*–coyote hybrid.

Recently, controversy has arisen over the status of the small eastern wolf, spp. *lycaon*, which shares some similarities with *rufus*. Some workers have joined these two as one new species *C. lycaon*. This small wolf has hybridized with coyotes to the east and grey wolves to west as its range has shrunk, making determination of its taxonomic status very difficult. This hybridization may be a good thing, however, and could enhance its survivability in human-dominated habitats.

The distribution of wolf species and hybrids in North America also reflects the repeated colonization of areas by different wolves after the retreat of glaciers, and evolution in glacial refugia. Some have argued that the eastern *rufus/lyacon* wolf is primitive, occupying the margins of the North American wolf range, and may represent a distinct, relict lineage. If so, it is clear that it is not the progenitor of the other wolf subspecies in North America today.

Biological Diversity: Exploiters and Exploited, First Edition.   Paul Hatcher and Nick Battey.
© 2011 John Wiley & Sons, Ltd. Published 2011 by John Wiley & Sons, Ltd.

This concern over wolf taxonomy is not just a nice academic exercise: it is crucial for continuing wolf conservation. The US Endangered Species Act recognizes the need to conserve valid subspecies, but not hybrids. Therefore, if the red wolf is a distinct species or subspecies then it will be protected, but should it be classified as a wolf–coyote hybrid then it is likely to be exterminated. Similarly, elevating *lycaon* to species rather than subspecies would immediately give this threatened wolf greater claim to protection (and also reduces the number of *C. lupus*, making this more threatened in eastern US). However, if *lycaon* and *rufus* were the same species, then the southern US red wolves suddenly become part of a much more abundant species.

Although several subspecies are recognized in North America, the larger wolves share far greater similarities than differences, and too great an insistence on rigid differences can lead to conservation difficulties. For example, the reintroduction of Canadian wolves into Yellowstone was once criticized for using the 'wrong' subspecies.

Box 19.1 Fig. 1   *North American wolves;* **(a)** Canis lupus *spp.* nubilus, *Great Plains wolf;* **(b)** spp. occidentalis, *northern wolf* [Wolves courtesy of the UK Wolf Conservation Trust]

destruction on the other. Here we will discuss part of this story, starting with aspects of wolf biology, leading through wolf mythology to the eradication of wolves in Britain and North America and subsequent recent conservation efforts.

THE REAL WOLF

Wolves are mammals of the order Carnivora (which now contains 271 diverse species), family Canidae. Cat and dog branches of this order diverged 50–60 mya; cats in the Old World and dogs in the New. The Canidae are the most ancient living family of carnivores, and are present in the fossil record from 35 mya as small animals living in North American forests. These canids later moved out of the forests, and by the late Miocene (5–7 mya) they had moved across the Bering Strait into Asia and Europe, arriving just after the last of the remaining large carnivores, the creodonts, had died out. Climate change, the replacement of forests by scrub and savannah, and the appearance of agile ungulates (hoofed mammals including deer, cattle and sheep) as prey provided opportunities for these fast predators, and species radiation took place. About 2 mya some of these canids re-entered North America across the Bering land bridge.

BOX 19.2   WOLVES AND THE EVOLUTION OF DOGS

It is now generally agreed that dogs evolved from wolves alone and not from any other *Canis* species (e.g. jackals), as has been suggested in the past. However, a number of questions over dog evolution remain, including when, where and how many times did the initial domestication occur?

The use of molecular genetics since the 1990s has started to clarify some issues. Studies have used mitochondrial DNA (mtDNA), short (16–18 000 base-pair) circular pieces of DNA that code for some mitochondrial proteins and some of the transcription apparatus. This DNA possesses a number of advantages for evolutionary study: each cell contains hundreds of mitochondria, so that mtDNA is much more abundant than nuclear DNA, and it has a higher mutation rate. However, as mitochondria are maternally transmitted one can only produce a history of maternal lineages, and the possibility of a non-wolf paternal ancestor remains.

Studies have found that the mtDNA of dogs contains four lineages, suggesting either that dogs arose from four separate wolf domestication events, or that one event was followed by backcrossing with female wolves. There is support for both arguments, and dogs have probably always been backcrossed with wolves by humans (there are currently over 100 000 wolf-dog hybrids in the USA alone), and recently coat melanism in North American grey wolves has been shown to have arisen from past hybridization with domestic dogs.

The date of domestication can be estimated by studying mutation rate of the mtDNA in the dog lineages. However, the same evidence can be interpreted in different ways: some suggest that dogs were only domesticated 15 000 BP, while others suggest a first domestication as much as 135 000 BP. It is hard to distinguish wolf bones from those of wolf-like dogs, and this has limited archaeological evidence: remains from 9–10 000 BP in Utah and 12–14 000 BP in Iraq, Germany and Israel are generally accepted to be the oldest domesticated dogs. There are also records from near this time from Star Carr, Yorkshire (about 9500 BP). This evidence could suggest that multiple domestication events occurred. However, mtDNA analysis again shows that the Native American dogs arose from Old World lineages that accompanied late Pleistocene humans across the Bering Strait.

Early dogs were probably not bred for particular traits, but bred freely and associated with humans only for food. Their sociality and ability to imprint on dominant animals would have aided domestication. Early dogs could have been indistinguishable skeletally from wolves, and may have diverged in morphology only when man started farming and imposing greater selective pressures on dogs. Thus the archaeological and DNA evidence are not necessarily in disagreement. Although, at least in the UK, greater dog diversity is noted from Roman times onwards, most breeds extant today were only developed in the last 200–300 years.

Dogs have usually been classified as a species, *C. familiaris*, separate from wolves, but there has been an increasing tendency to classify them as only a subspecies (spp. *familiaris*) of *C. lupus*. Scientists will probably never agree on which is correct, but dogs do show some considerable differences from wolves. In particular, the skull size in dogs has decreased, and its shape altered (shortening of muzzle, steeper forehead and more crowded teeth) compared to the wolf. As this makes adult dogs resemble young wolves of a similar size, some have suggested that paedomorphosis, or arrested juvenile development, has taken place during the evolution of the dog. In some behavioural characteristics dogs also resemble juvenile wolves. Paedomorphosis has been reported during the domestication of other mammals, and if this is the case with dogs, does it make them anything other than wolves who never grew up?

At this time there were up to nine wolf species in North America. The most striking was the massive dire wolf, *C. dirus*, which suddenly appears in the fossil record across all of southern North America 300 000 years ago. This species could be much larger than a modern grey wolf, with a massive head, short legs and large teeth, and became one of the only two wolf species left in North America at the end of the last glaciations, becoming extinct about 8000 BP. Possibly its prey had been eliminated by human hunters, or it was out-competed by the more agile *C. lupus* for the smaller and swifter prey then appearing. During the late Pleistocene (about 1 mya) it is estimated that there were about 33 million wolves world-wide, and this declined to about 1.2 million in the recent past and today only around 300 000 exist. The distribution of wolves and their taxonomy in the USA is given in Box 19.1 and Fig. 19.1.

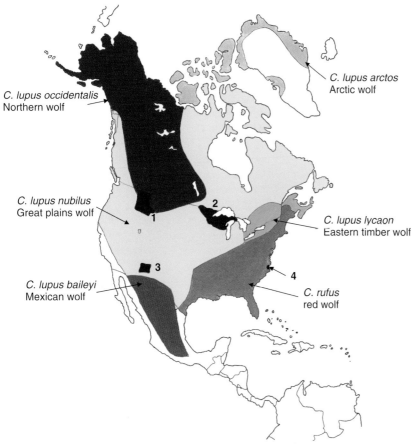

C. lupus occidentalis
Northern wolf

C. lupus arctos
Arctic wolf

C. lupus nubilus
Great plains wolf

C. lupus lycaon
Eastern timber wolf

C. lupus baileyi
Mexican wolf

C. rufus
red wolf

Fig. 19.1 *The historical distribution of wolves in North America. Wolves in Canada and Alaska still exist in 85% of this distribution. In the lower 48 states of USA, and Mexico, wolves are restricted to the black areas.* **1**, *Rocky Mountains;* **2**, *Great Lakes;* **3**, *Mexican wolf recovery area;* **4**, *red wolf reintroduction area. The previous distribution of lycaon and rufus is controversial. See Kyle et al. (2006) for other distributions [Based on Nowak (2003) and Grooms (2005).]*

Wolves are generalist, opportunistic hunters, preying on animals of all sizes from elk and bison down to rabbits, but they usually specialize on ungulates. The ability of wolves to tackle the largest herbivores is attributed to their teamwork and social organization into packs, maintained by a complex and flexible dominance hierarchy between individual, usually related, wolves. Pack size can vary depending on habitat and prey size from four or less, to over 40, and contains the alpha male and female, offspring from a number of years, uncles and aunts (who often have babysitting duties) and sometimes wolves adopted from other packs. The social structure, maintained by a complex set of visual gestures, scent marking, sound and displays, usually ensures that severe intra-pack violence is avoided and a dominant mating couple (the alpha male and female) exist. It is only this couple that usually breed, and after two months' gestation a litter of usually four to six cubs is born in a well-sheltered den in April or May. The cubs are blind for 12–13 days, suckled for 4–6 weeks, and may start to leave the nest when three weeks old. During this time they are not part of the pack hierarchy and are often tended by older relatives. As the cubs mature, starting to join the hunt at six months of age, they have to observe and fit into the social structure of the pack. Although they will become sexually mature in 2–3 years, unless the alpha pair are killed or deposed these young wolves stand little chance of breeding within the pack. A young wolf has to choose between biding its time for a chance to mate (if the alpha couple are unsuccessful other pairs may be allowed to produce cubs), deposing the alpha male or female, or leaving the pack to find a mate and set up a new pack. The latter is risky. The pack defends a territory containing sufficient prey animals for its needs (and thus territory size varies widely throughout the world: for example in elk-rich Montana some successful territories are only 50 km², whereas in the arctic they may extend over 2500 km²). Scent marking and pack howling signal to other wolves an occupied territory and serve to prevent fighting between packs, but a lone wolf is liable to be killed by another pack if it strays into its territory. Unless the area is sparsely populated with wolf territories, a new pack will have to carve out its territory from those of existing packs (Figs 19.2 and 19.3).

In the wild, most wolves die from starvation, disease or from attacks from other wolves, and rarely live more than 5–6 years. A wolf can go for many days without food, and when it is available can consume a large proportion of its body mass. The hunt for prey is all important. The wolf has a stiff-limbed gait; a steady rolling trot of about 8 km/h which it can keep up for hours when tracking prey. A pack may cover 50–100 km during a night's hunting and may spend eight hours a day on the move. The large ungulate prey have the potential to injure or even kill attacking wolves (many wild wolves examined have broken bones from such encounters), and a pack is not ensured success against a healthy individual. Thus, much of the encounter between predator and prey involves the wolves evaluating the prey and singling out a weaker animal, possibly younger or older, or diseased. The prey will often then be rushed at by the pack; a large animal standing its ground will often escape attack, but animals that try to escape can be singled out. Wolves can quickly accelerate to over 50 km/h for attack and can maintain a chase speed of 40 km/h for up to 30 minutes if necessary. Although all the adult members of a pack may take part in the stalking and rushing activities, the killing is usually performed by a couple of the wolves. Wolves have their preferred prey, for example elk form over 90% of the prey of the wolves in Yellowstone National Park, and in Europe red deer were favoured. However, when this prey is denied them, due to eradication, or management by humans, wolves invariably turn to attacking domestic livestock, bringing them into greater conflict with humans.

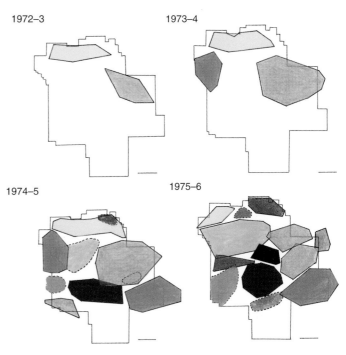

Fig. 19.2 *Wolf pack territories during recolonization of north-western Minnesota. Same colours between years indicates the same pack; solid lines indicate minimum area occupied by the pack; dashed lines the approximate area occupied by non-radio-collared packs. Scale bar = 10 km [Redrawn from Mech and Boitani (2003b).]*

THE MYTHICAL WOLF

Over the millennia there has been considerable feedback between mankind's attitude to wolves and the mythology that surrounds the wolf. This deep-seated mythology still has an underlying influence on people's attitudes to the wolf, so it is important to consider it here, before investigating the extermination and reintroduction of the real wolf. As Lopez (1978) says, 'the truth is we know little about the wolf. What we know a good deal more about is what we imagine the wolf to be.'[1]

Wolves featured heavily in Norse mythology. Giants in the shape of wolves were created by the Ironwood Witch and two of these chased the sun and moon across the sky:

> [F]rom this race shall come one that shall be mightiest of all, he that is named Moon-hound; he shall be filled with the flesh of all those men that die, and he shall swallow the moon, and sprinkle with blood the heavens and all the air; thereof shall the sun loose her shining, and the winds in that day shall be unquiet and roar on every side.[2]

This also introduces the concept of the wolf as a scavenger of the fallen in battle, which reoccurs often:

> The She-Wolf's evil Kindred
> Swallowed the corpse, harm-swollen,
> When the green sea was turnèd
> To red, with gore commingled.[3]

[1] Lopez (1978), p. 3.
[2] Gylfaginning XII; in Sturluson (1929).
[3] Skáldskaparmál LVII; in Sturluson (1929).

(a) (b)

(c) (d)

Fig. 19.3 *The typical wolf, Canis lupis.* **(a, b)** *Note powerful neck and forequarters.* **(c)** *Like dogs, wolves often engage in play. Due to complex dominance hierarchies, the darker wolf behind is unsure whether to join in (if allowed by the handlers!). In this case the lighter wolf is also engaged in scent marking.* **(d)** *The wolf has a mixed, herterodont dentition. Note the powerful canine teeth and the sharp molars behind for crushing bones [Wolves courtesy of the UK Wolf Conservation Trust, www.ukwolf.org.]*

The essential characteristics of wolves were given in their names, for example Geri (ravener) and Freki (glutton, greedy) were the two wolves always at Odin's side.

The most terrifying wolf of Norse mythology was the Fenris-Wolf, or Fenrir. This was the son of Loki, the arch-trickster of the Norse gods, 'First father of falsehoods, and blemish of all gods and men'[4] and Angur-boda, a giantess. Fenrir grew to an enormous size, even the gods were afraid to approach him. After two attempts Fenrir was bound by a fetter called Gleipner (Fig. 19.4), in the process biting off the hand of Týr. Later, the breaking free of Fenrir from his fetter signalled the start of Ragnarok, the Norse doomsday. The two wolves chasing the sun and moon catch up with and swallow them, and the stars vanish from the sky. Then:

> Fenris-Wolf shall advance with gaping mouth, and his lower jaw shall be against the earth, but the upper against heaven, – he would gape yet more if there were room for it; fire blaze from his eyes and nostrils.[5]

The gods were awakened and Odin was swallowed by the wolf before his son Vidarr managed to break the wolf's jaw. At the death of the wolf, fire was cast over the Earth and all was burnt, before renewal of a new world.

In the Bible, wolves, never named or personalized, symbolized the sinful side of human nature pitted against the Lamb of God, and disciples sent 'out like sheep among wolves'.[6] The wolf is portrayed as cunning, deceitful and greedy in medieval bestiaries – 'nothing on which they trample can survive'[7] – and summarized:

> The wolf is the devil, who is always envious of mankind, and continuously prowls around the sheepfolds of the Church's believers, to kill their souls and to corrupt them.[7]

[4]Gylfaginning XXXIII; in Sturluson (1929).
[5]Gylfaginning LI; in Sturluson (1929).
[6]Matthew 10:16.
[7]MS, Bodley 764, AD 1220–1250; see Barber (1992).

Shape-shifting into wolves was a minor feature of Norse mythology and also occurs in medieval contexts. This often involved wearing a wolf skin, and if this was removed the person would return to a human form. Werewolves were different and involved the metamorphosis of man into wolf. They appeared in Roman stories, but between the fifteenth and seventeenth centuries werewolves were written about in Europe as fact, not story, and there were a number of werewolf trials in Europe during this time, with the convicted usually being burned alive.

This negative image of the wolf can only have been reinforced in recent times by the Nazi cult of the wolf. They saw themselves as wolves or werewolves: the main underground headquarters near Rastenburg, East Prussia, was named Wolfsschanze (Wolf's Lair); 'wolf packs' of U-boats operated against Allied shipping (rich in wolf–sheep symbolism); and the largely unrealized plan to create Nazi terrorist cells in Europe as the Second World War ended was code-named 'Werewolf'.

However, not all wolf analogies were entirely negative. The wolf could become an almost sympathetic figure in some cases. For example, in the Latin beast-epic *Ysengrimus* written in the middle of the twelfth century and based around Ghent, the wolf Ysengrimus is the victim. In this brutal tale, redolent of medieval monastic allegory and satire, the wolf's greed is his downfall. 'While he feared for his stomach, he would undergo many fearful things',[8] and the fox Reynardus exploits this. The wolf is repeatedly beaten, flayed alive twice, deprived of his tail and a foot, his wife raped by the fox and finally is eaten alive by 66 pigs. In the later *Romance of Reynard the Fox* which developed from *Ysengrimus* over the next hundred years, the fox takes centre stage, and the wolf Isegrim is again one of the sufferers of Reynard's actions. There are many versions of this romance and they all end in a battle pitting the wits of the fox against the power of the wolf: the fox always survives, the wolf does not.

In some cases the wolf is an emblem of good. From the legend of Romulus and Remus

(a) (b)

(c) (d)

Fig. 19.4 *Wolves in Cumbria. The 4.5m tall stone cross in the churchyard at Gosforth, Cumbria, UK* **(a)** *carved in the first half of the 10th century, is one of the most beautiful of the Early English stone crosses. It has complex carved images on all four sides, the meaning of which has caused controversy, and many interpretations exist (Berg, 1958). The images probably contain both pagan Norse and also Christian scenes (including a crucifixion). The pagan scenes resemble those later written down in the Edda, and a number of wolves can be discerned:* **(b)** *Vidarr rending the jaws of the Fenris-Wolf;* **(c)** *the Fenris-Wolf bound by the fetter Gleipner;* **(d)** *the wolf (arrowed) and the hart above (both a pagan and Christian symbol), the tangle at the feet of the wolf may represent the broken fetter from which the Fenris-Wolf escaped*

[8]Mann (1987), II 324.

Fig. 19.5 *The wolf guarding the head of St Edmund: a wall painting in the nave of St Mary's church, Padbury, Bucks, UK, dating from the second half of the fourteenth century*

onwards there have been many tales and reports of wolves taking human foundlings into their packs and rearing them as one of their own, for example in Kipling's *Jungle Book*.[9] These stories still appear, and are sometimes reported as fact. For example, *Misha: a memoire of the holocaust years* by Misha Defonseca was published in 1997. The author tells of her travel across Europe, aged six, during the Second World War to find her parents. At one point she is rescued by a pack of wolves, which adopt her, protect her and save her from starvation. The book sold well, and was made into a film. However, early in 2008 the author admitted to the press that the story was fiction. It is interesting that many people, including some involved in wolf conservation, believed her story.

Good wolves were thought to be a demonstration of divine power in inciting a dangerous and unruly animal to behave. This is exemplified by the Wolf of St Edmund. Edmund, last king of East Anglia, was martyred by the Danes in 869 and beheaded. His head was hidden by the Danes in a forest so that the body could not be given a Christian burial. A wolf, supposedly under God's guidance, guarded the head from other animals until the king's followers could recover it. The cult of St Edmund grew in popularity during the Middle Ages (he was an early patron saint of England) and became a popular pre-reformation dedication for churches. Medieval wall paintings and wood carvings depicting scenes from Edmund's martyrdom can still be found in English churches (Fig. 19.5).

The symbolic wolf is still alive today. As an example one can do little better than examine the short stories of 'Saki' (Hector Hugh Munro). Ostensibly these humorous stories are similar to those of P.G. Wodehouse: dissolute young men, malicious children and appalling aunts drift through the long Edwardian house party inhabited by the English upper classes between the death of Victoria and the start of the First World War. Yet Saki's stories have something else; there is a very modern undercurrent of cruelty and misanthropy, and a sense that the animal kingdom is waiting to take revenge on mankind. Saki was fascinated, almost obsessed, with wolves. In 1904 he wrote to his sister, after the death of her pet dog:

> Have you thought of getting a wolf instead of a hound? There would be no licence to pay and at first it could feed largely on the smaller Inktons [a large neighbouring family], with biscuits sometimes for a change.[10]

This is probably only half in jest. Wolves appear regularly in Saki's short stories, often 'as sort of elusive undercurrent in the background that would never satisfactorily be explained',[11] but also as werewolves, eating or threatening to eat humans, or in folklore. In one of his most effective stories involving wolves, *The Interlopers*, two men intent on killing each other are drawn together when both are trapped under a tree. They reconcile their differences and shout for help. Figures appear through the forest, but they realize with 'the idiotic chattering laugh of a man unstrung with hideous fear'[12] that it is not men who approach, but a pack of wolves. Many of Saki's stories have this theme; the forfeiture of man's dominion

[9]Baden-Powell, the originator of the UK scouting movement, was a friend of Kipling. The cub–scout movement, with its pack leader Akela, was based on the *Jungle Book* wolf-pack. This has caused problems in exporting scouting to countries in which wolves are still feared.

[10]H.H. Munro, letter 1.4.1904; see Munro (1930), p. 682.

[11]*Reginald's Drama*, H.H. Munro, 1904.

[12]*The Interlopers*, H.H. Munro, 1923.

over animals due to man's inhumanity to man. This forms an important strand in later supernatural stories by other authors following the carnage of the world wars, for example, Arthur Machen's peerless *The Terror* (1917), and Daphne du Maurier's *The Birds* (1952).

EXTERMINATION OF THE WOLF

The first human recognition of wolves comes from Palaeolithic (13–11 000 BP) wall art in Europe, where wolves are represented along with much more frequent bison, deer, cattle and horses, in wall engravings and paintings in French and Spanish caves (Fig. 19.6).

By the Mesolithic (10–6000 BP), Maroo and Yalden (2000) have estimated that 2500 humans and 6600 wolves shared Britain with 950 000 wild boar (*Sus scrofa*), 830 000 roe deer (*Capreolus capreolus*), 65 000 moose (*Alces alces*), 1 250 000 red deer (*Cervus elaphus*) and 84 000 aurochs (*Bos primigenius*, wild cattle). There is likely to have been enough prey for both humans and wolves here. Smith (2006) has suggested that excarnation, whereby a corpse is exposed for stripping by animals prior to burial, took place in early Neolithic Britain, and has found canid teeth marks consistent with this on human bones from a round barrow on the Oxfordshire/Gloucestershire border. It is not possible to distinguish between wolf and dog as the cause, but if wolves were responsible, Smith wonders whether this is a sign that wolves were being actively appeased by being allowed to participate in this ritual.

Although we cannot be certain, it is likely that wolves had value to humans at this time, some being domesticated (see Box 19.2), and some providing pelts. However, wolf meat is inedible, and medieval people considered the wolf useless; even the pelt was considered of little value as it was difficult to tan and remove the odour.

Wolves were a noticeable feature in Britain in the first millennium AD: the *Anglo-Saxon Chronicle* reports, in the first patriotic poem in the English language, that in 937 after the battle of Brunnanburgh where Athelstan was victorious over a combined Scots and Irish Viking army:

> To enjoy the carnage, they left behind
> The horn-beaked raven with dusky plumage,
> And the hungry hawk of battle, the dun-coated
> Eagle, who with white-tipped tail shared
> The feast with the wolf, grey beast of the forest.[13]

Aybes and Yalden (1995) have recorded over 200 places named after wolves throughout England (Fig. 19.7), dating from this time. Eighteen per cent of the names recorded are for wolf pits, suggesting, although some disagree (e.g. Rackham, 1986), that pits for trapping wolves were common at this time.

Wolves have been trapped and hunted since ancient times for their attacks on humans and livestock, and from the Middle Ages were feared as carriers of rabies. The likelihood of wolf attacks on humans is controversial. Certainly in North America there seem to be no authenticated records of wolves attacking humans, but this is not the case in Europe. Kruuk (2002) states that wolves were regular predators of humans in Europe and Asia during the nineteenth and twentieth centuries, with documented records from Holland and Estonia of children being taken by wolves. Kruuk also notes people being killed by wolves during the

Fig. 19.6 *Part of a Palaeolithic wall engraving from Altxerri, Guipuzcoa, Spain, showing a wolf (~26 cm long) lightly engraved on the shoulder of a reindeer*

[13]Parker Chronicle AD 937; see Garmonsway (1954).

Fig. 19.7 *Woolmer pond in Woolmer Forest, Hampshire. In southern England, wolf place names are usually derived from the Anglo-Saxon Wulf, which usually becomes Wolf. Hence Woolmer is Wulf's mere = Wolf's pond*

1990s in India, Belarus and Russia, the latter where wolf attacks are said to be not uncommon. Wolf attacks on livestock, however, are incontrovertible.

Considerable numbers of wolves were hunted in Anglo-Saxon times. For example, in 957 King Edgar reaffirmed a treaty with the kingdoms of North and South Wales on receipt of payment of 300 wolf pelts. Apparently this tribute continued to be paid for three years after which no more wolves could be found. However, from the middle of the eleventh century the increasing importance of three factors combined to put the remaining English wolves under pressure from humans, and led ultimately to wolf extinction.

The first was the development of deer management after the Norman Conquest. Game was insignificant in the Anglo-Saxon economy and diet, but the conquering Normans brought with them a continental aristocratic hunting culture. This took two forms: Forests and Parks. The Forest was a legal term, and did not imply ownership of land, or that the land was wooded; rather that the king (predominantly) owned the deer and some other beasts therein. The Forests were managed for deer, and governed by a separate set of byelaws from common law, with potentially strict penalties for infringement:

> He [William the Conqueror] set apart a vast deer preserve and imposed laws concerning it.
> Whoever slew a hart or hind
> Was to be blinded.
> He forbade the killing of boars
> Even as the killing of harts.
> For he loved the stags as dearly
> As though he had been their father.[14]

This Forest law was first introduced by William the Conqueror to conserve the remaining red deer, and although the Domesday Book records about 25 Forests, at least 120 further ones were created up to the signing of the Magna Carta in 1216, which effectively ended Forest creation. Wolves are thought to have caused considerable losses in some Royal Forests and specialist wolf hunters, *luparii*, appeared in the twelfth-century royal household.

Although Forests were restricted to the King and a few nobles, anyone who could afford it could enclose land into a park. The park was a piece of private land, usually wood pasture, enclosed by a ditch, bank and fence. Although expensive to enclose, parks enabled much more effective deer management and more efficient production of venison than Forests. Parks became popular after the introduction of the fallow deer (*Dama dama*) by the Normans in the twelfth century; these rapidly replaced the native roe and red deer, and were eminently suitable for the managed park environment. By 1300 it is estimated that 2% of the country was enclosed as parks, from which wolves were actively excluded.

[14]Laud (Peterborough) Chronicle AD 1086; see Garmonsway (1954).

Concomitant with the management of deer was the rise of sheep husbandry in Britain. In Anglo-Saxon times sheep were primarily kept for milk production. From the ninth century sheep numbers steadily increased and woollen cloth production became more important. In the Domesday Book, sheep outnumber all other livestock combined, and their numbers continued to increase during the eleventh and twelfth centuries, along with a rise in wool prices. Sheep, even more vulnerable to wolves than cattle, were kept at night in thatched sheep houses, and during the day shepherds led flocks to pasture, walking in front with a mastiff or wolfhound, and the flocks herded by sheepdogs at the rear.

Lastly, wolves sheltering in unmanaged land were likely to come under increasing pressure as, due to the significant population increase in the thirteenth century, many 'waste' areas were cleared and put under cultivation.

All these changes generated considerable pressure for wolf removal. Land was granted to those who cleared wolves from an area and bounties given for dead wolves. Monarchs paid for wolf extermination. In 1210, King John gave 15 shillings to huntsmen who killed three wolves within Royal Forests; Henry III (1216–72) granted land in return for wolf extermination; and Edward I employed a hunter in 1281 to destroy wolves in several midland counties. In one of the last wolf records in England, a man was paid to guard cattle against wolves in the Royal Forest of Lancaster in 1295–6: seven calves were killed here by wolves in 1295 and eight in 1304–5. The final reliable record for the wolf in England is usually taken to be in about 1394 when the monks of Whitby were paid for 'tawing 14 wolfskins'.

The eradication of the English wolf was, according to Thomas (1983), 'a matter of some importance and the occasion of much self-congratulation',[15] as England became the first European country to be wolf-free, enabling easier and cheaper sheep husbandry, obviating the necessity of costly houses and protection for the animals (Fig. 19.8). The importance of sheep to the economy continued to increase, and from the beginning of the thirteenth century wool became the major British industry and at the centre of the economy until the sixteenth century. Higham (2004) suggests that wolves could have deterred graziers from developing large sheep flocks in Lancashire and Cheshire until the late thirteenth century, and some have suggested that the extermination of the wolf was crucial to the success of the British medieval wool industry.

In Scotland and Ireland, wolves persisted for longer, and consequently we have a better documentary record. Wolves were still a problem when James I passed a law in 1427 requiring every baron to organize his tenants to take part in four wolf-hunts in the spring, with rewards for those who killed a wolf. In 1457 this law was amended by James II to order local magistrates to gather locals three times a year in spring and summer to kill wolf cubs.

Wolves were common in the reign of Mary (1547–87) and she hunted wolves in the forest of Atholl in 1563. It is often stated that deforestation was deliberately practised during the sixteenth

Fig. 19.8 *Two early nineteenth-century wood engravings of mythological wolves:* **(a)** *Shepherd chasing away the Wolf, by William Blake, from Thornton's* The Pastorals of Virgil *(1821); note the fenced-in sheep;* **(b)** The Wolf, the Fox and the Ape, *by Thomas Bewick, from* The Fables of Aesop and others, *by Thomas Bewick (1818)*

(a)

(b)

[15]Thomas (1983), p. 273.

century in Scotland to hasten the destruction of wolves, but there is no evidence for this. A large bounty was paid for a wolf in Sutherland in 1621, which Rackham (1986) suggests is the last positive record of the wolf in Scotland. Others note that one of the last Scottish wolves was killed in Perthshire in 1680, but there are reports of wolves being killed in remote Sutherland in the 1690s, and wolves may have survived here into the eighteenth century.

Medieval visitors to Ireland mention that this was a land that produced wolves. Chieftains would hunt wolves by trailing a dead horse through the woods and dropping it in a clearing, wolf-hounds would then be set on the wolves that gathered to feed upon it at night and wolves were still common in Ireland after the Civil War. By 1652 the reconquest of Ireland after rebellion was complete, and Cromwell was concerned that the country needed to be cleared of wolves for the settlement of the 'adventurers' involved in the suppression to prosper. The records of the government of Ireland at this time give a fascinating insight into their concerns over wolves, culminating in June 1653 when an act was passed for general wolf destruction:

> Ordered that the Commissioners of Revenue in every precint do take all means for destroying wolves. A reward to be paid for every bitch-wolf of £6, dog-wolf £5, every cub, which preyeth for himself forty shillings, every suckling cub ten shillings.[16]

These high bounties caused problems and loans soon had to be given to some counties to enable the bounties to be paid. Up to 1656 almost twice as much was paid out in wolf bounties than paid for apprehending human outlaws in Ireland. It is not surprising that wolf numbers rapidly declined; they became scarce by the 1680s, although the last wolf is not reported to have been exterminated, in Co. Carlow, until 1786.

The early English settlers to North America in the seventeenth century, therefore, came from a land that had been without wolves for several centuries. They brought with them their superstitions and folklore, and found themselves pitched into a wilderness, populated again by real wolves. They also brought cattle. It was these cattle that fuelled the conflict between the settlers and wolves. Unlike in England, these cattle were not fenced in but allowed to roam free. Supervision of beasts was a low priority- people were needed in the fields—so livestock became easy prey for wolves.

During this early phase of colonization, wolf extermination was a low-key activity carried out by smallholders who used and modified European techniques. These included various hooks hidden in carrion, and community hunts. Again, severed wolf heads became symbolic, a totem nailed to meeting houses and displayed as for human outlaws. Legislation for bounties was enacted in many settlements from the 1630s; this led to some specialization of wolf hunters by the mid seventeenth century and Native Americans were also hired to kill wolves. Wolves were thus eliminated from some areas early, for example in Boston by 1657.

By the early nineteenth century, territorial expansion of settlers westwards from the original coastal settlements and an increasing sense of the need to tame, pacify and civilize the landscape (and America) led to a marked increase in wolf hunting: the wolf being 'the beast of waste and desolation'.[17] The payment of bounties was crucial to the development of a class of professional wolf hunters

---

[16]Dunlop (1913), p. 350.

[17]Theodore Roosevelt (1902) *Hunting the Grisly and other Sketches*; see McIntyre (1995), p. 108.

and in encouraging many to try their hand at this occupation. These bounties were quite generous and led to a strain on finances. For example, one year after a wolf bounty was introduced in Utah, its payment claimed 15% of the whole state's budget, and it had to be repealed.

During the 1840s, travellers moving west across America watched packs of wolves following the herds of buffalo across the Great Plains; there are records of herds of 8000 buffalo being followed by packs of 40 wolves. Wolves could only successfully attack small or weak buffalo, but Coleman (2004) suggests that wolves learnt to associate with human hunters: the slaughter of the buffalo herds from the middle of the nineteenth century led to a great excess of meat and wolf packs would wait until nightfall to feast on any dead buffalos. Wolves were largely ignored while there were sufficient buffalos; but when during the 1870s buffalo numbers crashed, and were exterminated during the 1880s, many buffalo hunters turned to wolf hunting as the wolves turned to livestock predation. Wolf hunting here was largely by the use of poison bait—meat laced with strychnine (see Box 17.3). Lopez (1978) suggests that in the period 1878 to 1895 this poisoning reached a peak and thousands of trappers were engaged in what seems a frenzy of poisoning. The strychnine-laced bait killed indiscriminately anything that ate it, and although no North American wolf has killed a child, wolf trappers poisoned several children. Lopez estimates that 500 million creatures, including 1–2 million wolves, may have died during this campaign.

Wolf extermination in North America was also encouraged during the eighteenth and nineteenth centuries by local legends reinforcing the demonization of the wolf as a child-killer, and attacker of defenceless travellers. This version of the wolf as threat to humans justified the enthusiasm with which wolves were destroyed. Coleman (2004) discusses this in detail: the activities involved in killing wolves were far more bloodthirsty and violent than needed for mere vermin control; it became a symbolic act. One manifestation of this was the late eighteenth- to early nineteenth-century grand circle hunt. This was highly organized: often several hundred armed participants would be involved and surround an area. The party would advance, moving all wildlife into the centre, where the animals would be slaughtered. If any wolves were caught (although this was the aim of the hunt, few were actually caught by this method) the bounty payments would go some way to offsetting the cost of the drink that fuelled the Bacchanalian celebrations that followed. In one such hunt, the famous Great Hinckley Hunt, on Christmas Eve 1818 in Ohio, 600 hunters killed a reported 300 deer, 21 bears and 17 wolves.

Ranchers grazing animals on federal land had long lobbied the state to control wolf populations, and in 1915 the Department of Agriculture's Biological Survey took over wolf control. The unit employed professional salaried hunters, who used wolf-traps with drag chains and denning along with strychnine.[18] This third phase of wolf elimination saw the start of a change of attitude to the wolf. In order to get sufficient government funds for their activities, the Biological Survey needed to market their activity, and in an early form of 'spin' publicised through books, posters and talks the remaining wolves as 'doomed-yet-heroic' desperados. The last wolves became heroes, which could only be caught by persistent effort from skilled hunters. The elimination of these wolves signalled the end of the old west, the march of progress and the close of the colonial epoch.

This wolf control programme, which killed over 24 000 wolves, was terminated in 1941. Another 46 000 wolves were killed by the Fishery and Wildlife Service (USFWS) between 1942 and 1970.

[18]Strychnine was used until 1931 for wolf control, and in large amounts. In 1926, enough strychnine to kill 14 million wolves was distributed.

Wolf control in Alaska and Canada, however, never succeeded in eradicating the wolf, although the same controversies exist. There are estimated to be 7–11 thousand wolves in Alaska, over most of the mainland and major islands (see Fig. 19.1). These hunt mainly moose and caribou and packs have large territories owing to scarcity of food. Wolves have been tolerated for longer in Alaska, and problems only arose in the 1950s with the advent of aerial hunting, which became especially popular in the 1960s, until banned. A decline in ungulate numbers in the 1970s was linked to wolf predation and ever since there have been legal battles between those who want increasing wolf control, and those who see any wolf control as anathema.

In Canada, 56 000 wolves (more than anywhere else in the world) currently occupy 85% of their original range. Wolf control here took the same form as in the US; however, with greater wilderness and lower human population density its effects were smaller. As a worker quoted in Grooms (2005) said: 'Canada has done all the same things to its wolves that we have in the US, just a little later and not nearly as completely.'[19]

The wolf was once abundant over most of North America (see Fig. 19.1). The general extermination of wolves through the late eighteenth and nineteenth centuries led to considerable habitat change and the spread of deer and coyotes (once restricted to the plains and deserts of central North America) bringing the latter in contact with wolves and allowing hybridization.

RETURN OF THE WOLF

The return of the wolf to the USA was a long time coming. Although increasing industrialization during the nineteenth century and the resulting human isolation from wildlife, along with the formation of a middle class led to a burgeoning animal welfare movement in the US from the 1860s, this welfare did not extend to the wolf. Likewise, the pioneering wolf-related eco-philosophical writing of Aldo Leopold in the 1940s, while influential, had little effect on public opinion at the time. Gradually, however, some attitudes changed. L. David Mech's 1970 book *The Wolf* was important in giving an accurate biological picture of the wolf, and when the US Congress passed the Endangered Species Act (ESA) in 1973, the wolf was one of the first animals listed. During the next 15 years the wolf became revered by many, rather than excoriated, as a symbol of wilderness, and also became seen as a symbol of mankind's cruelty to animals. This has led to the complicated and contentious situation of three different wolves now being 'seen' in North America: the old demonic wolf of some livestock owners and hunters; the wolf of the biologist, neither good nor bad; and also the revered, uncritically adored wolf of much of the public. It is possible that the balance of opinion has shifted too far in the direction of this latter wolf, which could ultimately damage efforts at wolf population recovery.

By the early 1970s there were fewer than 1000 wolves left in the lower 48 states of the US. About 700 of these lived in north-eastern Minnesota, especially in the Superior National Forest, bordering Ontario, Canada. Wolves were never exterminated in this hard-to-hunt land as it was not entirely dependent on ranching. After the great white pine forests were clear-felled in the late nineteenth century, a dense brush forest of aspen, maple and birch arose; both deer and wolves flourished in this environment, until this stand reached maturity in the 1950s. A

[19]Grooms (2005), p. 143.

management plan was set up in 1978 to increase wolf numbers, at a time when wolves had been reclassified here as threatened rather than endangered (thus they could now be killed if attacking pets or livestock), and when declining deer numbers could not support greater wolf numbers. However, foresters, who had considered the aspen as useless and thus left it in the woods, now found a use for it. The felling of aspen recreated the young forest ideal for deer and wolves, and populations increased to about 2500 wolves in the mid-1990s. Wolves can now be found throughout the top half of Minnesota, and the numbers have stabilized. As numbers increased in Minnesota, so wolves began to move into neighbouring states: from the mid-1970s into Wisconsin and Michigan, where by 2003–4 there were over 750 wolves resident.

The successful increase in wolf numbers in these three states bordering the Great Lakes shows that enlightened management and at least public tolerance can enable a wolf population on the brink of extinction to be rescued. However, in other US states the wolf has not been so lucky, either eradicated totally or a relict population squeezed into unsuitable habitat. But the greatest problem in the examples that follow has been lack of support from wide sections of the community coupled with vested anti-wolf interests from powerful bodies.

The first of the wolf populations to be conserved was the red wolf. This was the native wolf of the eastern seaboard of the US (see Box 19.1 and Fig. 19.1), and occupied a niche filled by the grey wolf to the north and the coyote to the west. By the late 1960s only about 100 individuals remained, restricted to less than 5000 km² of marshland along the coast of Texas and Louisiana—marginal land wholly unsuited to the wolf. The USFWS listed it as endangered in 1967, and captive breeding started in 1973. As hybridization with coyotes in the wild could not be prevented, the habitat was so unsuitable, and there was no place to translocate the animals to, it was decided to take all the red wolves into captivity—a brave and risky decision. Out of the over 400 wolves collected, only 17 were found to be pure red wolves and were sent to a zoo for captive breeding in Tacoma, Washington. At one time the number of red wolves declined to 14, and they bred only slowly, but by 1992 there were 200. Public opposition prevented the release of animals that would be protected, even if they attacked pets and livestock, under the ESA. However, in 1982 an amendment to the Act created the category of 'experimental and nonessential' animals. This allowed reintroductions under less rigorous conditions and treated as them as threatened, not endangered, allowing the removal or control of troublesome animals. This amendment was first used for the release of the red wolf. In 1987, four pairs were released in the Alligator River National Wildlife Refuge, a peninsular on the coast of North Carolina (see Fig. 19.1). This was the first time an extinct predator had been reintroduced into the wild in the US.

In the first decade 71 captive-born red wolves were released. Although 70% died, disappeared or had to be recaptured shortly afterwards, the first wild-born pup was observed in 1988, and by 2003 there were about 100 adults in the wild and 160 in captivity. In part this programme has succeeded because the wolves are well behaved, and rarely attack livestock. Hybridization with coyotes is still a problem, and a programme of coyote and hybrid control at the edge of the refuge has been necessary to enable red wolf range expansion. However, the number of wild red wolves has not increased, and numbered only 78 in spring 2010.

Probably the greatest, longest running and most acrimonious wolf restoration project in the US has centred on the Yellowstone National Park. Viable wolf populations mostly disappeared from the western states by the 1920s, with the last wolf killed in Wyoming in 1943. After the ESA was passed the government was mandated to prepare a plan for wolf reintroduction to the Rocky Mountains. A powerful livestock industry and shifting national government support during the 1970s and 80s prevented progress, which only began with the establishment of a fund to compensate for livestock losses due to wolves.

Arguments within pro-restoration organizations as to whether to reintroduce the wolf or wait for them to reintroduce themselves as in Wisconsin and Michigan also slowed progress. However, given that wolves that appeared naturally would be fully protected under the ESA, while those translocated would receive lesser protection under the 'experimental and nonessential' clause, this was a valid debate. As plans developed for reintroduction, the situation became more acrimonious.

Progress was made only when a local champion for the wolves emerged. The Nez Perce tribe of Native Americans volunteered to be the managers of any wolf reintroduction programme, educating the public, tracking wolves and controlling misbehavers. Although some wolves had naturally arrived from Canada in 1979, and the first breeding was noted in 1986, in December 1994 wolves were trapped in Alberta and brought to Yellowstone and Idaho and a total of 33 were released in 1995 and 1996, with 35 released in Idaho. These wolves prospered, and there was no need for further reintroductions. By 2004 there were 169 wolves in Yellowstone and over 400 in Idaho and north-western Montana. The effects of wolf introductions on the complex ecosystem of Yellowstone have been intensively monitored, and are starting to yield important information (Box 19.3).

Important also has been the enhancement of the Yellowstone economy by the wolves. The wolves here, as opposed to those in Idaho, have become very visible and wolf tourism has boomed – with over 130 000 visitors by 2004, and estimated to have added US$ 35 million to the local economy during 2006.

Some think that wolf conservation in the US is currently in a golden age – wolves are returning, and there is increasing public support for them. However, this may be temporary. The Great Lakes wolves were delisted in 2007, when they topped 4000. In 2008, there were 1645 Yellowstone wolves, in 217 packs, and 1500 of them were outside the National Parks. In February 2008, wolves in Idaho, Montana and Wyoming were removed from the endangered species list, and each state was required to maintain only 100 wolves, including ten breeding pairs. Idaho and Montana classified them as game animals and set quotas, while Wyoming classed them as vermin, allowing year-round virtually unlimited unlicensed killing. The delisting was challenged in the court, in particular the maintained populations were thought far too small to allow proper genetic mixing. In July 2008, a US federal judge overturned the delisting, and wolves were put back on the endangered list in all three states. However, in April 2009 this was lifted in Idaho and Montana, they were again classified as game animals and quotas for hunting were set. Wolves were still protected in Wyoming, and this state was asked to come up with a better management plan. The first legal wolf hunting season started in September 2009, and wolves that strayed out of the National Parks were at risk. At present it appears that many more wolves than expected are being killed – it is possible that they have become less afraid of humans. The converse is not true, and Yellowstone wolves need to relearn this fear, although in August 2010 they were given a respite when a US federal judge ordered the Montana and Idaho wolves to be put back on to the endangered species list.

BOX 19.3   THE YELLOWSTONE TROPHIC CASCADE

The importance of trophic cascades – a transmission of effects through the ecosystem from carnivores through their prey to alter the abundance and distribution of plant species on a community-wide basis – is debated in terrestrial systems.

Workers predicted that a trophic cascade might occur when wolves were reintroduced into Yellowstone. However, the Yellowstone ecosystem is very complicated and the wolf would be joining five other large predators: (humans; black bears, *Ursus americanus*; brown bears, *U. arctos*; coyotes, *C. latrans*; and mountain lions, *Puma concolor*), and eight ungulates (elk; big horn sheep, *Ovis canadensis*; bison, *Bison bison*; moose, *Alces alces*; mule deer, *Odocoileus hemionus*; Pronghorn, *Antilocapra americana*; mountain goats, *Oreamnos americanus*; and white-tailed deer, *Odocoileus virginianus*), making predictions difficult.

Elk represented 92% of the wolf kills during the winters of 1995–2000; 43% of these were calves, the remainder adults, with the median age of adult females killed of 16 years. After wolf reintroduction the elk population has declined from about 17000 in 1995 to about 8000 in 2004. However, it is not clear whether this is due to wolf predation (both bear species have also increased; they account for over 50% of elk calf deaths compared to less than 20% from wolves). Vucetich *et al.* (2005) produced a variety of simulation models based on elk populations before 1995, and suggested that climate change and human hunting has the greatest effect on elk numbers, with wolf predation primarily compensatory, targeting old and weak animals that would likely die anyway during the winter. Others disagree and argue that the effect of the wolves is additive to that of other factors. This is discussed in White and Garrott (2005), and mirrors disagreements over the effects of wolves on ungulates in other systems.

There is evidence of a wolf-induced trophic cascade in Yellowstone, but its effects are patchy, and evidence is largely correlative – the Yellowstone wolf reintroduction is a one-off event, it is neither replicated nor has controls and thus one must be cautious in ascribing causality to observed effects, however plausible. However, elk have changed their behaviour, and have started to avoid riparian habitats in high wolf-use areas, and this has led to the first regeneration of stream-side aspen in over fifty years (aspen regeneration stopped soon after wolves were eradicated in the 1920s) (Fig. 1). Grazing damage is still found in upland habitats, and thus it is likely that this effect is due to a change in elk behaviour, not just the reduction in elk numbers. It is thought that the elk are avoiding areas with poor visibility of and escape routes from wolves: the 'ecology of fear' as some workers put it. There has also been some release of willow in the same habitats from elk grazing. This regrowth is altering the species of song birds present. In time, the tree regrowth could consolidate the banks, reducing erosion and improving water quality.

Other effects are starting to occur. For example, the 50% reduction in coyotes after wolf reintroduction could favour small mammals: pronghorn fauns, highly predated by coyotes, now have a 4-fold higher density in sites used by wolves compared to areas containing coyotes.

The winter mortality of elk is related to the severity of the weather, especially snow depth. Recently the winters have become shorter and milder, causing reduction in the amount and duration of later-winter carrion. Increased carrion production by wolves is mitigating this effect and is thus buffering the effects of climate change on the scavengers, especially ravens, eagles, coyotes and bears.

However, the complexity of the system may mean that some hoped-for effects will not occur without assistance. It is thought that beavers could be favoured as they now have young trees to eat. Beavers would be an important part of the ecosystem, their dams creating habitats for mink and ducks. However, the high productivity of willow in low water-table areas is dependent on elk grazing. As the water table rises with beaver recolonization, so this elk-induced increase in productivity decreases. This positive feedback may inhibit the natural recolonisation by beavers; they may have to be reintroduced.

(a)

(b)

Box 19.3 Fig. 1 *Aspen regeneration in Yellowstone, August 2006.* **(a)** *3–4 m aspen regrowth along Lamar River;* **(b)** *lack of regrowth in adjacent upland. The dark, furrowed lower bark is from long-term bark stripping by elk* [Ripple, W.J. & Beschta, R.L. (2007) Restoring Yellowstone's aspen with wolves. *Biological Conservation* **138**, 514–519 © Elsevier.]

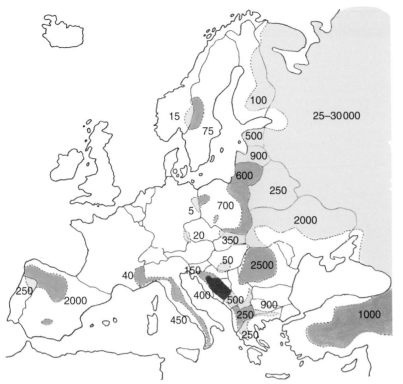

Fig. 19.9 *The range of the wolf in Europe and Eurasia in 2000. Numbers give a rough estimate of wolf numbers per country. Colours highlight the approximate distribution and status of wolves per country: green, increasing; yellow, stable; red, decreasing population; grey, unknown [Based on Boitani (2003).]*

Wolf reintroduction and conservation has also been successful throughout much of Europe, with notable conservation programmes in the Italian Apennines, attempts at reintroduction in Scandinavia, and wolves remain reasonably abundant in Eastern Europe and Eurasia (Fig. 19.9).

Can wolves be reintroduced into the UK? Several schemes have been proposed but as yet none has been given the go-ahead. Since the 1960s there have been repeated proposals to introduce wolves to Rhum, an uninhabited island National Nature Reserve 19 km off the coast of Scotland. This has a large population of red deer, which have to be severely culled every year. Yalden (1986) calculated that the cull of 250 deer a year could support 19 wolves, and a further nine might be supported if sheep and other herbivores were reintroduced. Although to some extent an ideal site, Rhum with an area of 10 700 ha is thought rather small to maintain a wolf population over the long-term, and attention has shifted to the highlands of Scotland, the only area in the UK that could conceivably support wolves. Wilson (2004) reviewed the possibilities for large carnivore reintroduction in the UK, and suggested that 200–250 wolves would be needed for a viable population; this would require at least 10,000 km^2 of suitable habitat, five times the size of the English Lake District. He concluded that wolf releases were unlikely to be approved owing to public opposition. The problems of mixed public opinion are illustrated by Sweden: in 1979 the first successful wolf breeding for 15 years was met with hostility, villagers applied for licences to kill wolves and a wolf destruction society was formed. By 1985 most of the wolves from this pack had been killed illegally.

Current plans for UK reintroduction hinge on the potential role of wolves in reducing the increasing wild ungulate population (mainly red deer) in the highlands, which cause damage to vegetation including extensive loss of heather cover and reduction of tree regrowth. Although much of the highlands of Scotland are managed for deer hunting, this does not control their population and an extensive and expensive cull is needed every year. By the early years of the twenty-first century this removed about 50 000 animals per year. Although this cull is insufficient to control red deer numbers, it is unlikely to be practically or economically feasible to increase it by much. This argument has been used by Nilsen *et al.* (2007) to suggest that wolf reintroduction would have a positive benefit to the ecosystem (as is becoming clear in Yellowstone; see Box 19.3). Their surveys suggest that attitudes to wolves have changed, with more favouring reintroduction to the wild. Given the experience in Yellowstone, it is thought likely that wolves would target the red deer first, but if deer numbers were significantly reduced, would wolves shift to livestock next?

Wolf reintroduction to the highlands may be first achieved on a private estate. In 2003, Paul Lister purchased the 10 000 ha Alladale estate to the north of Inverness. He plans to recreate the pre-human highland woodland ecosystem as a game reserve, with brown bears, lynx, and following from the Yellowstone results, wolves as an integral part. As a start he has enclosed the estate with an 80 km long 3 m high fence, halved red deer numbers, and planted 80 000 saplings. He has reintroduced wild boar to break up the blanket of heather and bracken to form a seed-bed, and in 2007 introduced two elk from Sweden.

It is thought that this estate is too small to contain a viable wolf population, and further land will have to be purchased or enter into the scheme before wolves can be reintroduced. The fence also may conflict with recent Scottish land reform legislation, which gives the public the right to unimpeded access to the countryside. However, there is a greater legal problem which puts the proposed plan in a 'Catch 22' situation: wolf reintroduction is very unlikely to be approved unless into a fenced site, and yet by erecting a fence this area might come under zoo legislation, which prohibits the keeping of predators and their prey in the same enclosure.

Wolf reintroduction in the UK still has some way to go. Russell and Russell (1978) concluded from previous proposals for wolf reintroduction into the UK that:

> To propose conserving them in the wild is therefore analogous to suggesting the conservation of desert locusts or malarial mosquitoes; it can surely only be explained by something like totemic survivals.[20]

This came from academics with no vested interest in the matter. We will leave the reader to compare relative merits of the conservation of the wolf as discussed in this chapter with that of the locust and mosquito discussed in previous chapters.

FURTHER READING

There is a huge array of material relating to the wolf in North America, rather less for wolves in Europe. Mech and Boitani (2003a) is the standard scientific work on the wolf. Grooms (2005) is a more popular account, with good photographs, which covers contemporary wolf controversies in the US, and Lopez (1978) is still useful, both for Native American attitudes to wolves and the state of wolf conservation at that time.

A useful introduction to wolf mythology is given by Pluskowski (2006). For Norse mythology it is best to consult one of the excellent modern translations of Sturluson's *Prose Edda*, or the *Poetic Edda*, rather than summaries, which can embellish the tales.

There is no one good comprehensive source for the extermination of the wolf in the UK, but it is covered by Yalden (1999), Cabot (1999, for Ireland), and also by Dent (1974). For an introduction to medieval hunting methods see Cummins (1988). Both Hampton (1997) and Coleman (2004) are recommended for general coverage of wolf extermination in the US, and McIntyre (1995) is an excellent source book of original documents relating to wolves in the US.

For reintroduction in the US, apart from the books above, see Fitzgerald (2006) for the lengthy legal battles over Mexican wolf reintroduction, which we

[20]Russell and Russell (1978), p. 178.

have not discussed, and Hedrick and Fredrickson (2008) for captive breeding of red and Mexican wolves. Reintroduction plans in the UK are covered by Yalden (1986), Wilson (2004), Nilsen *et al.* (2007) and Manning *et al.* (2009).

Wolf taxonomy is well covered in chapters in Mech and Boitani (2003a); see also Zimmer (2008) for a good introduction to the current debates on the species problem, relating in part to the wolf, and Kyle *et al.* (2006) for the genetics and distribution of the red wolf. The evolution of dogs is introduced in several chapters in Jensen (2007) and summarized by Honeycutt (2010). The return of the wolf to Yellowstone is covered by Smith and Ferguson (2005) and Morell (2008). There are many papers on the wolf-induced trophic cascade at Yellowstone; one of the first experimental tests of this is given by Berger *et al.* (2008), and Beschta and Ripple (2009) give a useful review.

REFERENCES

Aybes, C. and Yalden, D.W. (1995) Place-name evidence for the former distribution and status of wolves and beavers in Britain. *Mammal Review* **25**, 201–227.

Barber, R. (transl.) (1992) *Bestiary*. Folio Society, London.

Beschta, R.L. and Ripple, W.J. (2009) Large predators and trophic cascades in terrestrial ecosystems of the western United States. *Biological Conservation* **142**, 2401–2414.

Berg, K. (1958) The Gosforth cross. *Journal of the Warburg and Courtauld Institutes* **21**, 27–43.

Berger, K.M., Gese, E.M. and Berger, J. (2008) Indirect effects and traditional trophic cascades: a test involving wolves, coyotes, and pronghorn. *Ecology* **89**, 818–828.

Boitani, L. (2003) Wolf conservation and recovery. In: *Wolves: behavior, ecology, and conservation* (eds L.D. Mech and L. Boitani), pp. 317–340. University of Chicago Press, Chicago, USA.

Cabot, D. (1999) *Ireland*. HarperCollins, London.

Coleman, J.T. (2004) *Vicious: wolves and men in America*. Yale University Press, New York.

Cummins, J. (1988) *The Art of Medieval Hunting: the hound and the hawk*. Weidenfeld & Nicolson, London.

Dent, A. (1974) *Lost Beasts of Britain*. Harrap, London.

Dunlop, R. (ed.) (1913) *Ireland under the Commonwealth, Being a Selection of Documents Relating to the Government of Ireland from 1651 to 1659*. University Press, Manchester.

Fitzgerald, E.A. (2006) Lobo returns from limbo: *New Mexico Cattle Growers Ass'n v. U.S. Fish and Wildlife Service*. *Natural Resources Journal* **46**, 9–64.

Garmonsway, G.N. (transl.) (1954) *The Anglo-Saxon Chronicle*. Dent, Everyman's Library, London.

Grooms, S. (2005) *Return of the Wolf: successes and threats in the US and Canada*. Nova Vista, USA.

Hampton, B. (1997) *The Great American Wolf*. Henry Holt, New York.

Hedrick, P.W. and Fredrickson, R.J. (2008) Captive breeding and the reintroduction of Mexican and red wolves. *Molecular Ecology* **17**, 344–350.

Higham, N.J. (2004) *A Frontier Landscape: the north west in the Middle Ages*. Windgather Press, Macclesfield, UK.

Honeycutt, R.L. (2010) Unraveling the mysteries of dog evolution. *BMC Biology* **8**: 20.

Jensen, P. (ed.) (2007) *The Behavioural Biology of Dogs*. CAB International, Wallingford, UK.

Kruuk, H. (2002) *Hunter and Hunted: relationships between carnivores and people*. Cambridge University Press, Cambridge.

Kyle, C.J., Johnson, A.R., Patterson, B.R. *et al.* (2006) Genetic nature of eastern wolves: past, present and future. *Conservation Genetics* **7**, 273–287.

Lopez, B.H. (1978) *Of Wolves and Men*. Dent, London.

McIntyre, R. (ed.) (1995) *War Against the Wolf: America's campaign to exterminate the wolf*. Voyageur Press, Stillwater, MN, USA.

Mann, J. (ed.) (1987) *Ysengrimus*. Brill, Leiden, The Netherlands.

Manning, A.D., Gordon, I.J. and Ripple, W.J. (2009) Restoring landscapes of fear with wolves in the Scottish Highlands. *Biological Conservation* **142**, 2314–2321.

Maroo, S. and Yalden, D.W. (2000) The Mesolithic mammal fauna of Great Britain. *Mammal Review* **30**, 243–248.

Mech, L.D. and Boitani, L. (eds) (2003a) *Wolves: behavior, ecology, and conservation*. University of Chicago Press, Chicago, USA.

Mech, L.D. and Boitani, L. (2003b) Wolf social ecology. In: *Wolves: behavior, ecology, and conservation* (eds L.D. Mech and L. Boitani), pp. 1–34. University of Chicago Press, Chicago, USA.

Morell, V. (2008) Wolves at the door of a more dangerous world. *Science* **319**, 890–892.

Munro, E.M. (1930) Biography of Saki. In: *The Short Stories of Saki* (H.H. Munro), pp. 635–715. John Lane, The Bodley Head, London.

Nilsen, E.B., Milner-Gulland, E.J., Schofield, L., Mysterud, A., Stenseth, N.C. and Coulson, T. (2007) Wolf reintroduction to Scotland: public attitudes and consequences for red deer management. *Proceedings of the Royal Society B* **274**, 995–1002.

Nowak, R.M. (2003) Wolf evolution and taxonomy. In: *Wolves: behavior, ecology, and conservation* (eds L.D. Mech and L. Boitani), pp. 239–258. University of Chicago Press, Chicago, USA.

Pluskowski, A. (2006) *Wolves and the Wilderness in the Middle Ages*. Boydell Press, Woodbridge, UK.

Rackham, O. (1986) *The History of the Countryside*. Dent, London.

Ripple, W.J. and Beschta, R.L. (2007) Restoring Yellowstone's aspen with wolves. *Biological Conservation* **138**, 514–519.

Russell, W.M.S. and Russell, C. (1978) The social biology of werewolves. In: *Animals in Folklore* (eds J.R. Porter and W.M.S. Russell), pp. 143–182. Brewer/Rowman and Littlefield, Ipswich, UK.

Smith, D.W. and Ferguson, G. (2005) *Decade of the Wolf: returning the wild to Yellowstone*. Lyons Press, Guilford, CT, USA.

Smith, M. (2006) Bones chewed by canids as evidence for human excarnation: a British case study. *Antiquity* **80**, 671–685.

Sturluson, S. (1929) *The Prose Edda* [transl. A.G. Brodeur]. American-Scandinavian Foundation/Humphrey Milford, Oxford University Press, New York/London.

Thomas, K. (1983) *Man and the Natural World: changing attitudes in England 1500–1800*. Allen Lane, London.

Vucetich, J.A., Smith, D.W. and Stahler, D.R. (2005) Influence of harvest, climate and wolf predation on Yellowstone elk, 1961–2004. *Oikos* **111**, 259–270.

White, P.J. and Garrott, R.A. (2005) Yellowstone's ungulates after wolves: expectations, realisations, and predictions. *Biological Conservation* **125**, 141–152.

Wilson, C.J. (2004) Could we live with reintroduced large carnivores in the UK? *Mammal Review* **34**, 211–232.

Yalden, D.W. (1986) Opportunities for reintroducing British mammals. *Mammal Review* **16**, 53–63.

Yalden, D. (1999) *The History of British Mammals.* T. & A.D. Poyser, London.

Zimmer, C. (2008) What is a species? *Scientific American*, June, 48–55.

Index